THE NATURAL HOUSE

THE REAL GOODS SOLAR LIVING BOOKS

Wind Power for Home & Business: Renewable Energy for the 1990s and Beyond by Paul Gipe

Real Goods Solar Living Sourcebook: The Complete Guide to Renewable Energy Technologies and Sustainable Living, 9th Edition, edited by John Schaeffer

The Straw Bale House by Athena Swentzall Steen, Bill Steen, and David Bainbridge, with David Eisenberg

The Rammed Earth House by David Easton

The Real Goods Independent Builder: Designing & Building a House Your Own Way by Sam Clark

The Passive Solar House: Using Solar Design to Heat and Cool Your Home by James Kachadorian

A Place in the Sun: The Evolution of the Real Goods Solar Living Center by John Schaeffer and the Collaborative Design/Construction Team

Hemp Horizons: The Comeback of the World's Most Promising Plant by John W. Roulac

Mortgage-Free! Radical Strategies for Home Ownership by Rob Roy

The Earth-Sheltered House: An Architect's Sketchbook by Malcolm Wells

Y2K and Y-O-U: The Sane Person's Home Preparation Guide by Dermot McGuigan and Beverly Jacobson

Wind Energy Basics: A Guide to Small and Micro Wind Systems by Paul Gipe

The New Independent Home: People and Houses that Harvest the Sun, Wind, and Water by Michael Potts

The Natural House: A Complete Guide to Healthy, Energy-Efficient, Environmental Homes by Dan Chiras

Real Goods Trading Company in Ukiah, California, was founded in 1978 to make available new tools to help people live self-sufficiently and sustainably. Through seasonal catalogs, a periodical (*The Real Goods News*), a bi-annual *Solar Living Sourcebook*, as well as retail outlets, Real Goods provides a broad range of tools for independent living.

"Knowledge is our most important product" is the Real Goods motto. To further its mission, Real Goods has joined with Chelsea Green Publishing Company to co-create and co-publish the Real Goods Solar Living Book series. The titles in this series are written by pioneering individuals who have firsthand experience in using innovative technology to live lightly on the planet. Chelsea Green books are both practical and inspirational, and they enlarge our view of what is possible as we enter the next millenium.

Stephen Morris
President, Chelsea Green

John Schaeffer
President, Real Goods

THE NATURAL HOUSE

A Complete Guide to Healthy,
Energy-Efficient, Environmental Homes

Daniel D. Chiras

CHELSEA GREEN PUBLISHING COMPANY

White River Junction, Vermont / Totnes, England

Printed in the United States.
First printing, June 2000.

03 02 01 00   1 2 3 4 5
Printed on acid-free, recycled paper.

Library of Congress Cataloging-in-Publication Data
Chiras, Daniel D.
      The natural house : a complete guide to healthy, energy-efficient, environmental homes / Daniel D. Chiras.
          p.   cm.
      Includes bibliographical references and index.
      ISBN 1-890132-57-8 (alk. paper)
      1. House construction—Amateurs' manuals. 2. Green products.
3. Sustainable development. 4. Architecture and society. 5. Dwellings—Energy conservation. 6. Construction industry—Appropriate technology. I. Title.

TH4815 .C485 2000
690'.837–dc21
                                                 00-022725

Green Books Ltd
Foxhole, Dartington
Totnes, Devon TQ9 6EB
44-1-803-863-843

Chelsea Green Publishing Company
Post Office Box 428
White River Junction, VT 05001
(800) 639-4099
www.chelseagreen.com

SELECTED TITLES BY THE AUTHOR

Beyond the Fray: Reshaping America's Environmental Response (Johnson Books)

Lessons from Nature: Learning to Live Sustainably on the Earth (Island Press)

Voices for the Earth: Vital Ideas from America's Best Environmental Books (Johnson Books)

Environmental Science 6th ed. (Jones and Bartlett)

Natural Resource Conservation 6th ed. (Prentice Hall)

Study Skills for Science Students (Brooks/Cole)

Essential Study Skills (Brooks/Cole)

Human Biology: Health, Homeostasis, and the Environment (Jones and Bartlett)

Biology: Web of Life (West)

Dedicated with love and affection to Linda
for her kindness, patience, and love

CONTENTS

ACKNOWLEDGMENTS

MANY THANKS TO ALL the pioneers in the natural building movement who have dedicated their lives to offering a sane alternative to modern, unsustainable architecture. My deepest appreciation also goes out to the folks at Chelsea Green and Real Goods, who have played an integral part in the dramatic rise of the alternative building and sustainable living movements. You deserve medals for your tireless efforts, often against great odds. Many thanks to Stephen Morris for his willingness to publish my book, and tireless efforts to make it a success. To Jim Schley, my articulate, supportive, and perceptive editor, words can't express my gratitude. It's been a pleasure—indeed an honor—working with you. A special thanks to Bob Moran, who read an early draft of the manuscript and whose candid comments helped me whip it into shape, and to my copyeditor Mary Twitchell, who helped fine-tune the manuscript. I'd also like to thank Alan Berolzheimer for his invaluable assistance in editing the manuscript, locating artwork and photographs, and acquiring permissions. His patience and attention to detail in a hurried production schedule made the project run smoothly. Thanks to my incredibly talented artist, Michael Middleton, for his phenomenal sketches. Thanks also to the many folks at Chelsea Green who helped in one manner or another.

I'd also like to gratefully acknowledge people who selflessly took the time to read the manuscript in its early stages, while it was still very much a work in progress—looking more like a lump of clay than a finished cob home. Kit Cohan and Stuart Habel, you're terrific. Thanks also to some of the experts in natural building who took time to provide detailed reviews of chapters. Steve

Berlant, Rob Roy, Cathy Wanek, and Mark Piepkorn, your critiques proved invaluable. Thanks, too, to the pioneers who sat for interviews: David Easton, Bill and Athena Steen, Michael Reynolds, Paul McHenry, Jr., Ianto Evans, Rob Roy, Robert Chambers, Michael G. Smith, David Adamson, and Maggie Reynolds. And buckets of thanks go to all the homeowners, builders, suppliers, authors, and video producers who spent hours on the phone teaching me about natural building or who sent me samples of their work, let me view their products, or toured me through their homes. I'm hoping that I'm not missing anyone, but here's the list as best as I can remember it: Todd Stewart, Stan Huston, David Zornes, Tom Wuelpern, Scott Ely, Paul Linville, Bill Christensen, Marc Richmond-Powers, Tim and Sheryl Pettet, Steve Berlant, Cedar Rose, Keith Lindauer, Kaki Hunter, Doni Kiffmeyer, Scott Andrews, Ken Campbell, David Sorensen, David Johnson, Laurie Campbell, Gary Dillard, Ann Edminster, Mike Frerking, Dave Moshel, Gwyen Gorman, Art Ludwig, Kathleen Jarschke-Shultze, and Bob-O Schultze.

Finally, many thanks to my family for their love and support. To my brother Stan, who taught me a lot about building in his workshop, helped me build a few things for my house, and lent me the money to build my own home; to my brother Jim, who wired my ceiling fans; to my mother and father, who graciously donated cash to this ribald project; and to my sons, Skyler and Forrest, who make it all worthwhile.

A world of thanks to Linda, who worked with me on many weekends over the past three years to finish my house. Without her patience, knowledge, skilled hands, and keen ability to figure out complex problems, I'd still be trying to wire my wind generator! Thanks for sharing my enthusiasm and interest in natural building and for putting up with me when long hours at the computer strained my back and nerves. And many thanks to Linda's dad, John Stuart, for answering our many questions over the phone as we worked our way through various electrical projects. Finally, I'd be remiss not to thank the inventor of the screw gun! You should receive a Nobel Prize for this tool!

INTRODUCTION

N THE FALL OF 1994, I was faced with a major life decision. My marriage had broken up and I needed a place to live. One option was to rent. Unfortunately, the only homes I could afford to rent were small, dark, dingy, run down, and often at the end of dirt roads that would be impassable in the snowy winters. I also pondered the idea of buying a home. Dedicated as I was to the concepts and practices of sustainability, I knew that the environmental performance of even the best-built homes is—let me be frank—pretty awful. That left me with my last option: to build a home.

If I was going to build a home and office, however, I knew that I had to build according to my convictions. A quote by the late Edward Abbey rang in my head: "Sentiment without action is the ruin of the soul." A quote by Richard Lamm, a three-term governor of Colorado, also called to me: "It is not enough to have a handful of heroes, what we need are generations of responsible people." I reasoned that the divorce could be a positive thing, as it opened me up to creating an environmentally sustainable shelter. Having been an activist for many years and having burned out on trying to change people and public policy, I rather liked the idea of living responsibly and helping others see that we can live well with little impact on the planet. Maybe life could be a bit more enjoyable for me, too—not having to pound on my podium all the time.

Drawing on my work in environmental science and policy, I set out to build a state-of-the-art environmental home. In my writings and teaching, I emphasize an understanding of the principles of sustainability, and using them

What is the use of a house if you haven't got a tolerable planet to put it on?

HENRY DAVID THOREAU

1

as guidelines to reshape human systems. This project would give me an opportunity to see if these principles could be applied to housing.

Not having built a home before, however, I found myself in unfamiliar territory. One thing that became evident was that the design phase of the house was crucial. If I was going to build an environmentally sustainable home, I would need a design consistent with the principles and practices of sustainability. I knew that a house, like most things we humans construct, is a long-term contract with the environment. It commits its inhabitants to a level of resource demand, environmental pollution, and monetary outlay that must be sustained until the structure collapses or is torn down. And, lest we forget, in most cases, design decisions commit us to the burden of a huge mortgage and high utility bills. Dream homes often become nightmares, obligating us to much more work than is necessary. We end up working hard for our houses when it should be the other way around. Our houses should work for us.

I decided the design would minimize environmental destruction, economic outlay, and labor, letting me live lightly in every sense of the word. In keeping with my fiercely independent spirit, I also wanted my new home to be as free of the power companies as possible. Solar power would provide heat and electricity for the boys and me. Earth sheltering would help maintain internal temperatures year-round, keeping the structure warm in winter and cool in summer. I would also capture and purify rainwater and snow melt for drinking, bathing, cooking, and watering plants; no wells or water company hookup would be necessary.

Being one who would rather write, play music, hike, canoe, and camp than maintain a home, I vowed that my new home would require as little maintenance as possible and would be supremely efficient in its use of water, electricity, and heat. I would recycle graywater, decrease electrical demand, pack the house with insulation, and build with earth-friendly materials, or I would hand in my environmental badge.

So began my journey, a journey that provided a wealth of information for this book. Along the way, I learned a lot about building, builders, and building departments. I learned about alternative building materials, catchwater systems and graywater systems, alternatives to septic tanks, and much more. I talked to builders, subcontractors, and building department officials. I talked with the general public, schoolchildren, and friends to learn about their concerns and objections to alternative building. And I talked to people like myself who let ethics divert them from the beaten path. My goal is to share what I have learned to help you understand the full dimensions of natural, sustainable building—what it is and what it requires of us—so you, too, can join the handful of responsible citizens who are blazing a new path in sustainable housing.

Almost everyone I talked with who lives in a natural home shared with me the need to warn readers of potential pitfalls. Many extolled the necessity of being totally honest, by which they meant telling the full story. "Don't mislead

INTEGRAL DESIGN

It became clear to me at this stage in my life—because I was actually focusing on building and no longer just philosophizing about it—that the design and construction of a home are convergent forces. They can make our lives either environmentally sustainable or environmentally destructive. Design decisions influence how comfortably we live and how much it costs us to attain the lifestyle we aspire to, and thus deserve extraordinary consideration.

readers into thinking they can build a natural, sustainable home for a song and a dance," advised one builder. "Don't lead them to believe that building is simple. It isn't," advised another.

Taking their advice to heart, I will present the information as soberly as possible, pointing out costly and time-consuming mistakes that I and others have made, so that each new home can build on the knowledge of previous homes. I will also present a realistic view of costs. It is the only way we will succeed.

Our journey together will consist of three parts. In Part 1, we will explore principles of sustainable design and construction, guidelines vital to your future success. The principles will keep you on the path to sustainability. As you will soon see, it is much too easy to wander off the track and end up with a house that falls short of your goals.

In Part 2, we will explore natural building techniques such as adobe, cob, straw bale, cordwood, and Earthships. You will also learn about some new-comers to the natural building scene, such as earthbags and papercrete. I will discuss the essential characteristics, strengths, and weaknesses of each alternative. My goal is to give you enough of each technique so you gain a realistic understanding of what building with the technique will be like. In other words, these chapters won't provide enough information to enable you to build a natural home but they will give you an overview of each natural building method—enough information to decide which ones might work for you. At the end of the book there is an extensive resource guide that lists additional books, articles, Web sites, suppliers, builders, and more, so you can further your research.

In Part 3, I suggest ways to build a fully sustainable home. Experience has shown me that while many people want to build environmentally friendly homes, very few understand how far they must go to create a truly sustainable structure. In Part 3, we will explore ways to heat and cool a home naturally—with little, if any, fossil fuel energy. We will examine ways you can generate your own electricity from clean, renewable sources such as the sun, wind, and falling water. We will also explore unconventional, but environmentally sustainable, ways of obtaining fresh water and treating waste water. Finally, I discuss green building products and describe how site selection and landscaping can contribute to the sustainability of your natural home. The final chapter of the book focuses on the building process and how you can successfully navigate these largely unfamiliar waters.

Throughout the book I will present conversations with innovators in the natural building field and sustainable home construction. These men and women who are leading the charge will share their experiences, knowledge, and wisdom.

So, let's get started. . . .

FOR MORE INFORMATION

A vast body of literature exists on each of the specific topics I explore in this book. A guide to these sources, organized by chapter, begins on page 441. Resources that are relevant to multiple chapters are listed with the chapter of their primary topic.

THE NATURAL HOUSE / PART

The Sustainable Imperative

STRIKING OUT IN A NEW DIRECTION

E ACH YEAR, MILLIONS of new homes are built to house the world's population. In the United States alone, the National Association of Home Builders estimates, 1.3 million homes are constructed each year. Although some of the world's new housing is built to replace shelter that has become obsolete, most homes are constructed to house the growing human population, a trend with no end in sight. Except in Europe and Japan, where the population is projected to decline over the next few decades, most other countries are expected to post steady growth well into the future.

If you are a business economist the influence of population growth on housing demand is no doubt a thing of exceptional beauty. If you are tuned into the environmental scene, however, these trends are anything but welcome news. The millions of new homes needed to accommodate the world's burgeoning population will require massive quantities of raw materials and energy. These resources, of course, are extracted from the Earth's crust or harvested from forests already straining under current demands.

To understand the impact that this massive growth in housing will have, consider the resources required to build a single-family American home, admittedly one of the most lavish structures being built these days to shelter human beings. A typical American home measuring a little over 2,000 square feet will require more than 13,000 board feet of framing lumber. If laid end to end the 2 x 4s and 2 x 10s used for framing walls, floors, doors, windows, and roofs would extend nearly two and a half miles. Sheathing material (plywood

Why not go out on a limb? That's where the fruit are.

WILL ROGERS

and other similar products) for exterior walls, roofs, and floors would cover an area approximately the size of two tennis courts—or about 6,200 square feet. Drywall would cover an area of equal size. Roofing material, insulation, exterior siding, and floor coverings (rugs, tile, and such) would cover an area of 10,000 square feet. But that's not all. This small American house would also require 14 tons of concrete, 15 windows, 12 interior doors, 3 toilets, 3 bathroom sinks, 2 bathtubs, 2 garage doors, and much more (see sidebar). If you multiply this resource demand by 1.2 million (the number of new homes built in the United States each year) you may begin to appreciate the extraordinary resource demand U.S. home construction places on the planet's already overtaxed resources. The framing lumber for 1.2 million new homes alone, if laid end to end, would extend 3 million miles—to the moon and back six and a half times. Add to this the new homebuilding taking place in Canada, Australia, Japan, Europe, Central America, and other countries and you begin to comprehend the impact homebuilding has on the planet.

The building materials required, however, are just part of the total demand our homes place on the environment. Intense demand for energy and a host of other natural resources continues long after the hammers, saws, and screw guns are silenced as the occupants consume the food, water, electricity, fuel, and household products they need. The construction and maintenance of households are clearly on par with other resource-intensive activities, such as transportation and industrial production. In addition, there's another aspect of modern housing that we need to be concerned with: the dependency and vulnerability our new homes create.

Michael Reynolds, a New Mexico builder of alternative homes called Earthships, sums up the situation with typical candor. In his book *Earthships* (Volume I), he likens the modern home to a patient in a hospital's intensive care unit. Reynolds observes, "A person on a life support system in a hospital has to be always within reach and 'plugged in' to the various systems that keep him/her alive." The modern home is connected by pipes and wires to its own life-support systems, consisting of the electrical power grid, gas lines, water lines, sewage lines, and so on, which provide energy, water, and a host of vital services. As Reynolds points out, there is a dark side to this dependence. When systems fail due to some natural event, such as a hurricane, tornado, or earthquake, existing housing becomes nonfunctional. The modern home is as vulnerable as any intensive-care patient.

Our homes are also a source of enormous waste, the most common being solid waste. Americans alone produce over 220 million tons of garbage every year—enough to fill the Superdome in New Orleans more than 2.5 times per day. Although more and more household waste is being recycled, the vast majority still ends up being burned in incinerators or buried in landfills. Modern homes also generate considerable amounts of air pollution from furnaces and fireplaces as well as from power plants that produce electricity for our homes. According to the U.S. Department of Energy, American homes consume 20 percent of the nation's energy and produce approximately 20 percent of the greenhouse gas carbon dioxide. Rising levels of greenhouse gas may be causing a major shift in the climate, a change that could be catastrophic to the planet—and us.

Although most of us do not think of our homes as a source of pollution, they are. Our houses—and the many systems that support them with power and other resources—are slowly poisoning our planet with sewage, industrial chemicals, household cleaners, radioactive wastes, greenhouse gases, and acids. Moreover, the acquisition of resources to meet household needs is destroying the ecosystems that are vital to our survival and to the survival of millions of other species that share the planet with us. We accept such travesties as the price for making our homes functional, says Reynolds, but we ignore them at our own peril.

As Michael Potts points out in his book *The New Independent Home*, modern homes are something of an embarrassment to enlightened builders. Not only do they contribute to pollution and habitat destruction, they may cause unnecessary illness to the occupants. Research has shown that many conventional building materials, carpets, paints, and finishes release potentially toxic fumes, such as formaldehyde, that pose a health risk to many people. Especially vulnerable are those individuals suffering from chemical sensitivity, for whom toxic substances from modern products can be severely debilitating.

Today, a new generation of builders is emerging to challenge the 20th-century notions of shelter. The central question they are posing is this: Can we build, power, furnish, and operate homes that are aesthetically appealing, convenient, and comfortable, yet healthy, nourishing to the human spirit, *and* easy on the environment? And can we do all this without spending a fortune?

The answer is an emphatic yes.

WHY BUILD A NATURAL, SUSTAINABLE HOME?

I have no doubt that we can build homes that meet our needs while incurring only a small fraction of the impact on the environment of conventional homes. We can live well while protecting, even enhancing, global ecosystems, the life-support system of the planet. In so doing, we will ensure future generations and the millions of other species that make their home on this planet the resources they need to prosper. And, we can easily build and finish homes with products that safeguard human health.

What may be surprising to many readers is that we can achieve these goals without going bankrupt. Natural, healthy, environmentally sustainable homes, far from being prohibitively expensive, can be built at the same or lower cost than conventional structures. And once they are built, these homes can operate at a fraction of the cost of a conventional home.

In the sweep of history, the twentieth-century American house will probably be regarded as a temporary aberration, an embarrassment to enlightened builders and planners. It will be called the "out of place house" or "the utterly dependent home."

MICHAEL POTTS,
The New Independent Home

In a bitter irony, modern homes not only threaten the health of the planet, they threaten the health of those they are intended to shelter.

Stick-frame construction gets a bad rap in this book. To satisfy a sense of fairness, I do need to point out that while standard wood-frame construction has its cons, it does offer many advantages. Bear in mind, too, that with the growing number of green building products, traditional stick-frame construction can be made more sustainable. In that regard, pay particular attention to chapter 14, which lists products you can use to build or remodel a conventional home to make it more sustainable.

PROS

- Wood is a fairly strong, versatile material.

- Wood is produced from a renewable resource and can be grown and harvested in a sustainable manner. It is readily available in many parts of the world.

- Wood-frame construction is widely practiced in many countries. So, finding someone to build a house, add on to, or repair an existing structure or assist you in a project of your own is relatively easy.

- Books and videos on wood-frame construction are widely available. There's no shortage of information!

- Wood can even be harvested and milled on site to reduce transportation and processing costs.

- Wood can be used for a variety of applications and thousands of products and techniques have been devised to meet a wide range of building needs.

- Wood is a lightweight material that lends itself to a modular approach. With modern building techniques, components can be readily assembled into complex structures.

- Building codes cover wood construction in detail.

- Building officials are familiar with wood-frame structures. Getting a permit to build is easy, so long as you're proposing a structurally sound building.

CONS

- Wood-frame construction requires special skills and knowledge. You need to know what you're doing to build a strong, durable home.

- Wood-frame construction is labor intensive. It requires lots of cutting and nailing or screwing. It also requires a lot of energy.

- Wood-frame construction requires enormous amounts of lumber, far more than most natural homes. It therefore contributes significantly to global deforestation.

- Much of the wood in modern stick-frame homes comes from huge clear-cuts, often many miles from the building sites. Clear-cuts can destroy valuable wildlife habitat, foreclose on recreational opportunities, increase runoff and erosion, and devastate views.

- The processing and transportation of wood and wood products requires a lot of energy and produces a lot of pollution.

- Standard building materials often contain glues made from toxic materials, which are released into our homes.

- Wood-frame homes are not very resistant to fire.

- Wood-frame homes are highly vulnerable to termites, tornadoes, and hurricanes.

One of the keys to success in sustainable building is the use of natural, locally available materials. Countless examples scattered across the globe, from Japan to Germany to Sweden to Australia to the United States, demonstrate that attractive, comfortable, healthy, and environmentally sustainable residences can be fashioned from a wide assortment of natural materials, such as straw bales, clay and sand, and adobe, as well as waste products, such as used automobile tires. These materials can be acquired on site or from nearby fields, or forests, or the waste stream, greatly lowering construction costs.

Catherine Wanek

Using local resources minimizes and localizes one's impact. As veteran builder Michael Smith notes in an article in *The Art of Natural Building*, "Digging a hole in your yard for clay to make a cob house may look ugly at first, but it's a lot less ugly than strip mines, giant factories, and superhighways. Nature has enormous capacity to heal small wounds; that hole in your yard would make an excellent frog pond."

Another key aspect of the new generation of environmentally sustainable homes is that they employ building materials that have a low embodied energy (figure 1-1). Embodied energy is the amount of energy that is required to manufacture a product. It includes the energy needed to harvest or extract a resource and transport it to processing facilities. And it includes the energy needed to transform raw materials into finished products and then transport them to the user. Locally acquired adobe or cob (a mud used to build walls), for instance, has extremely low embodied energy, compared to many conventional building products such as steel studs. Using materials with low embodied energy reduces global energy demand as well as local and global air and water pollution created during the production and distribution of building materials.

The new generation of homes is also being fashioned to be more energy independent—in some cases, totally energy self-sufficient (figure 1-2). This goal is achieved by increased insulation, proper siting, and a reliance on renewable energy resources such as the wind and sun. As many readers already know, solar energy can be used to heat homes in almost any climate. Although backup heat may be required in a solar home, the demand for home heating fuels is a tiny fraction of that required by a conventional home. Electricity can also be generated by renewable resources to power lights, fans, computers, stereos, and appliances, once again without the need to extract and burn fossil fuels.

EMBODIED ENERGY

If the relative energy intensity of lumber is set at 1, brick and cement are 2, glass is 3, fiberglass is 7, steel is 8, plastic is 30, and aluminum is 80.

FIGURE 1-2.
A wind generator and photovoltaic panels (right) supply electricity to this home.

Sustainable homes are sited, designed, and built so they are much less vulnerable to seasonal climate fluctuations than standard wood-frame houses. They fulfill the nearly universal dream of staying warm in winter and cool in summer—as well as the fantasy of being emancipated from the tyranny of high utility bills. While people living in conventional houses typically shell out $100 to $300 each month to pay the local utility, those living in solar homes made from straw bales, adobe, or other natural building materials pay $10 to $25 per month. In some cases, they pay nothing at all! In my 2,600-square-foot home and office, 8,000 feet above sea level, for example, the monthly bill for natural gas to supply my cooking stove and water heater runs about $12 per month during the spring, summer, and fall. During the winter, approximately five months of the year, gas bills are $25 to $35 per month higher to supplement solar heat. Electricity is supplied free of charge year-round from the wind and sun.

Many natural homes are being made from recycled building materials. Diverting materials from the waste stream to build houses greatly reduces the amount of trash that is burned or buried in landfills. My home and office in Evergreen, Colorado, for example, used 800 discarded automobile tires, gathered locally. My house also used 20 square yards of carpeting made from recycled plastic pop bottles, discards that would have ended up in a landfill had there not been a market for them. The ceilings are packed with 12 inches of cellulose made from newspaper and cardboard diverted from the waste stream. Each reused or recycled product makes a small but significant contribution to the betterment of the planet (figure 1-3). Using recycled building products also reduces air pollution, because products manufactured from recycled materials require much less fossil-fuel energy to make than those made from virgin materials. The energy savings range from 30 to 95 percent, depending on the material. And any time you burn less fossil fuel, you reduce air pollution.

Finally, natural building tends to rely in large part on simple, easy-to-learn techniques. Construction of exterior walls and floors often relies more heavily on human labor and creativity than on capital, high technology, and specialized skills. As Michael Smith points out in *The Art of Natural Building*, natural building is necessarily regional and idiosyncratic. There are no "right answers, no universally appropriate materials, no standard designs. Everything depends on local ecology, geology, and climate, on the character of the particular building site, and on the needs and personalities of the builders and users."

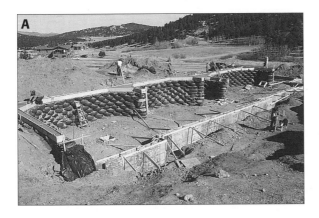

FIGURE 1-3 A, B.
The author's home under construction. Stucco was applied to packed tire walls (A) to create an attractive interior finish (B).

SUSTAINABLE DESIGN AND CONSTRUCTION

Before we begin our exploration of natural building and sustainable systems, we need a set of guiding principles and a set of operating principles. The guiding principles answer the question: Why build a natural, sustainable home? Why not go out and buy a tract home? The operating principles answer the question: How do you make a home from natural building materials that is healthy, comfortable, nourishing, environmentally sustainable, and affordable?

Guiding Principles

In this age of modern building materials and energy-intensive homes, those of us who build alternative structures are often asked why. What makes us veer from the conventional path?

The answer is threefold. The first reason is that the Earth contains a limited supply of resources. Fossil fuels and minerals are finite. There is only so much and it can only last so long. Even renewable resources like fisheries and forests can be depleted by rapacious overuse.

The second reason we veer from conventionality is because we feel an obligation to ensure future generations the resources they will need to reach their full potential. If you genuinely care about the future and want to manifest your feelings, there is no choice: You must avoid the resource-intensive path that is leading us to ecological and economic ruin. Concern for future generations is called *intergenerational equity*. Think of it as fairness to future generations.

We who stray from conventionality also believe that humans have a moral obligation not just to people but to other species as well. That is, present decisions must consider the welfare of the millions of species of plants and animals that share in the Earth's diminishing bounty. This is called *ecological justice*.

Embracing intergenerational equity and ecological justice places a moral burden on present generations, one that requires many conscious decisions during the design, construction, and operation of a home. It changes our

The sustainable homeowner gains the satisfaction of living lightly on the land while helping to create a brighter future for our children and the millions of plants and animals that depend on the life-support systems of the planet—forests, fields, air, and water—that provide us with food, fiber, water, clean air, and countless free ecological services.

thinking in fundamental ways. The question is not, "Can I afford this or that?" but "Can future generations and the Earth afford my extravagance?" If I build a 4,000-square-foot home, what will be the repercussions on the planet?

For too many years, the central operating principle of our frontier society has been "There's always more—and it is all for us." William McDonough put it well when he wrote, "We were operating as if Nature was the Great Mother who never has any problems, is always there for her children, and requires no love in return." Recognizing that the Earth is not a smorgasbord for gluttonous consumption and that future generations and other living creatures depend on the Earth for survival compels many of us to choose a different path.

There's a third reason for leaving the beaten path. It is time we recognize that to survive on this planet we must learn to cooperate with nature. For years we humans have striven to dominate and control nature. The 18th-century philosopher John Locke said that we must become "emancipated from the bonds of nature."

Today, virtually every system we depend on, from water supply to agriculture to housing, is based on the notion of domination rather than cooperation. But this conquest attitude may actually be making the task of living on the planet much harder. Ecological, or more appropriately sustainable, design, on the other hand, seeks to fashion human systems that work in concert with nature. Designing homes in harmony with the natural environment, for example, means conceiving of structures that are cooled by the Earth or by shade trees and warmed by the sun rather than expensive, fossil-fuel-consuming cooling and heating systems.

Contrary to Locke's beliefs, the key to success in life and building is to learn to avail ourselves of natural forces. Our job will be easier and less costly—not just in terms of dollars, but in terms of environmental impact and labor. We have been making this all a lot harder than is necessary.

Principles of Sustainable Design and Construction

Creating a sustainable home requires ingenuity, persistence, and forethought. The guiding principles outlined in the previous section can assist in this process, but to implement them, we need some practical design guidelines. What I present here are those operating principles; I think of them as a mental filter through which all decisions must pass during the design, construction, and operation of a home.

Make It Small. In the natural world, extravagance is a rarity. As a rule, organisms survive on what they need—no more, no less. In my classes, I like to remind my students that you won't find a robin with two nests, one for the chicks and the other to hold all the stuff. They don't have a lot of stuff. A juicy worm and a morning sunrise are quite enough for these delightful birds.

Health for the body, peace for the spirit, harmony with the environment—these are the criteria of the natural house.

DAVID PEARSON,
The New Natural House Book

Paul Lacinski

In contrast, in the human world frugality is an endangered species. If a 130-horsepower engine will do, we go for 200. If a 2,000-square-foot home is enough, we build 4,000 square feet. We don't know when enough is enough. In this age of bigger is better, the trophy home seems to be the pot of gold at the end of the house-buying rainbow. In such a climate, advice to build a small home may seem like heresy. Real estate agents will roll their eyes. Loan officers will scoff. Fellow homebuyers will decide you have lost your mind. "Did you actually say you wanted a smaller house?"

As Dianna Lopez Barnett and William D. Browning point out in their book *A Primer on Sustainable Building*, "Our cultural assumption is that we should buy as much house as we can afford." This conventional wisdom is clearly out of step with social and environmental realities (figure 1-4). On the social front, today's families in most developed countries are small, averaging about two children each—or fewer in some places. In addition, many homes are occupied by single adults, a divorced parent with his or her children, or an unmarried adult who may live with a roommate. In many developed countries, the number of retired couples is rising. They, too, need smaller dwellings, but many builders insist on building large, spacious homes.

Because of the resources required to build and operate a large home over its lifetime, many builders are contributing unnecessarily to environmental deterioration. The great irony in this is that many large homes don't always yield a lot more usable space. In fact, many large homes waste enormous amounts of space—for example, in wide hallways or single-purpose rooms like music rooms. Elegant and grand they may be, but they are exceptionally wasteful. Another source of waste is the combination of a formal living room and a family room, the former being reserved for rare occasions, the latter being real living space. Do we really need both?

The truth is that we can live in a small home. For many people, 1,200 square feet will do. Some may need a bit more. Others may need less.

Small, efficiently designed homes conserve resources, lessen your impact on the Earth, and ensure adequate resources for future generations and other

FIGURE 1-4 A, B.
Trophy homes (A) consume an inordinate quantity of resources to build and maintain. A small, natural home like this cob structure (B) uses a fraction of the resources and represents a major contribution to sustainable housing.

Large homes not only require more energy and raw materials to build, they require more raw materials to paint, finish, and furnish. Bigger furnaces, more furniture, and more flooring material add substantially to the drain on the Earth's natural capital.

species. Small houses not only use less energy, wood, and other building materials, they also require less land. In addition, a smaller home is a more inviting project for owner-builders to tackle.

Small homes need not be cramped and restrictive, either. By careful design, you can get a lot of use out of a little space. In fact, many design features make smaller spaces function well without making you feel closed in. Windows, a nook or a bay window with a seat, for example, can be used to open up a room. A large common area will make a house feel bigger. Recessed storage and built-in furnishings, such as tables, can optimize floor space. For more on the subject, see *New Compact House Designs, Small Houses,* and other books on the subject, which are listed in the resource guide at the end of the book.

Make It Efficient. Efficiency is a cornerstone of sustainability. We can't create a sustainable future without it. Thankfully, many steps can be taken to achieve efficiency in new homes. For example, a designer can save a considerable amount of wood by selecting a simple rectangular design over a more complex one. "Simpler shapes and volumes mean fewer pieces to cut and fewer framing errors," write Ann Edminster and Sami Yassa, authors of *Efficient Wood Use in Residential Construction,* a must-read for anyone using lumber in a building project.

Wood use can be slashed by adjusting measurements to conform with the dimensions of standard building materials. "Since wood products are typically milled or produced in increments of 2 feet, designs based on a 2-foot module will result in the least waste from off-cuts," write Edminster and Yassa. Therefore, drawing a room with even dimensions, say 10 feet by 18 feet, rather than 9 feet by 20 feet, will reduce the amount of wood you waste but yield the same square footage. You'd be amazed at how little scrap lumber will be produced by following this advice.

Energy is another prime target for efficiency. One of the most effective ways of saving energy is to properly insulate the walls and ceilings of a house and to install energy-efficient windows. These simple, cost-effective measures help a house retain heat during the cold winter months. As an added benefit, insulation reduces the amount of sunlight required to heat a solar home and lowers the need for oil, natural gas, or electricity for primary heat in a non-solar home. Insulation also makes the task of cooling a home much easier, a fact that escapes many home designers in hot climates. In many areas of the world, such as the desert Southwest or arid parts of Australia, cooling costs actually exceed winter heating costs.

Sealing a house against leaks is another efficiency measure worth its weight in wool sweaters. Building airtight homes or retrofitting existing homes with caulk and weatherstripping dramatically reduces heat loss over the winter and slashes cooling costs in the summer. According the American Council for an Energy-Efficient Economy, each year approximately $13 billion worth

Large, wasteful rooms; careless design of entries and hallways; vaulted ceilings; and other design "mistakes" make for houses that are larger and more resource-intensive than necessary. The rule is simple: Build a house only as large as you need and make the design as efficient in its use of space as possible.

of energy is lost in American homes through cracks and holes in the exterior of houses, around windows, near foundations, and around doors. This averages about $150 per household. It's not much on an individual basis, but adds up to a rather significant sum when you take into account all the leaky houses in existence today. And remember, all energy consumption results in pollution. Thirteen billion dollars worth of wasted energy is responsible for a lot of pollution that just doesn't have to be.

Daylighting, the appropriate placement of windows and skylights to provide natural interior lighting, is another efficiency measure that can dramatically reduce daytime electrical demand while creating a more aesthetically appealing atmosphere. Working in well-lighted spaces may increase personal productivity and enhance mood, as well. Daylighting provides a welcome connection to nature.

Water-efficient showerheads, low-flush toilets, and water-conserving dishwashers and clothes washers are all essential to furnishing a sustainable home, as are a variety of energy-efficient appliances. As energy and water efficiency have become more "mainstream," the quality, performance, and availability of these fixtures and appliances have greatly improved.

By beginning with a small, efficient design and including efficient lighting, appliances, and electronic equipment, you can significantly lower your demand for electrical power, whether from the sun or wind or the local utility. Efficient homeowners find that such measures reduce electrical energy consumption by 50 to 75 percent, making smaller solar electric systems suffice, and saving thousands of dollars in the process. My previous house, a standard passive solar home, would have required a $40,000 to $50,000 photovoltaic system; my current house, personally designed to be as electrically efficient as possible, does nicely with a system that cost $12,000. I use about one-fourth of the electricity of a similar home. For more on efficient appliances, see the *Consumer Guide to Home Energy Savings* by Alex Wilson, Jennifer Thorne, and John Morrill. This is a remarkably useful little book.

Another efficiency measure that pays huge dividends is the use of products with lower embodied energy. Energy is required to harvest or mine raw materials, process them, and transport them to stores and end users. But some materials require more energy than others. Steel and aluminum, for example, are much more energy intensive than wood. Mining and processing of the ores requires more energy than cutting trees and processing them into lumber.

Buying locally produced goods also lowers the embodied energy of a house because less energy is needed to deliver them to customers. Buying locally also reduces pollution and supports the local economy. Natural building materials, which are often locally available, have some of the lowest embodied energy ratings of all building materials.

Another consideration is the efficiency of design for the occupants. How efficient will the house be as you move about? Will traffic flow nicely? Or will

BUILDING NOTE

To reduce wood waste from off-cuts, adjust overall house dimensions, as well as roof pitch and/or overhang length, to obtain multiples of two.

ANN EDMINSTER and SAMI YASSA, *Efficient Wood Use in Residential Construction* walls and furniture obstruct movement? How quickly and conveniently will one be able to get from one location to another? Will you have to hike a half a mile to get from one end of the house to the other to retrieve a stamp, or will you have to run down two flights of stairs every time you want to get a snack from the refrigerator? How efficient are the locations of outlets and cabinets? Will you have to get on your hands and knees to plug in an appliance? Will you have to climb onto a small step ladder to reach food in your cupboards? Will countertops be situated conveniently for all family members to participate in food preparation?

In most homes, designs are created for the average adult—that is, for mid-height, active, young, healthy adults. But at little or no extra cost, most commonly used light switches, outlets, and cabinets can be placed in an Optimal Reach Zone, making everyone's life easier, not just the imaginary person for whom most houses are designed.

The study of how people do things and ways to make daily functions more efficient is called *ergonomics*. Although specific design features such as these are not the main thrust of this book and not characteristically within the purview of sustainable design, they are worth considering. They promote an efficient use of your time. They also promote optimum health by reducing unnecessary strain on backs, arms, and legs. You can learn more about accessibility and ergonomics in Sam Clark's book, *The Real Goods Independent Builder: Designing and Building a House Your Own Way.*

Cohousing, cluster development, and ecovillages—all of which are ways of fitting multiple families on the same piece of land—represent yet another means of conserving the planet's resources. Consider cohousing. A cohousing development typically consists of single-family homes or condominium-like structures on a commonly owned piece of ground. Many cohousing developments feature a common house where participants sometimes share meals, hold meetings, or house guests. Guest rooms and washing machines in the common house reduce needed floor space and appliances in each home. In the cohousing arrangement I've toured, for example, the common house contained a large kitchen for community meals. As a result individual homes can be built with much smaller kitchens. If homeowners want to entertain a large number of guests, they can book the common house for the evening. Many cohousing communities have a community garden. Heartwood cohousing near Durango, Colorado, has a greenhouse, workshop, and barn for use by the members.

Cohousing developments typically devote a portion of their land to open space and gardens. Children play freely on common lawns and the fields and forests that surround the homes (figure 1-5). Some private outdoor space is often provided as well. In many cohousing developments, people share equipment. Harmony Village in Golden, Colorado, for example, has a single lawnmower for 27 units!

BUILDING NOTE

The Center for Renewable Energy and Sustainable Technologies sells an interactive CD-ROM program that helps individuals and builders identify actions that will reduce the environmental impact of a building project. It's listed in the resource guide.

Most cohousing communities plan their homes and the surrounding grounds around people, not the automobile. Parking and garages are often in a central location. In Harmony Village, parking is restricted to the periphery. Groceries and other large items are delivered to individual units on the walkway, but cars are meant to be parked a short stroll from the housing for aesthetic reasons and to encourage community interaction. Contrast this to the typical U.S. suburb where people drive up to their house, hit the control to open the garage, then disappear from sight. Some cohousing communities are even experimenting with jointly owned vehicles.

FIGURE 1-5.
Although abandoned on this cold, wintry day in Colorado, the common area between the homes of this cohousing development is often bustling with children engaged in play and adults engaged in discussion.

Use Recycled Building Materials and Recycle All Waste. It is often said that "in nature, there is no waste." The truth be known, as soon as there is life, there is waste. All organisms produce waste, sometimes lots of waste. In nature, however, waste generally doesn't go to waste—it is used and reused over and over. As countless nature programs remind us, the waste of one organism serves as the food source of another. Dung beetles have a heyday when the rhinos arrive on the African savanna and begin depositing tons of fecal matter that serves as a food source for the hungry beetle.

In building a sustainable home, domestic and industrial waste are one of your greatest assets. Used automobile tires, straw, pop bottles, drywall, clothing scraps, newspapers, cardboard, plastic milk jugs, and a host of other secondary materials are all being used to make good-quality, reliable, durable products for homebuilding. Chapter 14 describes many products made from recycled waste and the resource guide lists places where you can purchase them.

Recycling should be built into each and every new house. Convenient recycling centers in kitchens and garages are essential and will become more important as cities and towns the world over divert more and more waste from landfills and incinerators into the recycling network. In building a home, select materials that can easily be recycled. You might, for example, choose steel roofing, which is readily recyclable, over asphalt shingles, which are not currently recycled. Do the planet and the next owner a favor.

And don't forget to recycle construction material. You will be shocked at the stream of waste paper, scrap metal, lumber, bottles, and the like that is generated at a construction site. According to one source, up to 25 percent of the garbage in U.S. landfills is construction and demolition waste. Unless you or your builder is environmentally minded, most of this perfectly usable material will end up in a dumpster and be hauled off to a local landfill. Let's recycle it. As a T-shirt of mine (made out of fibers from recycled plastic) says, "Recycle or Go Extinct."

Tap into the Earth's Generous Supply of Renewable Energy. The economy of nature depends almost entirely on renewable resources: the sun, soil, water, and plants. The human economy, however, is largely built and maintained by tapping into the Earth's supply of nonrenewable resources, such as coal, oil, natural gas, steel, aluminum, and a host of other metals. The differences between the two are many and profound. One of the most important is that nonrenewables are finite; they cannot be regenerated by natural processes. Reliance on them is dangerous, especially in light of the relentless expansion in both the human population and the global economy—and our unwillingness to do anything about these pernicious forces. Increased demand and finite resources are a train wreck in the making.

FIGURE 1-6.
Photovoltaic modules like these photographed in Rico, Colorado, produce electricity and are growing in popularity worldwide. These modules are mounted on a tracking device that keeps them in line with the sun from sun up to sun down.

Renewable resources such as trees and soil are regenerated by natural processes. They could provide a steady stream of energy and building materials *ad infinitum* if only we are smart enough to manage them properly and protect them from the avarice of many. Unfortunately, we seem unable to do this. Wind and sun are also renewable resources that cannot be depleted; the only limitation to their use is our own imagination (figure 1-6).

The potential of renewable energy resources is immense. One study published by the U.S. Department of Energy compared the potential of nonrenewable fuels, such as coal and oil, to renewables, such as wind, hydro power, and solar energy. The author found that nearly ten times more energy *annually* is available from renewable sources than from *all* of the fossil fuel remaining in the Earth's crust. That's right, *the annual renewable potential is nearly ten times greater than the total potential of the Earth's finite fuel base.*

Restore the Earth. In nature, organisms and ecosystems possess many mechanisms that permit them to repair damage. These natural restorative mechanisms are vital to the sustainability of ecosystems and nature's rich web of life. A sustainable human presence is possible as well, but only if we learn to restore the massive damage we have created.

Restored systems, if properly managed, have the potential to contribute significantly to future supplies of renewable resources, including many building products. Restoring previously devastated landscapes has many secondary benefits, including a potpourri of free services such as flood control, air purification, water pollution control, oxygen production, and even pest control.

How can you contribute to the restoration of the planet? First and foremost, you can assist by building a smaller home. Not only does it use fewer resources, it results in much less land disturbance. Fewer trees will have to be cut and fewer minerals will have to be mined. Smaller homes also create a smaller footprint—less disturbed land. As in human health, prevention is the first line of defense in ensuring planetary health.

Another preventive measure is to regulate construction activities on a building site to limit peripheral damage caused by workers' vehicles and heavy equipment. Although it is often difficult to reign in overzealous equipment operators, the effort is well worth it. One technique is to cordon off areas with fluorescent orange tape. If you don't, you may find a large area of devastation around your home.

After completing a building project, restoration work must begin in earnest. Replanting native vegetation increases the chances of speedy recovery and supports native wildlife populations. Immediately restoring disturbed areas also limits the establishment of hardy, invasive weed species that thrive on denuded land.

If a lawn is required, choose grasses and shrubs that are native to your region. They are better adapted to local soil conditions, temperature, and rainfall, and are more likely to survive. Planting native species minimizes time spent pampering delicate unacclimated foreign species and reduces demand for irrigation water and fertilizer. (For more on this subject see chapter 15.)

You might even consider earth sheltering your house, that is, burying large parts while maintaining a bright, comfortable interior, or installing a living roof—a roof of soil and vegetation (figure 1-7). If properly constructed and waterproofed, a living roof will provide a lifetime of protection and enjoyment. And it ensures that the surface of the Earth is used as intended, as a place for plants rather than buildings and pavement. Elk periodically graze on the

Restoring ecologically impoverished landscapes is not just an environmental do-gooders' project, it is essential to our long-term survival. Planet care is the ultimate form of self-care.

FIGURE 1-7.
The front roof of the author's tire and straw bale home in Colorado is covered with steel roofing but the back of the roof is underground. Dirt on the roof, now sporting a rich layer of grasses and wildflowers, reduces noise, provides insulation, and gives back part of the site to vegetation.

grasses growing on my roof, and wildflowers adorn the roof throughout much of the spring, summer, and early fall. Earth sheltering and living roofs also restore more of the house site to vegetation. (The living roof is described in more detail in chapter 3. Earth sheltering is described in more detail in chapters 11 and 15.)

Another way to promote restoration is to support suppliers like lumber companies who buy their wood from companies that have sound harvesting and restoration policies (see chapter 14).

Create a Safe, Healthy Living Space. A sustainable home is powered by renewable energy and is as efficient as humanly possible. It is built in a way that promotes restoration, and is made out of recycled and reused materials. It is also a healthy place to live. In other words, it is as good for the people who will live in it as it is for the environment.

A healthy home is warm and free of drafts in the winter and cool in the summer. It is well lighted and pleasing to the eye, both inside and out. And, of course, it is free of toxic substances.

Make It Easy to Operate and Maintain. Sustainable homes should be easy to operate and should require as little maintenance as possible. Passive solar homes are a good example.

A passive solar heating system has only one moving part, the sun. Although this is technically incorrect—the sun is stationary and the Earth is the moving part—it underscores the technological simplicity of a structure that meets human needs with little impact on the environment and with little effort by people. Passive solar energy is about as low maintenance and easy to operate as any heating system I can imagine. Passive solar heating relies on the sun as the source of energy for late fall, winter, and early spring heating; south-facing windows (in the Northern Hemisphere) let the sunlight stream into the home; thermal mass stores solar energy in the interior of the house; and insulation and curtains or shades retain the heat. It requires the brainpower of a dairy cow to operate, so if you can open a curtain or a window shade, you have all the necessary skills.

Plaster or cement stucco exteriors, if properly applied, reduce maintenance. Many natural building products also provide durability and low maintenance. Stone is the ultimate in long-lasting materials. Rammed earth, a building technique described in chapter 2, is a close second. Native vegetation and preserved natural areas reduce yard maintenance, the bane of modern suburban existence.

Another way to reduce repair and maintenance is by eliminating pumps and other electrical equipment that are typically used to move air or water to compensate for poor design. Gravity-fed catchwater systems, for example, harvest rainwater from the roof with far less energy than traditional water systems, such as wells, as you will see in chapter 13. Proper placement of thermal mass to absorb sunlight stores and disperses heat within the house, eliminating the need for fans and ducts. Properly placed windows and openable skylights provide ventilation and eliminate the need for air conditioners. Earth sheltering and shade trees make expensive, energy-consuming cooling systems unnecessary. You get the idea.

Your job is to devise as many simple, creative ways as you can to make your house work for you without pumps, fans, air conditioners, heaters, and, of course, electricity. As you read this book, you will discover many ways to work with nature to achieve the same ends we achieve through technology, energy, and, of course, our hard-earned salaries.

Design Your Home to Be Accessible. When most people, including professional architects, design homes, they design them as if the present were all that mattered. That is, they create floor plans that are based on a static state, ignoring aging and injuries that will occur over time. Little thought is given to the occasional mishap that relegates a person to a wheelchair or crutches. But injuries do happen and our lack of foresight becomes all too clear at such times.

The good news is that small changes in design can accommodate inevitable changes in our bodies as well as the occasional accident. To promote wheelchair access in a home, build slightly wider doorways and make bathrooms a bit larger—spacious enough to park a wheelchair so one can comfortably transfer from the chair to the bathtub or the commode. Special turn-around zones can also be designed into hallways and other high-use areas to facilitate wheelchair movement. Sinks and cabinets at wheelchair height and with sufficient room for a person's legs are simple design adjustments. Access through the front door is another extremely important consideration. Inconspicuous ramps or ground-level access that allow one to enter right through the front door without having to climb steps are easily designed into a home.

The type of handles and knobs you specify for doors and cabinets and the type of faucets installed in the kitchen and bathrooms can greatly facilitate the lives of people who develop arthritis as they age. These small considerations also make a home more manageable for aged parents or grandparents or friends with disabilities. What is more, it doesn't take much thinking or much work to make a house disability friendly. Check out the *Directory of Accessible Building Supplies* published by the Research Center of the National Association of Home Builders for a list of accessible building products. For more information on design with accessibility, see Sam Clark's book, *The Real Goods Independent Builder: Designing and Building a House Your Own Way.*

Make Your Home Economical in Every Sense of the Word. The popularity of natural, sustainable housing depends, in large part, on economics—how affordable they are to build, furnish, and operate. Contrary to the myth that building an environmentally responsible home is a costly venture reserved only for the well-to-do, I have found that this option can be quite cost competitive. Even with some unfortunate cost overruns, for example, my home cost $5 to $15 per square foot less than most other new spec homes in my area. It also operates at a fraction of the cost of these homes. But that is not to say that every aspect of building and furnishing my house was cheaper than a more conventional route. Not at all. I found that some environmentally friendly products, such as a super-efficient refrigerator and compact fluorescent light bulbs, cost more, sometimes a lot more, than conventional materials. However, other green building products, such as recycled tile, carpeting, carpet pad, insulation, and paint, cost the same or sometimes less than the environmentally unfriendly products they were designed to replace.

Some products and features of a natural, sustainable home produce immediate economic savings (reduced operating costs) and additional comfort. Installing extra ceiling insulation, for example, keeps a house warmer in the winter and cooler in the summer, lowers fuel bills over the lifetime of a house, and lessens economic burdens. Some investments create additional savings in reduced equipment and construction costs. For example, a well-insulated house will require considerably less heat than a poorly insulated one. Although additional insulation costs more, it enables one to install a smaller and less expensive furnace, saving you money right away. Installing low-flush toilets and showerheads slashes water demand and reduces the size of the leach field you will need, if your house has a septic system. Because low-water toilets and showerheads are no more expensive than standard fixtures, you save money both in the short and the long run.

Not all purchases result in immediate cost savings. A solar electric system initially may cost $10,000 or more, but you will reap its benefits for many years to come. When the cost of the system is amortized over its useful life and when avoided costs, such as meters and hookup to the utility, are factored in, the system often turns out to be a pretty good investment, especially if you are some distance from a local utility line. In my case, solar electricity ended up costing a little more than a conventional electrical hookup. An investment of this nature, however, buffers one against utility price hikes and saves thousands of dollars in damage to human health and the environment because solar electricity is so much cleaner than electricity from coal-fired and nuclear power plants. Decreasing these "external costs" is part of the benefits package offered to the sustainable homeowner.

Consider Life-Cycle Costs of All Materials and Products. All building materials have a cost that exceeds the sticker price. This cost, called its life-cycle cost, takes into account the social, economic, and environmental price tag of a product from cradle to grave. That is, it includes all costs incurred from the extraction of the raw ore, the harvest of the wood, or even the collection of secondary (recycled) materials through the manufacturing, to the end use, and then to disposal or recycling.

During a life-cycle cost analysis, economists attempt to tally *all* costs, then try to place an economic value on them—including the very tangible costs, such as the cost of labor and fuel prices, as well as the very-real-but-hard-to-assess impacts, such as a product's contribution to air pollution, illness, habitat loss, and the like. Converting environmental losses into dollars and cents may seem heartless and stupid, but it does have a purpose. It enables us to understand the full impact of our lifestyle and to compare various products. As such, it is an important tool for sustainable building. I'll mention lifecycle costs of various products and materials in the following chapters to help you make decisions.

These are the key principles of sustainable design. I strongly advise you to study them, discuss them with friends, and incorporate them into your decision-making process. I guarantee that they will help you steer a steady course through the process of sustainable design and construction.

A PLACE THAT SOOTHES THE SOUL

In his book *The New Natural House Book*, David Pearson writes, "To support personal and planetary health, we need healthy and conserving homes; homes that help us to lead a new lifestyle; homes that are designed not to damage the environment but to bring positive regeneration to it; homes, in fact, that are not sick, but are healing places for body, mind, spirit, and planet." He goes on to note that a home is the largest investment most people will ever make. For many people, Pearson notes, it will become a lifelong focus, as they make repairs, modify existing systems, care for the grounds, add on as needs dictate, and, in many cases, make monthly payments. Our homes should be a true source of beauty and well-being, not just physical well-being, but also spiritual and mental health.

Our homes provide refuge, and by working carefully with natural materials and design we can make them comforting and nourishing. Many natural materials lend themselves to organic designs that replenish the human spirit. Design strategies that link the inner and outer space of a home create outside living spaces that flow gracefully into the interior of our homes, joining us more firmly with the basis of life, our environment. The pliability of natural

---

PRINCIPLES OF APPROPRIATE TECHNOLOGY

While searching the Web, I found a set of criteria for thinking about appropriateness of technology—criteria that could easily be applied to all building products and techniques. I thought it might help readers remember the principles I've outlined in the preceding paragraphs. Here is my adapted version, with thanks to the author, Jan Sturmann, who posted it on the Internet:

Does it harm
the producer, the worker, the user, the biosphere?

Does it improve
my work, my life, my awareness, my community, our environment, our collective future?

Is it affordable
to purchase, power, repair, and maintain?

Is it graceful
to use and to behold?

Does it make redundant
human skill, creativity, and lives?

Is it simple and convenient
to make, use, repair, and maintain?

Is it durable,
capable of lasting a long time?

Does it connect
me with you with us with our environment?

Is it inheritable
for children to enjoy and benefit from?

Bill Steen

materials also leads to greater artistic expression, allowing us to create a personal habitat that expresses our individuality while permitting us to break out of the oppressive influence of the rectilinear world we are imprisoned in. This organic design, often drawn from patterns of nature, permits us to create dramatic structures with curved walls that express our individuality and nourish our spirits (figure 1-8). A home is also a defining statement. By turning to sustainable design and natural materials, we express our commitment to environmental stewardship and good health.

Creating a home that will soothe the soul is an extraordinarily complex task. In modern society, convenience tends to dictate unimaginative rectilinear structures. Even some natural homes leave much to be desired. In conventional or natural homes, homeowners who have not designed beauty and grace into their structures often end up spending a fortune trying to decorate a home to give it a soul, often to no avail.

The art of creating a home that feeds the human spirit is the art of creating a house that not only works for people at the most basic levels of convenience, but whose features "are in sync with the psychology and physiology of how people thrive, or don't thrive," according to Sam Clark. This task is not just about creating a visually appealing home. While vision is

FIGURE 1-8.
Many natural building materials allow construction of sensuous, curved walls that help us escape the confines of our increasingly rectilinear world. This is a straw bale house.

important, it is only one sense through which we experience a home. We must not forget warmth, comfort, textures, solidity, and the play of sunlight on the floors and walls. We are participants in a structure. When we move from one room to the next, our senses shift. The size and shape of a room, its color, warmth, and light give us a bigger sense of the whole. It is these subtleties and interactions that give a house a soul of its own.

For assistance in this complex, yet exciting artistry, I strongly recommend reading *A Pattern Language* by Christopher Alexander, Sara Ishikawa, Murray Silversteen, and colleagues at the Center for Environmental Structure. They present a detailed analysis of use patterns, ways to optimally design homes around various functions, and ways to optimize the aesthetic and spiritual values of a house.

SOME PARTING ADVICE

Sustainable housing is not a new invention. For the most part, people have always built their homes out of natural, locally available materials in the most climate-sensitive ways. It was not until the modern era that we became divorced from these natural homebuilding methods. But, today we are witnessing a resurgence. People are tapping into the wisdom of important, but discarded building methods and melding them with the best modern practices of today. There is much to be learned as we experiment.

Because the new sustainable home movement is still in an experimental phase, you, your architect, and your builder will surely make some mistakes. I know, I made more than a few myself!

This book is my attempt to help you make as many correct decisions as possible, eliminating costly and embarrassing mistakes, while fashioning a house that serves the spirit, the planet, and the body. As you proceed, you will find that many so-called experts will express divergent views on various issues—for example, whether it makes sense to place rigid foam under a concrete slab. People just plain differ in their views. Always seek competent advice and be wary of those who can't back up their assertions with hard facts. If you feel in your gut that something isn't right, trust your intuition. Some misinformation comes from proponents of particular homes, who often promote "their technology" as if it were the ultimate in homebuilding. Well meaning and vital to the advancement of natural homebuilding, these pioneers sometimes suffer from a bit of zealousness that we must accept, understand, and lovingly forgive.

Beware of rosy construction cost estimates, too. Tim Pettit, an Earthship, straw bale, and adobe builder based in Ouray, Colorado, makes a living largely helping people who have gotten into trouble, either financially or technically, while building a natural home of their own. Many of his clients, he says, have read an article or two in a popular magazine or newspaper about people who have built Earthships for some ridiculously low price, then set out to replicate the experience. Sadly, many of them find that they can't build the home of their dreams for $40 per square foot. Others run into difficulty, being swayed by wrongful assertions that it is "easy to build a natural home." Low on money or in over their heads, they call Tim to bail them out. (Tim is listed in the resource guide for chapter 4.) If you are building your own home, hire a consultant early on, before you start construction. As Cedar Rose, a consultant and natural builder from Carbondale, Colorado asserts, "People get into trouble because they don't spend enough time planning. A few thousand dollars spent on planning can save you tens of thousands of dollars in the long run."

It is important to use common sense at all times. Check out new ideas and new products. "Some revolutionary designs and products that swept through the industry have been swept right out again a few years later—usually after many people have tried out the new approaches with mixed results," says Sam Clark. My advice is that it is okay to be on the cutting edge, indeed desirable, but don't let builders experiment with you and don't experiment with far-out ideas that really don't hold up to scrutiny. Discuss ideas with others and listen to honest feedback. If it sounds too good to be true, it probably is. You can

We learn most from our mistakes. People with a hands-on approach are prepared to go ahead and make mistakes, learn from them, and achieve results.

RENE DALMEIJER

build a sustainable home without going so far out on a limb that you will come tumbling down. Keep in mind the basic principles presented in this chapter and elsewhere.

By adhering to the principles of sustainable design and common sense and learning from the builders and the experts who live in various forms of alternative housing, you can make the best choices available at this time. And, of course, your choices will be influenced by economic decisions. Do you install aluminum roofing made from nearly 100 percent recycled scrap or do you install steel roofing, that may or may not be made from recycled material, but that costs half as much? My advice is keep your eye on the principles and look for the economic trade-offs. Although you may spend more for one product, you may save a little on another, or you may save substantially over the long run in reduced maintenance and operation costs. Ultimately, it's the long term that matters.

As Sam Clark notes, "Innovation is important, but even new methods that seem well engineered and brilliant have hidden costs, maintenance problems, and unintended consequences." Be bold, but be careful. Be a good earth steward, but be wise. A recycled window may seem like a great bargain for you and the planet, but if it leaks and rots out in a few years, what have you gained?

A house is more than a shelter. It is more than protection from cold winds or summer heat. It is more than a liberating factor in our lives. A home express a point of view. It express our values and our commitment. It expresses our willingness to act upon our values and our commitment.

But, lest we forget, one key value in forging a sustainable future is wisdom. In natural and sustainable building movements, unfortunately, many of us get caught up in thinking of a home principally as a moral choice. Our self-righteousness can taint our decisions, leading us to make unsound choices. Many fall into a trap of thinking that if a new product or idea promises to promote sustainability, it must be good. More than once I have heard people express the mistaken view that if a product is natural, it must be perfect. Many people in this movement shun conventional building practices, as well. If conventional builders are doing it, they think, it must be ill advised. Nothing could be further from the truth. Conventional approaches have been refined over decades of trial and error, and good practices often make good sense. As you will see in chapter 14, many products are now available to make our homes sustainable.

If nothing else, conventional practices help us understand building principles. It is for this reason that I recommend reading a book or two on conventional home design and homebuilding. You may well need to build a portion of your home, for example, interior walls, using more conventional techniques. *How Buildings Work: The Natural Order of Architecture* by Edward Allen, and *American Building: The Environmental Forces That Shape It* by James M. Fitch and William Bobenhausen, are good sources of information, the former being a more basic treatise on building techniques. Look again at the sidebar "Stick-Frame Construction: Pros and Cons" on p. 10.

Our goal in natural and sustainable building is to produce a home that nourishes the spirit and the body. Our goal is to produce a home with minimal impact on the environment by using natural and sustainable products and by tapping into solar energy and other natural forces that greatly reduce our burden on the planet. In a sense, our goal is to achieve harmony between people and their homes and between the home and the natural environment. But we must invite common sense and sound information to be our guides and companions on this fascinating journey. And we must do so with sobriety to temper our enthusiasm and desire to create a gentler form of architecture as we forge a sustainable future one house at a time.

THE NATURAL HOUSE / PART

Choices

RAMMED EARTH HOMES

WHEN PEOPLE FIRST HEAR about rammed earth homes, structures whose walls are built of dirt packed between forms similar to those used to pour concrete, they express a variety of emotions, among them skepticism, amazement, and incredulity. "What a novel idea!" one enthusiast remarked to me. "Who would ever have thought you could build a house out of dirt!"

The truth is that earthen construction is no newcomer to the art of building. In fact, humans have been building shelter from packed dirt for nearly 10,000 years. Over this enormous expanse of time, numerous structures—even entire cities such as the world's oldest city, Jericho—were constructed from packed earth. And this technique continues today in timber-poor, labor-rich countries. Few technologies can boast as long and as successful a history as rammed earth.

The buildings constructed of rammed earth over the millennia weren't simple, primitive mud huts, either, but often rather large and impressive buildings, including elaborate temples, churches, and mosques that sometimes stretched many stories high.

OVERVIEW OF RAMMED EARTH CONSTRUCTION

Rammed earth building is a rather simple natural technique. Solid, earthen walls are built on a variety of foundations (as you shall soon see). Workers begin by erecting wooden or steel forms like those used in concrete work on the foundation. They then shovel moistened earth containing the right mix of

Man has always managed to find ways to use earth for building shelter in all parts of the globe, with variations to suit each particular climate.

PAUL G. McHENRY, JR.,
The Adobe Story

33

sand and clay or cement into the forms. Special tamping devices are used to compact the first layer. Once it is properly tamped, another layer of soil is shoveled into the form and then compacted. This process is repeated until the form is filled. Soon after the wall has been completed, the forms are removed and another section is built.

Rammed earth hardens like stone. It is durable and weather resistant. As described below, it can be coated with stucco for further protection.

HISTORY OF RAMMED EARTH BUILDING

Rammed earth building technology took root in North Africa and in the Middle East in the cradle of civilization, where it is still practiced today. In North Africa, rammed earth buildings date back to the time of the pharaohs. This technology can be traced back many centuries in the Far East as well. Archaeological digs in China have revealed structures built of rammed earth dating back to the 7th century B.C. Parts of the core of the Great Wall of China, the construction of which began approximately 5,000 years ago, are also made of rammed earth.

In arid climates, rammed earth structures lasted for centuries, providing shelter from the wind and a cool respite from oppressive desert heat. As testimony to its strength and durability, ancient earthen buildings still remain long after the civilizations responsible for their construction have disappeared. The Biblical dust-to-dust lesson, it seems, applies to people, but not to their earthen buildings.

Contrary to what one might think, rammed earth building is not restricted to arid climates. It has been successfully used in more temperate zones, such as Europe, where it was introduced by the Romans and Phoenicians in their many conquests. The Romans, for example, introduced this building technology to the Rhone River Valley of France, where it became the dominant form of architecture for an estimated 2,000 years. A traveler in the region today delights in seeing hundreds of rammed earth structures, approximately 15 percent of the rural buildings. Many are still in use after centuries of inhabitation. The French called this new (to them) building technology *pisé de terre* (translated "packed earth").

Rammed earth was introduced into North and South America by the Spaniards, who built structures out of a mixture of soil and crushed seashells called *taipa*. Packed between removable wooden forms, this material formed extraordinarily durable walls. Many of them, like those in St. Augustine in northern Florida (the United States' oldest city), are still standing today. Taipal buildings were erected in South America, where rammed earth construction became much more popular than in the U.S. and has persisted until very recently as a dominant form of architecture. A traveler to Brazil, Peru, or Chile will find centuries-old taipal structures still in use today.

In Europe and North America, rammed earth building faded from prominence after its early successes, but experienced a resurgence in the 1840s and again in the 1930s during the Great Depression. Both upswings in popularity, though modest by most standards, were stimulated by a desire to develop a low-cost means of building homes that would be simple enough for would-be homeowners to construct their own shelter. During these periods, several key figures both in Europe and the United States played a significant role in furthering knowledge about rammed earth building and popularizing the technique. Most noteworthy was Frank Lloyd Wright. Wright saw rammed earth as a means for people to build affordable homes themselves—one of his passions. Numerous rammed earth buildings were built during these resurgences in popularity, including a housing development in Gardendale, Alabama, and the Church of the Holy Cross in South Carolina (figure 2-1). Despite its early success, this earth-friendly building technology declined in prominence as a result of mass-produced building materials that allow workers to build homes, especially exterior walls, in much less time than the slower and more cumbersome technique of ramming earth between forms. Like many great ways of doing things, rammed earth construction has fallen victim to convenience, speed, and economics. It is a story you will hear repeated over and over again in this book as we explore natural building techniques.

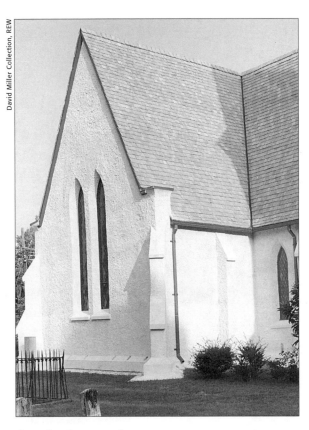

David Miller Collection, REW

FIGURE 2-1.
The Church of the Holy Cross in South Carolina was built of rammed earth in the 1800s. It stands today as proud testimony to the durability of this natural building technique.

In recent years, concern over the environment and dwindling natural resources has resulted in yet another, albeit still modest, upsurge in interest in this time-tested construction technology. The rebirth began in the 1970s as a handful of architects, contractors, and individuals in various countries independently began to explore the potential of rammed earth building. One of their goals was to find ways to meet demand for housing in an environmentally compatible manner. As author and builder David Easton points out in *The Rammed Earth House*, this enthusiastic group of visionaries was "inspired by the apparent simplicity of the system and the inherent logic of building with such a basic material." Today, most of the activity in rammed earth building is concentrated in Australia, Arizona, California, New Mexico, and France. Although there is great interest among some people, "earthwall construction, which holds such promise for affordable, durable, efficient, healthy housing, still struggles for respectability," writes Easton. Despite millennia of proven performance in hot, desert climates as well as more temperate

FIGURE 2-2.
Rammed earth structures vary in size and complexity. This 100-room, four-story resort in Australia is one of the most elegant structures I've ever seen.

climates, rammed earth structures in the U.S. desert Southwest capture only a tiny fraction of the new housing market, unable to compete with Southwest-style homes mass-produced in housing tracts from lumber and stucco.

Nonetheless, several U.S. rammed earth builders are thriving, giving encouragement to those interested in sustainable building. Tom Wuelpern's Rammed Earth Development, Inc., in Tucson, Arizona, builds a dozen rammed earth houses each year. Stan Huston of Huston Rammed Earth in Edgewood, New Mexico, builds an equal number of homes in New Mexico, Colorado, Utah, Texas, and Arizona.

One place where rammed earth construction is more than treading water is western Australia. In some parts of this arid region, rammed earth homes capture 20 percent of the new housing market. In Queensland, on the east coast of Australia, architect David Oliver designed and constructed an exquisite $26 million, 100-room, four-story resort hotel out of rammed earth (figure 2-2).

The remarkable success of rammed earth in Australia results from a variety of factors. One of the driving forces is the scarcity of trees. The second reason is that this area is home to many termite colonies that can do significant damage to a wood-frame house. In such cases, rammed earth building provides a logical, indeed intelligent, alternative. A third factor is that many houses in the area are already built out of brick and other forms of masonry. Rammed earth walls, which are as hard as concrete, are therefore viewed not as an anomaly, but rather as another form of masonry wall. This environment of acceptance is aided by the fact that rammed earth walls are less expensive than limestone or double brick walls.

In other nations, houses made of dirt, although beautiful and elegant, carry a negative stigma harkening back to the days of sod houses. Such a view is unfortunate, given the many advantages earthen structures have over "more civilized" and considerably less environmentally friendly modern building techniques.

AN INTRODUCTION TO FOUNDATIONS

As in any structure, one of the first details we must consider is the foundation. How is the foundation of a rammed earth home made? Are there any special requirements? Before we explore the various types of foundations commonly used in rammed earth homes, a few general words on foundations are in line. This information is relevant to just about any natural house you may build.

The foundation of a home—or any building for that matter—is a unified, stable base on which the building sits. Foundations come in many forms and are built from a variety of materials, including concrete, stone, and gravel. Although most of us give little consideration to the foundations of our homes, they are vital to a house's success and durability. The foundation is the structural component on which the walls rest. It also provides a point of attachment, anchoring the building to the ground and thus limiting the movement of the walls—an important task, especially on sloping property.

The foundation must also support all of the loads a building presents to it. The word *load* is an engineering term synonymous with force; there are two kinds of forces or loads, live and dead. The dead load placed on a foundation consist of forces resulting from the weight of the whole house, including the walls, floors, and roof. Live loads are movable, variable forces such as the weight of furniture, appliances, and even the pet ferret. Live loads also include the transitory weight of guests who drop by to share a bottle of wine or snow that periodically accumulates on the roof.

A third category are the *lateral loads*, a type of live load, but so important that they merit a name of their own. Lateral loads are forces, such wind or earth pressure, that push against the walls of a house.

The foundation of a building has to be solid enough and deep enough to handle *all* of these loads with some margin of safety built in for unexpected events such as a hurricane, an unexpectedly thick blanket of snow, or the weight of a new water bed. The foundation must also be rigid enough to resist cracking as parts of the ground beneath it settle a bit here and there. This is especially important for rammed earth homes. As Paul McHenry, Jr., points out in his book *Adobe: Build It Yourself*, in a masonry structure, such as rammed earth or adobe with plastered walls, even a 1/8-inch shift will create large cracks in the wall. In earthquake-prone areas, foundations and walls may require special precautions.

Finally, the foundation must provide a moisture break between itself and the walls so that soil water does not seep from the ground into the walls of one's home. This is achieved by coating the outside surface of the foundation with a waterproofing material. It is vitally important for rammed earth walls, which can deteriorate if they get too wet.

Few people pay much notice to the foundation of a house, other than structural engineers, builders, and building department personnel. But its importance cannot be overemphasized. As Easton points out, "An earthwalled building, even more than a lightweight frame building, demands the strength and protection afforded by a well-designed foundation."

FOUNDATIONS FOR RAMMED EARTH HOMES

Rammed earth homes can be built on a variety of foundations. The simplest and least costly is a foundation of earth achieved by starting the earthen walls below grade—that is, just below the ground's surface. Beware: This technique, while inexpensive, is only effective in dry climates where freezing is a rare

> *Any type of ground can resist only a certain amount of load, dead or live, before it cracks, slides, caves in, or slips.*
>
> NADER KHALILI, *Ceramic Houses and Earth Architecture*

occurrence. Also keep in mind that the soil under the earthen wall/foundation must be dense and stable enough to prevent settling and cracking. Ideally, buried earthwalls should be stabilized with cement to provide additional strength and resistance to moisture. Translated, this means you should add cement to the earthwall you are building to enhance its strength and water resistance.

Another type of foundation consists of a stone wall that extends beneath the earth's surface. Builders begin by excavating a trench slightly wider than the wall, around the perimeter of the proposed home, usually with a backhoe. The trench extends below the frost line, the deepest level to which the soil freezes during the winter, and is then filled with large and small stones. A cement-based mortar can be used to stabilize the stones. The earthen wall is then constructed on top of the stone foundation. More information on stone wall foundations is presented in chapters 6 and 9.

Rubble Trench Foundations

Trenches can also be filled with gravel, road base, and crushed stone with concrete poured over the top to form a grade beam (see figure 2-3a). Such systems, referred to as rubble trench foundations, require drainage when they are installed in wetter, colder climates, usually porous pipe at the base of the trench. This prevents water from accumulating beneath the foundation. Drainage is vitally important, because any freezing of water and soil below the foundation can cause heaving (lifting) of the foundation. Remember: Water expands when it freezes. Heaving, in turn, can result in severe cracks in the foundation and the overlying walls. The grade beam should be poured so that it is raised off the ground to prevent water from seeping up into the wall it supports.

The depth of the trench for these foundations varies considerably, depending on where you live and the type of climate you are cursed or blessed with. In Arizona, where the soil rarely freezes, you may only need a 12-inch-deep trench. In Iowa, where winters are considerably colder and the ground routinely freezes and stays frozen for long periods, the trench may need to be 4 feet deep to be safe. According to some sources, insulating a rubble trench foundation permits one to create a much shallower foundation.

Although rubble trench foundations are an economical option, they do not present as much resistance to wind as a foundation with a concrete stem wall and footing anchored in the earth below grade. In addition, they may not be sufficient in seismically active areas because they don't anchor the house in place. In the absence of sound anchorage, a home can literally be rocked off its foundation by an earthquake. Finally, this type of foundation is insufficient in weaker soils—those with a load-bearing capacity of less than one ton per square foot.

A foundation filled with rubble or other materials may seem problematic, but if the truth be known, rubble trench foundations are an age-old practice and have been used extensively by such architectural dignitaries as Frank Lloyd Wright. Attesting to their soundness, similar foundation systems are used for railroad tracks and highways. In places where the frost line extends halfway to the center of the Earth, this type of foundation is often cheaper to construct than a concrete foundation.

Concrete Foundations

The most common foundation in use today in the United States is made of concrete. Concrete foundations typically consist of two parts, the spread footing, or simply footing, and the stem wall. As shown in figure 2-3b, the spread footing is much like your foot, which besides providing a handy place for your socks, distributes your weight on the ground. The stem wall is akin to your legs. It connects the rest of your body to the feet.

In strong, dense (well-packed) subsoils, the footing needs to be only as wide as the stem wall. In weaker, less dense (less well-packed) soils and subsoils, the footing needs to be wider—sometimes two or three times wider than the stem wall itself. To understand why, think of snowshoes and how well they support one's weight even on a loose layer of freshly fallen snow.

Builders use several techniques to pour stem walls and footings, well beyond the scope of this book. But what you do need to know is that stem walls and footings are generally poured between wooden forms in excavated trenches and are reinforced with steel rods, known as rebar (reinforcing bar). Rebar is placed lengthwise in the form near the bottom and the top throughout the entire foundation. Rebar is wired in place before the concrete is poured in the form. Details of this process can be found in many of the more detailed books on natural building techniques listed in the resource guide.

FIGURES 2-3 A, B.
Rubble trench foundations (A) are inexpensive and generally acceptable in most regions. Insulation along the outside of the foundation reduces the depth of the structure and saves money and resources. Concrete foundations (B) are more commonly used in rammed earth construction. They consist of a footing and stem wall.

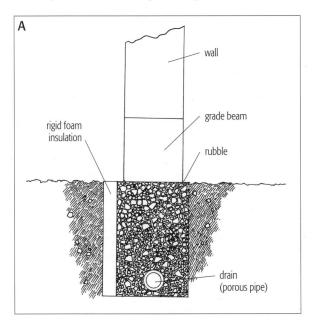

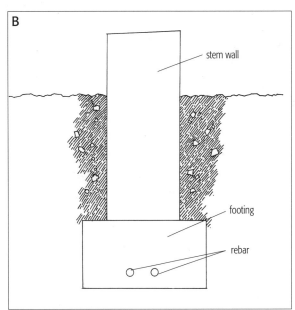

Steel reinforcing is required to enhance the tensile strength of concrete. Although you may think of concrete as a rather strong material—after all, they build bridges and tall buildings out of the stuff—it's really not that strong. On its own, concrete has natural strength in one area: compression. In other words, it resists compression; you can support a lot of weight on a concrete pillar. However, don't bend concrete because this deceptively strong-looking material is exceptionally weak in tension. If you do, it will break apart. Steel, on the other hand, possesses a lot of tensile strength, which means that you can bend it a lot before it breaks. When added to concrete, steel provides tensile strength, greatly increasing the uses to which concrete can be put.

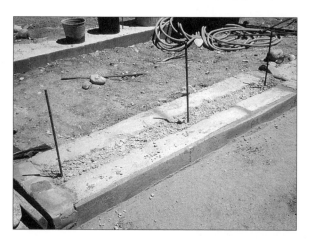

FIGURE 2-4.
Rebar anchors the wall to this foundation for a rammed earth home built by Stan Huston. Notice also the roughened surface of the top of the stem wall to increase adhesion, and the depression, which serves as a keyway, further locking the rammed earth wall to the foundation.

Rebar typically runs lengthwise in the stem wall and footing. In building a foundation, the deeper the footing and stem wall, the more horizontal steel is required. In rammed earth construction, especially in earthquake-prone regions, rebar may be placed vertically so that it extends above the uppermost level of the stem wall, as shown in figure 2-4. This technique, also used in straw bale construction, provides a way to anchor rammed earth walls to the foundation.

Before you or your architect or builder can design a foundation, you will need a soil test to determine how dense the soil is and how much weight it can bear. This information will determine how wide the footing should be. In addition, you will need to know the depth of the frost line—that is, how deep the soil freezes. This information permits the designer to determine how deep a foundation must be to avoid frost heave. The deeper the foundation, the more steel reinforcing is required.

For those who have never built a home, remember this: Although an architect or a builder can design a set of plans that typically includes a foundation drawing, a structural engineer usually has to approve your plans before you can obtain a building permit (this process is discussed in more detail in chapter 16). If changes are required by the engineer, the architect or builder will need to make them, then resubmit the plans to the engineer for his or her stamp of approval. After the engineer has stamped and signed the blueprints, they can be submitted to the building department for approval.

The foundation of a house is so important to its structural integrity that building departments typically require an inspection prior to pouring concrete—that is, after the forms have been set and the rebar has been placed, and after the pour to be sure it went well. This may seem like overkill, but it's not. A faulty foundation can ruin a home. Don't cut corners here. You do so at your own peril.

Yet another popular and cost-effective system used to create foundations is known as the monolithic slab. In this technique, the stem wall and footings are formed and then poured along with the concrete slab. This technique is used on houses that have no basements and results in slab-on-grade—that is, a slab of concrete with thickened edges (that serves as the stem wall and footing). The concrete slab is the subflooring over which you can lay tile or carpet—or just about anything else.

To create a slab-on-grade, a trench is first excavated around the perimeter of the house and the forms are set up for the stem wall and footing. The floor is also prepared for a concrete pour. Floor preparation includes most, if not all, of the following steps, depending on local conditions and local code requirements: (1) excavation for the slab and foundation, (2) installation of plumbing (both water and sewer lines and floor drains), (3) proper tamping or compaction of the soil so the floor is on solid ground, (4) installation of sand to give the concrete a more solid base, if the subsoil is inadequate, (5) installation of rigid foam insulation and steel reinforcing mesh (typically 6 inch by 6 inch, 10-gauge wire mesh laid flat on the floor), (6) placement of electrical conduit, and (7) installation of a subslab system for venting radon gas.

In the excavation (step 1), be certain to remove all grass and vegetation from the site of the future slab. This material will decay over time and cause the slab to settle and crack. Also, be certain that the slab depth is fairly uniform. If it is deeper in some places, you may not have ordered enough concrete. While you can get more, it is expensive to have a small amount delivered. Also be sure to insulate hot water lines that will run in your slab. If you don't, you will find that it takes a century or two to get hot water at distant faucets and once you turn a hot water faucet off, the water in the line will cool down almost instantly.

Concrete foundations for rammed earth homes require some modifications to make wall construction easier. For example, to facilitate the assembly of the wall forms, many builders create a small ledge known as a ledger on the top of the stem wall. David Easton uses 1 by 2-inch lathing or 3/4 by 1½-inch strips of plywood nailed to the inside of the form boards (both sides) at the top of the future stem wall. The ledger makes it easier to clamp the forms to the stem wall.

Leaving the concrete on the top of the stem wall very rough improves the mechanical bond between the top of the stem wall and the earth wall above it. Rebar that extends from the foundation through the lower portion of the rammed earth wall enhances the strength of the bond as well, and is recommended for areas of high wind or seismic activity. Although the rebar creates a stronger bond between the wall and foundation, it also poses a risk to workers who will later descend down into the forms to pack the earth. Another technique used by rammed earth builders is to carve out a depression in the top of the stem wall before the concrete has cured. This creates a keyway, which further enhances the bonding of the rammed earth wall to the foundation.

BUILDING NOTE

When building with rammed earth, be sure that your foundation is made precisely to the dimensions of the rammed earth wall. If you don't, forms will not fit and you will increase the cost of wall construction enormously!

PAUL LINVILLE,
Koala-T Construction

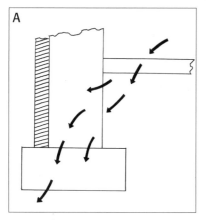 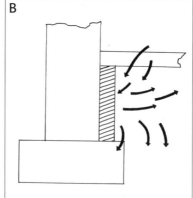 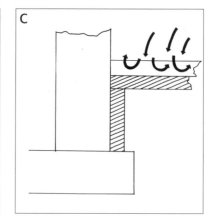

FIGURE 2-5 A, B, C.
*Placement of rigid foam
insulation greatly affects the
flow of heat (arrows) from a
structure into the surrounding
ground. The first option (A)
is better than no insulation.
Inside placement (B) is better
yet. Optimal insulation
is shown on the right (C).*

BUILDING NOTE

R-value is a measure of resistance to
heat. It is a relative measure of how
fast heat travels through a material.
The higher the R-value, the lower the
rate of heat flow.

Insulating and Protecting Foundations

Many builders in cold climates insulate the exterior of the stem wall of a foundation using 2 to 4-inch rigid foam (figure 2-5a). The insulation helps keep cold out of a house and is highly advisable for those living in cold or temperate climates. Studies show that up to 17 percent of the heat loss in a building may occur through the foundation. (I even recommend rigid insulation for garage foundations, as it keeps the concrete warmer and reduces damage.) Paul Linville, a builder with many years experience in rammed earth and conventional construction, prefers to place the insulation along the inside of the stem wall. This method offers several advantages. One of the most important benefits is that it physically separates the stem wall from the heated interior of the house, including the floor. This, in turn, prevents heat from escaping into the ground through the stem wall, as illustrated in Figure 2-5b.

A variety of foams with different R-values are available for insulating foundation walls. (See sidebar for a definition of R-value.) One of my favorite rigid foams is Cellotex, an insulation that provides considerably more R-value per inch than standard white or blue foam.

Adding insulation to a project is easy and relatively inexpensive. The additional labor and cost are well worth it in the long haul, as your house will be warmer and a lot easier and cheaper to heat and cool. According to some sources, foundation insulation can reduce the depth of the stem wall—cutting costs that easily make up for the additional cost of insulating. For example, if code calls for a three-foot-deep foundation, proper insulation can reduce the depth to one foot. Builders call these frost-protected shallow foundations. Be sure to check out frost-protected shallow foundations with your building department and with a competent engineer.

After the concrete forms have been removed, rigid foam insulation can be anchored to a foundation wall with ripple or brick ties or with a suitable adhesive. Insulation can also be placed in the concrete forms prior to pouring.

Another innovative way to insulate your foundation, which is popular in Sweden, is to bury underground-rated rigid foam insulation around the perimeter of the house. As shown in figure 2-6, the insulation extends out horizontally from the structure, sloping slightly away from the house, for drainage. This keeps the ground underneath from freezing and helps retain heat in the house. For more on foundation insulation, get a copy of *Home Insulation* by Harry Yost, an extremely useful book that should be required reading for all homeowners. In chapter 14 I'll describe innovative ways to make concrete foundations less resource and energy intensive, by using fly ash concrete or insulating concrete forms.

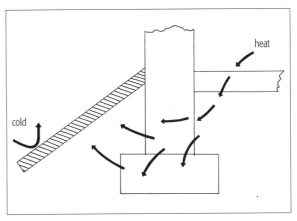

FIGURE 2-6.
Perimeter insulation popular in Sweden reduces heat loss from buildings.

As a final note, all foundations require good drainage to prevent water from accumulating next to the house. Drainage is usually achieved by grading the property so the ground surface slopes away from the house. Additional measures that help are extenders on downspouts that transport water away from the house or underground pipes attached to downspouts that perform the same task but are out of sight and out of the way of children and lawnmowers. Underground pipes can also be used to collect water from a roof for irrigating gardens, trees, shrubs, and grass, and for providing household water (see chapter 13).

As Paul McHenry, Jr., points out in *Adobe: Build It Yourself*, "Foundations and slabs built by nonprofessionals frequently are either more substantial than necessary" and thus more costly, or "inadequate." Creating foundations and pouring slabs is demanding work and, in my view, best left to professionals.

BUILDING RAMMED EARTH WALLS

When I visited my first rammed earth home, I was amazed by massiveness of the 24-inch-thick outside walls of the house. What solid structures! I also found the houses to be quiet and warm, almost like a sanctuary. The experience reminded me of my many peregrinations over the past twenty-five years through the canyon country of the West. And well it should, for rammed earth walls are nothing more than artificially made rock with the look and feel of rock. In fact, building a rammed earth wall essentially replicates the geological process of sedimentary rock formation, albeit in a much shorter time span. Let's look at how rammed earth walls are made. First stop on this journey is a brief exploration of the soil used to make them.

Creating the Proper Mix for Building Walls

One lesson that rammed earth wall builders have learned over the years is that for a wall to endure the vagaries of weather and geology, it has to be made of the right stuff, a mixture of mineral matter that when compacted can withstand wind, rain, snow, and even earthquakes. One of the first rules for creating the right stuff is that organic soil, such as the topsoil in your garden, is

Think of rammed earth as a sort of "instant rock." The natural process of creating sedimentary rock occurs over a span of thousands and millions of years. An earth rammer, on the other hand, creates it in a matter of minutes.

DAVID EASTON,
The Rammed Earth House

strictly forbidden. Builders either exclude it entirely or limit it to 1 or 2 percent of the mass of the wall to ensure the adequate strength and permanence. Organic soil is structurally weaker than mineral matter. It also contains varying amounts of organic matter, notably plant material, present in various stages of decomposition. If decay continues in a wall, it will leave tiny air pockets that weaken the structure.

Historically, rammed earth walls have been made from a variety of mixtures of mineral matter, reflecting the widely varied conditions of the world's subsoils. Over time, though, the mix that seems to have "won out" over all others consists of 70 percent sand and aggregate (small stone or gravel) and 30 percent clay, which acts as a binder, a substance that "cements" all of the other particles together. Clay comes in many varieties and not any old clay will do. The correct clay has to be a nonexpansive, that is, one that does not expand when wet and then crack when it dries. These expansion/contraction cycles stress the wall internally. Although clay is used as a binder in many parts of the world, it is best suited to arid regions that experience minimal seismic activity, such as western Australia, New Mexico, northern Africa, and the Middle East.

In places where moisture and earthquakes are a problem, or where building codes require it, Portland cement is added in small amounts (3 to 12 percent by volume or about one to three 50-pound sacks per cubic yard of soil) to the earth mixture as a stabilizer. Adding cement renders a wall more water resistant and increases its compressive strength and longevity. Several builders I've talked to use cement routinely, choosing a mix that has no clay at all.

According to David Easton, very few soils contain the optimum mix of sand and clay. Moreover, soils vary markedly from one region to the next and may vary substantially on a given building site! Because you need the correct mix of sand and clay, test your soil before you start building. A quick field test you can do yourself is to add a handful of soil and some water to a glass jar, then mix it and let it settle. Particles settle out according to size and mass. Fine particles such as clay and silt end up on top, while coarse particles such as sand and gravel end up on the bottom. A visual inspection of the settled material tells you whether your proposed soil mix is at least close to being acceptable. Details of this test are described in Easton's book.

If your soil passes this pretest, the next step is to send a sample to a soils lab for further testing. Soils labs perform tests to determine the precise composition of the soil and other important features such as shrinkage, moisture absorption, density, and the compressive strength of the soil when it has hardened. The information you obtain from a soils lab is essential for your structural engineer, and may even be required by your building department.

If the soil on site is suitable for building rammed earth walls, you could save hundreds of dollars in materials costs. However, if your soil isn't right, don't give up. You may be able to augment the soil by adding sand or clay from local quarries to make it suitable for wall building.

BUILDING NOTE

Not all builders agree with the use of clay in rammed earth building. Sam Huston of Huston Rammed Earth in Edgewood, New Mexico, and Tom Wuelpern of Rammed Earth Development, Inc., in Tucson, Arizona, for example, prefer a mix with little if any clay. Clay sticks to the tools, they say, and expands and contracts in concert with changing moisture levels, which causes cracking.

If your soil is still not amendable for rammed earth construction and your heart is set on this building technique, you may be able to obtain a suitable mix from a local quarry. Most quarries sell road base and structural backfill that are suitable for rammed earth wall construction. Road base consists primarily of sand and gravel with 10 to 20 percent clay, which is a relatively decent mix, although a little coarse for many people's tastes. Because of the low clay content, it may require the addition of a little Portland cement as a stabilizer.

Quarry fines are another decent wall-building material available from local sources. Quarry fines consist of mineral matter left over from the screening of decorative rock. They contain sand, clay, silt, and small gravel. The clay concentration is higher than in road base and usually falls within the acceptable range for rammed earth construction. Quarry fines have another benefit: They often cost half as much as road base. Even though quarry fines are less expensive, the cost of supplying wall-building material from an external source is largely determined by transportation costs, which usually far outweigh material costs. If you can transport your soil, you can save some money.

Another problem with quarry fines is that the composition may change if the quarry starts extracting from a different stratum after you have obtained and tested samples. In a project in the Napa region of California, engineer and author Bruce King found that when they were ready to start construction, the quarry that had supplied the initial material on which they performed their preconstruction tests had begun excavating from a different vein. The new material not only had a different appearance, it had different structural characteristics. The story ended happily, however, as they were able to obtain enough from the initial vein to complete the project.

Besides finding the right mix of materials for proper structural stability and resistance to weather, one must also consider aesthetics—what color the wall will be. After all, aesthetics is a major reason why people choose this particular natural building method. Color is provided naturally by the contents of your mix—clays, sand, aggregate, road base, quarry fines, or whatever material you use. Builders modify colors by adding different colored clays or sand or cements—white or tan cement, instead of gray cement, for instance. Or, various colorizers can be added to achieve an aesthetically appealing wall.

After you or your builder has determined whether the local subsoil will be suitable for building walls—alone or with modifications—or that a local quarry can provide you with the right mix, and after you have chosen a color or settled on the natural color of your soil, you will need to construct some test blocks and test cylinders. These tests allow you to see what your walls will look like, how strong they will be, and how resistant they will be to wind and rain. Even if your soil has passed all previous soil tests, subject it to this last battery of tests so there are no shocking surprises awaiting you after the walls have been built.

A NOTE OF CAUTION: RADON

Some builders suggest that you test your soil for radon before you build so you don't end up with radioactive walls. Contact a local radon mitigation company for a test kit.

BUILDING NOTE

One advantage of buying structural backfill or road base from a local supplier is that the quarry typically has had the mix tested by a soils lab and will provide you with a copy of the test data, saving you a little money.

FIGURE 2-7.
Workers spread and tamp dirt delivered to the tall steel forms in this home under construction in Buena Vista, Colorado. This technique pioneered by Stan Huston allows rapid construction of walls but is only suited to contractors.

BUILDING NOTE

Different sources make different recommendations for water content, ranging from 8 to 13 percent. Basically, the water you add should produce a slightly wet mixture that forms a ball. When dropped from the waist, the ball should break into several coherent clods. In the words of engineer Bruce King, the clods "should not shatter or splatter."

Assuming the test blocks look good and will hold up well under different weather conditions, it's time to start building. You will need to build or purchase forms, which are assembled on the foundation and packed with soil. Thorough mixing and crushing of the earthen material is required to eliminate lumps of clay. If the clay is not adequately broken into the smallest particles possible during mixing, soft lumps will remain in the mix. Interspersed in the otherwise solid earthen matrix, they can substantially weaken earthen walls, so much so that some builders won't use soil if it has any clay in it at all; they insist on using cement stabilizers.

Soil used for wall building is mixed on site in one of several ways, depending on your budget and your preferences for machines or hand labor. Hard-core enthusiasts with a meager budget and an aversion to mechanical assistance prefer to mix the soil by hand, which is painstakingly slow. Those who have a more generous building budget and only a modest distaste for machinery may opt to mix their soil with a garden rototiller. Or, if you have an even larger working budget and don't mind machinery, you can mix the soil with the bucket of a backhoe or Bobcat. Experts warn that even this is rather slow, especially compared to the system used by Stan Huston (figure 2-7). Huston's workers feed sand and cement into separate hoppers of a pug mill mixer using a Bobcat. The two are mixed and fed onto a conveyor belt that carries the mix to a second conveyor that delivers the mix to the wall forms. This technique is fast and economical, but it is only suitable for professional builders who can afford the $85,000 worth of equipment.

After the soil, stabilizers, and colorizers are mixed, water is added. In Huston's conveyor system, water is sprayed on the mix just before it is dumped onto the conveyor. If your mix passes the hand-drop test, described in the sidebar and designed to test for the proper water content, you are ready to start ramming earth into the forms. At this point you have to work fairly quickly, as the mix will set up in an hour or two. A word to the wise: Don't mix the soil until the forms are in place and don't mix more than you can use in any given form.

Form Systems for Rammed Earth Walls

The success of rammed earth building depends on having the right soil mix and proper forms. Although every builder seems to have his own way of forming rammed earth walls, most forms are made out of 2 x 4 or 4 x 8 sheets of plywood. Because the walls are subject to enormous force, builders use either

standard ¾-inch plywood, 1⅛ inch plywood subflooring, or ¾-inch HDO (high-density overlay) plywood similar to that used in concrete form work. The forms also contain an external support system (figure 2-8).

Rammed earth homes are built using one of three techniques: (a) the panel-to-panel system, (b) the freestanding wall system, or (c) the continuous wall system.

The Panel-to-Panel System. In the panel-to-panel system, alternating sections of wall are created one at a time to full ceiling height, as shown in Figure 2-9. After these sections have cured, the forms are removed and intervening sections are constructed, thus creating a continuous wall.

Forms for the walls are laid on either side of the stem wall along the ledger. The ends are then closed off with pieces of plywood (the end forms) to create a box that is open at the top. To prevent the walls from bulging as the earth is packed into the forms, 2 x 10-inch or 2 x 12-inch planks or steel bars, known

FIGURE 2-8 A, B.
Forms used in rammed earth construction. On the left are wooden forms used by David Easton. On the right are steel forms used by Stan Huston.

BUILDING NOTE

Because of the extensive form work required for rammed earth construction, rammed earth building of any serious nature is best suited to contractors and subcontractors, not owner-builders.

FIGURE 2-9
This home was built by David Easton using the panel-to-panel system. Notice the vertical lines demarcating individual panels and the concrete bond beam holding the walls together.

David Easton

How long have you been involved in rammed earth construction?

Twenty-five years.

What got you interested in rammed earth building?

The opportunity it afforded for building a less expensive house. In those days I had an excess of personal energy and a shortage of capital. I had always been attracted to adobe construction, having been familiar with it from my youth in southern California. Earthwalls seemed a way to build for less money. I read about rammed earth as opposed to adobe in Ken Kern's *The Owner Built Home* and in George Middleton's book on rammed earth. The technique seemed to be more direct than making adobe bricks. So, it was building my own home that first stimulated my interest. As soon as I built my very first wall, I felt a passion to improve the technology. That kept me going, and has kept me going to this day.

What changes have you seen in this building technique over the years?

The changes are massive, and they are the reasons that rammed earth has gained such popularity. First, forming systems. The traditional rammed earth forming system was extremely cumbersome, heavy, and slow to set up and take down. It was difficult to maintain plumb, straight walls, and the walls themselves were so unsightly as to always need plaster cover-ups. Second, mixing and delivery. Mixing was slow with shovels and garden tillers, and delivery was equally slow and tedious with shovels, wheelbarrows, and buckets. Thirdly, not enough was known about soil composition, and quality control was inconsistent. Finally, architects were not designing correctly for the material, rather they were thinking frame house and just using a thicker pencil for the exterior walls. Much had to be learned about construction detailing and interfacing thick, monolithic walls with the other trades.

Over the years, the few veteran earth rammers in Arizona, Australia, and California have made improvements to the technology so that it is not only acceptable to building departments and banking institutions, but attractive to the market as well.

How are rammed earth homes accepted?

The banks, at least in my area, like rammed earth houses because they retain their value and are typically built by financially secure clients. The building departments accept the building technology because there is now sufficient engineering proof of its load-bearing capacity and design parameters. Engineers accept it for the same reasons. Architects like it because clients are asking for it and they like to design with thick walls. Environmentalists like it because it reduces the number of trees that need to be cut to build a house and because a house built with rammed earth walls will last many times longer than a wood-frame structure, thus preserving resources. Well-designed rammed earth homes also use less energy to heat and cool. The marketplace, certainly here in the wine region of northern California, has warmly embraced thick earth-walls because of the connection architecturally to the style of southern France and northern Italy. People lucky enough to own a house with thick earthwalls become strong advocates because of how uniquely quiet and comfortable the houses are.

Is rammed earth construction growing in popularity?

Yes, although not to the degree I had originally hoped, at least in my own small geographic niche. In Australia, rammed earth has caught on to such a degree that in certain parts of the country, it represents 20 percent of all new house construction. This is primarily because wood frame is not the principal alternative, but rather brick; rammed earth is less expensive than brick. It is more expensive than wood frame, which is the primary reason, I think, why rammed earth has not been more successful in the United States.

What are the biggest barriers to making rammed earth a more popular building technology?

Cost and market perception. It is more expensive than frame construction because it is a better product. Unfortunately, buyers are controlled by the purchase price and without any tax incentives or utility company participation, frame walls will continue to dominate.

Is it as costly as some critics argue? Is it just something for the rich?

Rammed earth houses are slightly more expensive than frame houses, but that is because they are better than frame houses, just as a Volvo is better than a Toyota. They are safer, last longer, and are more enjoyable to be inside. They are not, however, only for the rich. Rammed earth can be economical if designed appropriately and simply. It will never be as cheap as frame and stucco until the price of lumber reflects its true environmental cost.

Are costs going down?

Our market is demanding more elaborate and exciting houses, but the cost of the wall system is going down. The introduction of our PISE technique (pneumatically impacted stabilized earth) reduced the price of monolithic earth walls by 30 percent. If there were widespread competition, I think you would see the cost of "thick PISE" walls only slightly higher than

frame. If some very large builder were to use PISE in a well-orchestrated major development, I think PISE could actually beat frame.

How are your homes holding up?

Our houses are holding up well. The oldest ones, built when we were still learning about soil types, have problems with accelerated surface weathering. Houses with inappropriate design details, such as parapet walls in high rainfall areas, have more significant weather-related problems. All of these problems are correctable with topically applied water sealers and earth patching. The houses built in the past ten years are holding up very well.

What comments do you get from your customers? Any negatives?

Our homeowners love their rammed earth houses. Even the ones who had trouble with the process and became anxious and distrustful of their contractor, grow to love them once the construction crises have faded into memory. The houses are comfortable and quiet. Restful, dramatic, calming. The negatives? I've heard people say they can be cold. I haven't heard anyone say they felt damp.

What mistakes do people who build their own rammed earth houses commonly make and how can they be avoided?

The mistake they make is in soil selection and stabilization. They can avoid it by thoroughly testing

the soil and sample blocks well ahead of time. They must also design a house appropriately for the climate.

What's your prognosis for rammed earth twenty years from now?

As recently as five years ago I would have said that rammed earth (actually PISE) had a very bright future. I expected the environmental movement to have more strength in getting legislation passed that would place a higher value on "green" building systems and that somehow the marketplace would be moved along to demand more responsible housing. In response to that demand, I thought production homebuilders would start building environmentally responsible housing units and that gradually we would see a plethora of young, bright, idealistic architects and builders revolutionizing the way people live. Now I don't know. I have a suspicion that thick walls such as rammed earth and straw bales will lose to thin, synthetic wall systems. Engineered studs from wood derivatives or even recycled plastic may dominate. Space is so expensive that there may not be any room in the housing industry for thick walls.

DAVID EASTON is a California-based builder and author of the best-selling book *The Rammed Earth House*. He has built 100 rammed earth homes and has pioneered many new techniques to improve rammed earth building.

FIGURE 2-10.
Steel frames are extremely heavy and must be moved with mechanical assistance like this Bobcat.

as walers, are placed horizontally against the form and the whole assembly, including planks, end forms, and plywood is held together by pipe clamps, as shown in figure 2-8a.

For durability, some builders use commercially available concrete forms, the support structures of which are made of steel and water-resistant material in place of plywood to enhance the longevity of the form. Although durable and long lasting, these forms are heavy and expensive. They also require a Bobcat to set in place and remove (Figure 2-10).

Soil is delivered to the forms until they are filled to a depth of 7½ to 8 inches. The soil is then tamped down to a depth of 4½ to 5 inches. Tamping is performed either by using a hand tamper or with a special pneumatic tamping device powered by a compressor. Both techniques require a strong back, strong arms, and great physical endurance.

Tamping is not only difficult, it is potentially dangerous because workers must climb down or reach down into tall forms to compact the first few layers. Working eight feet down inside a form, they also have to be careful not to impale themselves on the rebar protruding from the foundation.

After the soil is compacted, the next layer of untamped mix is added and the process is repeated until the form is full. Each layer must be carefully tamped to ensure the structural integrity of the wall. If a worker doesn't compact the soil sufficiently, the wall will lack the strength it needs to support the roof load. If, on the other hand, a worker overcompacts the soil, the forms may blow out. Too much compaction can also lead to fracturing and reduced wall strength, according to engineer Bruce King.

With experience, builders find the right balance between too little and too much tamping. They know the consequences of making a mistake, which are difficult or usually impossible to rectify. By their very nature, notes author Doug Eure, rammed earth walls defy alteration. The only thing to do when a wall is defective is to knock it down with a tractor and then break it up with sledgehammers, a process that takes many hours.

After the walls of the house have been completed, a concrete bond beam is poured to hold the wall together (figure 2-11b). A bond beam is a structure that allows a secure attachment between the roof and the walls. (You'll learn more about bond beams in chapters 4 and 6.

Bond beams can be poured directly into the upper 8 or 10 inches of the form. Doing so will require many forms, all of which must be left in place until

the exterior walls of the structure are completed. Bond beams can also be created by pouring concrete into forms made of 2 x 10s erected on top of the finished wall, an option that requires additional labor, lumber, and cost (figure 2-11b). Because of their massive weight, bond beams need not be anchored in the wall, although you must anchor the roof to the bond beam.

The Freestanding Panel System. In seismically active regions, rammed earth builders use a slightly different system of forming, known as the freestanding panel method. Like the panel-to-panel system, individual segments of the wall are built one at a time. However, the gaps between the wall segments are only 10 to 12 inches wide (figure 2-12). After the forms have been removed and the walls have cured, forms are erected for the gaps and the bond beam, then reinforced with steel, and filled with wet concrete.

FIGURE 2-11 A, B.
The concrete bond beam (top stratum) on the wall on the left (A) was poured in the upper 8 to 10 inches of the rammed earth wall forms. Bond beams can also be poured separately in specially constructed forms, as on the right (B).

FIGURE 2-12.
This home is being built using the free-standing panel system. Concrete will be poured in the section gaps between wall segments after the forms have been built.

FIGURE 2-13.
Commercial builders may build enough homes to be able to afford to erect forms for an entire building. This eliminates the vertical joints common in other systems. Panels can be built to the entire height of the wall or a few feet at a time.

Cynthia Easton

In this system, the concrete columns and bond beam absorb most of the forces generated by earthquakes and reduce the stress on the rammed earth walls. Because it is more labor and material intensive, the freestanding panel system costs more than the panel-to-panel system.

The Continuous Wall System. The last technique is known as the continuous wall system. As shown in figure 2-13, forms can be laid along the entire foundation to the full wall height. Smaller forms, measuring only two feet high, can also be laid along the foundation. Soil is then added and tamped as in other systems. After the first course has cured sufficiently, the forms are raised two feet, secured, then filled with dirt and tamped. After this section has cured, the process is repeated until the entire wall is built. When the form is approximately 9 inches from the top of the wall, a concrete bond beam is poured.

This technique requires smaller forms that are much easier to erect and much easier to fill and tamp. They're a lot safer, too. Some builders consider rammed earth walls built this way to be the most aesthetically appealing, for they lack the concrete columns of the freestanding panel system and lack the vertical joints of the panel-to-panel system. Like most things in life, this method has its drawbacks. For one, it requires a lot of form work and therefore a lot of lumber. Because of this, it is generally recommended only for contractors, who can afford to invest in the forms that they can use over and over again to make their investment pay off.

How High Can Walls Go?

If your plans call for a two-story building, rest assured that rammed earth will work fine. In fact, many multi-story buildings made from rammed earth have survived admirably for centuries. At least three methods are used to add a sec-

RAMMED EARTH FACTS

• Thickness of typical rammed earth walls is 18 to 24 inches.

• Amount of earthen material required to build a 2,000-square-foot home out of rammed earth is 200 tons.

• Rammed earth walls weigh about 2 tons per running foot.

• An 18-inch-thick wall can safely run about 18 feet high.

TOM WUELPERN, Rammed Earth Development, Inc.

ond story (see more advanced treatises on the subject, such as *The Rammed Earth House*). Here's a quick overview for the curious reader.

The first method of constructing a two-story building is to build the first-story wall, top it with a bond beam, then erect a slightly narrower second-story wall on top of the wall and bond beam. As shown in figure 2-14, the second-story walls are 3 to 6 inches slimmer. Because the outer surface of second-story wall is flush with the exterior surface of the first-story wall, a ledger is created on the interior surface at the interface of the two walls. This is used to support the floor joists for the second story.

The second method is to build the walls full height without an intervening bond beam, but still making the second-story wall 3 to 6 inches narrower than the first-story wall, to create a ledger for the floor joists. The third method is to build the walls to full height, keeping them the same thickness throughout. If this method is used, anchor bolts must be embedded in the walls to attach 2 x 12 or 4 x 12 ledgers needed to support the floor joists.

Unfortunately, building two-story rammed earth increases the complexity and costs of a project. Tom Wuelpern says that while "it is structurally not a problem to go two stories high, labor costs escalate dramatically." In fact, to build second stories his company rents an oversized Bobcat with an extension or uses a conveyor to move the dirt to the second-story forms. Although a second story adds floor space without the need for additional roof, and thus saves a little money over a single-floor design, the foundation must be more substantial to support a two-story building. This, combined with additional labor and equipment costs, negates any savings you might generate.

Doors and Windows

In rammed earth construction, doors and windows are installed in rough openings in the exterior walls. One way to make these openings is to insert a removable wooden box, called a volume displacement box, or VDB, the exact size of the rough opening into the form as it is being filled and tamped (figure 2-15). To do this, pack dirt up to the bottom of the future window (or door), and insert the VDB, making sure it is level. Then pack dirt around the box. After the forms are removed, the box is removed and *voilà!* a rough window (or door) opening exists.

Simple. Yes. But experienced builders warn that the process is very time consuming. The reasons for

FIGURE 2-14.
A cross section through a two-story rammed earth wall showing the ledger needed to support floor joists for the second story.

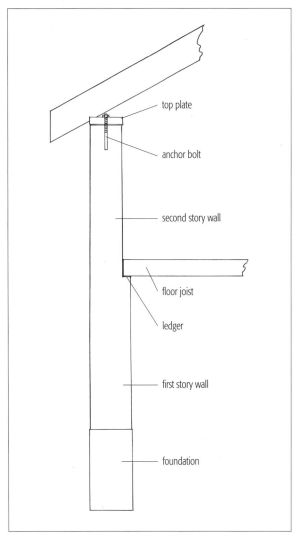

- top plate
- anchor bolt
- second story wall
- floor joist
- ledger
- first story wall
- foundation

this are many. First of all, you can't insert any old, haphazardly built box into a rammed earth wall. The VDB has to be made to the exact dimensions of the rough opening. If the rough openings are not right, that is, you mess up and build one that is too big or too small, you will have to reduce or enlarge the opening. Even the slightest error can take a lot of time to fix, and your alterations, unless skillfully carried out, can detract from the aesthetics of the opening.

VDBs must also be sturdy to resist collapse as you pack dirt around them. At the same time, they need to be easy to remove after the forms are removed. To do so may require partial disassembly, but because you may be using the box again, disassembly doesn't mean demolition. Packing dirt around the window box also requires special care and that takes more time. Furthermore, because windows in a house often vary in size, separate boxes must be built for each window size.

Rough openings can be created without VDBs—but only if you are using the panel-to-panel method. In this technique, the walls are built in sections with intervening gaps, which is where windows are placed. To make a window opening, you erect partial forms in the intervening gaps. These forms rise to the bottom of the rough window opening. Add dirt to the form and tamp down. At this point, you have created three sides of the rough opening. After the bond beam is poured on top (spanning the panels and the window opening), you have a completed opening. Door openings can be made with VDBs and by carefully planning the placement of panels.

The tops of doors and windows may be spanned by lintels or headers. The lintel supports the weight of the wall and the roof above the door or window. Lintels accept the load imposed from above and distribute it to the walls on

FIGURE 2-15 A, B.
Volume displacement boxes are placed in forms to create rough openings for windows.

Cynthia Easton

FIGURE 2-16 A, B.
Lintels support the wall and roof over doors and windows and can be made from a variety of sturdy building materials (a wooden timber, A, poured concrete, B). They also add a touch of class to rammed earth homes.

either side of door and window openings. Without a support structure spanning the top of an opening, door frames will shift, making it impossible to open and close your doors. Windows frames bend and glass can even break. In some cases, the wall could collapse.

In standard wood-frame homes, lintels or headers are built from framing lumber, usually two 2 x 4s or 2 x 6s. In rammed earth homes, lintels are typically made out of precast concrete, timber, steel, poured concrete, or packed, reinforced earth (the latter, however, only when openings are small), some of which are shown in figure 2-16. More advanced treatments are described in books on rammed earth construction.

Small arched openings can be made using rammed earth. They add a lot of character to a home. Arches are made by inserting specially built wooden frames (figure 2-16). Dirt is packed around the frame, and after the soil has cured, the arch frame is removed.

Special forms can be used to create niches in walls for design or functional purposes such as shelving. During wall construction, wooden forms are lowered into the wall forms, leveled, and attached to the surface of the form. Dirt is then carefully packed around the niche form. After the wall has set up, the niche forms are removed, creating a wonderfully creative space.

The Finished Wall

The appearance of a finished earthen wall varies considerably, depending on the color and texture of the materials and the forming method. Some builders produce a fairly uniform surface, which they achieve by using a uniformly colored and textured soil mix. Jim Wilson, a rammed earth homebuilder in Texas, however, specializes in building walls that show lines or strata much like the natural layering of rock strata. He achieves this effect by using materials of different color as the forms are filled. He also adds traces of lightly packed, poorly mixed earth, which form honeycombs on the surfaces of the walls, creating an ancient look. Australian builder Giles Hohnen builds walls with an

BUILDING NOTE

Some form systems leave noticeable lines in the surface of the wall. These can be sandblasted off in a day or so, depending on the size of the house, creating a nice, uniform wall surface.

Rammed Earth Homes / **55**

even more stunning array of colors, created by different colored clays available in his area. To author Doug Eure, these walls are reminiscent of the shifting hues of the American Southwest's Painted Desert. Paul Linville once built a house in which he painstakingly matched the color of the wall strata to the nearby mountains, using different colored soils as the wall grew in height.

Forms often leave lines on the walls, which builders can sandblast off. Others prefer to cover interior and exterior walls with plaster or stucco. In Tucson, Arizona, builder Tom Wuelpern finds that most of his customers prefer a plaster or stucco exterior finish so their homes fit in better with their neighbors' homes. Tint added to plaster or stucco gives virtually any color you'd like. Plaster and stucco also protect the walls, but applying wall finishes of this nature is labor-intensive and can add substantially to the cost of a home. If that's what you'd like but you don't have the budget, you can wait until funds become available.

RAMMED EARTH WALLS AND PASSIVE SOLAR HEATING

Builder Quentin Branch, of Rammed Earth Solar Homes in Tucson, asserts that "Although many [people] choose rammed earth for its sculptural potential, most want it for its passive solar capabilities." Rammed earth walls have a high thermal mass. Thermal mass refers to any material that absorbs heat, either from sunlight striking it or by being in contact with warm air (see sidebar). In passive solar designs, heat from sunlight or from sunlight-warmed air is absorbed by the rammed earth walls on sunny days, then dissipated into the room at night or on cool days. To achieve this effect, careful placement of solar mass walls is essential, as described in chapter 11. One method, suggested by Jim Chiaro, a straw bale builder from southern Colorado, is to use rammed earth to build some of the interior walls of a house, especially those in close proximity to south-facing windows. This way sunlight entering the house strikes the mass directly, increasing the efficiency of heat absorption. Figure 2-17 shows a suitable solar design and traces the path of the incoming solar radiation.

Although solar heating is described in detail in chapter 11, a few words are important to allay common fears and concerns. One of the most frequent concerns I hear is that solar homes will overheat in the summer. This fear is based on the erroneous assumption that a solar house allows so much sunlight to enter that the house becomes unbearably hot in the summer.

As you may know, in the northern hemisphere the summer sun rises in the northeast, crosses overhead, then sets in the northwest. Very little sunlight can penetrate the south-facing windows of a properly built passive solar home. Therefore, although it is bright and sunny outside, the interior of a passive solar house is not bathed in direct sunlight. The sun is too high in the sky. Only

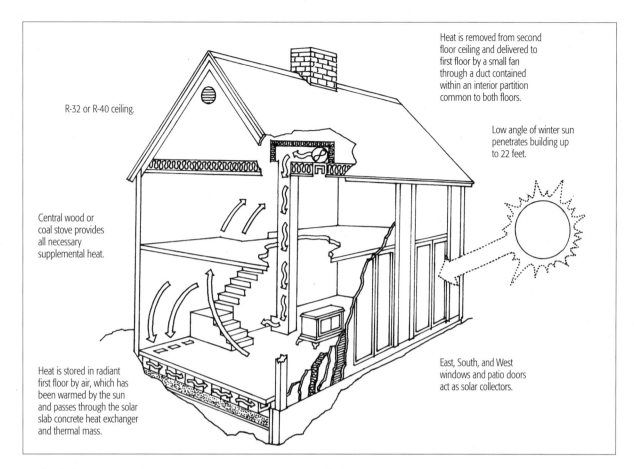

Heat is removed from second floor ceiling and delivered to first floor by a small fan through a duct contained within an interior partition common to both floors.

R-32 or R-40 ceiling.

Low angle of winter sun penetrates building up to 22 feet.

Central wood or coal stove provides all necessary supplemental heat.

Heat is stored in radiant first floor by air, which has been warmed by the sun and passes through the solar slab concrete heat exchanger and thermal mass.

East, South, and West windows and patio doors act as solar collectors.

in the fall, winter, and early spring, when heat is needed, does sunlight enter the structure, and it only comes in then because the sun is lower in the sky.

Although exterior walls heat up in the summer and transfer heat to the interior, this occurs with any house. To avoid the problem, include appropriately sized eaves (or overhangs) or other architectural features, such as vine-covered lattices, in your plans. These provide shade that keeps walls cooler. Trees and shrubs also reduce sunlight striking exterior walls, as explained in chapters 11 and 15.

In hot desert climates, rammed earth homes tend to stay cool during summer months even if sunlight strikes the exterior walls. Sunlight is converted to heat (radiant energy), which begins to migrate inward. In such a climate, however, by the time the sun sets, the heat has penetrated only nine inches into the wall. When the air cools, heat reverses its path and exits to the cool evening air.

Although rammed earth homes function well in hot, arid climates, they do not perform well in hot, humid climates because nighttime temperatures do not fall as dramatically as they do in the desert. Because the walls can't cool off at night, heat accumulates in the thick earthen walls, making the interior of the house hot and uncomfortable.

FIGURE 2-17.
In this passive solar home designed by James Kachadorian, the floor is the main thermal mass. Heat is channeled through concrete blocks under the cement slab and through vents in the floor at the back of the house. Heat absorbed by the blocks and slab is then released into the room air at night or during cold weather.

The suitability of rammed earth in cold climates is uncertain. Some builders advise against the use of rammed earth in cold climates because the thick earthen walls provide little insulation—surprisingly little. In fact, an inch of rammed earth has an R-value of 0.25. Therefore, an 18-inch wall has an R-value of 4.5, not much better than double-pane, low-E glass.

Other builders note that with a few modifications, rammed earth homes could be made suitable for colder climates. One solution is to build thicker walls. The mass will absorb heat and radiate it back into the room. Another is to add rigid foam insulation, either on the exterior of the walls or inside the walls themselves. Rigid foam insulation with a fairly high R-value can be placed on the outside wall after the forms have been removed. A layer of cement stucco applied over the insulation will protect it. Rigid foam can also be inserted inside the wall as it is being built.

Rigid foam creates a thermal break, a barrier to heat and cold. In the winter, the thermal break helps hold heat inside the building, keeping it from seeping through the walls into the great outdoors. In the summer, the thermal break helps prevent heat from migrating into the structure and retains cool temperatures inside the structure.

ROOFS

Rammed earth construction lends itself very nicely to a wide variety of roofs. The most important advice is to select a roof that suits your climate. If you live in a wintry climate, choose a steeper roof that sheds snow to prevent excessive load, potential leakage, and possible roof collapse. Don't erect a flat roof in snow country. You'll regret it when the roof begins to leak or, worse yet, fails.

If you are collecting snow melt from your roof to recharge your household water supply (described in chapter 13), you will want as much south-facing

FIGURE 2-18.
Anchor bolts to be placed in the bond beam of this house are used to attach the top plate, to which the roof is attached.

Cynthia Easton

roof as possible (if you live in the northern hemisphere, that is). This design promotes melting and contributes a considerable amount of water to your annual supply. On north-facing roofs, snow may linger for months on end. If it does melt, the water tends to freeze along the edge of the roof and in the gutters, creating ice dams that can damage roofs by causing leakage or collapse. South-facing roofs are also suitable for mounting solar panels, both photovoltaics that generate electricity and solar hot water panels for domestic hot water and space heating (discussed in chapters 11 and 12).

Let your imagination go wild. A good roof adds character to a house. But remember, the more complex your roof, the more it will cost.

How do you attach a roof to a rammed earth home? The process is actually quite simple. The concrete bond beam on the top of the walls acts as the point of attachment for the roof, as long as you have installed anchor bolts in the concrete prior to the pour (figure 2-18). Anchor bolts are used to attach a top plate, also known as a sill or roof plate. Top plates typically consist of 2 x 6s bolted to the top of the bond beam. Wooden trusses or rafters are attached to the top plate and the rest of the roof is assembled as in standard wood-frame construction.

The roof frame can be made from a variety of products, but one I recommend to those wishing to reduce their wood consumption and protect the forests is the I-joist, which I like to call wooden I-beams, shown in figure 2-19 and discussed in more detail in chapter 14. Made from oriented strand board and plywood, I-joists are as strong as the solid lumber typically used for roof rafters. I-joists are lightweight, easy to use, and consist of 40 to 60 percent less wood that the solid dimensional lumber they are designed to replace. In addition, they are constructed from wood derived from small trees, so you don't have to cut down an old-growth Douglas fir to create 2 x 10 or 2 x 12 roof rafters. I-joists are widely available and come in depths ranging from 9½ to 16 inches, providing adequate space for superinsulating a ceiling.

Prefabricated wood trusses are another choice for roof framing. Wood trusses come preassembled or can be built on site, usually from 2 x 4s. Although they use a lot more wood than an I-joist, they utilize smaller trees than those necessary to make 2 x 10s or 2 x 12s.

The bottom line is: you can get a deep roof to stuff with insulation without using much wood or without cutting down old-growth trees. Alternative roofing materials are described in chapter 14. For more information on roof insulation see chapter 14 and Harry Yost's book *Home Insulation*.

Willamette Industries

FIGURE 2-19.
I-joists are used to build roofs and floors. They contain 40% to 60% less wood than solid dimensional lumber and are strong, competitively priced, and widely available. They also allow for the construction of thick insulated spaces.

BUILDING NOTE

A truss is a system of rigid, interconnected triangles that utilizes small individual members (dimensional lumber) to create a structural element capable of spanning much greater distances and carrying much greater loads than the constituent pieces.

ANN EDMINSTER and
SAMI YASSA, *Efficient Wood Use in Residential Construction*

FLOORS

Rammed earth homes are amenable to a variety of floors. You can, for example, install tile, carpet, wood, marble, slate, bricks, flagstone, tinted concrete . . . whatever. Perhaps the most environmentally sound floor is made of earth, which I'll describe in chapters 3 and 4.

Tinted concrete is an interesting option. Pigmented concrete is poured, then steel-troweled to create a smooth finish or stamped to create marvelous realistic patterns resembling flagstone or various types of tile. Stamped concrete floor can be beautiful, although they tend to crack. However, if a concrete floors is covered with tile or carpet, cracking isn't a problem, except that radon gas may seep into the room if you haven't taken precautions to avoid this problem. In exposed floors, cracking creates an aesthetic problem, which can be minimized, sometimes prevented, by pouring slabs in smaller sections and placing expansion joints between them. (Concrete can also be tinted after it has been poured. Special chemicals are applied after the floor has cured.)

Another environmentally sound option is soil cement. As its name implies, soil cement is a mixture of soil and cement, not unlike concrete, only a little softer and less expensive. Soil cement is not a new idea; builders have been using it extensively over the years to line ponds, pave driveways, and stabilize steep slopes along highways. Some builders use soil cement for countertops, window sills, hearths, and garden pavers.

Soil cement cracks more than concrete and more unpredictably, so special precautions must be made to prevent it from fissuring in wild and weird patterns. Score lines, like those you see in concrete slabs, can help prevent uncontrolled cracking. Or, you can let the floor crack and when it is done simply grout and seal the cracks. Soil cement is also stickier and harder to create a smooth, flat surface with than concrete. Because soil cement is more porous than concrete, it stains if it hasn't been surface sealed. On the positive side, soil cement takes longer to solidify and dry, which gives you more time to get it right. As David Easton points out, "Soil cement is intriguing, but very tricky to work with." He adds, "You may think you have poured a good floor, only to see it soon ripped to shreds by Rover's claws or the legs of the dining room chair." So be wary. If you need to save money and want an environmentally sustainable floor, try soil cement, but seek counsel from those in the know and proceed with caution.

ELECTRICITY AND PLUMBING

Rammed earth construction poses a few special challenges when it comes to running electrical and plumbing lines—problems that are easily overcome. Wiring, for example, can be run in the exterior walls, in the concrete pillars (in homes built using the freestanding panel system), or in the floor. In all three instances, wiring must be run in conduit, typically a flexible metal or plastic pipe that protects the wire from contact with cement or rammed earth. Wiring is generally pulled through the conduit after the shell of a house (roof and walls) is complete, using an electrician's fish tape. Wiring can be run without conduit in the ceilings and interior walls.

In rammed earth construction, conduit and outlet boxes for light switches and receptacles need to be rated for use in masonry. Check your local building code for exact details. Electrical boxes and conduit that travels through rammed earth walls must be put in place after the forms have been erected and before workers start filling and tamping them. Electrical boxes are typically screwed into the form, open side facing out. Workers must take special care not to dislodge electrical boxes and conduit when packing dirt around them.

TV cable, phone lines, wires to speakers, and fiber optics can be run in conduit in exterior walls if you don't want unsightly wires running down your walls. This conduit must also be installed prior to tamping. As a precaution, you may want to add a few extra conduits; you may need them later.

To minimize the hassle, many builders run the majority of their electrical service through more convenient locations, such as ceilings and interior walls or under floors. This saves a lot of time for the dirt crew!

Like electrical lines, plumbing can run through exterior walls of a rammed earth house. Beware: A small leak or condensation on pipes can cause severe damage to earthen walls. In addition, because water pipes can easily be damaged when workers ram earth in a form, many builders find other, safer places to run water and sewer lines. The slab or space between floor joists in a wooden floor are good options. Interior walls, usually framed by conventional means, are another. Besides protecting the pipe from damage, these options make the job of packing earth in the walls a lot easier and faster. If you have to install water lines in the earthen wall, use a hand tamper when packing dirt around them to avoid breakage.

One exception to the rule is vent pipes from sinks and toilet lines, which carry noxious odors to the outside. Because these lines run vertically in the walls of a house, are relatively easy to work around, and carry no water, they can be located in exterior walls.

WHAT DOES IT COST TO BUILD A RAMMED EARTH HOME?

The cost of a rammed earth house varies. According to builder Quentin Branch, his customers pay from $65 to $125 per square foot, depending on a variety of factors, among them, how much labor the owner supplies, the complexity of the project, and the level of detail and refinement.

On the lower end of the scale, one client designed her own home, prepared all of the construction documents, and chose a simple design with unplastered unstuccoed walls, brick-on-sand floors, and a steel roof. She installed the plumbing and wiring herself under supervision. Finish work was minimal. On the high end of the scale at $125 per square foot, the customers did none of the work and chose a much more elaborate home with fireplaces, niches, and other more costly details, including interior plaster, stucco over exterior insulation, sandstone and tile floors, handcrafted cabinetry, tile roof, and viga-and-latilla ceilings. Expensive appliances, lighting, and plumbing fixtures can send the final cost even higher. You could easily spend $150 per square foot if your tastes run toward the elaborate end of the spectrum. Expensive tastes and fancy details are, however, inconsistent with the goals of living lightly on the land. Simplicity is the key. You can live well without Italian marble. I guarantee it! And the world will be a better place for all of us, except maybe the Italian marble importers.

SHOULD YOU BUILD A RAMMED EARTH HOME?

As with most alternative building technologies, enthusiasts and avid supporters—including builders and owners—sing the praises of rammed earth. That's okay. But a sober, objective look at the pros and cons of rammed earth construction is far more beneficial than the uncritical chorus of praise from its proponents. In the spirit of objectivity, let's examine the advantages and disadvantages of rammed earth construction.

Advantages of Rammed Earth Homes

- Rammed earth homes use local materials, reducing our impact on the planet.
- Building homes with earthen walls reduces our demand for lumber and helps protect forests.
- Rammed earth homes can be extremely attractive. They are some of the most inviting natural homes I've seen.
- Rammed earth homes lend themselves to passive solar design.
- In many climates, these homes are energy efficient and thus reduce our dependence on fossil fuels, which reduces air pollution and other impacts associated with the production, transportation, and consumption of fossil fuels.
- Rammed earth homes can be built incorporating a wide variety of styles, including contemporary, classical, and Southwestern.
- Rammed earth walls are very strong and capable of resisting earthquakes, hurricanes, and tornadoes.
- Rammed earth walls are resistant to decay. If well designed and properly built, they can outlast wood-frame structures, which deteriorate over time and need serious renovation after 50 to 75 years.
- Rammed earth walls are safe in fires. In fact, fire hardens and strengthens the exterior walls.
- The thick walls of rammed earth are soundproof, a benefit of immense value in this increasingly noisy world.
- Rammed earth walls are impenetrable to insects and rodents, unlike standard wood-frame homes.

Although natural building principles have been applied to homes of all sizes and price ranges, genuine economy and its concomitant—financial freedom—are available to those who adopt a culture of modesty.

BRUCE A. SILVERBERG,
Introduction to *The Art of Natural Building*

- Rammed earth building technology may be less expensive than brick, stone, adobe, and even ferro-cement construction.

Disadvantages of Rammed Earth Homes
- Like many forms of construction, rammed earth is fairly labor intensive and, in places where labor costs are high, may be more costly than conventional (though less environmentally friendly) techniques.
- Building rammed earth walls, while magical, is very hard work.
- Rammed earth construction requires careful analysis of soils. Without it, costly errors can be made.
- On-site mixing, wetting, and compacting of soil requires careful attention to detail. Mistakes are difficult, if not impossible, to repair once a wall has hardened.
- Rammed earth building requires special skills in form construction to make walls, windows, doors, and niches.
- Rammed earth construction is not widely understood, creating barriers to its acceptance, especially by other builders, homeowners' associations, and building departments.
- Rammed earth homes require a lot of wood to build forms, although some wood can be reused after the form work has been completed.
- Rammed earth construction is best conducted by contractors and not generally well suited for owner-builders. Contractors can use the same forms over and over again, saving money and resources.

This chapter has given you an overview of rammed earth construction. While greatly simplified, it should be sufficient to enable potential architects, builders, and homeowners to decide if they would like to learn more about rammed earth construction. If you like what you've heard and are undaunted by considerations of soil, form work, and tamping, continue your research by starting with David Easton's book, *The Rammed Earth House.* You may want to explore cast earth homebuilding as well, described in chapter 10; see sidebar for a brief description. View the videos on rammed earth construction, attend a workshop or two, and contact the people listed in the resource guide who are building rammed earth homes, and, if possible, visit some homes under construction.

The information these sources provide can help you further your understanding of this unique natural building option, as well as the many challenges you will encounter along the way. You may also want to read about other building technologies presented in this book. As you will soon find out, there are many other natural building techniques, and some might suit you even better.

Working with earth is simple in one sense, but very complex in another. Soil is so diverse, its properties and reactions so widely variable, that to fully understand its uses and limitations, and to build with it successfully, takes years of study and experience.

DAVID EASTON,
The Rammed Earth House

CAST EARTH ALTERNATIVE

Cast earth homebuilding is a modern spinoff of rammed earth home construction. In this technique, a slurry of sand, lime, and water are poured into concrete forms, greatly accelerating the wall-building process and cutting down on costs. For more on this emerging technology, see chapter 10.

3

STRAW BALE HOMES

Straw bale building is not just about building. It's a more sustainable approach of living together in peaceful and healthy conditions amongst the increasing number of humans and the decreasing number of plants, animals, and other expressions of natural life.

MARTIN OEHLMANN

THE IDEA OF BUILDING a house out of straw may seem laughable, perhaps even foolish. I know. I had the same reaction the first time I heard of straw bale construction. You may have the same reaction yourself.

A month or two after my first encounter with this alternative building technique, I saw a news report on a straw bale house being built in southern Colorado. Even though I could see how bales were used to build a house, the idea still seemed pretty strange. I decided to run the idea by a friend of mine who has built and remodeled many houses over the past twenty years. He shared my skepticism and quickly rattled off a list of issues that made this preposterous idea seem even more ridiculous, among them potential problems with moisture, insects, rodents, and fire. His final words on the subject were: "Forget it."

One of the many lessons I've learned in my life is to remain open to new ideas, even in the face of skepticism, my own included. Sitting in my living room pondering how I should begin this chapter, surrounded by walls built of straw bales, I had to laugh. Yes, I was wrong. Straw bale is a viable building technique—one that offers numerous benefits to people and the planet.

Unlike rammed earth building, straw bale construction is a relative newcomer, emerging in the late 1800s. Although straw bale is a recent innovation, straw and grasses have been used to create shelter for centuries in many parts of the world. The earliest grass and straw shelters were made from loose or bundled straws and grasses. In some cases, mud was used as a mortar to stabilize the walls. In Asia and Europe, homes were constructed using tied bundles

Courtesy of Simonton family

FIGURE 3-1.
*Simonton House, Purdum,
Nebraska, 1908.*

of straw stacked in mud mortar. This practice stretches back several centuries. Straw bale construction, the next step in this evolutionary process, was made possible by the invention of the baler in the 1880s.

The emergence of baling equipment powered by horse or steam engine opened up a whole new realm of possibility in straw and grass building. Experimentation began in a region in northwestern Nebraska known as the Sand Hills. As Bill and Athena Steen write in their book, *The Straw Bale House*, "It took only a slight stretch of the imagination for early homesteaders in the timber-poor region of the Great Plains of North America to think of using bales as oversized bricks." Their visions soon bore fruit as early settlers began using hay bales to build homes (figure 3-1). These early pioneers quickly discovered that baled grass—and later straw—made a surprisingly good building material. Inexpensive and easier to work with than sod, straw and hay bale buildings were also a lot more pleasant to live in than sod homes.

In 1904, the U.S. Congress passed the Kincaid Act, a law that offered settlers full sections of land, measuring a mile long by a mile wide, rather than the quarter sections previously granted by the government. This generous offer brought large numbers of immigrants into the Sand Hills, and they too built homes, barns, and churches out of baled hay and straw. However, many of the early farmers viewed their new homes as temporary structures that would suffice until they could afford to build "real" houses out of sawn lumber with clapboard siding, like those of their civilized aunts, uncles, and cousins living "back East."

Often they left the exteriors exposed while finishing interior walls with plaster to create a more pleasant environment. But something strange happened along the way. Many folks found that their straw bale homes were not only durable, they were exceptionally comfortable, even in the coldest winters

*Over the years,
bale homes have been
built out of baled
meadow grass, kudzu,
tumbleweed, and straw
from a variety of grain
crops such as wheat,
oats, rice, rye,
and barley.*

or the hottest summers. As a result, many families plastered the exteriors and made their straw or hay bale homes permanent residences.

Despite its early success, straw bale construction never caught on. Even in Nebraska, it fell out of favor as railroads penetrated the frontier, bringing with them milled lumber and other modern building materials. It was only a matter of time before imported wood products replaced straw bales as a building material and straw bale construction entered a period of dormancy.

In late 1970s and early 1980s, straw bale building experienced a rebirth in the United States that continues to gain momentum today. The impetus for this re-emergence was growing concern over energy and environmental problems, such as pollution, deforestation, species extinction, and resource depletion. These new pioneers viewed straw bale as an environmentally sound alternative to standard wood-frame houses that could reduce timber harvesting. Thick insulated walls made from straw bales could also slash our nation's energy demand, especially our demand for fossil fuels that are used for heating and cooling homes. Advocates quickly realized that straw bale construction, if widely adopted, would also reduce global air pollution caused by farmers each year who burn straw after harvest season, a practice that generates millions of tons of carbon dioxide, carbon monoxide, and nitrogen oxides worldwide.

For the new generation of straw bale builders, the appeal of this technique goes far beyond resource and environmental issues. Straw bale construction was, and still is, seen as a simpler way of building homes and thus more amenable to the owner-builder. Some advocates view straw bale as a more affordable alternative to conventional forms of construction. Although this may not always be the case, the straw bale homes that are being built in Belarus to house people displaced by the disastrous accident at the Chernobyl nuclear reactor are three to four times cheaper than standard brick houses—and are four times more energy efficient! With these kinds of savings possible, straw bale seems to be catching on. One house was built in Belarus in 1996, three the following year, and nearly forty in 1998. One popular book on straw bale construction, *Build It with Bales* by Matts Myhrman and S. O. McDonald, has even been translated into Russian.

Despite the encouraging results in Russia, straw bale construction does not always save huge sums of money. Unless you do a lot of work yourself, scrounge materials, rely on a simple and efficient design, and use inexpensive finish products, don't expect huge financial savings. As those of you who build

STRAW VERSUS HAY: WHAT CITY SLICKERS DON'T KNOW

Straw and hay are not synonymous. Hay consists of meadow grasses with seed heads still attached. It is used as animal feed and is popular among mice because of the presence of seeds. Nonetheless, you can use hay for a bale home, but it's generally not recommended.

Straw is the shaft of rice, wheat, rye, and other grain crops, and includes everything from the root crown to the seed head. Seed heads are snipped off by harvesting equipment prior to baling and seeds are used to make animal feed and food for humans. Although straw contains seeds and seed heads, it doesn't have many. This makes straw less desirable to rodents and thus the preferred building material.

That said, it is important to note that not all straw is the same. Rice straw is rather difficult to work with and generally avoided. David Eisenberg, a leader in straw bale construction, says it was "like working with Brillo pads or steel wool." Most people build with wheat straw.

For more information on buying straw, see Mark Piepkorn's article "Buying Your Bales" in vol. 23 of *The Last Straw.*

Courtesy of Burritt Museum

FIGURE 3-2.
Straw bale is suitable for a wide variety of climates, including hot, humid ones, as this mansion turned museum illustrates. Burritt Museum, Huntsville, Alabama, built in 1938.

a natural home will soon find out, the cost of building a wall represents only a small fraction of total building costs—about 10 to 20 percent. Therefore, even if you build straw bale walls for half the cost of standard construction, you will only save 5 to 10 percent on the house. But costs aren't everything. As Joseph McCabe writes in an article entitled "The History of Straw Bales Used for Construction," straw bales "don't provide great reductions in construction costs compared with conventional building practices," but they "do provide society with a reduced cost associated with environment and health." This alone is justification for their construction.

Because of dedicated individuals and builders who want to make a positive contribution to the betterment of our planet, straw bale home construction is experiencing a meteoric rise. Gary Duncan of the Smart Shelter Network in Montrose, Colorado, has documented 180 straw bale homes in relatively sparsely populated southwestern Colorado. Today, straw bale homes can be found throughout the United States in every state except Hawaii, according to Catherine Wanek, straw bale videographer and managing editor of *The Last Straw.* The largest concentration is in New Mexico, Arizona, and California. Outside the United States, straw bale homes have popped up in Canada, Mexico, Russia, England, Ireland, France, Germany, Denmark, Holland, Norway, Sweden, New Zealand, and Australia.

One of the most impressive straw bale homes I've encountered is a huge mansion built in Huntsville, Alabama, in 1938 (figure 3-2). Surviving well in Alabama's humid climate, this structure now houses the Burritt Museum. Straw bale advocates point to this structure to allay people's concern over moisture—notably, the fear that humidity will cause straw in the bales to mildew and perhaps even rot (see sidebar on p. 68).

The oldest known straw bale structure is a one-room schoolhouse built near Bayard, Nebraska, in 1896 or 1897. Unfortunately, no report of its con-

Moisture is one of the key concerns in straw bale construction. Fortunately, there are many ways to protect straw in a wall from becoming wet or saturated. Four of the most important are (1) suitable overhangs to protect walls from driving rains, (2) toeing up the foundation so that the first course of straw bales is at least 6 to 8 inches above grade, (3) moisture-proofing the surface of the foundation that bales are laid on, and (4) creating breathable walls that permit moisture to escape.

There's a lot of controversy in the straw bale building movement as to how one should deal with moisture. And there's a lot of experimentation going on. To learn more about the subject, look at *The Last Straw*, vol. 22. You may want to subscribe to this quarterly journal to stay abreast of the latest developments in this and other areas as well.

dition is available. The Burke home, built in 1903 in Alliance, Nebraska is still in relatively good condition, even though it has not been inhabited since 1956. This structure, and others like it, illustrate the durability of straw bale construction.

FOUNDATIONS: SPECIAL REQUIREMENTS FOR STRAW BALE WALLS

Construction of a straw bale home begins with the foundation. If you haven't read the section on foundations in chapter 2, take a moment to read it now. This information will help you understand the material presented in this chapter and others that follow.

One of the most common foundations found in straw bale homes is the concrete stem wall and footing (see chapter 2). The uniqueness of straw bale homes, however, requires several crucial structural adaptations. First of all, because the walls are three to four times wider than a standard wood-framed wall, the stem wall for a straw bale home must be wider. Because the thickened foundation requires more concrete, it will also cost more. (However, because the stem for a straw bale home is wide, footings are typically unnecessary.)

Steel reinforcing bar, or rebar, is added to the stem wall and footing to provide additional strength (figure 3-3). In parts of the world where rebar is difficult to obtain or too costly, builders use split bamboo rods in place of rebar. Bamboo is a natural, renewable resource with low embodied energy (see chapter 1) and is readily available in many parts of the world. Its most notable disadvantage as a substitute for rebar is that it doesn't bind as well as steel to the concrete in foundations. According to Mark Piepkorn, editor of *The Last Straw*, this problem can be overcome by splitting the bamboo, then twisting it to unlock the fibers to which the concrete will adhere. (For more on bamboo as a building material, see chapter 10.)

Rigid foam insulation applied to the inner or outer surface of the stem wall is essential to create an energy-efficient home, as noted in chapter 2. Two-inch rigid insulation is generally adequate. If you are installing insulation on the outside of the foundation, size the foundation so that the rigid insulation will be flush with the straw bales. This makes it a lot easier to plaster the wall (figure 3-3).

FIGURE 3-3.
A cross section of a straw bale wall showing placement of rebar and rigid insulation.

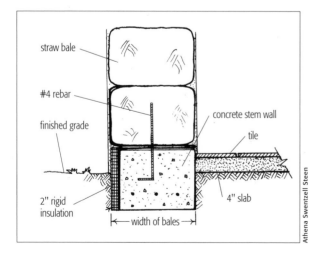

straw bale

#4 rebar

finished grade

concrete stem wall

tile

2" rigid insulation

4" slab

width of bales

Athena Swentzell Steen

If you are using a cement stucco finish, a practice now frowned upon by many natural builders, you will need to embed two nailing strips in the top of the stem wall to attach stucco netting, usually chicken wire, to hold the cement stucco in place. Nailing strips are made by placing 2 x 4s or smaller strips of pressure-treated lumber along the inside and the outside edges of the form after the concrete is poured. Bolts or nails are driven through the nailing strips so that they will adhere to the stem wall.

Unlike cement stucco, earthen plasters adhere very nicely to straw bales without any assistance from stucco netting. Earthen plasters not only reduce labor costs, they are more aesthetically appealing, and they allow air and moisture to move through the walls. Builders like to say that they allow your walls to breathe. In so doing, plasters prevent moisture from building up inside the straw bale wall, which reduces the chance of mildew forming in the wall and the straw rotting out.

Special precautions must be taken to protect straw bale walls from moisture that may wick through the foundation. One technique that is commonly used in virtually all concrete foundations is to apply a waterproofing material (foundation sealer) to the outside surface of the stem wall to repel groundwater. Because concrete is hygroscopic (that is, it absorbs moisture) and because the straw bales sit directly on the top of the stem wall, builders also apply a layer of waterproofing along the top surface of the stem wall after the concrete has set up.

Several techniques are currently used to create a waterproof barrier on the stem wall. In the straw bale portion of my house, the moisture barrier consists of a commercially available concrete sealer painted along the top of the stem wall just before bales were set in place. A better choice, I learned at a later date, is a nontoxic, organic sealer called DynoSeal. The product is extremely easy to apply and very pleasant to work with, especially when compared to the standard foundation waterproofing agents. DynoSeal is relatively odor free, and easy to clean off skin and clothing. It costs more—a lot more—than mass-produced concrete waterproofing agents, but if it will prevent me or one of my family members from being sick or developing cancer later in life, then to me it is worth the extra money!

A waterproof seal between the foundation and the first row of straw bales can also be created by applying plastic, roofing felt, or galvanized metal flashing on the top of the stem wall. In New Mexico, the state building code for straw bale homes requires a fairly elaborate galvanized metal flashing, which also serves as a barrier to insects, especially termites. As the authors of *The Straw Bale House* point out, there is no evidence to indicate that termites pose a threat to straw in straw bale structures. They can, however, damage the wooden support posts, door and window frames, roof framing, and other wood components.

BUILDING NOTE

The term *stucco* refers to cement-based and synthetic products for wall coating. *Plaster* refers to lime, adobe or mud, and gypsum. In Europe, coatings applied to the inside are called plaster and those applied to the outside of a wall are called renders.

BUILDING NOTE

Plastic and roofing felt should be laid flat under your bales against the top of the foundation stem wall. They should probably not be extended upward onto the straw bale wall, as some codes require. If you do this, you risk trapping moisture in the wall.

Another means of protecting straw from moisture is to build a raised ("toed-up") stem wall that is 6 inches or more above grade (figure 3-3). A standard practice in all construction, this keeps water that may pool on the ground from coming in contact with the straw bale walls. (Proper grading should prevent pooling.)

As I researched this book, I was surprised at the creativity of builders. One idea that I found particularly intriguing was the use of straw bales, rather than wood, to create forms for stem walls. As shown in figure 3-4, bale forms are made by lining straw bales along the perimeter of the house in place of wooden concrete forms. Stakes are then driven into the bales (a couple per bale) to prevent blowout.

Straw bale stem wall forms may sound pretty wacky and may produce a somewhat irregular stem wall, but as long as the foundation meets structural standards imposed by the building department, they're a perfectly reasonable alternative to wood forms—and they're relatively easy to install. In addition, they provide the builder with a relatively easy means of making curved foundations to support curved walls—a refreshing variation in an otherwise rectilinear human-built world. Rigid foam insulation can be added to the form, either along the inside or outside edge, depending on your preference.

Another special requirement of straw bale construction is the placement of vertical rebar in the stem wall, as illustrated in figure 3-3. Rebar is placed along the centerline of the stem wall, every two feet, and serves to anchor the first row of bales. The most common method is to use 18-inch #4 L-shaped rebar pins, placed to a minimum depth of 6 inches with 12 inches extending above the stem wall. In New Mexico, where much of the contemporary pioneering work on straw bale homes has been performed, state building code requires the use of rebar that extends halfway into the second row of bales. Two pins per bale are required by code. For structural stability, rebar is placed within a foot of all corners.

Straw bale foundations can also be made of rock and gravel (see chapter 2). As in rammed earth construction, a concrete grade beam is poured on top of the rubble trench foundation for added stability. In most cases, an 8-inch-high grade beam will suffice. The grade beam elevates the straw bale wall off the ground and consequently prevents water from contacting the walls. The grade beam also creates a unified foundation and provides a solid anchor for vertical rebar (needed to anchor the bottom row of bales). Rigid foam insula-

FIGURE 3-4.
Straw bales can be used to form a foundation wall, cutting down on skilled labor, time, and costs. Bales must be firmly anchored with rebar to prevent blowout. This technique allows the creation of curved walls, as shown here.

BUILDING NOTE

A word of caution: Don't cut corners when building a foundation. The foundation of the house is too important to mess up. If the foundation is inadequate, your whole house will be put in jeopardy.

tion installed on the inside or outside of the rubble trench foundation will reduce energy loss. As noted in the last chapter, insulation may substantially reduce the depth of the foundation as well.

Rubble trench foundations are very inexpensive, durable, and code approved. They are a lot easier for the owner-builder as well. By minimizing the use of concrete, they also cut down on the energy required to make a house (cement has twice the embodied energy of wood).

If you are strapped for cash or simply want to try an alternative foundation that uses fewer materials than a concrete or rubble trench foundation, I encourage you to explore your options by reading some of the authoritative books on straw bale construction, especially *The Straw Bale House* and *Build It with Bales*. The Uniform Building Code permits all manner of foundations, provided they safely support the loads imposed on them. However, building officials are used to standard foundations and may balk at foundations built from alternative materials such as railroad ties, wooden pallets, or tires packed with dirt. If you are proposing such an alternative, get a structural engineer's approval first.

One of the most common questions I get asked is, "Can I build a straw bale home with a basement?" The answer is a qualified "yes." If you have the money, and your water table is deep enough to permit construction, and you're not on bedrock, you can build a home with a basement. However, you will have to use cement blocks or poured concrete (or alternative products discussed in chapter 14). Alternative foundations such as rubble trench, railroad ties, and wooden pallets won't work.

In colder climates, basements can be a fairly economical option, as you have to construct a rather deep stem wall anyway to get the foundation below the frost line, unless of course you're building a frost-protected shallow foundation. If you are digging a deep foundation and want extra square footage, it may be worth the additional expense to excavate a basement.

BUILDING STRAW BALE WALLS

After the foundation is completed, it is time to build the walls of your straw bale home. You have two options: the load-bearing wall, or the non-load-bearing wall, more commonly called the in-fill method.

Load-bearing walls consist of straw bales stacked in a running bond (overlapping pattern) like bricks. The bales are either pinned or, less commonly, mortared like bricks. In load-bearing structures, the roof typically rests on a wooden beam known as a top plate, which is attached to the top of the load-bearing straw bale wall. The bales not only support the roof, they provide the insulation.

In the second technique, the in-fill method, a frame is built to support the weight of the roof and additional floors in a multistory building. The frame is made out of posts and beams and a host of materials, including concrete

BUILDING NOTE

The use of foam insulation in rubble trench foundations allows one to create a shallower foundation, known as a frost-protected shallow foundation, that saves money. More and more builders such as Terra Designs in Boulder, Colorado, find them quite suitable for straw bale construction.

BUILDING NOTE

Straw bales can be purchased in many parts of the world. For a list of U.S. suppliers, see the annual resource issue of *The Last Straw* or check with a local straw bale association. Readers outside the United States should check with national straw bale organizations.

For tips on buying directly from farmers, see Mark Piepkorn's article "Buying Your Bales" in vol. 23 of *The Last Straw*.

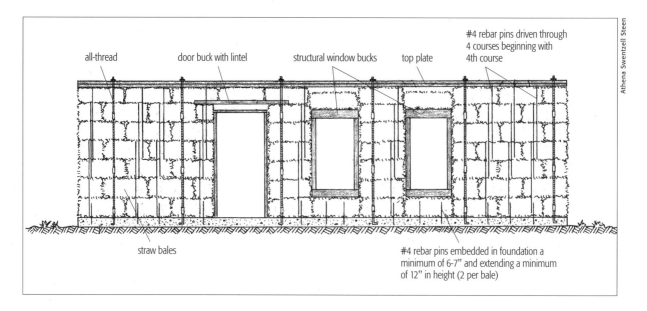

Athena Swentzell Steen

#4 rebar pins driven through 4 courses beginning with 4th course

all-thread

door buck with lintel

structural window bucks

top plate

straw bales

#4 rebar pins embedded in foundation a minimum of 6-7" and extending a minimum of 12" in height (2 per bale)

FIGURE 3-5.

Details of a load-bearing straw bale wall. All-thread anchors the top plate to the foundation. Window and door bucks are inserted in the wall and anchored to bales. Rebar pins embedded in the foundation anchor the bottom course of bales. Notice half bales in the wall needed to create a running bond pattern.

blocks. The bales are placed between the vertical supports where they provide insulation but no support.

Hybrid forms also exist containing features of both the load- and non-load-bearing walls. If you want to learn about hybrid structures and other innovative approaches, see volume 25 of *The Last Straw*.

Load-Bearing Walls

Many people find it difficult to believe that a wall built out of bales of straw can support the weight of the roof and resist lateral loads imposed by winds and earthquakes. But it's true. In fact, the early Nebraska homes were load-bearing structures and many have withstood the test of time.

All of this wizardry is made possible by the remarkable characteristics of the straw bale itself. As described in detail in *Buildings of Earth and Straw*, engineering tests show that bale walls are remarkably strong. For example, hydraulic pressure tests, which test the compression strength of a material, show that a straw bale (laid flat) can support a load of 10,000 pounds per square foot. Pretty remarkable, especially when you consider that a standard 2 x 4 wall supports 1,500 pounds per square foot. Furthermore, straw bale walls with rebar reinforcement and a unifying top plate attached to the foundation experience minimum deflection when exposed to lateral forces (loads) equivalent to a 100-mile-per-hour wind. These and other tests lead to an astonishing conclusion: straw bale walls are extremely strong and can easily support the weight of the roof and additional floors, not to mention the live (transient) loads such as those imposed by heavy spring snows.

Figure 3-5 shows the detailed structure of a load-bearing wall. As illustrated, straw bales are anchored to the stem wall by rebar pins. The first row of bales is positioned over the rebar, then pressed and pounded in place. Constant attention must be given to the alignment of the bales so that their outer surface is flush with the exterior of the foundation wall. The next row of bales is placed on top, overlapping the ones below by a minimum of 12 inches to create a running bond that gives additional strength to the wall. In order to achieve a running bond, half bales are used where needed. Half bales are specially made on site, a process discussed in more detailed books on the subject.

Another feature of load-bearing walls are the pins. After the third or fourth row of bales is in place, rebar or bamboo "pins" are driven into the bales, usually two per bale. Pins are also sometimes used to provide lateral stability. Although rebar is commonly used, bamboo is an economic and potentially sustainable alternative, as are wooden dowels, wooden stakes, or even stout tree branches.

Driving pins into the bales is a lot of fun and is suitable for family members and folks who come out to help build your walls as part of a workshop. As I was building my home, numerous people stopped by or called to volunteer, donating their labor in exchange for the experience.

Pins are driven directly into bales with a sledgehammer, but the job is a lot easier if you use a rebar driver, a device that fits over the pins to increase the "strike" surface. I made a rebar driver by screwing a cap onto a six-inch-long (threaded) steel pipe large enough to slide over the rebar. Rebar drivers can also be used with wooden dowels and bamboo pins. Besides making it easier to pound the pins into the wall, they reduce splitting.

With load-bearing walls, it is important to provide a strong connection between the top plate and the foundation. This prevents the roof from being blown off in a strong wind, and it makes the house more rigid and resistant to lateral loads. Several methods are currently used. One of the earliest methods, now falling out of favor among straw bale builders, is the all-thread connection.

All-thread is a fully threaded rod connected to ½-inch anchor bolts embedded vertically in the foundation prior to the pour, 6 feet apart and about 12 inches from the corners of the building. After the foundation has set, coupling nuts are attached to the anchor bolts. These allow you to attach the all-

ORGANICALLY GROWN STRAW?

Pesticides are often used to produce grains such as wheat. But do the pesticides pose a threat to straw bale home builders and occupants?

According to Catherine Wanek, managing editor of *The Last Straw*, pesticides are applied while the crops are young and should have deteriorated or been washed off by the time straw is harvested, thus posing little or no threat to straw bale builders or occupants.

For those who want to be absolutely sure, you might check out the Northern New Mexico Organic Wheat Project. They're working with wheat growers trying to promote organic wheat production. Organic straw suppliers are also listed in the annual resources issue of *The Last Straw*.

FIGURE 3-6.
All-thread (foreground) was once widely used to secure the top plate to the foundation. Notice rebar pins protruding from the foundation.

thread, usually in 3-foot sections. Straw bales are then impaled on the all-thread and vertical rebar protruding from the stem wall (figure 3-6).

After the first and second rows of bales are in place, additional three-foot sections of all-thread are attached—again using coupling nuts. Bales are added to the wall until the desired height is reached. The top plate is then set in place and attached to the all-thread. All-thread is run through holes drilled in the top plate, then secured by washers and nuts.

This process is fairly time-consuming. Special care must be taken to be sure that the walls remain plumb. When combined with pinning, it slows straw bale wall construction down considerably. Because of this, builders have begun to explore faster ways of anchoring the top plate to the foundation, including wire, aircraft cable, polyester strapping, wood strips, and external pins (figure 3-7). Wire, cable, and strapping are tightly looped over the top of the wall and anchored to the foundation on either side of the bale wall by 6-inch-long eye bolts embedded vertically in the top of the stem wall every 2 to 6 feet.

Another means of attaching the wall to the roof plate and enhancing the lateral stability of a straw bale wall is external pinning—using wood strips or rods. External pinning was introduced by Bill and Athena Steen as they worked on low-income housing in Mexico, where rebar was too costly. Other builders, such as Bob Bolles of Poway, California, have adopted the system and

BUILDING NOTE

Straw bale walls are irregular by their very nature. You can remove the irregularity when you apply plasters. However, the smoother the surface prior to plastering, the easier your job will be and the less material you will need.

FIGURE 3-7.
External strapping or cabling can be used in place of all-thread.

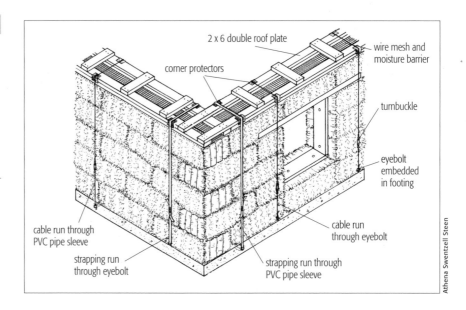

helped to perfect it. External pins can be made from #4 rebar or strips of wood that extend from the foundation to the top plate (figure 3-8). Eight-foot sections of rebar are attached by galvanized fence staples to the top plate and to a bottom plate, a piece of wood attached to the top of the stem wall. Polypropylene twine is then looped around pins, coursing from one side of the wall to the other, and tied.

Compared to internal pinning, external pinning reduces labor and greatly increases the stability of straw bale walls. It also makes it easier to produce plumb walls. For more details on the external pinning system, read Bob Bolles's article, "Exterior Pinning Put to the Test" in volume 25 of *The Last Straw*.

Although some people view straw bale wall construction as the easiest part of building a bale home, it requires special care and attention to detail—especially if you want level, plumb walls. Keeping your walls plumb is extremely important for aesthetics and structural stability, especially when building load-bearing walls.

After each row of bales is in place, check them using a line strung between corner braces (discussed shortly) and a carpenter's level. Bales that stick out too far should be pounded in line using your gloved fists or a mallet.

Vigilance is especially important when your walls are being built in workshops by participants who are still learning. If you are planning to raise your walls in a workshop, be certain that everyone is trained to stack bales properly and be sure that someone is in charge of monitoring progress—checking and rechecking the walls as they are being erected. I can't emphasize this point enough. Straw bale walls go up very rapidly; if you overlook a mistake in the first row of bales, you may have to tear the wall down and start over or live with the consequences.

Before construction of a load-bearing wall begins, rough door frames (also known as door bucks) should be constructed and attached to the foundation in appropriate locations. Rough window openings (window bucks) should also be built and readied for installation (figure 3-9). As the bale wall is built, window frames are inserted in proper locations. Straw bales are then stacked around them and the bucks are secured to the bales.

In load-bearing walls, straw bales are generally laid flat, that is, on their "wide side." This gives the wall the greatest strength. Two-string bales laid flat produce 17-inch-wide walls; three-string bales produce 23- to 24-inch walls.

Straw bales may vary in length by as much as four inches. If your bales vary, gaps will appear in your walls. One-quarter- to one-third-bale-width gaps are filled by breaking apart large bales and (tightly) retying them to create

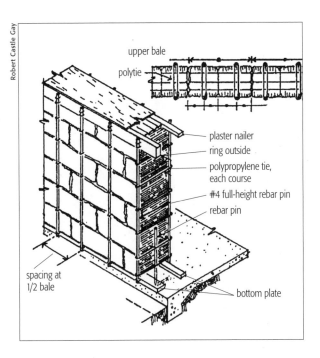

FIGURE 3-8.
External pinning is one of the most popular ways of attaching the top plate to the foundation. In this instance, #4 rebar pins are attached to the bottom plate and top plate.

Robert Castle Gay

upper bale
polytie
plaster nailer
ring outside
polypropylene tie, each course
#4 full-height rebar pin
rebar pin
bottom plate
spacing at 1/2 bale

BUILDING NOTE

Many individuals have observed a kind of wall-building frenzy that occurs at workshops and parties, as people's pent-up excitement is unleashed. The frenzy can lead to crooked, poorly pinned, tilting walls, mistakes that can jeopardize their structural stability and take several days to correct. So watch your helpers like a hawk.

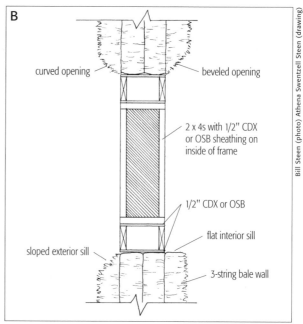

Bill Steen (photo) Athena Swentzell Steen (drawing)

In the drawing (B):
- curved opening
- beveled opening
- 2 x 4s with 1/2" CDX or OSB sheathing on inside of frame
- 1/2" CDX or OSB
- flat interior sill
- sloped exterior sill
- 3-string bale wall

FIGURE 3-9 A, B.
Window bucks are made from wood (A) and inserted in the wall (B).

minibales that fit into the odd spaces. If the gaps are smaller, simply insert bale flakes (untied sections). Even smaller gaps are stuffed with loose straw.

Top Plates. After the wall has attained full height, the top plate is set in place. Also known as the roof-bearing assembly, the top plate is a means of attaching the roof to the walls of a house. Because it forms a rigid structure along the top, it helps to stabilize the straw bale walls. In addition, the top plate bears and distributes the roof load to the walls of a house, which is extremely important in load-bearing bale walls. Uneven loading (more pressure in one place than another) can cause sections of the wall to bulge and eventually fail.

Straw bale builders use a variety of materials to make top plates, including 2 x 12s, 2 x 10s, wooden ladder-style plates, logs, wooden box beams, bamboo, and concrete bond beams, similar to those used in rammed earth. For details, refer to *Build It with Bales* by Matts Myhrman and S. O. McDonald. This book does a particularly good job of illustrating the options. Architects, builders, and structural engineers may also have ideas. Check out *The Last Straw* for innovative top plate designs, too.

Letting the Roof Settle. After you have built the load-bearing walls and installed the top plate, the walls must be compressed. This can be achieved naturally by the weight of the roof. Under the weight of the roof, load-bearing walls compress approximately 1 to 2 inches, sometimes more, over a period of 4 to 8 weeks. If winter is fast approaching or if you have given up your lease and need to move into your straw bale home as soon as possible, you can hasten the process by precompressing the walls.

BUILDING NOTE

A load-bearing wall built out of three-string bales can only safely rise to a height of 10 feet 8 inches, while a wall built of their smaller cousins, the two-string bale, can only be 8 feet 4 inches high.

ATHENA SWENTZELL STEEN,
BILL STEEN, and
DAVID BAINBRIDGE, with
DAVID EISENBERG,
The Straw Bale House

Precompression is achieved by tightening the connections attaching the top plate to the foundation—for example, all-thread, wire, or strapping—before the roof is constructed. All-thread is probably one of the easiest systems for tightening. (It's otherwise difficult to work with.) After the top plate has been bolted on to the wall, tighten the nuts using a ratchet or a torque wrench. Torque wrenches are preferable because they allow you to control and monitor the force you are applying to each nut. This, in turn, allows you to tighten each nut uniformly to keep the top plate level, which is essential for roof construction.

Equal compression is achieved by proper spacing of all-thread rods and by tightening the nuts uniformly. Walls built with straps and wires to attach the roof plate to the foundation can also be precompressed by increasing tension on the connectors. Buying bales from farmers or suppliers who ensure uniform compaction and density helps prevent uneven compression. For a source of bales in your area, check out the annual resource issue of *The Last Straw*.

WHAT KIND OF STRAW BALES SHOULD YOU USE?

Straw is widely available and can be found at most feed stores, where it is sold as animal bedding or mulch for gardens. Unfortunately, commercially available straw bales are often fairly loosely packed and may be mildewed, which renders them useless for straw bale construction.

To build a wall out of straw bales, you need to start with the best possible bales. The authors of *The Straw Bale House* sum it up best: "Use good bales that are dry, solid, and compact. Poor-quality bales will affect the structural stability of the building." Treat them well, too. Rough handling can damage bales.

If you hire a contractor who specializes in straw bale construction to build your house, he or she will inevitably have a supplier. If you are doing it yourself, contact local builders to see where they get their bales.

What you want is a bale that is dry and properly compressed. The compression setting on a New Holland baler should be set between 250 and 500 pounds. Bruce King, author of *Buildings of Earth and Straw*, recommends using bales with a dry density of 7 to 10 pounds per cubic foot.

Non-Load-Bearing Walls: The In-Fill Method

The most popular way of building straw bale walls is the in-fill method with a post-and-beam structure. Post-and-beam construction is widely used in many countries for barns, sheds, and homes; it is fast, economical, and relatively easy. As the name implies, a post-and-beam structure consists of vertical posts (made from 4 x 4s or larger lumber) spanned by wooden beams. The beams serve the same function as the top plate. In a post-and-beam structure, this "skeleton" supports the weight of the house. In a straw bale home, bales are stacked between the beams to provide insulation. They provide no structural support whatsoever.

Builders use all manner of wooden posts for the vertical supports in post-and-beam straw bale structures, including 4 x 4 posts, 4 x 6 posts, 6 x 6 posts, utility poles, lodge poles, and large-diameter bamboo. Some builders use concrete posts or, more commonly, posts constructed from concrete blocks or some similar block material like Faswall block (see p. 79). Builders are even experimenting with elaborate steel frames in their straw bale structures, in part to reduce pressure on the world's forests. Steel frames work well and are strong, readily available, and easy and quick to assemble. But steel is a nonrenewable resource with much higher embodied energy than lumber. When

BUILDING NOTE

The new rotary combines chop straw into very short lengths. Although these bales are compact, straw can be pulled out of the bales by the handful. So check your source carefully.

CATHERINE WANEK,
The Last Straw

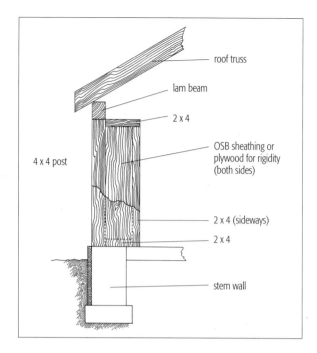

FIGURE 3-10.
Details of panels used in my house.

roof truss
lam beam
2 x 4
OSB sheathing or plywood for rigidity (both sides)
4 x 4 post
2 x 4 (sideways)
2 x 4
stem wall

BUILDING NOTE

If you need help getting a building permit, check out volumes 23 and 27 of *The Last Straw*. They contain a wealth of information on building codes, including contacts at other building departments who have approved straw bale structures, and a comprehensive list of research and testing on straw bale walls.

If your building department is dubious about straw bale construction, ask them to contact other jurisdictions and talk to code officials there or give them videos and technical reports that will help them answer questions they have about this novel means of building homes.

deciding, choose a product that makes the most sense from an environmental standpoint.

Rather than building with large posts, I installed wooden panels made from 4 x 4s, 2 x 4s, and oriented strand board as illustrated in figure 3-10. You could also use paneling made of straw. We replaced the solid beams with a microlam, an engineered beam that is made from much smaller trees (see chapter 14). This technique, known as modified post-and-beam construction, worked extremely well and was relatively easy to master—and it greatly reduced wood consumption and costs.

To ensure that your roof stays in place, the panels in a modified post-and-beam building must be anchored to the stem wall of the foundation. We used Simpson connectors, shown in figure 3-11b, attached to hurricane bolts driven into the concrete after it has cured.

In addition to being relatively easy to construct, post-and-beam systems are familiar to structural engineers and building department officials who must approve your plans before you are granted a building permit. You may also find that insurance companies, lenders, and building inspectors are more amenable to straw bale homes with post-and-beam construction. In contrast, this cast of characters is often dubious about load-bearing straw bale walls and more inclined to reject your plans or make it difficult for you to get a permit or services such as the financing and insurance you may require.

One of the chief advantages of post-and-beam and modified post-and-beam wall systems is that they use very little lumber compared to a framed wall built out 2 x 4s or 2 x 6s. If you are interested in the post-and-beam system, consider the less lumber-intensive methods, and look at the article by Bob Bolles in volume 25 of *The Last Straw*. You may even want to consider using recycled posts and beams. But be careful. An inefficiently designed post-and-beam straw bale home may use more wood than a stick-frame house.

Another building support structure for a straw bale home is Faswall block. Recently introduced into the United States from Europe, where they are widely used, the Faswall block is a hollow-core block much like a concrete block. Composed of 88 percent wood chips (made from discarded wood pallets) and a low-toxicity cement, Faswall blocks offer several important advantages over concrete blocks. Faswall blocks are made from a lower-embodied-energy material; they are 75 percent lighter than concrete blocks and therefore easier to handle; they can also be cut with handsaws and circular saws with carbide-tipped blades; and nails can be driven into them, as can

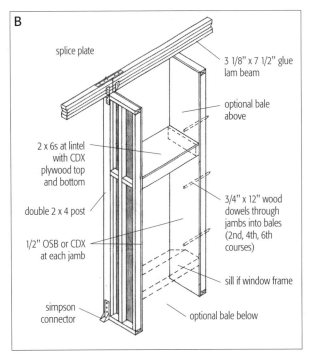

splice plate

3 1/8" x 7 1/2" glue lam beam

optional bale above

2 x 6s at lintel with CDX plywood top and bottom

double 2 x 4 post

3/4" x 12" wood dowels through jambs into bales (2nd, 4th, 6th courses)

1/2" OSB or CDX at each jamb

sill if window frame

simpson connector

optional bale below

screws and bolts. Don't try any of these things with a concrete block! In addition, plaster can be applied directly to Faswall blocks. The blocks come in several sizes and can be used to construct grade beams, stem walls, and bond beams. For more on this product and a block made from recycled foam known as the Rastra block, see chapter 14.

FIGURE 3-11 A, B. *Wooden panel with lam beam (A). Details of the construction (B).*

Finishing Straw Bale Walls

Perhaps one of the most appealing aspects of straw bale homes is their thick walls that give a solid "earthy" feeling when plastered or stuccoed that so many people like. In many ways, they resemble walls made from adobe bricks (see chapter 5). Builders use the term *plaster* (as do many others) to refer to earthen, lime, and gypsum plasters. In Great Britain, plasters are called renders. Cement or synthetic coatings are called stuccos.

Plaster and stucco serve many useful functions in a straw bale home. For example, they protect the bales from rain and pests such as mice and insects. Plaster and stucco also help to make walls more rigid and fireproof. In fact, a good coat of cement stucco will make a straw bale wall virtually impervious to fire. In one test, it took nearly two hours for a flame at over 100 degrees F to crack the stucco and reach the interior. Even if a fire comes in contact with the straw, combustion is thwarted by the lack of oxygen in the bales. I tell folks who are worried about fire safety that trying to ignite a straw bale is much like trying to burn a phone book. Try throwing one in a fireplace or woodstove sometime and see what happens. In my house, the combustible materials (the

EMBODIED ENERGY

Required to produce:

1 ton of straw	112,500 BTUs
1 ton of concrete	5,800,000 BTUs

Criteria/Property/Characteristic	TYPE OF PLASTER				
	Cement-based	Lime-based	Gypsum-based, Interior use only	Clay-based, Natural	Clay-based, Asphalt-stabilized
Low embodied energy in binder	□□□■	□□□■	□□■■	■■■■	■■■■
Chemically benign binder (noncaustic)	□□□■	□□□□	□■■■	■■■■	■■■■
Availability of binder or prepackaged mix	■■■■	■■■■	■■■■	□□□□ to ■■■■	□□□□ to ■■■■
Workability (good cohesion and adhesion)	□■■■	□□■■	□■■■	■■■■	■■■■
Likelihood of success on straw bales without reinforcement	□■■■	□■■■	□□■■	■■■■	■■■■
Resistance to erosion by water	■■■■	□■■■	□□□■	□□□□	□□■■
Rapid development of strength	□■■■	□□□□	■■■■	□□■■	□□■■
Eventual hardness	■■■■	□□□□	□■■■	□□□■	□□□■
Breathability	□□□■	■■■■	□□■■	□■■■	□■■■
Low maintenance	□■■■	□□■■	■■■■	□□□□	□□■■
No moist curing needed	□□□□	■■■■	■■■■	■■■■	■■■■
Friendliness to novices, overall	□□□■	□□■■	□□□■	■■■■	□■■■

■■■■ = best, greatest   □□□□ = worst, least

Source: Myhrman and McDonald, *Build It with Bales*, p. 99.

FIGURE 3-12.
This table offers a somewhat subjective comparison of plasters and stucco (labelled as cement-based plaster).

BUILDING NOTE

Unplastered straw bale walls can burn and have with great success. Until the walls are plastered, fire is a serious consideration.

MARK PIEPKORN,
The Last Straw

wood-frame face and roof) would burn out well before the fire posed any danger to the cement-stuccoed straw bale walls.

Straw bale builders differ considerably in their preferences for plaster or stucco (Figure 3-12). Earthen, lime, and gypsum plasters are pleasing to the eye, pleasant to touch, easy to repair, and easy to drive a nail through. Of these, I think earthen plasters are the most beautiful and inviting. Cement stuccos, on the other hand, are not as pleasing to view or touch. They are difficult to repair, and hard as nails. Because they are so hard, cement stuccos have poorer acoustic qualities than plasters. The most important difference, however, is breathability—plasters permit air and moisture to penetrate more readily than stuccos.

Although opinions vary widely, the consensus among natural builders is that you want a breathable wall. That is, you want a coating that will allow moisture to pass through a straw bale wall with ease. Coatings that block moisture cause it to accumulate in the walls, leaving the straw susceptible to mold and decomposition.

In post-and-beam structures, sheetrock (drywall) may be secured to the frame on the interior or black board attached to the exterior surface of the wall. These materials make plastering easy and efficient, although the result is a wall surface that looks pretty conventional. More often than not, plasters and stuccos are applied directly to the straw bales, usually in three coats.

Cement Stucco. For cement stucco, chicken wire is often applied to support the weight of stucco, although it is probably not really needed (figure 3-13). Even though chicken wire is inexpensive, it takes a lot of time to install and therefore adds to labor costs. Most builders use 1-inch chicken wire that comes in long rolls. Stucco netting is applied row by row to the inside and outside surfaces of the straw bale wall. It is first held in place by wire staples—U-shaped pieces of wire that are driven into the straw. After the chicken wire is in place, two workers sew it into a wall. Working with two long steel "needles," specially made for the project, and a roll of wire as "thread," one person passes the needle through the wall to his or her partner waiting on the other side (figure 3-14). That person detaches the wire from the needle, loops it around a section of chicken wire, then sends the wire "thread" back through with a second needle, all the time being sure to keep the thread tight so that the chicken wire adheres very tightly to the bales. If the chicken wire isn't tight, it is useless.

In the post-and-beam or modified post-and-beam method, roof felt paper is applied over any lumber that will come in contact with cement stucco to prevent cracking. Exposed wood draws moisture out of the stucco so rapidly that it causes the stucco to crack. The felt paper is covered with chicken wire or

FIGURE 3-13.
Bottom left. Chicken wire attached to straw bales and wood of the window frame.

FIGURE 3-14.
Needle and wire used to sew bales to secure chicken wire to both sides of the bale wall.

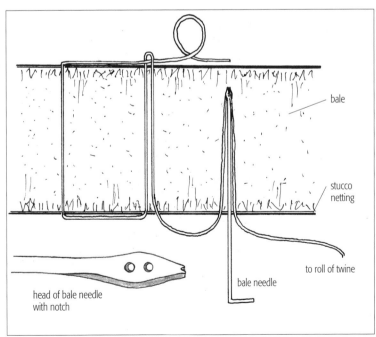

metal lath, a mesh with smaller angular openings specially designed to adhere well to stucco.

Cement stuccos are typically applied in three coats: a base coat, known as the scratch coat, a brown coat, and a finish coat. The base coat is the thickest layer and smoothes out irregularities in a wall. After the base coat has been applied, a scratch tool resembling a small hand-held rake is scraped across the surface—hence the name *scratch coat*. The grooves it produces form small keyways that help the next coat, the brown coat, adhere. After the brown coat is in place, it should be smoothed using a short piece of 2 x 4. This eliminates irregularities. A trowel can also be used to smooth out the surface. When the surface meets your approval, the final coat, known as a color or finish coat, is applied. This may be either cement stucco or synthetic stucco (a latex and sand mix).

Cement stucco is often applied by hand, using trowels. No matter how strong you are, this process is hard on wrists, elbows, and shoulders. To minimize labor and wear and tear on joints, some builders prefer to spray cement stucco on walls, especially the first coat, which is the thickest and therefore requires the most work.

However, if you want to do the job by hand, you can sponsor a plastering party (figure 3-15). A group of five or six friends, relatives, and would-be straw bale homeowners can plaster a wall with a minimum of training.

FIGURE 3-15.
Plaster is applied to straw bale walls using a hawk and trowel or by hand.

Bill Steen

Although this may save you some money, it can create problems if you are hoping for a smooth, flat wall. Volunteer laborers rarely have the skill to create a wall surface of professional quality.

Earthen Plasters. For a more natural-looking and longer-lasting home, consider earthen plasters, also known as adobe (clay-based) plasters. Earthen plasters are applied directly to the bales without chicken wire or stucco netting—saving time, money, and resources. Earthen plasters adhere extremely well, especially if the bales are laid flat—that is, with their wide surfaces down. This positioning presents a fairly irregular surface (consisting of straw ends) that increases the bonding surface for plasters.

Like cement plasters, earthen plasters are applied in three coats. However, some applicators like to spray a thin layer of clay slip (a watery mixture of mud) over the bale wall to increase adhesion before the scratch coat is applied (figure 3-16). After this, they pack crevices in the wall surface with loose straw soaked in clay slip, to ready it for the scratch coat.

Earthen plasters consist of mud, straw, and sand mixed with water. Some workers add manure and other materials such as flour paste to the scratch coat to increase the stickiness of the wall. At least one expert on plaster I know advises against the addition of flour for fear that it will mold and make people sick.

The scratch coat of an earthen plaster is often referred to as the first rough coat, because there's no need to run the scratch tool over it. Most people apply the first rough coat by hand, although there are machines that can spray a mud plaster on the wall much more rapidly. Mud is applied one handful at a time by pushing it on to the wall with relatively short upward strokes. The next coat, the brown coat or second rough coat, is also applied by hand. It fills the irregularities in the wall and produces a relatively flat surface. Many plasterers run a trowel over the second rough coat to achieve a flat, smooth surface. After the wall has been smoothed, the finish coat is applied, either with bare hands or trowels.

Earthen plasters can be used for interior and exterior surfaces. Interior surfaces require very little care and upkeep. However, exterior surfaces require constant vigilance and occasional repair. Exposure to the hot sun and freezing temperatures may cause the walls to crack. Because tiny cracks will allow moisture to enter and because small cracks will enlarge over time, they must be promptly repaired to prevent moisture from penetrating the wall and damaging the bales. Earthen-finish coats may need to be resurfaced or recoated from time to time to compensate for thinning caused by wind and water.

To reduce wear and tear on exterior walls and to cut down on maintenance, some homebuilders apply a lime plaster finish coat over two coats of earthen plaster. Lime plaster is extremely durable, adheres well to earthen plaster, and can be tinted. Although it is a bit more difficult to work with than earthen plasters, it may be more acceptable to building department officials, and its durability more than makes up for its difficulties. Lime plaster is discussed in more detail in chapter 6 on cob and in Steve Berlant's book *The Natural Builder*, Volume 3: *Earth and Mineral Plasters*.

Rainscreen Siding. Although most straw bale homes are stuccoed or plastered, some builders apply siding on bale walls. This practice is most common in wet climates—places where driving rains soak exterior walls and there are not enough warm, sunny days in between storms to permit the walls to dry out. In

FIGURE 3-16.
Clay slip can be sprayed onto straw bales before plastering to improve the adhesion of mud plaster to the wall. This device is powered by compressed air.

such instances, siding prevents driving rains from penetrating the walls and saturating the straw. In this technique, builders leave a small gap between the siding and the exterior of the bale walls that allows moisture to escape, as shown in figure 3-17. Otherwise, moisture emanating from people, pets, fish tanks, plants, and showers will accumulate on the inside surface of the siding and drip onto the bales, causing them to mildew and rot.

Rainscreen siding is mounted on wood strips embedded in the surface of the wall in a thin layer of earthen, lime, or gypsum plaster. This layer is known as backplaster. The air space between the backplaster and the siding opens into the soffit (overhang), allowing moisture to exit via the soffit vent.

If you are building in a drier climate, use a plaster or cement stucco. As noted earlier, finish coats come in many colors. The most beautiful ones I've seen are natural; that is, they rely on the natural pigments found in earthen plasters. Plaster and cement stuccos can be painted. Chapter 14 describes environmentally friendly paints made from a variety of materials such as milk protein and chalk.

ROOFS

The roof of a building serves several important functions. Besides protecting the structure and its occupants from rain and snow, it provides structural rigidity. In fact, engineers consider the roof to be a rigid diaphragm that enhances the structural rigidity of a house. The roof also defines the home. That is, the roof contributes in a major way to the "character" of a house.

Like rammed earth homes, the straw bale house can accommodate many different types of roofs. The roof should reflect the climate in which you live. Consider rain and snow very carefully before designing or selecting a roof type. A "flat" roof that may work in Arizona's dry climate may be a prescription for disaster in Vermont.

Even though straw bale homes are suited to a wide range of roof options, the possibilities are greatest if the walls are constructed using the non-load-bearing or in-fill method (post-and-beam and modified post-and-beam systems). The reason for this is that the wooden support structure used in these building techniques carries the load of the roof and can be adjusted wherever the load increases, say, because of a dormer or some other feature that adds to weight to the roof.

If you build a home with load-bearing walls, your roof options are much more limited. The reason for this is that the load-bearing capacity of straw bale

FIGURE 3-17.

A cross section through a straw bale wall with rainscreen siding. Notice the backplaster and the space between the backplaster and rainscreen siding to permit moisture to escape through the soffit vent.

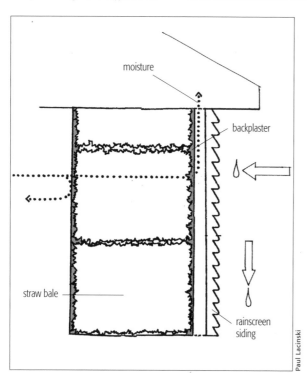

moisture

backplaster

straw bale

rainscreen siding

Paul Lacinski

FIGURE 3-18.
The hip roof distributes the weight of the roof on all four bale walls, unlike the more commonly used gable roof which distributes the load on two walls.

walls is pretty uniform throughout the house. Architectural features such as dormers may cause uneven compression.

One of the best choices when building a house with load-bearing straw bale walls is the hip roof, shown in figure 3-18. Hip roofs distribute the load evenly over all four walls. Other roofs can also be built, but you should consult with a qualified architect or structural engineer to be certain your walls will support the roof of your choice. For more on roofs, see *The Straw Bale House* and *Build It with Bales.* These are the most coherent and thorough treatments of the subject.

As you consider the type of roof you want, remember to choose a style that provides adequate space for insulation. I recommend ceiling insulation of R-38 or more for everyone, whether you live in a hot or cold climate. In cold climates, ceiling insulation as high as R-50 is advisable. Don't skimp on insulation, especially if you want comfort and freedom from utility bills. Proper house design, combined with generous insulation, will produce a home that is warm in the winter and cool in the summer, even in the most brutal heat wave. (Keeping your house cool naturally is discussed in more detail in chapter 11).

Insulation exists in a variety of forms; some are more sustainable than others. Chapter 14 describes the many options for insulating your home in the most environmentally and economically sustainable manner. Harry Yost's *Home Insulation* also provides a great deal of information on this topic.

Vaulted or cathedral ceilings, although they are beautiful, greatly add to the volume of air space in your house—air space that you must heat or cool. If you want the exalted feeling of a cathedral ceiling, try it in one room, like your living room, then bring the rest of the ceilings down below the ozone layer. You'll be a lot warmer if you live in a cold climate. Remember to insulate the

GET PROFESSIONAL HELP

Most roofs must be built to precise dimensions to ensure proper pitch and alignment with the underlying structure. Roofs are fairly complex to design and build and thus require skill in design and construction. Get professional help if you are a do-it-yourselfer. Like the foundation, the roof is too important to mess up.

Bill Steen and Athena Swentzell Steen

What sparked your interest in straw bale construction?

Many things sparked our interest, among them, a lack of housing throughout the underdeveloped world, the overbuilding of houses here in the States, the excessive use of materials, manufactured materials, and the like. Basically seeing how consumptive we are in our lifestyles and the need to live a simpler, more meaningful life.

How long have you been building with straw bales?

Athena built a small cottage for her family when she was 19 years old in 1981. She had never heard of straw bale building before and got the idea from her father. I heard about it for the first time in 1986, and in 1990 she and I put up the first building, where we live now.

What are some of your most interesting projects? Why?

The work we have done in Mexico, because it is not based on any predetermined architectural style. We used predominantly local materials, many of them waste materials. Primarily we have relied on straw (it is abundant); clay for plasters, blocks, and roofs; bamboo-like reeds; and discarded pieces of concrete. The form of the buildings evolved from what was available. This is quite different from the States, where people pick a style, move materials all around the world, and build irrespective of climate and conditions. In Mexico our work has included men, women, and children. We have relied on little more than shovels, plastering trowels, and other hand tools. Power tools are rarely needed. We built a large office building for the Save the Children Foundation using these methods. The building is unique, beautiful, and elegant. We have also been working on a project with a group of women who are working together to build their own houses. We have been helping organize, train, and work along with them to build small 300-square-foot one-room houses using the same techniques and materials. They cost around $500 each. The money for the houses has come from small donations from groups around the United States, Europe, Canada, and Australia.

Worldwide, housing is in short supply. How could straw bale construction alleviate this chronic and serious problem?

It is not so much straw bale that could alleviate housing shortages, as it is the recognition that it is possible to build good shelter using materials other than concrete and manufactured materials. We have to throw away the models that we have been using to define what constitutes quality housing and look for something completely different. That's where straw bale helps, because it teaches people that you don't only have to build using wood-frame, concrete block, etc. Straw bale is just one of a great many options that people are thinking about.

The other idea is that people in the developed world can't continue to build houses as large as they want, even if they can afford it. There are not enough materials on this Earth for everyone to be building houses over 2,000 square feet. Nor is there the energy to build and operate them. We have to scale way down and use a lot less. It is time to start thinking about the implications of how we build and what it means to the rest of the world.

What role do you see for straw bale construction in the coming years in the more developed countries?

Straw bale is a relatively simple way of building that allows participation of friends and family. It can be used in conjunction with other natural building components, meaning that one can rely more on locally available materials. People need to take a larger role in the construction of their homes, and straw bales can help them do that. Again, what is built needs to be small and simple. By keeping it that way, we can rely on people who aren't professional contractors and methods that may be more time consuming. This is not to say that professional builders aren't necessary, but rather that others can assume part of the work. It is one way of building that helps us be more than just consumers.

What are the major changes in straw bale construction since you began working and experimenting with it?
There have been improvements and simplifications. On the other hand, there is the danger that straw bale construction will become just another modern method of building in which the final end product is little more than a tract house. The walls end up being thicker, but the feel of the place is essentially the same. The argument is made that one is getting a more energy-efficient house, but it would seem that a home ought to be much more than an energy-efficient box.

Do people generally have problems getting building department approval for straw bale homes? Why or why not?
As with anything new, straw bale construction can be a struggle if the building inspector has never heard of it before. However, if the person seeking the permit is willing to accept the responsibility that they will have to do a little educating, then it is usually possible. There are enough code-approved buildings around the country so that if the local building inspector is willing to engage in a little dialogue with an inspector from another area, it can simplify the process.

What steps can people take when dealing with building department officials to allay concerns and obtain a permit?
There is an organization in Tucson called DCAT that specifically deals with code issues related to straw bale construction. It would be advantageous to contact them and view their Web site, which has a lot of code-related information (520-624-6628; www.azstarnet. com/~DCAT).

What are the most common mistakes people make when building with straw?
Not protecting the bales from moisture before and during the construction process. People get their bales wet from rain while waiting to build and then put them in the walls anyway. The same happens while building. We never build with bales that have gotten rained on — that includes in the field, during storage, or during construction.

The next mistake is not paying attention to good building practices. Good overhangs and window detailing are extremely important.

What recommendations do you have for someone interested in building with straw?
Take it slow, build something very, very small to start with just to learn about the material. Take a workshop, if possible, and learn everything that you can. Then go slowly—don't be in a hurry. In the end, when it is time to build, design carefully, build small, consume wisely, and have fun during the process. Don't be careless; learn what building details are important in the area where you live, because much of it will be the same for a straw bale building. Finally, remember that this is a very young building system. It is still evolving. By joining the ranks, you become part of a process of redefining the way we build and the way we look at the world.

ATHENA SWENTZELL STEEN and BILL STEEN make their home in the oak woodlands of southeastern Arizona, where they teach workshops, write, and raise their family. They utilize their diverse backgrounds of mixed Anglo, Pueblo Indian, and Mexican cultures to direct the activities of The Canelo Project, a small nonprofit corporation dedicated to "connecting people, culture and nature." Much of their time is spent working in Mexico to develop simple and inexpensive straw bale and clay building methods that are applicable in both the developed and underdeveloped worlds.

vaulted ceiling and add a ceiling fan to force the warm air back down. Extra thermal mass may also be necessary to absorb sunlight needed to heat the room.

Another consideration is the type of roof material you use. As the authors of *The Straw Bale House* point out, "Until a creative new roof system for bale structures is developed, long horizontal spans of roofs will be most appropriately met by lumber and steel roof systems." Although wood may be one of the few options you have, there are ways to build a roof that minimize the amount of lumber without sacrificing strength. One product worth considering is the wooden I-beam or I-joist. As noted in chapter 2, I-joists use a fraction of the lumber of 2 x 10s and 2 x 12s, which are often used for roof construction. I-joists are also made from smaller trees, unlike 2 x 10s and 2 x 12s that come from old-growth trees.

Another option that cuts down on lumber is the prefabricated wood truss or parallel chord truss. Made from 2 x 4s cut from smaller-diameter trees, wood trusses are often delivered to the building site preassembled and ready for installation. If you make your own roof trusses, use lumber from producers who engage in sustainable forest practices. There's a lot more on roof wood options in chapter 14.

The Living Roof

After the roof is framed, it is covered with sheathing, felt paper, and shingles made of wood or asphalt or some other roofing product, such as steel roofing.

FIGURE 3-19.
Workers from ArchiBio prepare a living roof. Straw bales are placed on the decking, over a waterproof membrane, then cut and exposed to the elements. After the bales decompose a bit, manure is added. Seeds can be broadcast on the manure and soon sprout, creating a living roof that returns a portion of the site to vegetation.

Francois Tanguay

However, as the nonprofit group ArchiBio of Quebec, Canada, points out, there are older traditional methods of roof building that can also be used. One of them is the living roof.

A living roof is made of soil and vegetation placed atop a wooden roof support system. The roof can be freestanding as in figure 3-19, or built into the slope of the land. The three most important considerations in designing a living roof are providing adequate support, waterproofing, and proper drainage.

Living roofs can be nearly flat or sloped, up to 6 to 8 inches per foot. To ensure that the decking, the wood laid on the surface of the roof trusses, is kept dry, a waterproof membrane is applied over the decking. Polyethylene has been used. Although it is inexpensive and can be doubled over to reduce chances of developing a leak, polyethylene punctures easily and will deteriorate over time. It is pretty slippery, too, and hard to walk on, posing a danger to workers. A much wiser choice is a reinforced waterproofing materials such as Bituthene, Volcay, Armorplast, or Torchdown. According to ArchiBio, "These membranes have been used for many years on commercial and industrial buildings and are long lasting."

I used Bituthene on the underground portion of my home, but applied a thin (½ inch) layer of rigid foam insulation over it to protect it against puncture. Although this added to the cost of construction, it was more than justified. A small leak can do a lot of damage and can be extremely difficult to locate—and very costly to repair. The cost of a repair, which might require all of the dirt to be removed from the roof, would far exceed the extra money I spent on insulation. The peace of mind it creates is well worth the extra expense.

After the waterproofing and protective foam were in place, we pushed dirt back onto the roof using the bucket of a tractor and shovels. We then planted seeds gathered from the property the preceding spring and large clumps of sod dug up by the excavator when he prepared my driveway. Within a year, my living roof was overgrown with lush grasses, wild roses, and wildflowers.

The folks at ArchiBio have experimented with another approach to living roofs. Instead of dirt, they apply a thick layer of straw over the waterproof membrane—usually bales laid on their sides. After the bales are in place, the strings are cut and the straw is left uncovered for about six months. During this time, the straw absorbs moisture and begins to decompose. A couple inches of well-aged manure or compost is added next. Seeds are sown and in a year or so the roof comes alive with grasses, wildflowers, strawberries, whatever you want.

Native vegetation—that is, wildflowers and grasses that are indigenous to your region—are well adapted to local conditions and more likely to thrive on your roof. Because they are hardier than introduced species, they minimize your work and increase the likelihood of success. You'll find more information on the subject in chapter 15.

BUILDING NOTE

No matter how you design your living roof, be sure water that seeps down into the soil, if not used by the plants or retained in the soil, eventually drains off the structure. Proper sloping and peripheral drainage are vital to the success of a living roof.

Be careful when you load dirt onto a roof. Even with a carefully designed one, and that's the only kind to have, says Malcolm Wells in his book, *The Earth Sheltered House*, construction loads can be far more severe than the long-term loads. Dropping a load of dirt on a roof, or driving a front-end loader onto the roof, can cause severe strain, even roof failure. So, be careful at this stage!

There are many other roof options that border on the fringe, when compared to "modern" standards, but are really better known as "old favorites" once you look around the world or scan the pages of history. Among these is the thatch roof. "Contrary to the commonly held opinion that thatching is an antiquated, obsolete roofing system," say the authors of *The Straw Bale House*, "it has evolved into a modern craft." They go on to point out that thatch roofs share many of the advantages of other modern roofing materials and are still highly desirable in many parts of the world. A visitor to Mexico will see lots of thatch structures.

Thatch roofs are durable. If properly built and constructed from the best materials, a thatch roof can last 50 years or more. Getting one approved in the United States, however, could take nearly as long unless you live in Hawaii, where thatch roofs are code approved. For more information on thatch roofs, check out volume 26 of *The Last Straw*.

FLOORS

Straw bale homes are amenable to a variety of floors. The one most often used is the concrete slab-on-grade floor (described in chapter 2). Concrete floors are relatively versatile and can be colorized, carpeted, or tiled. They can even be stamped or scored to look like flagstone, tile, or brick, with stunning results.

If "economical" is the byword of your project, try an earthen or a soil cement floor (the latter is described in chapter 2). Earthen floors have been popular for centuries and can be remarkably solid. Anita Rodriguez of Taos, New Mexico, who builds floors for clients, guarantees them for life. "They will never crack, they hold up to excessive wear—from high heels, children, or even puddles of beer," she says, then adds, "You can do just about anything on them but chop wood." Of course, for this kind of durability, you have to know what you are doing.

Earthen floors are made of 20 percent clay and silt and 80 percent aggregate and are typically laid on a layer of crushed rock. Additional hardness is sometimes provided by adding other substances to the mix, such as cement.

Surface hardeners and sealers like boiled linseed oil are vital, too. Without them, your floor may not hold up very well. Besides being relatively inexpensive and extremely attractive, these floor systems provide good thermal mass for passive solar designs (chapter 11). Earthen floors are discussed in *The Earthbuilders Encyclopedia, The Straw Bale House, Earthen Floors,* and many of the books on adobe and cob construction, which are listed in the resource guide.

Another option for the economically minded is the use of brick and tile on a layer of sand. Easy to install, durable, economical, and attractive, these floors provide good thermal mass and are ideal in high-traffic areas. They also provide easy access to subfloor plumbing or electrical lines. Tile and bricks are laid on a 1½-inch-thick layer of sand. Although this may seem like a dubious way of making a floor, if the sand layer rests on a firm foundation of dirt and care is taken when laying bricks and grouting the tiles, bricks and tiles will not shift and the floor will remain level for a long time.

ELECTRICITY AND PLUMBING

Electrical layout for a straw bale home is very similar to that of a standard home. Electrical wires run through interior walls, through ceilings, and even through the slab or under a wooden floor. If wiring must course along an exterior wall, you must use wire rated for burying or run conventional wire through metal conduit laid in the "joints" between bales or pushed between the bales. Using a chain saw, you can even "carve out" grooves in straw bales for conduit.

Electrical boxes are attached to wooden wedges that are subsequently hammered into the straw bale walls. Boxes typically protrude 1 to 1½ inches from the surface of the bale to separate the electrical connections in the box from the potentially combustible straw and to leave room for the plaster. Although this system works well, innovators have come up with newer ways. Take a look at volume 24 of *The Last Straw* for an interesting alternative, also shown in figure 3-20.

Building inspectors may insist on further measures to isolate the electrical box from the straw bale wall and thus lessen the danger of sparks igniting a wall (which is a pretty remote possibility). Some builders spray a fire retardant, such as clay or borate, on the straw behind each electrical box.

As in rammed earth construction, plumbing is generally restricted to interior walls, slabs, and floor joist cavities for convenience but also to avoid water damage caused by leakage or condensation on pipes. The rule of thumb when building a straw bale home is to run plumbing where it can be easily accessed in case of a leak, such as interior partition walls. If you must run plumbing in

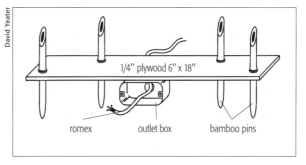

David Yeater

1/4" plywood 6" x 18"

romex   outlet box   bamboo pins

FIGURE 3-20.
An outlet board is used to insert electrical boxes in straw bale walls. The flat wooden plate lies between bales and the bamboo stakes secure the structure to the wall. The lower bale must be notched to accommodate the electrical outlet box.

a straw bale wall, place water pipes in plastic sleeves, larger-diameter pipes that prevent moisture from contacting the bales. Plastic pipes should be daylighted (run to a location outside the home) in case there's a break in the water pipe.

WHAT DOES IT COST TO BUILD A STRAW BALE HOME?

As I noted earlier, some proponents of straw bale construction point to its affordability as a major benefit. However, cost savings may be rather small. No matter what natural building technique you use, remember this: Costs vary enormously and are influenced by many factors, including the complexity of your floor plan, the amount of labor you and your friends contribute, the economy of your area, the quality of the finish work, the type of appliances you buy, and a host of other factors.

Richard Hoffmeister of Taliesin West estimates straw bale home costs ranging from $5 to over $80 per square foot. The lowest category, $5 to $20 per square foot, assumes a very simple structure with the owner, friends, and family providing all of the labor. These costs are possible only if you use salvaged materials. Bear in mind, though, that tile can run you $3 per square foot. Installing it can cost another $1 to $2 per square foot. It doesn't take much to drive the cost skyward!

The next category of homes runs from $20 to $50 per square foot. Once again, this price range is possible only if the owner, friends, and family do the bulk of the work, subcontracting the more technical parts, such as electrical and plumbing. Basically, though, walls, roofs, and finish work are the responsibility of the owner. This price range does permit a few more amenities such as tile, doors, and cabinetry on the lower end of the price range.

Moving up on the cost scale, a $50- to $80-per-square-foot home is contractor-built using the in-fill method with a minimum of custom features such as expensive tile, cabinetry, and built-in entertainment centers. I personally think that this would be on the very low end of contractor-built homes.

The final category is $80 per square foot and above. According to Hoffmeister, this would get you a custom-built home with a significant number of features. Expensive tile, fancy appliances, elaborate window coverings, a complex floor plan, breaks in the roof line by dormers, solar electric systems, and a host of other features can easily drive the cost to $125 per square foot.

These numbers are only rough guidelines. I recently toured a straw bale home that fit the description of the $80-per-square-foot category. But because of problems with the foundation, difficulty stringing an electrical line to the house, and general cost overruns, the house ended up costing the owners $125 per square foot. The foundation for this 2,000-square-foot home cost nearly twice the builder's estimate. Electrical service cost $15,000, rather than the builder's original estimate of $1,500.

BUILDING NOTE

Conventional stick-frame homes vary considerably. Tract homes may cost $50 to $80 per square foot. Custom homes may run from $80 to $150 per square foot, depending on local labor costs and other factors.

So remember that prices vary. Check around. Ask builders and owner-builders what it cost them. Then see what they gave or got for the money. Be realistic.

As a final note on cost, realize that the real cost of a home is not just the price of building it. The real cost, known as the full cost by economists, includes the day-to-day operating costs and the environmental costs (often called external costs)—damage to the biosphere and damage created in supplying the resources needed to build and operate a house.

One useful comparison is the comparative lifetime costs of various homes (see sidebar). The lifetime cost shown here includes the costs of construction, financing, and operating a home made from straw bales compared with the conventional stick-frame home. (It doesn't include environmental costs.) As you can see, lifetime costs range from $171,000 for a conventional house over a 30-year period to $29,625 for a straw bale structure, built entirely by the owner with salvaged and scavenged materials, and equipped with super-efficient appliances.

Don't get carried away with these numbers. They are meant to show differences, not to be an absolute guide. Costs add up very quickly. Many an owner-builder has gotten into trouble thinking that they can build a home for $5 to $50 per square foot, only to find themselves out of money with only half a house built. If your house has any of the niceties you've grown accustomed to, it will more likely run in the upper price range. However, straw bale will save you a lot of money over its lifetime by reducing energy costs.

HOW MUCH WILL IT COST IN A LIFETIME?

This estimate is based on a 30-year period in a moderate climate, for a 1,375-square-foot home with three bedrooms and two baths.

Conventional	$171,000
Straw bale, contractor-built	$153,000
Straw bale partially owner-built with super-efficient appliances	$74,000
Straw bale, owner-built (no financing) with efficient appliances	$29,625

See *The Straw Bale House* (pp. 38–39) for more details.

SHOULD YOU BUILD A STRAW BALE HOME?

Like all natural building technologies, indeed every building technology known to humankind, straw bale has both pros and cons that you should consider.

Advantages of Straw Bale Homes

- Straw bale homes are produced from an abundant and highly renewable resource, a by-product of grain production that is often burned after the grain is harvested, creating enormous amounts of air pollution.
- Straw bales are relatively easy to acquire in most locations (at least in more developed nations). (For a source near you, see the annual resource issue of *The Last Straw*, which even lists organically grown straw.)
- Straw bale construction requires much less lumber than traditional wood-frame construction and could reduce deforestation.
- Straw has low embodied energy. In fact, properly built straw bale homes require much less energy (about 30 times less) than standard wood-frame walls insulated with fiberglass batting.

- Properly built straw bale walls are safe, strong, durable, and long lasting.
- Stuccoed and plastered straw bale walls are fireproof, rodent- and insect-proof, and, if kept dry, resistant to decay.
- Straw bale homes are suitable for a wide range of climates.
- Straw bale walls provide an extraordinary measure of insulation, helping to ensure a thermally stable environment. By keeping homes cool in the summer and warm in the winter, they conserve energy.
- Thermal properties make this construction technology suitable for passive solar heating and cooling systems.
- Straw bale homes require less energy to heat and cool and thus decrease our dependence on fossil fuels, thereby reducing air and water pollution, oil spills, and land disturbance.
- Lowered heating and cooling costs reduce the size of backup heating and cooling systems, saving money.
- The heavy insulation of bale walls protects against outside noise.
- Straw bale walls can be built with relatively simple tools and relatively untrained workers. Individuals can learn what they need to know to stack walls in a two-day workshop. (But remember, walls are only part of the structure. You still have to build a roof, install doors and windows, and so on.)
- Straw bale construction is conducive to community participation and owner-builder projects.
- Straw bale construction is flexible, allowing one to build in a variety of different styles. Straw bales can be used in conjunction with other natural building methods, such as rammed earth, or even conventional building techniques.
- As straw bale construction becomes more widely known, it becomes easier for the owner-builder to obtain building department approval, financing, and insurance.
- When covered with plaster, straw bale walls resemble thick adobe walls.

Disadvantages of Straw Bale Homes

- Despite its growing popularity, straw bale construction is still considered unconventional, which may make it difficult to obtain a building permit, financing, and insurance.
- Straw bale walls may require a thicker foundation than wood-frame homes, which adds to the project cost.
- Although straw is a waste product that is often burned in fields or plowed under, these practices help replenish agricultural soils by adding vital organic and inorganic nutrients. Diverting straw to build houses could damage soils.
- Walls of straw bale must be protected from moisture to avoid mildew and rotting (but so must walls of conventional wood-frame homes).

- Bale walls are relatively easy to erect, but they represent only a fraction of the total construction project. Don't be lulled into thinking it is easy to build straw bale homes. Electricity and plumbing, interior framing, tile, and other features of modern homes will increase substantially the complexity of the project.
- Straw bale walls cannot be built underground or earth bermed, although there are exceptions.
- Straw bale building is in a state of rapid evolution. As information changes, current building methods may become obsolete or prove inadequate. Therefore, there is some risk involved in building straw bale homes.
- Opinions on different aspects of straw bale construction vary considerably, making it hard for the neophyte to know which path to follow.

So that's the scoop on straw bale construction. If you want to learn more, see the resource guide. Attend a workshop or two. Read books. Watch the videos. And, by all means, check out *The Last Straw*. Their annual resource issue contains a comprehensive list of resources, including a list of straw bale buildings you can visit.

4

EARTHSHIPS AND BEYOND

YOU CAN'T MENTION the word "Earthship" without conjuring up memories of actor-activist Dennis Weaver and his sprawling 10,000-square-foot Earthship mansion made of tires and aluminum cans in southern Colorado. Although the charismatic actor has done a great deal to promote and legitimize Earthships and alternative building technologies, he is not the originator of the Earthship idea. That honor goes to maverick builder Michael Reynolds, of Taos, New Mexico, who began building Earthships about thirty years ago.

These funky-looking structures made from used automobile tires packed with dirt and stacked on top of one another like huge bricks, consist of a series of U-shaped rooms. The curving walls are covered with stucco or earthen plaster, which hides any evidence of the tires and makes the walls resemble those of adobe homes. The front face of many Earthships consists of sloping glass that gathers up the sun to heat the house (figure 4-1). Sunlight also supplies the indoor planters that provide food and purify wastewater.

Typically earth sheltered and equipped with photovoltaics to generate electricity and heat water, the Earthship sports a unique roof design that collects water from rain and melting snow. Stored in large cisterns, then purified in a series of filters, this water supplies all domestic needs.

As is evident from this brief description, the Earthship is more than a house, much more. In the words of Michael Reynolds, the Earthship "is a completely independent globally oriented dwelling unit made of materials readily available in most parts of the world." In my view, the Earthship is the

FIGURE 4-1.
Built from tires packed with dirt, the Earthship is the ultimate dream machine—that is, if your dreams run toward helping to build a sustainable future.

Salvaging what is left of the forests, the waters, and the air was once a noble issue for environmentalists, intellectuals, and various scientists as they thought of future generations and what they might be facing. Now, it is clear that we who will live on into the 21st century are, in fact, "the future generation."

MICHAEL REYNOLDS

epitome of sustainable design and construction. No part of sustainable living has been ignored in this ingenious building. Although it is not perfect, and is evolving as fast as any building technology on the planet, the Earthship is probably the model of a sustainable home unrivaled in the modern world.

ANATOMY OF AN EARTHSHIP

As shown in figure 4-2, the Earthship consists of two parts: the U-shaped living spaces or Earthship rooms, and the environmental interface corridor. The rooms of an Earthship are delineated by plaster-covered tire walls, which can be closed off for privacy. Flanking the building are two cisterns, monstrous water-storage tanks that collectively can hold up to 10,000 gallons of water.

Earthship "Us"

environmental interface corridor

cistern

cistern

Solar Survival Architecture

graywater planter

kitchen

systems package

bathroom

graywater planter

FIGURE 4-2.
Earthships consist of a series of Us built out of tires packed with dirt. The front section is the environmental interface corridor. It houses many of the essential components of a sustainable lifestyle.

In many Earthships, the kitchen is located in the sunny environmental interface corridor. Beware: Food preparation during the period of maximum solar gain—that is, when the sun is low in the sky—may become intolerably hot. The refrigerator is also bathed in sunlight much of the year, making it work much harder than necessary. Shades can be installed to prevent you from overheating, and refrigerators should be shaded by a partition wall—or by some other means—for optimal efficiency.

Americans discard an estimated 253 million automobile tires every year. Nearly half of them end up in stockpiles or in landfills.

Scrap Tire Management Council

(In the most recent versions, the walls of cisterns are typically made from packed tires plastered with cement.)

The environmental interface corridor, the "front" of an Earthship, houses the independent living systems: the solar hot water system, the water purification system, and the batteries for storing electricity from the photovoltaic panels. It also houses the kitchen, bathrooms, and planters that purify waste-water from sinks, showers, and bathtubs.

THE EARTHSHIP AS A MODEL OF SUSTAINABILITY

In one of those anonymous Internet articles, someone wrote that Michael "Reynolds believes that Earthships could change the nature of the human mind itself." In other words, "the concept of the Earthship extends beyond that of a structure independent of any outside inputs. It is also a structure operating as a catalyst, influencing things beyond its physical bounds." I agree. It has changed my thinking profoundly.

The Earthship is a demonstration of the wisdom of fitting in. Its success proves that we humans can live sustainably on the Earth by using natural materials and waste to build our homes and by cooperating with nature and tapping into the sun and nature's forces.

To me, the Earthship is a reflection of a deep and abiding understanding by its developers that the Earth's resources are limited. It reflects the knowledge that we humans are a part of nature. It shows us that we can live well without putting Earth's life support system at risk. The Earthship helps us live in harmony with the natural world, honor natural limits, and protect the ecosystems that supply us and a host of other species with the resources we need to survive on a finite planet.

The Earthship is a structure that valiantly "obeys" the laws of sustainable design, construction, and living outlined in chapter 1. In keeping with the efficiency principle, the Earthship is often nestled in the ground and insulated to the hilt to conserve energy. Locally available dirt is used not only to pack tires but often to make the earthen wall plasters and earthen floors. Tapping into such a local resource cuts down on transportation energy demand. In keeping with the reuse and recycle principle, much of the structure is composed of waste, notably, automobile tires, an abundant waste product of modern society. Recycled tile, carpeting, wood, and other products can further the Earthship's sustainable attributes. Like other buildings described in this book, the Earthship is a solar home. It obtains its electricity from photovoltaic panels (PVs) powered by the sun. Sunlight also serves as the major source of heat. Earthships are healthy, affordable, and easy to maintain and operate—and they're easy on the environment. The Earthship is just about everything an aspirant of sustainable living could want. And, what is more, Reynolds's company has made it easy to equip an Earthship with independent living systems such as solar hot water, solar electricity, and catchwater. They will, for example, sell you a systems package that contains the preassembled components of a water purification system to cleanse drinking water, a solar hot water system (panels and tank), a pressure tank, an on-demand hot water heater to provide backup heat, and a solar electric system (batteries, vented battery vault, photovoltaic panels, inverter, etc.). In addition, the systems package comes in a 5-foot-by-8-foot crate that is installed on the front of the Earthship. Pretty amazing.

If you want to live lightly on the land and achieve your independence from the local power company, and those aggravating bills they insist on sending you every month, but are thinking about building straw bale or rammed earth or adobe, study the Earthship, if for no other reason than to become a better systems thinker. Look carefully at the Earthship's systems, so that you can see the full range of options. Incorporate these systems into your design. You may even want to consider purchasing a systems package from Reynolds's company. The price is reasonable. And it will save you a lot of time and trouble chasing down the parts of solar and water systems. It will also reduce installation time; the components of the water and solar electric systems, for instance, are preassembled and mounted on a board. They simply need to be attached to the wall and then connected to pipes or wires.

Another feature that makes the Earthship concept even more exciting is that it is not static. Over the past decade, the Earthship design has undergone a thoughtful evolution. Its basic structure and the configuration of the systems that provide inhabitants with food, water, energy, and environmentally sound wastewater disposal have undergone dramatic improvements, which are documented in Earthship books, the *Earthship Chronicles*, and a newsletter available from Reynolds's company (all listed in the resource guide at the end of the book).

The appearance of the Earthship has also undergone considerable change. If you don't find the exterior appealing, maybe it's a bit too funky for your tastes, there are many new options to choose from (see figures 4-3 and 4-4). *The Earthship Menu: How to get an Earthship* lists them. If you would like to see an Earthship, even live in one for a night or a month or two, contact

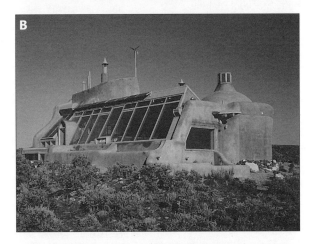

FIGURE 4-3 A, B, C.
Three new Earthship designs: Shell House (A), Nautilus House (B, straw bale), and the Hut House (C).

Reynolds's company, Solar Survival Architecture (SSA), in Taos. They have rental units in Greater World Community just outside of Taos, New Mexico. They also list units for sale. If you are interested in building one, Solar Survival Architecture provides training, blueprints, consultation, and experienced work crews who will travel to your site to build tire walls, install graywater systems, whatever you want—even build your entire home. SSA sells blueprints for a relatively small fee. After you pick out a style you like, they will supply you with three sets of architect-stamped blueprints, often suitable for submission to your building department! And, as noted earlier, components such as pre-assembled solar electric systems and water filtration systems can be purchased from SSA. They have taken much of the work out of sustainable design, in an effort to make it more accessible. As building technologies go, the Earthship has a relatively short history, only three decades or so. Its greatest history may yet lie ahead of it as these structures are built throughout the world and undergo the ultimate test, living in them.

OTHER TIRE HOMES

Earthships have inspired some builders, such as Dave Johnson and Dave Sorensen of Vanguard Homes in Colorado Springs, Colorado, to strike off in new directions. Using the basic concepts of the Earthship, they are building homes that may have a much wider appeal. My home is one of them (figure 4-4). Located in a mountain subdivision, my home resembles a modern passive solar home, and fits fairly well into the neighborhood. Although it may look like other homes, it isn't. Inside you see curved walls made from used automobile tires—about 800 of them. Compact fluorescent lighting and the super-efficient refrigerator reduce my use of electricity to about one-fourth of my neighbors. Unlike them, I have no electric lines running to my house. I get all

FIGURE 4-4.
If your tastes run toward a more conventional-looking solar home, you can still build with tires. This is a photograph of my home in the middle of winter. Most of the walls were built with packed tires.

of my electricity from the sun and wind. The roof captures snow melt and rain, filling a 5,000-gallon cistern that supplies all of our domestic water needs. There is no well. Water from the washing machine is used to nourish plants indoors and outdoors. The list goes on.

I could not have received approval for an Earthship in this subdivision. It would have stuck out like a wart, and the architectural review committee would have sooner voted Democrat than grant permission to build it. But an Earthship derivative, well, that was another story. Because it looked a lot more like the homes in my neighborhood, they granted me permission to build. If your tastes wander more toward mainstream architectural design, but you like the Earthship concept, this hybrid may be the house for you.

A WORD OF CAUTION

As noted previously, the Earthship is an evolving building technology. Since its inception, numerous mistakes have been made. For example, early in the development of drinking water systems, roofing materials were used that turned out to be less than optimal. Some even fouled the water. And the slope of the entire roof had to be changed to collect water. Builders who didn't know as much as they needed to know also built structures that fell short of expectations—some didn't perform well and others were structurally defective. Some people modified plans—against the advice of Michael Reynolds—which they lived to regret. Be careful if you are buying an older Earthship. And be sure to check out the newest Earthship books and publications.

Michael Reynolds writes: "We are attempting to establish a business of producing and aiding the production of Earthships in the flailing economic arena of the time." The struggle to create the perfect sustainable living system has been difficult. Reynolds and those who have joined him have had to fight antiquated laws, beliefs, and building codes. They have made mistakes, had clients turn on them, and suffered criticism from skeptics. They have been "chastised for not having the whole process of Earthship production up to par with the production of conventional energy-hog housing that has been around for many years."

Builder Gary Dillard (listed under Red Pueblo, LLC, in the resource guide) has found that the process of obtaining a building permit for an Earthship in Colorado is fraught with frustration and disappointment. Some counties are receptive, but the majority of them are not. They usually object to the graywater systems, which treat water from showers, sinks, and washing machines, fearing that pathogenic microbes in the water (if present) might spread to people when, for example, the water is used for outdoor watering. Other building department officials object to the solar electric systems, fearing that they won't work adequately when struck by lightning.

You may have heard about Earthship failures, but as Reynolds points out, "After twenty-five years of research and development, the Earthship and all its systems have evolved to a point where we are sure of what it takes to get a home that will independently and reliably take care of you." But that is not to say "it won't continue to improve and evolve." Given the creative minds who are leading the development of this technology, I think you can count on it.

FOUNDATIONS AND WALLS

One of the advantages of an Earthship is that its tire walls require no foundation. Walls are built directly on the ground. Soils must be well drained and sufficiently strong to support the weight of the structure.

To optimize solar gain, Earthships are typically built into hillsides facing south (in the northern hemisphere). Topsoil is first removed, then stockpiled for reapplication after the house is complete. To build the walls, tires are laid out in U's with small squares of cardboard placed over the bottom opening of each tire to prevent dirt that will be packed in the tire from escaping. Tire packing is done in teams of two. One person shovels dirt into the tire and the other tamps it.

Tamping or "tire packing" has been performed by hand for years and is Reynolds's preferred method, especially in developing countries where labor is abundant and more affordable than machine tamping. Some builders, however, prefer a faster but more energy-intensive method known as pneumatic tamping. This technique requires a special tamping device, described in chapter 2, and a compressor, usually powered by diesel fuel. Although this technique is noisy and uses fossil fuels, it greatly accelerates tire packing.

During the packing process, tires expand dramatically. You won't believe a tire could get that big! When the tire is full and level across the top, the next one is packed. When full, it should be level with the first tire and so on down the line. After the first course of tires is packed and leveled, the second row is set in place in a running-bond pattern to increase the strength of the wall. Cardboard is placed over the bottom openings and the tires are filled with dirt and tamped. A carpenter's level works well to keep the wall plumb.

Tire walls of Earthships and Earthship derivatives are built one row at a time and usually one U at a time. The time required to build a wall depends on the size of the house, the technique (hand tamping or pneumatic tamping), and the number of workers. In my experience, a crew of four can pack 100 tires a day using the pneumatic tamper; an entire 2,000-square-foot home may take 7 to 10 working days. By hand, the same job may take a month or two, although the folks at SSA say their crews can pack 100 tires a day by hand.

BUILDING NOTE

Tires tend to shift as they are packed with dirt, especially when a pneumatic tamper is being used. To keep the walls of my house plumb, my builders ran deck screws into abutting tires. They also watched over the workers like hawks, using a carpenter's level to check their work. Like other tire homebuilders, they used tires of the same size throughout the project to achieve uniform wall height and level courses. Aware of the dangers of half-hearted packing, they also made sure that all tires were packed fully and equally. If they're not, a tire wall may contain weak spots and may crack or, worse yet, collapse.

Because tire packing by hand is grueling work that is hard on your lower back, arms, and shoulders, workers have a tendency to skimp. Check each tire in a row before the next row is started to see if it is hard and sufficiently filled (it will bulge considerably). There's too much riding on these walls to do a shoddy job. For more details on tire packing, and there are a lot more, see Reynolds's book, *Earthship*, volume 1).

In addition to yielding interesting curved walls, the U-shaped design of Earthship rooms provides walls with maximum structural stability. Like the stones in an arch, the tires of a U-shaped wall lock in place, creating a wall that is fairly resistant to lateral loads. Although most builders don't believe it is necessary to reinforce the walls, some drive rebar vertically through the centers of the tires to provide additional strength. Another means of increasing the strength of the walls is to terrace them (figure 4-5).

Earthships and derivatives are typically built as a series of U-shaped modules of varying size. Although a module can vary in depth and width, the maximum recommended width is 18 feet and the maximum recommended depth is 26 feet.

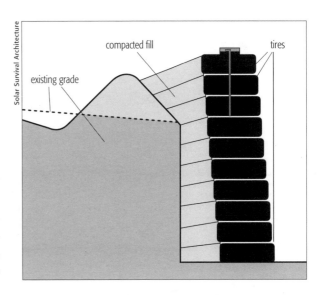

FIGURE 4-5.
Stepping tires into the earth.

As the walls of an Earthship rise up from the ground, dirt is backed behind them, usually with a backhoe. The Earthship is therefore nestled into the ground, or earth sheltered, and thus protected from cold and wind. Earth sheltering, as noted in chapter 1, makes a house much warmer in winter and cooler in summer. Earth sheltering also helps to protect the wall/foundation from frost heave.

Any time a structure is placed underground, care must be taken to prevent water from seeping into it. To repel moisture, Earthship builders apply a layer of 6-mil plastic over the tires where they will come in contact with the ground. Rigid foam insulation is often placed against the tires as well to create a thermal break, and a French drain is installed to draw water away from the structure. Dirt is then pushed against the tires. (There is more on the French drain later in this chapter.)

BUILDING NOTE

Plastic should be placed against the back of the tire wall to prevent water from seeping into the house. Thick foam insulation and a layer of pumice are often used to provide a thermal break between the tires and the Earth. This reduces heat loss from the structure.

SECURING THE ROOF TO THE TIRE WALL: BOND BEAMS

When the tire wall is completed, a bond beam must be constructed to secure the roof to the tire walls. Bond beams can be constructed in one of several ways. In early Earthships, anchor bolts were set in a stiff concrete mix poured in depressions carved out of the centers of the tires in the top row. A wooden top plate (also known as a sill) was then secured to the bond beam via the anchor bolts.

In recent iterations of the Earthship, builders have begun experimenting with concrete bond beams that are poured in "forms" made from aluminum cans and cement mortar (figure 4-6). Forms can also be made from dimensional lumber. (Details of the concrete bond beam are outlined in *Earthship*, volume III.) These days Solar Survival Architecture generally uses wood plating rather than concrete bond beams.

BUILDING NOTE

Horizontal rebar is inserted in the concrete form to provide tensile strength. Vertical rebar is driven into the tires to provide a means of attaching the bond beam to the tire wall. Anchor bolts are also placed in the concrete shortly after it is poured to provide a means of attaching the top plate to the concrete bond beam.

FIGURE 4-6.
Aluminum cans and concrete are used to build a bond beam to attach the top plate and roof.

Although aluminum cans are widely used in Earthship construction, remember that aluminum has a very high embodied energy— 80 times greater than wood!

FIGURE 4-7 A, B.
PVC pipe (A) laid in the bond beam will serve as a chase to run electrical wire to a wall. Otherwise you may end up with unsightly conduit protruding from your walls (B).

When framing a bond beam, it is a good idea to insert PVC chases for electrical conduit that may run through the ceiling down to an outlet (figure 4-7a). If you don't, you may have unsightly conduit running along the inside edge of the bond beam and tire wall, as shown in figure 4-7b. A few pieces of PVC with a 45-degree angle should suffice. If you don't need them, there's no harm done. If you do, you will be glad you took the time to add them. Builders also install knockout blocks in the forms prior to pouring the concrete. These 2 x 4 blocks are inserted against the inside edge of the form. After the concrete has set up and the forms have been removed, the blocks are literally knocked out of the bond beam, leaving behind channels in which run to your electrical conduit. When the walls are finished, a layer of stucco or plaster is applied to cover any evidence of the conduit.

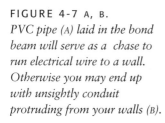

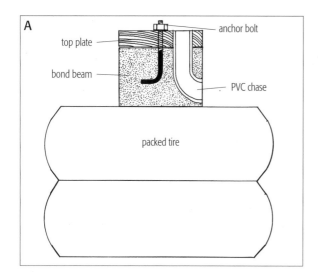

Some modifications of the bond beam may be required, depending on the type of truss you use. Trusses are wooden or steel assemblies that form the structural support for your roof. In New Mexico, many Earthships have been built using vigas (pronounced vee-gahs), peeled logs. They can be obtained from local forests and create a rustic-looking ceiling that appeals to many homeowners. For those interested in using vigas, see *Earthship*, volume I, which provides a remarkable level of detail on the subject.

Earthship roofs can also be built from 2 x 4 trusses or I-joists, discussed in chapters 2 and 14. Trusses such as these are generally easier to handle, being much lighter than vigas. Unlike vigas, they don't taper. They also create a roof depth that is suitable for superinsulating a house. *Earthship*, volume 3 discusses two types of trusses that you can make yourself, one of which is very similar to I-joists.

THE FRONT FACE

The front wall of the Earthship, known as the front face, is largely glass inserted in a wood frame. For years, sloped glass was the preferred method in Earthship construction because it maximizes solar gain in the winter. Although details of framing the front of an Earthship are beyond the scope of this book, a few comments are in order.

As shown in figure 4-8, the front face rests on a foundation built out of packed tires. (This is the only foundation you will need.) The front face is framed with dimensional lumber and the windows are then set in place (figure 4-8b). To protect the wood framing

Solar Survival Architecture (photo)

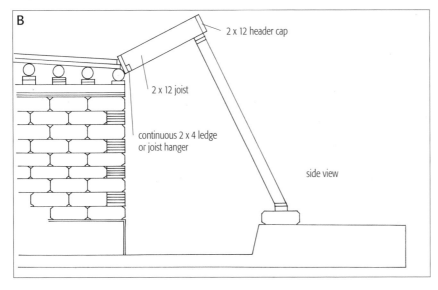

2 x 12 header cap

2 x 12 joist

continuous 2 x 4 ledge or joist hanger

side view

FIGURE 4-8 A, B.
Sunlight penetrates the front face of the Earthship, providing heat (A). The front face is framed with conventional lumber set on a tire foundation (B).

FIGURE 4-9 A, B.
The V-shaped roof of the Earthship (A) is ideal for capturing rainwater and snow melt which is then funneled into cisterns located on either end of the house. This gutter (B) is used on Earthship Nests, which don't have the typical V-shaped roof to collect rainwater.

from punishing sunlight, a steel cladding is applied to the frame. (Unprotected wood is no match for intense sunlight on the south side of a solar structure!)

In old-style Earthships built in the northern hemisphere, roofs sloped north (away from the sun) and were covered with a thick blanket of insulating earth, almost completely burying the house in the ground. While this was great for site restoration, it did not lend itself to catching rainwater. Recognizing this, Michael Reynolds has redesigned the Earthship roof, dramatically changing the orientation as well as the shape of the roof, turning it into a rain-catching machine unrivaled in the building industry.

The modern Earthship sports an odd **V**-shaped roof, shown in figure 4-9. Water falling on the roof drains into the apex of the **V**, then drains laterally to the cisterns, one on each end of the house. Although it is more difficult to build than earlier roof designs, the extra effort is amply rewarded in a generous supply of water.

KEEPING WATER AWAY FROM YOUR WALLS AND OUT OF YOUR HOUSE

Because Earthships are nestled in the ground, special effort must be taken to protect them from groundwater. Proper siting on a lot is the first and most important measure. Avoid swampy locations and never place a home in natural water drainage, even if it appears to be dry. If you must locate a house in or near a natural water drainage, divert the water around the house. Natural drainage courses may quickly fill with water in heavy downpours. Be especially careful in desert country. A sudden downpour can turn a dry and apparently harmless arroyo into a raging torrent within seconds. A house built in the path of what may seem like an innocuous, ancient watercourse could be flooded in no time under heavy rainfall.

Draping a layer of 6-mil plastic on the outside of the tire walls helps protect an Earthship or tire home from water infiltration. The plastic should be integrated into a peripheral drainage system, consisting of a plastic-lined trench filled with gravel and porous pipe located at the base of the wall along the outside perimeter of the house (figure 4-10). This is known as a French drain and is especially useful in wet climates.

After the house is completed, be sure that the site is properly graded. The earthen berm around the Earthship and the ground in front of the structure should be sloped away from the house to promote maximum drainage.

FINISHING THE WALLS

The interior walls of tire homes are typically covered with cement stucco or earthen plaster. This turns the ugly, raw tire wall into a thing of great beauty. If you haven't read the section on plastering and stuccoing straw bale walls in chapter 3, now would be a good time to do so, because much of this information pertains to the tire home as well.

Earthen plaster for interior walls can be made from local soils or can be imported from local suppliers or neighboring sites. Before you can start plastering, you must fill the deep V-shaped crevices in the wall formed where tires abut. Reynolds recommends using a "stiff" mud (one with not much water). Workers toss the mud against the wall to fill its deep recesses. Some Earthship builders also use aluminum cans as filler, as shown in Figure 4-11. As a side note, although aluminum cans are an abundant resource, aluminum is extremely high in embodied energy (80 times greater than wood). It is also highly recyclable. Because it is so economical to recycle and because many places have programs to recycle aluminum, many folks prefer to use other materials—for example, rocks. Aluminum is just too valuable to be locked up in walls of a home, especially when other widely available materials can be mortared in place with mud!

After the canyons in your walls are filled, the first coat (scratch coat) of mud plaster or cement stucco is applied. Before a cement stucco scratch coat dries, the wall must be raked with a "scratcher" to increase the adhesion of the second layer. After the scratch coat has dried, a second coat of cement or earthen plaster (the brown coat) is troweled on. Once you have achieved the desired thickness, removed any irregularities in the walls, eliminated all signs of tires (bulges), and the second coat is dry, the finish coat is troweled in place. To create a smooth finish, spray a fine mist of water on the wall, then trowel until you are satisfied with the result.

For readers who want to learn more about plastering, check out *Earthship*, volume 1 for more details. It describes the proper mix of sand and dirt

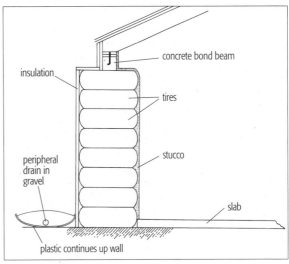

insulation

concrete bond beam

tires

stucco

peripheral drain in gravel

slab

plastic continues up wall

FIGURE 4-10.
A drawing of the back wall of my house shows the placement of a peripheral drain—porous pipe in a bed of rocks that runs along the periphery of the house daylighting at the end of my property.

BUILDING NOTE

In my home, the builders also believed that aluminum should be reserved for more important tasks than filler, like holding beer, so they took a shortcut that worked out marvelously well. They stapled 1-inch chicken wire to the face of the tires by nails, then troweled cement onto the wall, filling the spaces. The chicken wire held the concrete in place and speeded up the process.

FIGURE 4-11.
Aluminum cans and mud are used to fill the gaps between the tires of this wall. Some individuals (myself included) feel the use of cans is a waste of a high-embodied-energy resource that is readily recycled.

BUILDING NOTE

Cement stucco is often used on the interior walls of Earthships and exposed exterior walls, which are few. Cement stuccos work well in Earthships because the walls do not have to breathe as they do in straw bale homes.

Still, earthen plasters should be given special consideration because they use indigenous materials and have a lower embodied energy than cement. They are also very attractive. Solar Survival Architecture generally does not use cement stucco anymore.

---

and other important aspects of this step. You should also read Steve Berlant's book, *The Natural Builder*, volume 3, *Earth and Mineral Plasters*. It contains a gold mine of information!

Common Concerns: Fires and Odor. When many people first hear about Earthships, their eyebrows knit up as they express concern over fire breaking out in the walls, an unfounded fear that comes from news reports showing billowing clouds of black smoke issuing from tire dumps. While spontaneous combustion in tire dumps, containing millions of tires exposed to the air, is a very real concern, fears over fires breaking out in tire walls are unfounded. In an Earthship, used automobile tires are entombed in concrete or earthen plaster, a layer three to six inches thick. No air can enter to promote combustion, and even if air could reach the tires, bear in mind that they are packed with dirt. The chances of a fire are about as good as a frog flying to the moon.

Another concern I commonly hear is about odors. Won't you smell rubber all of the time? I've lived in a tire home for four years and have never detected the slightest odor of rubber. I've visited a dozen tire homes in New Mexico and Colorado, even ones whose walls haven't yet been plastered, and I've never detected the faintest smell of rubber. So don't sweat it. You won't smell the tires. For those who are chemically sensitive, stuccoed or plastered tire walls may be a great option.

CAN WALLS

Aluminum cans figure prominently in the construction of an Earthship. Besides containing the beverages that are consumed after a hard day of tire packing, aluminum cans are used to fill the V-shaped recesses in walls, a practice many people frown on. Aluminum cans can also be used to make bond beams and structures such as cisterns, interior walls, planters, vanities, shower stalls, stairways, and room partitions—the possibilities are endless and the results can be stunning (figure 4-12). Can walls require very little skill, and open up all kinds of opportunities for innovative design and aesthetic achievement. (As you will learn in chapter 6, however, cob could be used just as well as aluminum cans to shape many of these structures.)

Embedded in a stiff cement mortar, cans are stacked "on top" of one another like bricks. After the wall has been completed, it is plastered to cover evidence of the cans. Details of this unique form of construction can be found in *Earthship*, volumes 1 and 2.

CUSTOMIZING YOUR EARTHSHIP

One of the major advantages of the Earthship is the availability of a complete set of plans, costing $1,500 to $8,000. The variation in cost reflects the variation in size and architectural complexity (one-story vs. two story). The larger and more complex the structure (like split level), the greater the cost.

The engineer-stamped plans, thirty pages in all, contain most, if not all, of the drawings needed by the building department to issue a permit, including an overview, floor plan, building sections, framing plan, and a few details. (Chapter 16 clarifies what building departments typically require.) The remainder of this massive set of drawings includes details on virtually every aspect of Earthship construction, from excavation to the installation of mechanical systems. To have an architect draw up such plans would cost thousands of dollars more than you will pay the folks at Solar Survival Architecture.

Although SSA's plans have been accepted by building departments throughout the United States, local building departments may require some modifications or separate engineering approval for specific parts of the plans. Soil tests and other documentation may also be required by your local building department.

If none of the many plans suits your liking, you can customize your Earthship. For example, you can shorten interior walls, add doorways and other openings in the tire walls, hook up to conventional water and power, whatever. But beware. Many of the problems that arise in Earthships, says Reynolds, come from changes in the plans. Change at your own risk. Changes will also add to the cost of the structure. As Reynolds notes, if your "trek toward independent living must be made with a limited budget, customization is bound to make the trip perilous." The additional plans, labor, and materials needed to make changes will jack up the price tag. So if your dreams outstrip your budget, don't think that a customized Earthship will solve the problem.

FIGURE 4-12 A, B.
Aluminum cans and concrete can be used to make a wide assortment of things including this stunning dome in an Earthship in Taos, New Mexico.

BUILDING NOTE

Solar Survival Architecture can provide a detailed engineering report to your engineer and building department, for a small fee. SSA will even obtain a permit for you, as it is easier for them, with all of their accumulated knowledge, to deal with building department personnel than it is for novices.

ELECTRICITY AND PLUMBING

Installing a photovoltaic system requires specialized knowledge and experience. As you will learn in chapter 12, a solar electric system requires PV panels, an inverter, a battery charger, batteries, a vented battery vault, a control panel, and a few other odds and ends. The PV panels generate direct current (DC) electricity, which travels to the control panel where it registers on a current meter so you know how much electricity you are generating at any one time. The electricity is then stored in the batteries for later use. When an appliance, light bulb, or an electronic device is turned on, DC electricity is drawn from the batteries. Current travels to the inverter, a device that converts DC electricity to alternating current (AC) and bumps the voltage up, usually from 12 or 24 volts to 110 volts. AC electricity then travels to the main circuit box in your house and is distributed via standard AC circuits to run typical household appliances and lighting.

In my house, I bought these components separately for $9,300, and hired an electrician to install it all, which cost another $3,400. Had I purchased a solar system from SSA, the whole package would have arrived at once, mostly prewired and mounted on a board for easy installation. I could have saved a few thousand dollars. Even the most solar-illiterate electrician can do this job—and quickly.

The remainder of the wiring in an Earthship is pretty straightforward as well. Electrical wire is often run in code-approved conduit along tire walls, prior to plastering or stuccoing. (Check your local codes for the type of conduit they require if you are doing your own wiring.) Electrical wiring can also be run in conduit in the floor or between rafters or roof trusses. It then runs through the tire walls or interior framed walls to switches and electrical outlets.

Prior to plastering or stuccoing, electrical boxes for outlets and switches are mounted on small plywood plates that are screwed into the tires over the V-shaped spaces. Electrical conduit is then run along the grooves between rows of tires.

Plumbing in an Earthship consists of water lines that run from the cistern to a series of filters and a pressure tank, located in the front center of the house (in the environmental interface corridor). From the pressure tank, cold water flows to the water heater or through cold water lines that run to faucets and showerheads. Water lines from the water heater run alongside the cold water lines. Hot and cold water lines run through the frame wall along the front face and through the floor, if necessary. As a side note, hot and cold water lines are typically not run through tire walls, which would make them pretty inaccessible in case a leak developed and needed repair.

In most modern homes, water from sinks, showers, and tubs, which is known as graywater, and wastewater from toilets (and the kitchen sink), which is known as blackwater, are drained to a municipal sewer system or to a septic tank along a common set of pipes. In Earthships, graywater and blackwater are handled separately. That is to say, they flow in separate pipes. Gray-water is piped to a biological filtration/purification unit, a planter that purifies the water while supporting a lush jungle of plants. It is located along the front of the house in the environmental interface corridor and adds extraordinary beauty to the Earthship. Blackwater is drained into a solar septic tank located outside the house, usually right in front of the structure. The septic tank releases leachate (liquid) into an outdoor planter, which purifies the liquid runoff from the septic tank. These biological wastewater purification systems are described in chapter 13.

ROOFS

The roof of an Earthship is a relatively simple structure with vigas or trusses that rest on the top plate, which is secured to the top of tire walls. Trusses and vigas are placed on blocks to achieve the slope needed to drain water from the roof (figure 4-13).

Decking is nailed or screwed directly to trusses or to the tops of the vigas. Vigas must usually be planed to create a flat surface. Although builders use a variety of materials for roof decking, such as plywood, oriented strand board (OSB), and dimensional lumber, you'd better like its appearance. In this roofing system the decking also serves as the ceiling. Any planking wider than 6 inches will suffice, but Reynolds recommends the use of rough sawn 1 x 12s, because they are inexpensive and nice to look at. (If you take this route, I'd strongly recommend that you consider using salvaged or recycled wood or wood from sustainably harvested forests, a topic discussed in chapter 14).

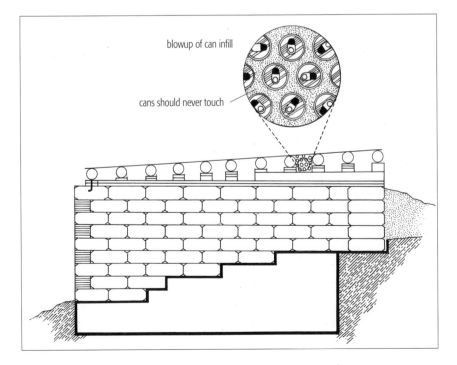

blowup of can infill

cans should never touch

FIGURE 4-13.
Vigas (peeled logs) rest on wooden blocks of varying size to anchor the roof to the tire wall and achieve the slope required to gather rainwater.

Skylights are typically installed in the roof near the back of each U to enhance interior lighting. Details of skylight construction are presented in *Earthship*, volume 1.

Earthship roofs have been specially designed to catch rainwater and snow

INTERVIEWS WITH INNOVATORS

Michael Reynolds

What gave you the idea of building houses out of automobile tires?

I was always more interested in homes for people than in "architecture." My thesis was on housing. I graduated and left the Midwest for New Mexico. In 1970, Walter Cronkite did a piece on the "clear-cutting of timber" in the Northwest. He predicted skyrocketing wood prices and a housing shortage. Immediately after, Charles Kuralt did a piece on beer and soda cans being thrown all over the streets and highways around the country. This was before recycling. Kuralt predicted a garbage problem. I began designing the first beer can house within two weeks after this news program.

Next, the energy crunch hit and I was already building houses of cans with permits and bank loans. I needed thermal mass to store energy. Other architects and engineers were using concrete and water in very expensive designs to store heat. I looked around with my "garbage" eyes and saw tires as abundant as trees. I stuffed

them with dirt. The result is the Earthship.

When did you build your first tire home?

I built my first tire home in the mid-1970s. It is now a rental unit and is never empty.

What were the early reactions to this novel idea? Has public opinion changed over the years?

Early on, many people thought I was nuts and some people actually got angry because I was bringing garbage into their neighborhood. It has always been uphill. I am a runner—I only like to run uphill. My work is always uphill.

Every so often, someone would congratulate me and tell me to keep going—that building with tires really was a great idea and someday everyone would realize it.

Now, with energy problems, wood prices, tree awareness, what to do with tires becoming almost a crisis, and recycling being very "in," the acceptance of building with tires is steadily growing. Others are doing it who have no connection to me other than seeing one of our books. Other

architects, developers, and institutions are picking it up. I do believe it will grow into being such a popular building method that tires will cease to be a problem. This will result in many people taking advantage of thermal mass for temperature stabilization— some without even knowing it.

What are the major changes that have occurred over the past 25 years in Earthship design?

The Earthship is like the automobile—it keeps getting better in performance and more "packaged" to build. The systems—power, water, sewage— have vastly improved in performance and price. Many components, such as gravity-operated skylights, front face dormers, doors, and battery boxes, have become mass-produced and are available to promote the "assembly" of an Earthship versus building everything from scratch. This allows inexperienced homebuilders to have the difficult parts prefabricated and ensures that these difficult items are executed correctly. The idea is to make sustainable buildings easily available, much like automobiles.

melt, which is funneled into huge cisterns to supply domestic water needs. To ensure the cleanest water possible, most Earthship builders now install metal roofing, typically steel. Steel roofing outlasts most other roofing materials, is easy to clean, and presumably releases no toxins. To install a steel roof over

Much about the automobile as a production item is very valid and has inspired us to bring the Earthship into this realm of availability. Even though the automobile creates many of the problems of the planet, it ironically produces the building block of Earthships and is the inspiration for their production methods.

When something like the automobile is invented, we should ask ourselves, "If there were millions of these inventions on the planet, what would their impact be?" Maybe if that question were asked, the evolution of the automobile would have been different. But now it is deeply rooted in our economic stability and seriously affecting our planet.

The Earthship concept as an invention, on the other hand, would actually benefit the planet if millions of them were around. The Earthship is inspired by trees. There are never too many trees. There are too many automobiles. Earthships will continue to evolve to be more like trees—impacting the planet in a positive way. It is necessary that we have shelter for people. This shelter should be as integral with the processes of the planet as are trees. Conventional shelter is carving away at the very essence of the planet while trees are constantly participating in the essence of the planet, i.e., they are part of the planet itself.

How suitable are Earthships for urban locations? Housing developments?
The Earthship is now, and growing even more, to be a cellular/modular concept. This independent, sustainable cell can grow in urban as well as rural areas. Towns and cities could be built without infrastructure—just like moss growing on the north side of a tree. We have been working toward the urban application for a long time and we are ready for it.

Do you have any additional thoughts?
Many people all over the planet know of Earthships. Few really know what they are and fewer still know what the effort is really all about. Some think they are just buildings made of old tires. "What a neat idea," they think. Others think they are solar buildings made of recycled materials—great. Still others are aware of the fact that they are an attempt (getting quite successful) at making totally sustainable homes from by-products of our society—still an incomplete understanding. The Earthship concept is an effort to slowly evolve humanity toward a method of life that continuously improves that life. This involves changing building methods, changing utility methods, changing living methods, and ultimately changing our thinking and understanding. This is a journey whose destination cannot be perceived, only imagined.

The Earthship concept is now only a seed being planted everywhere. As this seed grows, we will find that we are making homes and sanctuaries for plants instead of kings and politicians, because it is the plants that will truly be guiding us and nurturing us. We have recreated the world because it doesn't work. Now we must find the humility to follow and emulate other forces on this planet. The Earthship is a vessel toward that end.

MICHAEL REYNOLDS is an architect and builder of Earthships who lives in Taos, New Mexico. A world leader in sustainable design, Reynolds has written several books on Earthship construction and travels widely, speaking about and building Earthships.

The architectural design of the Earthship doesn't leave many options for roof type. You don't have the same options you would have with nearly every other natural home discussed in this book.

If your taste in building is more conventional or you need to place your home among other more conventional structures, you may want to consider an Earthship derivative (described earlier in the chapter). This option permits many other home styles and many different types of roofs.

vigas and decking requires some skill, however. A vapor barrier must first be installed on the top surface of the roof decking to prevent house moisture from escaping and soaking the insulation. Six-mil plastic is sufficient. Thirty-pound felt roofing paper is then laid down over the vapor barrier and secured with staples or nails. Its job is to prevent water that may leak through the roof from entering the house.

Next, 8-inch rigid foam insulation is placed over the roofing felt, then secured with long deck screws. Finally, metal roofing is screwed in place. The finished roof has an R-value of about 30.

Most of the Earthships I've seen use roof trusses or I joists rather than vigas. When asked why, homeowners and builders told me that roof trusses and I joists are lighter and easier to handle than vigas. They also make it easier to increase the insulation. Rigid foam insulation is often placed between the trusses, but fiberglass, cotton batting, and blown materials (for example, recycled cellulose or newsprint) can also be used. (A vapor barrier is especially important if you are using the latter types of insulation, which lose their R-value when wet.)

Trusses are nailed into wood blocks secured to the concrete bond beam and the decking is then nailed to the trusses. Roofing felt is applied to the decking and steel roofing is then screwed into the decking. After the insulation has been installed, a vapor barrier is stapled to the underside of the trusses to prevent water from coming in contact with the insulation. The ceiling is built when the house enters the final stages of construction.

FLOORS

Earthships are ideally suited for some of the more "earthy" types of floors. Earthen and flagstone floors are two candidates for those people seeking a more rustic look. (Earthen floors are described in chapter 3 and many other books listed in the resource guide, including *Earthen Floors, The Earthbuilders Encyclopedia, The Straw Bale House, The Rammed Earth House, The Cob Builders Handbook*, and *Earthship*, volume 3.)

Brick on sand is another good choice. It's inexpensive and, if done right, holds up very well even in high-traffic zones. A concrete slab can also be poured and covered with tile or carpeting. To promote sustainability, use tile and carpeting made from recycled materials. They are becoming much more widely available and are pretty cost competitive with standard flooring materials. (Chapter 14 covers this topic in more detail.)

THE EARTHSHIP NEST, THE WOODLESS HUT, AND THE HURRICANE-PROOF EARTHSHIP

Like the coyotes that roam the desert around the center of Earthship operations in Taos, New Mexico, the Earthship has proven to be an adaptable crea-

A

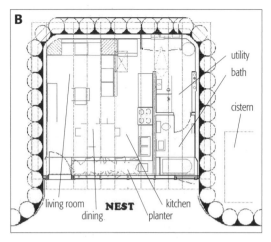

B

utility

bath

cistern

living room   **NEST**   kitchen

dining   planter

ture. One of its adaptations is the Nest, a one-room, oval-shaped Earthship (figure 4-14). The Nest was designed to provide basic human comfort: a "warm bed, a hot shower, a toilet, limited electricity, a place to grow some food and a place to prepare it, and a place to hang out and read or watch TV." Moreover, it was designed to be built rapidly, in a little over a month, and economically, for around $35,000 (without plastering). Like the mother Earthship, the Nest is a form of shelter that is easy on our planet. Furthermore, the Nest is available as a kit that comes with components for the energy and water systems, roof framing, front face framing, and so on. The parts are assembled on site like a log home kit. The Nest can also be expanded into a full-fledged Earthship, as one's means and needs grow.

The Earthship Nest is to homes what the Volkswagen was to automobiles. That is, both are built "for the people." Although we all know that the VW Beetles would not accelerate to 60 miles per hour in six seconds, we didn't expect that of them. By the same token, the Nest has all of the benefits of an Earthship—catchwater, solar electricity, passive solar heating, and a growing space—but it will not make you totally self-sufficient. The solar electric system, for example, consists of 4 panels and 4 batteries—which is minimal for survival but hardly enough power for luxury or unconscious power use. Therefore, even though the Nest won't do everything a full-fledged Earthship will do, it will get you by. For more details, see *The Nest Concept*, one of Solar Survival Architecture's *Earthship Chronicles*.

Working on a sustainable community in Israel, a region with a limited supply of wood, Solar Survival Architecture devised another Earthship offshoot, known as the Woodless Hut (figure 4-15). A circle of rammed earth tires with a concrete bond beam formed of cans or bottles, the Woodless Hut has a dome roof made of a "birdcage" of rebar draped with stucco netting. To seal the roof, rags and scrap cloth were dipped in a slurry of cement, then draped over the stucco netting. Two or three layers provided a solid roof that

FIGURE 4-14 A, B.
The Earthship Nest, small, affordable, and quick to construct.

The Nest takes the Earthship "self-sufficient with ecological awareness" concept and combines it with the Volkswagen "economically available with limitations" concept.

MICHAEL REYNOLDS,
The Nest Concept

FIGURE 4-15 A, B.
The Woodless Hut is a domed structure made from dirt-filled tires laid out in a circle, then bermed to reduce heating and cooling demand. Photograph on the left shows work in progress in Honduras. Photo on the right is a completed structure in Taos.

was then plastered and coated with an acrylic roof coating to make it waterproof. The space between the inner and outer domes was filled with insulation. SSA's next project (in Honduras) introduced a wooden inner dome to enhance the aesthetic appeal and a skylight to provide more light and ventilation.

Like other SSA creations, the Hut lends itself to modular building. This feature makes it perfect for those just starting out in the housing market. Further adding to its appeal, the Hut is relatively simple to build and is now being developed into a full-fledged Earthship with all systems. This unique structure could provide basic shelter in many parts of the world. For those of greater means, the modular Hut could be a home or a studio—or simply a peaceful sanctuary to escape the hustle and bustle of life.

Although Earthships are virtually indestructible and provide protection against hurricanes and tornadoes, Reynolds and his cohorts have produced a hurricane model that is even better. The hurricane-proof Earthship is nestled even more securely into the ground to withstand fierce winds and blowing debris. In fact, this unique design ensures that winds blow right over the structure. To protect occupants from flying glass, the hurricane-proof Earthship's windows are designed not to shatter. What is more, the hurricane-proof Earthship is also available as a prepackaged kit ready for assembly in any remote or urban area of the globe.

WHAT DOES IT COST TO BUILD AN EARTHSHIP?

As with any alternative construction technology, cost is a major consideration. The cost of Earthships varies dramatically depending on size, complexity, sweat equity, labor costs, and the level of detail and extravagance you want. Put more bluntly, those who want Italian marble tile and waterfalls and a contractor to build an Earthship for them, are going to pay a lot more than those who select earthen floors and do a lot of the work themselves.

According to one source, owner-built Earthships range in price from $11 to $50 per square foot, most commonly $30 to $50 per square foot. As Michael Reynolds points out, however, these prices pertain to totally owner-built houses (no subcontractors involved) without the independent utility systems—solar electricity, solar hot water, and water filtration, for example. The Earthships built for this price are basically passive solar structures without a power source.

Please don't be misled into thinking you can build a 2,000-square-foot home for $60,000 to $100,000. Costs mount up very quickly. Take time to perform a detailed estimate and be realistic, before you plunge into a project such as this. Many a homeowner has saved up the $30- to $50-per-square-foot cost only to find that he or she is out of money long before the structure is habitable!

To build an Earthship with the independent living systems, expect to pay at least $65 to $75 per square foot, if you do a lot of the labor yourself and keep the interior simple. But that's still quite a bargain. If you want a contractor to build the house, prices can rise to $75 to $100 per square foot. Again, the more elaborate your tastes, the higher the cost of the construction.

If Solar Survival Architecture builds the unfinished shell, consisting of tire walls, roof, front face, and systems package, it will cost $75 to $85 per square foot, but only if you live in the Taos, New Mexico, area and only if you complete the finish work yourself. See the annual resource issue of *The Last Straw* for information on roving plaster crews. If SSA constructs the finished building, it will cost approximately $95 to $100 per square foot, again, however, only if your live in Taos. If the folks at SSA have to send work crews to your site, the cost goes up. And, if you deviate from the blueprints and want a lot of extras, say expensive tile floors, you will, of course, pay more.

You may be able to hire a local Earthship builder to do the job. Make sure you tour homes he or she has built and talk to the people for whom they have built Earthships. The folks at SSA can recommend people in your area.

If you pine for the romantic experience of building your own home, see a trained therapist immediately. If he or she can't cure you of this dangerous affliction, Survival Solar Architecture sponsors three-day and week-long workshops or will let you join them on a project in exchange for training. SSA also provides technical support for a fee and will even send specialized work crews to your house to install the systems package, pack tires, or whatever. Don't skimp. Get advice on the project and hire help on those parts of the project that are beyond your abilities. The money is well worth it in the long run!

If you are interested in an Earthship, I'd recommend contacting the folks at Sustainable Solar Survival Architecture or Earthship Global Operations, listed in the resource guide at the end of the book. By all means, get ahold of their pamphlet, *The Earthship Menu: How to Get an Earthship*, which is a part of their *Earthship Chronicles*, a publication designed to keep potential buyers and builders up to date on the latest innovations.

INNOVATIONS IN TIRE HOME TECHNOLOGY: EARTHSHIP DERIVATIVES

If you like the idea of building an Earthship, but can't, perhaps because your neighbors would hang you by your toes from a tree, or if you don't like the appearance of an Earthship but like many of its features, don't despair. As noted earlier, contractors are now building homes out of packed tires using Earthship design ideas, homes that offer many of the benefits of an Earthship but don't resemble these decidedly offbeat forms of architecture. For legal reasons, they don't call their creations "Earthships," which is a trademarked name. Some folks refer to them as Earthship derivatives. Others call them tire homes or packed tire homes. Still others call them rammed earth tire homes. No matter what you call them, they're an exciting new architectural offering in the sustainable building movement.

By combining Earthship technology with more conventional modern architecture, the tire home offers people a choice that looks and feels more conventional but has many, if not all, of the independence and environmental benefits of its more earthy cousin. In fact, a lot of folks who have lived in standard wood-frame colonial and ranch-style houses all their lives say, "I could live in a house like this!" after they've toured my house.

To me, the Earthship derivative may be a bridge to the future. It offers housing that is more familiar and thus potentially more marketable to a wider audience. And lest we forget, to create a sustainable world we need rapid change in large numbers. A few converts living lives of quiet independence and sustainability are not enough. We need the masses to join in. Instead of less than 1 percent of the housing market dedicated to building environmentally sustainable homes, we need 99 percent. Tract homes need to go the way of the dinosaur, but much more quickly. The tire home is part of the answer to achieving a globally sustainable human presence, as are many other forms of natural building such as straw bale, rammed earth, and adobe.

Tire homes are built much the same as Earthships—in adjoining Us. Tires packed with earth form the foundation, the mass walls, and many of the partition walls. As in Earthships, a bond beam sits atop the tire walls and anchors the roof through a top plate. The rest of the building is framed in 2 x 6 lumber and I-joists (for roofs). The front face and roofs of these homes are where you see one of the major differences. Roofs are more conventional looking. The south-facing wall can be vertical or sloped, or some combination of the two, just as in the Earthship (figure 4-16). It could also be made of straw bales, rammed earth, or some other natural product. My house has a clerestory roof, like a modern solar home. The clerestory roof emits light to the back walls, where the mass is located. Other kinds of roofs can be built with this form of architecture as well. Plaster and cement stucco can be used to cover the front wall, but any kind of siding will work. Remember, though, unprotected wood

To optimize solar gain, sunlight needs to strike thermal mass. Although mass walls will heat up room air during the day, the process is not as efficient as when sunlight strikes the walls directly. When using clerestory windows to emit light to back walls, do not to make your ceilings too high in cold climates. This creates an enormous volume that is difficult to heat on cold days. If you want a vaulted ceiling keep it under 12 feet and use ceiling fans to drive heat back down.

should be avoided on the south face of solar buildings. Wood can't withstand the daily punishment of the sun.

Interiors of tire homes can be framed to look more normal. And rooms can be closed off for more privacy and sound control.

The Earthship holds no corner on sustainability. Any house, including the modern tire home, can be equipped with solar panels for hot water, photovoltaics and wind generators for electricity, a catchwater system for domestic water, and passive solar for heating. Mine is living proof that you can live well yet be independent and save a lot of money.

SHOULD YOU BUILD AN EARTHSHIP OR TIRE HOME?

Before you decide whether an Earthship or tire home is right for you, let's look at the pros and cons of this building technique.

Advantages of Earthships and Tire Homes

- Earthships are one of the most environmentally sustainable forms of housing available in the industrial nations of the world, especially given the independent living systems that they incorporate.
- Earthships reduce our demand for natural resources by using indigenous materials available near the building site (vigas and mud plaster and packing dirt) and "waste" (discarded automobile tires).
- The use of tires reduces stockpiles of tires that can catch fire and smolder for months on end.
- By relying on passive solar heating and photovoltaics for electricity, Earthships reduce our demand for fossil fuel energy to heat and light our homes.
- Earthships greatly reduce our impact on the environment by reducing resource demand and pollution.

- Earthships decrease our dependence on utility power and save money in reduced energy bills.
- Earthships capture rainwater and snow melt, recycle wastewater, and grow food. By recycling graywater, they protect groundwater and surface water supplies from depletion and pollution.
- Earthships are resistant to fire, termites, earthquakes, tornadoes, and hurricanes.
- Earthships protect the inhabitants from outside noise.
- Materials required to build an Earthship can be found worldwide in abundance.
- Earthships are relatively easy to build on one's own, especially with the detailed blueprints, workshop experience, and advice from Solar Survival Architecture. Tire wall building requires relatively unskilled labor and can be speeded up using pneumatic tamping devices.
- Earthships are a durable form of housing that outlasts conventional housing, while providing an array of other benefits.
- Earthships could be used in developing countries to promote self-sufficiency.
- The interiors of Earthships are relatively humid and easy on nasal passageways.
- Earthships are a proven technology that are often accepted by building departments.
- Earthships compete well economically with other forms of earth-friendly housing.

Disadvantages of Earthships and Tire Homes
- Getting approval for Earthship construction varies by state. Some states are quite receptive and easy to work with, such as New Mexico and Colorado. Officials there often understand the logic of solar systems and graywater and blackwater systems. Acceptability in other states varies from county to county. Most counties in Utah are very uncooperative.
- Although Earthships tend to be warm when the sun is out, they can be cold during cloudy periods, because the mass is situated too far back in the house to absorb incoming solar radiation directly. Many people find that backup heating is required for those days.
- Even though tire walls are relatively easy to build, costly mistakes can be made by those with little experience.
- Building tire walls is labor intensive and labor is expensive in many parts of the world.
- Slanted-glass designs can overheat, especially in the fall. Curtains or window shades should be considered to avoid this problem.
- Kitchens are typically located in the environmental interface corridor and, because they are in the direct path of the sun during the fall, winter, and

spring, they can become unbearably hot during the day. Refrigerators have to work harder if not shielded from the sun in these locations.

- As in any solar house, sunlight in an Earthship can be intense during much of the year, creating eye-straining glare on computer screens.
- Earthships offer less livable space than a regular home because of the planters and curved walls of the Us. Although planters do provide a means of recycling graywater, keep your house humidified, provide food, and create a pleasing atmosphere, they consume a lot of floor space. So, if you want 2,000 square feet of actual living space, you will need to build a larger Earthship.
- Earthships typically lack the storage space and closets needed by most families. Be sure to include storage in your design.
- Humidity can be a big problem in Earthships. Plants give off lots of moisture. If you dry clothes inside, humidity levels can become intolerable. A vapor barrier placed between the ceiling and the insulation is essential!
- The open design of an Earthship, which facilitates the movement of air to all parts of the house, reduces privacy and makes interior noise a problem. Sound from televisions and stereos and, oh yes, rug rats, can be unbearable. Rooms can be closed off to avoid this problem, although that alters the flow of warm air.
- Insects like white flies and aphids can be a big problem to those growing vegetables inside. Some plants, such as tomatoes, grow well inside but become heavily infested with white flies, which in my experience tend to kill the plants just after they begin producing tomatoes. Tropical plants are much more resistant to insects and grow well, although the bright light in an Earthship can be harmful to many tropical species. Choose species carefully. The tropical plants that have worked well for me in bright light include dieffenbachia, hibiscus, croton, bougainvillea, rubber plants, dracaenia, banana trees, poinsettias, and oranges. Cacti and succulents such jade plants and aloe vera also grow extremely well in bright light.
- Cats often prefer to use planters rather than litter boxes. Mine sure does! This creates a problem with odor and flies. Some say it is not good for soils where you're growing food. Making matters worse, cats may carry dirt on their paws after exiting a planter, soiling the floors.

PERSONAL NOTE

I've battled white flies and aphids with soap sprays and other natural pesticides, including lady bugs, but have found it very difficult to win against these pests.

Earthships and other tire homes represent a viable option for those interested in living by sustainable principles. But as Michael Reynolds writes in *The Earthship Menu*, "The pursuit of a new direction is rugged, whether it is through untracked snow, uncharted jungle, or out of a crystallized prevailing human condition." By which he means, building an Earthship and charting a relatively independent lifestyle is no picnic. You will fight dogma, mistakes, antiquated beliefs, laws, and asinine codes. And you may not get it all right. You will make mistakes that come back to haunt you. We all do.

Remember, though, that by building a natural, sustainable home, you join a group of new pioneers, and all pioneers face hardships. Expect to make mistakes and to find yourself in strange territory. But if you prepare yourself by reading and study, and learn the latest developments, it will make your experience as painless as possible. And remember, your experience will help our society evolve into a sustainable human presence.

ADOBE HOMES

ADOBE, LIKE RAMMED EARTH, is an ancient building technique, using what may be the oldest manufactured building material, the adobe brick. Adobe bricks or blocks consist of mud sun-dried in wooden forms. Once dried, adobe blocks are stacked in the walls in a running bond. They are cemented together with mud mortar and then either coated with earthen plaster or left unplastered.

The term *adobe* is often credited to the Spanish, who coined the term *adobar,* meaning to plaster. However, the word may actually have come from an Arabic word, *attubah,* which means "a brick." In modern parlance, adobe refers to the bricks, the mortar, *and* the mud plaster. It is even used to describe the flat-roof style of homes that adorn the tree-lined streets of Santa Fe and Taos, New Mexico. However, adobe bricks can be used to fashion a wide range of buildings in a variety of architectural styles in many climates. In keeping with popular misunderstanding, Marcia Southwick writes in her book *Build with Adobe,* "If you are planning to build an adobe house in New York State, forget it! Or in Illinois or Iowa or Washington. The only way you can join the club is to live in an arid or semi-arid region." Not so. The house shown in figure 5-1 is located in Geneva, New York. This magnificent home is not made from fired clay bricks, as you might suspect, but rather from traditional adobe bricks. In New York state alone, researchers have documented 40 adobe brick buildings in a nine-county region that spans nearly half of the state—a state not known for particularly dry weather. Adobe homes are also found in Massachusetts and Nebraska—even in China and Japan.

Adobe is the ideal material for the beginner. It is a warm, kind material that is forgiving of mistakes, and amenable to change. If you don't like what you have wrought, it is a simple matter to take it down and try again.

PAUL GRAHAM McHENRY, JR.,
Adobe: Build It Yourself

FIGURE 5-1.
Although adobe construction tends to be found in hot, dry climates, this house in Geneva, New York, demonstrates that adobe may be suitable to other climates as well. The main constraint in adobe construction is having a sufficient period of dry, warm weather to make blocks. In cold climates, precautions must be made to insulate walls.

Today, it is estimated that nearly half of the world's population live in earthen structures, adobe, rammed earth, and other more primitive structures.

Orlando Romero and David Larkin sum it up best in their book, *Adobe: Building and Living with Earth*: "Adobe is not something exotic found halfway around the world, nor is it confined, in the United States, to the romance of the Southwest." So, my friends, discard your thoughts about adobe being a building material that is restricted to desert climates.

HISTORY OF ADOBE CONSTRUCTION

Humans have used natural building materials to fashion homes since the dawn of time. Three of the most popular materials were grasses, sticks, and mud, with mud often used as a binder for the sticks and grass. Many early builders erected shelters for their families by weaving sticks together, then plastering them with mud to protect occupants from the elements, a procedure known as wattle and daub. The leap from wattle and daub to adobe construction was a small but significant one; it probably started independently in more than one location as early people began experimenting with mud bricks. How convenient the bricks must have seemed!

Although no one knows where adobe construction actually began, the earliest adobe architecture may have appeared around 6000 B.C. in the region now known as Iraq. Adobe structures have also been uncovered in Egypt, dating back to about 5000 B.C. From here, adobe building traveled to Spain, carried by the Moors, Arab and Berber soldiers who conquered Spain. (Who said nothing good ever comes of greed?) From Spain, adobe construction techniques traveled to the western hemisphere around 1600 A.D., starting in South America, then spreading north into what is now known as the desert Southwest of the United States. Like the Moorish transfer of technology to Spain, the dissemination of adobe to the New World resulted from conquest—this one spurred by the Spaniards' seemingly insatiable lust for gold.

The arrival of adobe architecture in the desert Southwest may not have marked its first presence. Recent evidence suggests that native people were already using earthen materials to build homes. Archaeologists have uncovered structures made from molded adobe bricks prior to the arrival of the Spaniards in several locations in the western hemisphere. Casa Grande, a prehistoric ruin in Arizona, is made from baskets of mud called "turtles." (It may be more similar to cob, described in the next chapter.) Mud mortar and plaster were also used on stone structures built by the Pueblos, Anasazi, and other early cultures.

This is not to discount the Spaniards' influence on the spread of adobe building. Even though adobe construction was occurring, the Spaniards, especially the Jesuit priests, disseminated it as they sought to Christianize the native Americans "fortunate" enough to be in their path. The Spanish legacy continued for many years. In 1850, an estimated 97 percent of all the homes in New Mexico were made from adobe bricks.

As this brief history illustrates, adobe is not the legacy of any one culture. It was an outgrowth of necessity and local availability. However, despite the fact that adobe building emerged independently in several locations, the techniques and appearance of buildings are remarkably similar. Adobe homes in China, for example, bear an uncanny resemble to those found in the U.S. desert Southwest.

Adobe continues to be a popular building material worldwide. The traveler will find modern adobe homes in Japan, China, northern Africa, southwestern United States, Mexico, and Latin America. Today, nearly 60 percent of the homes built in Peru are made from adobe or rammed earth. According to one source, there are an estimated 200,000 adobe homes in the United States alone, the vast majority of them located in the desert Southwest. However, like so many other environmentally sound natural building techniques, including rammed earth and straw bale construction, adobe fell victim to that pernicious predator of environmentally sound ideas: progress. In the United States, the railroad and the advent of mass-produced building materials caused a dramatic change in construction methods in the heart of adobe country. Even today, adobe construction falls victim to the key operatives of progress, speed, and cost. These forces have paved the way for the fake adobe home. With fake vigas and wood-frame walls covered on the outside by a thin veneer of cement stucco, these homes resemble adobe only to the untrained eye. To contemporary adobe builders, they are a mockery of an age-old practice. Although they may appeal to cost-conscious retirees living in Arizona or New Mexico, they can provide comfort in the desert climate only because of costly air conditioning, an energy-intensive amenity provided naturally by thick-walled adobe homes.

Adobe also suffered an identity crisis as people's evolving concept of what constituted suitable housing changed. Over time, adobe came to be viewed as a "poor man's" shelter, shunned in favor of contemporary building materials. Romero and Larkin sum up the problem: "At first glance, adobe appears to

ANCIENT FINGERPRINTS

One of the highlights of my life was to discover an Anasazi ruin while kayaking a remote section of Dolores River in southwestern Colorado. Made of stone and mud mortar, this structure was perched high on the canyon wall. When we entered and began studying our find, we discovered the perfectly preserved finger impressions of the early builders in the mud mortar!

Traditional adobe bricks are not as vulnerable to moisture as one might assume. Air, humidity, moisture, and rainfall have a negligible effect on fully cured bricks. Clay will limit moisture penetration, only the surface will be wet. Concentrated streams of water, however, must be kept off bricks.

PAUL G. McHENRY, JR., *Adobe and Rammed Earth Buildings*

The walls of a 2,000-square-foot ranch house built with a wooden frame weigh about 10 tons. The walls of an adobe home the same size would weigh nearly 340 tons.

the eye and the mind as a contradiction. How is it possible to build, out of mud, a simple one-room shelter, let alone a magnificent residence, mosque, or monastery?" They go on to ask, "How is it possible that this fragile composition of soil, straw, and labor has lasted centuries, nurturing and sustaining countless cultures . . . ?"

Despite these problems, adobe construction still persists. In regions of the world where money is in short supply, adobe construction represents a viable building option for the less fortunate. It is a home a family can build themselves. In wealthier regions, custom homebuilders erect million-dollar adobe homes for their wealthy clients. To them, adobe is anything but a poor man's shelter. It is elegant, efficient, and highly desirable. It blends well with the arid desert environment, too.

Adobe is also a potentially sustainable form of housing. Adobe mud is locally available, strong, and durable. Thick adobe walls have excellent thermal properties ideal for passive solar heating and passive cooling.

FOUNDATIONS

Adobe walls are massive yet easily damaged by standing water, streams of water, and seepage from the ground. Therefore, adobe must be carefully protected from water. While occasional rain will damage the surface slightly, groundwater can destroy it. For this reason, a solid waterproof foundation is essential.

Foundations for adobe homes, like many other natural homes, may be made from a variety of materials (see sections on foundations in chapters 2, 3, and 4).

For those interested in saving money and resources, rubble trench foundations are a good choice for an adobe home. A concrete grade beam provides a level point of attachment and a solid building surface; it also raises the earthen walls off the ground, protecting them from standing water. A grade beam may also protect a structure from earthquake damage, although not all engineers agree. Dissenters argue that a grade beam will not isolate a building sufficiently from seismic activity because the beam may move unevenly during an earthquake, causing the building to crack or even collapse. Before you close the book on this debate, however, consider one more bit of information. After studying buildings damaged by earthquakes, the Chinese government concluded that a layer of sand between the rubble trench foundation and the grade beam allows the grade beam to ride out an earthquake, effectively isolating the building from the strong lateral ground movements that are so damaging. With so much riding on this decision, I recommend that you consult with a structural engineer you trust and check your local building codes.

Rock foundations are also well suited to adobe architecture and represent a sustainable alternative to standard concrete foundations. Many builders recommend the use of cement mortar to hold the rocks in place. Mud can be used instead of cement, but the rocks should be tightly interlocked in case the mud washes away. (See chapters 8 and 9 for further discussion of stone foundations.)

Adobe brick foundations are another possibility. Although many buildings constructed on adobe brick foundations are still standing today, this option is suitable only in extremely arid countries, such as those in the Middle East, where rainfall is as elusive as peace. Some building codes prohibit adobe brick foundations, so check your local codes before you start building.

Poured concrete foundations are commonly used in adobe home construction in wealthier nations. In less developed nations, the cost of this option may be prohibitive. A concrete foundation consists of a footing and stem wall. Footings are typically made from poured concrete, but stem walls can be constructed from poured concrete, or concrete blocks filled with concrete, or compacted adobe dirt capped with cement to prevent water from seeping into the wall. (A variety of earth-friendly building blocks are currently available for use in foundations. See chapter 14.) As a general rule, the footing should be 6 inches wider than the stem wall, and the stem wall should be at least two blocks high.

Rebar is required in footings and stem walls to increase the tensile strength of the foundation. Rigid insulation and a coat of waterproofing material are also vital, to reduce energy loss and prevent water from seeping into the foundation, respectively.

Another foundation option is the monolithic slab (see chapter 2).

THE BUILDING BLOCK OF ADOBE WALLS: ADOBE BRICKS

Adobe bricks or blocks vary in size from one country to the next. In the United States, adobe bricks typically measure 10 x 14 x 4 inches and weigh 35 to 40 pounds, as much as a full-grown beagle. In contrast, blocks in countries where adobe homes are frequently built with the aid of children, elderly men and women, and mothers, are typically smaller, lighter, and thus more practical. In Iran, for instance, adobe blocks measure 8 x 10 x 2 inches. If you'd like to make your own bricks and involve more than your weight-lifting friends from the local gym, make your adobes smaller rather than larger. Remember: what matters is not how easily you can heft a single brick, but how well you can handle the hundreds—heck, thousands—of adobes required to build a house (figure 5-2). For more on the subject, see Steve Berlant's book, *The Natural Builder*, volume 1: *Creating Architecture from Earth*. (This is a very useful reference for all adobe construction projects, along with Paul McHenry's books.)

Paul G. McHenry Jr.

FIGURE 5-2.
A stack of adobe blocks awaiting use.

BUILDING NOTE

Slaked lime (hydrated building lime) is also a valuable additive to adobe. It stabilizes adobe bricks and reduces the absorption of moisture. It also increases the compression strength of adobe bricks.

STEVE BERLANT, *The Natural Builder*, Vol. 1, *Creating Architecture from Earth*

Adobe bricks are made from mud containing approximately 20 percent clay and 80 percent sand, although U.S. adobe building codes state that blocks should contain between 25 and 45 percent clay. Clay is an important component of the mixture because it binds together the sand and other materials in the adobe brick. Achieving the proper proportions is very important. Too much clay causes cracking. Too little clay (or too much sand) renders the bricks more brittle and much less resistant to rain and other erosive forces such as wind, an important consideration if your walls are not going to be plastered.

Topsoil is unsuitable for adobe brick making, because it contains organic matter. Organic matter is structurally weak and may decompose, further weakening a wall.

To test the suitability of your soil—actually, your subsoil—you will need to perform several simple tests. The first is the "jar test," introduced in chapter 2. This highly scientific procedure requires water, a jar, and some soil. The jar is filled with water and the soil you hope to use. The lid is screwed tight and the jar is shaken and then set aside. Pebbles and coarse sand will settle out first (figure 5-3). Fine sand and silt form the next layer. Clay forms the top layer. If your soil sample contains the appropriate proportions of each, approximately 20 percent clay and 80 percent sand, silt, and pebbles, proceed to the next step of making some test bricks. These bricks you can test for strength and tendency to crack.

Bricks are made in wooden forms, then left to dry. Once they are dried, you must assess their appearance. You will notice some cracking, perhaps even some moderate warping. These are to be expected, but if cracking is excessive, your bricks are useless. They will break too easily.

Excessive cracking is caused by several factors. Wind, for instance, sometimes causes adobes to dry too quickly and thus crack. If you suspect this to be problem, shelter the bricks. If that doesn't work, there may be too much clay in the mix. Reducing the clay content (by adding more sand) should solve the problem. But don't add too much sand. This makes bricks less resistant to wind and rain, and also causes them to break easily. For more information on these tests and adobe mix ingredients, see Berlant's *The Natural Builder*, volume 1: *Creating Architecture from Earth,* or Nader Khalili's book, *Ceramic Houses and Earth Architecture.* (Berlant's book describes additional tests and should be consulted by all readers who are serious about adobe building.)

Adobe soil can also be purchased by the truckload from local suppliers. One dump-truck-full is enough to make 400 to 500 adobe bricks. Before you order the soil, ask for some samples to test.

FIGURE 5-3.

The jar test is used to determine the suitability of soil for building adobe blocks.

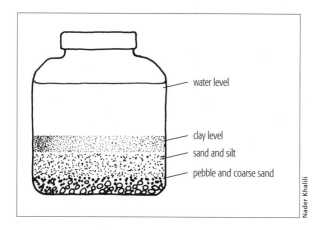

water level

clay level
sand and silt
pebble and coarse sand

Nader Khalili

Assuming your soil passes all of the tests and your foundation is ready, the next step is building the walls. To do so, you will first need to moisten and mix the soil. Hand mixing in an adobe pit may be appropriate for small operations. If your want to make a lot of adobe bricks quickly, you may want to mix your soil with a rototiller or the bucket of a tractor.

After mixing, the soil should have the consistency of bread dough. To facilitate the breakdown of large clumps of clay, most builders let their mud sit overnight. The next day, the mud is shoveled into wooden block forms, then smoothed by hand (figure 5-4). Some builders rake their fingers over the surface after the mud has been leveled to create a roughened surface that enhances bonding to the mortar.

Forms can be any size. In the hands of an experienced adobe brick maker with a trained assistant, even a small set of forms is sufficient to produce large numbers of bricks in a short period. When making forms, remember that adobe mud will shrink upon drying. Your forms should be slightly larger than the brick size you plan on building with.

As soon as bricks are dry enough to retain their shape, the forms are removed and the partially dried bricks are left on the ground to dry. A few days later, the bricks are laid on their sides, a measure that accelerates drying. Two days later, the bricks are carefully stacked. This makes room for additional brick making and also ensures thorough drying (figure 5-2). (Special stacking techniques are described in some of the reference books listed in the resource guide.)

Although adobe bricks reach full strength after 30 days, most people begin using them 11 to 14 days after the mud was shoveled into the forms, but only under optimal drying conditions. If the weather is cold or moist, drying may require considerably more time. To see if a brick is dry, simply break one in

BUILDING NOTE

Adobe blocks can be purchased from various suppliers. This cuts down on labor and reduces construction time.

BUILDING NOTE

Approximately 15,000 adobe bricks are needed to build a ranch house of 2,500 square feet.

Paul G. McHenry Jr.

FIGURE 5-4.
Adobe blocks are made in wooden forms and dried in the sun. If weather cooperates, they're usually ready for construction within 11 to 14 days.

two. If the color is uniform throughout, the brick is dry. If dark and light spots appear, drying is not complete.

Straw has traditionally been added to adobe to strengthen the bricks and to prevent cracking. However, many modern adobe brick makers have come to view straw as a superfluous additive. They argue that severe cracking is usually the result of too much clay and is remedied by altering the clay content—both the concentration and type of clay—rather than by adding straw. Cracking may also occur if bricks dry too quickly or if they are too large, both problems that a brick maker can easily rectify.

Some people think that straw added to adobe enhances its insulative properties. The straw is first chopped into finger-length sections, then added to the mud. The experts I've talked too, however, warn not to expect any increase in insulative value with the addition of straw. Straw may make blocks a bit lighter and a bit easier to work with, but it won't increase the R-value.

Another additive of dubious merit is asphalt emulsion, an oil by-product. According to Paul McHenry, 5 percent emulsion makes bricks water resistant, while 15 percent makes them waterproof. Besides being made from a nonrenewable resource, asphalt emulsion contains potentially harmful chemicals, which, Steve Berlant says, "smell really bad." The odors also persist and may render a home unhealthy to all occupants, especially to those who are chemically hypersensitive (see sidebar). Although I am not aware of any good evidence suggesting that asphalt emulsion is harmful to health, the precautionary principle should be applied here. That says, take the safe route. Let's not be building pioneers *and* guinea pigs!

Asphalt emulsion is not needed when exterior walls are plastered. Plaster will protect the blocks from moisture. However, asphalt emulsion could be used on unplastered freestanding walls in gardens and patios that are more likely to get wet.

Another commonly used stabilizer is Portland cement. Although more expensive than asphalt emulsion, it is less likely to present any health hazards. For a discussion of natural stabilizers, see *The Natural Builder*, volume 1.

In Mexico and the far desert Southwest, some builders construct adobe homes with kiln-fired adobe bricks. These bricks, known as a fired adobe or burned adobe brick, are much harder than sun-dried adobes. To make them, however, one must use a higher-clay-content mud and must remove impurities, especially rocks and pebbles containing lime, which will cause cracking if left in place. After the mud is "cleansed," it is placed in forms and then baked in the sun. When bricks can be handled, they are transferred to a kiln and baked for two days.

CHEMICAL SENSITIVITY

Many people suffer chemical sensitivity brought on by toxicants such as formaldehyde released from carpeting, engineered wood products such as plywood and oriented strand board, and finishes. The release of chemicals from a product is called off-gassing or out-gassing.

These chemical substances bind to naturally occurring proteins in the body. The immune system of the body recognizes these new chemical combinations as foreign substances, and attacks them. However, it also attacks the nonconjugated (normal) proteins. Subsequent exposure to the offending chemical can cause severe, often debilitating, reactions. The correct scientific name for this phenomenon is chemical hypersensitivity.

Although fired adobe bricks are harder than traditional adobe bricks, they absorb more moisture. Water accumulating inside bricks can freeze and expand, causing surface spalling, a fancy name for flaking. To prevent this problem, fired adobe bricks are coated with clear liquid waterproofer. Sounds great, except that my sources suggest it must be reapplied every two to four years to ensure continued protection. For more information on fired adobe bricks, see Khalili's book, *Ceramic Houses and Earth Architecture*, and McHenry's book *Adobe and Rammed Earth Buildings*.

Adobe bricks can be manufactured by machine. The machines range from small, economical hand-operated units to large, expensive, fully automated ones—all designed for on-site manufacturing. The Earth Press produced by Adobe International, Inc., puts out compacted blocks that, according to company literature, exceed the Uniform Building Code for compressive strength and "modulus of rupture"—which is engineering jargon meaning they're really strong!

Pressed earth blocks, as they are called, are made from lightly moistened soil with or without stabilizer. Unstabilized blocks—that is, blocks made by compression of soil alone—may be susceptible to erosion and may not be approved by local building codes.

Cement stabilization increases the strength of pressed earth blocks and enhances their resistance to water. They're also easier to get approved. Although stabilized pressed blocks may not appeal to natural builders, they do provide some advantages over traditional adobe blocks. For one, they require less water, which may be a big plus if you are working in an arid climate. Another key advantage is that because they are uniform in size and shape, they are easier to build with. Pressed earth blocks can even be stockpiled for long periods without protection. Don't try that with a conventional adobe block!

Another advantage of pressed earth blocks is that their production requires less space than traditional adobe-making operations. And, lest we forget, the blocks are ready for use almost instantly. No need to wait two weeks for them to dry. Many machines crank out blocks at a fairly rapid rate. Dave Moshel of Earth Uprising Adobe Block and Machine Company in Arivaca, Arizona, operates a machine he designed that produces four to five blocks per minute. Pressed earth block machines also make adobe construction feasible in wetter climates. If you can keep the mix from freezing, you can even make blocks in cold weather, something not possible with traditional adobe. Purchasing a machine, however, can be expensive, so unless you can rent one or borrow one from a contractor (good luck!), this option may have limited usefulness to the owner-builder. Alternatively, if you live near a block maker who owns a machine, you can purchase your blocks from him or her.

Some builders are also experimenting with a technique called poured adobe. In this technique, a very wet mud mix, about the consistency of thick cake batter, is poured into block forms placed on the walls or forms similar to

BUILDING NOTE

Although some books on adobe recommend cement plasters, evidence suggests that cement plasters will destroy an adobe wall over time by trapping moisture inside.

STEVE BERLANT

Pressed earth block machines represent the first significant advance in adobe construction in centuries.

Cement mortar is incompatible with unstabilized (traditional) adobe. The two materials expand and contract at different rates in response to changes in temperature. Use of cement mortar causes the adobe bricks to deteriorate. However, a soil-cement mixture containing mostly soil and a little cement with lime may be suitable mortar.

those used by rammed earth homebuilders. After the adobe has settled and firmed up, the forms are removed (as in the case of the block form) or slid upwards (as in the more standard form) so the next course can be poured. The end result of both forming methods is a wall similar to those found in cob homes (monolithic walls of adobe), which is discussed in the next chapter.

When building with adobe blocks, you will need to "cement" the bricks together with an earthen mortar made from the same mud used for making bricks. Like brick mud, the mortar must be free of organic matter. However, it must also be screened (before being mixed with water) to remove rocks and pebbles. Although these "impurities" are acceptable in bricks, they make mortaring difficult. Nader Khalili describes an old Persian custom of throwing coins in the mortar for workers to find. In the process, workers come across rocks and pebbles that they toss aside. (Greed comes into play again!)

Adobe blocks can also be "cemented" with gypsum mortars. Gypsum sets up quickly and is ideal for constructing arched window and door openings. Lime mortars are also suitable. For more details and additional references on the subject, check out Berlant's *The Natural Builder*, volume 1: *Creating Architecture from Earth*.

Another interesting variation on adobe construction is the ceramic house. Ceramic houses are made of adobe bricks and mortar. After construction is completed, the entire house is fired. Yes, that's right. The whole house is fired from within. To achieve this effect, an oil or gas heater is placed inside the house. The doors and windows are closed off and the heater is ignited. After a few days of firing, which brings the inside temperature to 1,000 degrees F, the adobe bricks and mortar turn hard as a kiln-baked brick. This technique has the potential to provide safe, affordable housing for millions of the world's less fortunate people. It is also a means of remodeling and strengthening previously built adobe and mud structures in China, Africa, and the Middle East.

Adobe offers many environmental benefits. In addition to its abundance, it has an extremely low embodied energy. Consider some simple figures to illustrate this point: According to Paul McHenry, eight ordinary red clay bricks require 125,000 BTUs of energy—the amount of energy contained in a gallon of gasoline. In contrast, manufacturing a bag of Portland cement requires four times as much energy—500,000 BTUs. But that's nothing compared to the energy required to build a 10-foot-high by 10-foot-wide wall stud wall using 2 x 4s. To cut, process, and transport the wood to build the wall consumes the energy equivalent of 6 gallons of gasoline. Frame the same wall out

EMBODIED ENERGY COMPARISON:

How Does Adobe Stack Up?

Eight clay bricks   125,000 BTUs
    —the energy equivalent to 1 gallon of gasoline.

One bag of Portland cement   500,000 BTUs
    —the equivalent of 4 gallons of gasoline.

Frame a 10 x 10-foot wall with wood   750,000 BTUs
    —the equivalent of 6 gallons of gasoline.

Frame 10 x 10-foot wall with steel   2,750,000 BTUs
    —the equivalent of 22 gallons of gasoline

Build a 10 x 10-foot adobe wall   6,000 calories
    —the equivalent of 24,000 BTUs or 0.2 gallons of gasoline.

PAUL G. McHENRY, JR., *The Adobe Story*

of steel studs and the energy investment climbs to 22 gallons. However, to build the same wall out of adobe, which requires two days of work, requires only about 6,000 calories or 24,000 BTUs—less than a quart of gasoline. That's the food energy required by the worker to accomplish this task. Not bad! As McHenry points out in *The Adobe Story,* "Of course there are better [building] materials than adobe. But at what cost?"

BUILDING AN ADOBE WALL

Laying up an adobe wall is fairly labor-intensive work, best performed by a pair of workers—one to lug blocks and mortar, while the other concentrates

INTERVIEWS WITH INNOVATORS

Paul G. McHenry Jr.

How long have you been working with or studying adobe?

I started working with adobe in 1960.

What first attracted you to this building technology?

I had been in the construction business for ten years and was bored. I thought it would be fun to design and build an adobe home. (Everyone has a secret desire to design the perfect home.)

You say adobe has a split image. What do you mean?

Adobe has a split image because it is available to the poor, who can build shelter with almost no cash outlay. However, it is so labor intensive that wealthy people can indulge in nostalgia and live in comfort and beauty. Some see it as a poor person's building material.

Others see it as something only the rich can afford.

What are the major myths about adobe?

The major myths are that it takes a special soil to make the bricks, and that the buildings will melt with the first big rainstorm. Both are wrong, but people have a vague sense of unease with a material that will melt in water.

Why do you consider adobe to be "the common bond of our global community?"

Adobe structures are made from a material (earth) that is found all over the world, and building patterns are basically the same.

The world population is growing and more and more people are without housing. How can adobe help rectify the situation?

Earth (adobe and the other forms) can alleviate the housing problem,

if we can get past the poverty image and the distrust of the engineering community who write building codes. Education and demonstration on a village level can help people see the contribution adobe can make to solving the world housing crisis. The building material is underfoot, just about everywhere, and people without work can have jobs. The several forms of earthwall construction can accommodate any climate.

What is it like to live in an adobe home?

Wonderful! It feels strong and secure, quiet, and not subject to the daily swings of temperature outside.

PAUL G. McHENRY JR. is an architect, builder, and author of numerous books on adobe building, listed in the resource guide. Over the years, he has gained considerable insight and knowledge on earthen building through his travels and research. He recently retired from his teaching position at the University of New Mexico's School of Architecture and Planning.

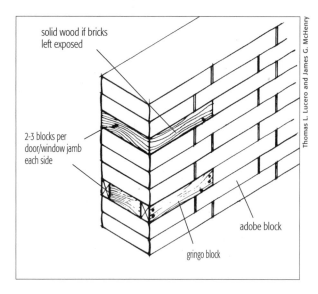

solid wood if bricks
left exposed

2-3 blocks per
door/window jamb
each side

adobe block

gringo block

Thomas L. Lucero and James G. McHenry

FIGURE 5-5.
Gringo blocks are adobe-block sized wood blocks set in the wall. They're used to secure door and window frames.

BUILDING NOTE

The height of an unsupported adobe block wall should not exceed ten times the wall thickness. In other words, a foot-wide wall should not be higher than 10 feet. Interior adobe walls can acts as buttresses, supporting long exterior walls.

his or her efforts on setting the blocks. Although you can build by yourself, two people create a better rhythm—and give you someone to complain to.

Adobe wall construction begins with the corners of the house. Take care that the corners are plumb and square, for the corners will serve as reference points. After they are built, the intervening walls can be laid up.

In building the corners and walls, the first row of bricks is laid on the foundation on a layer of adobe mortar approximately ¾-inch thick. Place mortar on the ends of blocks to seal the joints between adjacent bricks. The next course is laid on top of the first one in a running bond that increases the strength of the wall by eliminating continuous vertical joints. Care should be taken to ensure that each course is level and the walls are plumb. Adobe walls not built plumb have a tendency to tip over.

Most builders recommend laying up no more than six or seven courses of bricks a day. If you ignore this rule, you will pay dearly, for wet mortar in the bottom courses may compress and by the time you get to the final course of bricks, your wall will be too short. It may, for instance, fall short of the planned height of windows and doors. Wet mortar is also unstable. If your exuberance gets the best of you, your wall may come crashing down on you and your fellow workers.

As adobe walls go up, window and door frames are set in place. Wooden frames, like those used in rammed earth and straw bale construction, are made from construction grade lumber (2 x 6s and such) and are attached to the adobe wall via wooden nailer blocks, either thin strips of wood inserted between adjacent courses of adobe brick or larger adobe-sized "gringo" blocks, inserted in place of adobe bricks around window and door openings (figure 5-5). Obviously, careful planning and forethought must be given to their location.

Adobe structures are often viewed as a risky proposition when built in earthquake-prone regions. Architect, builder, and author Nader Khalili writes, "Towns and villages built with earth are not without problems. Sometimes an earthquake destroys an entire town." He tells a story of a town in his native homeland of Iran, built largely of adobe homes. After an earthquake rated 7.7 on the Richter scale, "practically the only things left standing were rug frames and gates. But then, the modern concrete and steel structures were also destroyed." Khalili notes that "the argument about the vulnerability of earth architecture to earthquakes is one of the greatest weapons used against it. Even though no one can deny the low seismic resistance of adobe structures, the argument itself is more destructive than the earthquake." And instead of improving its seismic resistance, Khalili complains, "almost all efforts are directed to replacing adobe with manufactured materials."

In earthquake zones, special precautions must be taken to reinforce adobe structures. Several techniques are available. Thicker walls, for example, render a structure more earthquake resistant. Rebar reinforcement also helps. According to Paul McHenry, the most successful adobe structures in earthquake regions are those reinforced with vertical rebar extending from the foundation to the bond beam on top of the walls. To accommodate rebar, workers drill holes in adobe bricks or split them in two and lay one on each side of the rebar (figure 5-6a). Because continuous pieces of rebar extending the full height of the wall are a pain in the neck to deal with, most builders use shorter pieces. These overlap by about 12 inches and are wired together to form a "continuous" vertical support.

Horizontal rebar is also recommended to enhance the structural stability of adobe walls and reduce their vulnerability to earthquakes. Many builders lay short sections of rebar in the horizontal joint spaces, approximately every third course. Others use horizontal tie beams, which are metal-ladder-type devices laid in the horizontal joint spaces every three feet of wall height (figure 5-6b). According to Steve Berlant, buildings with tie beams have survived repeated seismic assaults and are still standing after 85 years of earth shaking.

Horizontal rebar and tie beams are only two of a half dozen or so weapons in the builder's earthquake-fighting tool bag. Concrete bond beams make a home more earthquake resistant. And some engineers recommend the construction of single-story homes in earthquake zones, because they are less vulnerable to the shearing forces of an earthquake. Symmetrical designs, for example, circular or rectangular homes, are also effective against the often-violent rumblings of the Earth's crust, while asymmetric buildings (L-, T-, H-, or X-shaped) are more vulnerable. Light roofs are also recommended by many architects and engineers.

BUILDING NOTE

An air cavity left between inner and outer layers of adobes in a double adobe wall increases insulation. Vapor barriers should not be used so that moisture is not trapped in the wall. Moisture accumulating in the inside of the wall could cause it to deteriorate.

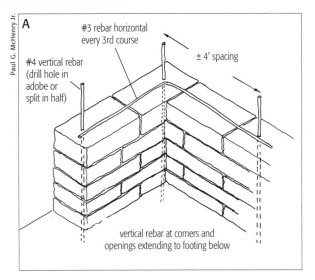

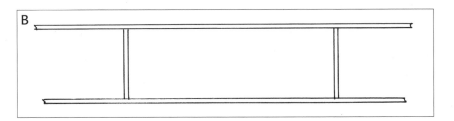

FIGURE 5-6 A, B.
Horizontal and vertical rebar in adobe walls enhances structural stability, especially important in earthquake-prone regions (A). Horizontal tie bars (B) are manufactured for use in adobe walls.

FIGURE 5-7.
Because adobe lacks compressive strength, vigas and rafters must rest on a bond beam to prevent buckling. In this home, the bond beam is made from poured concrete. The bond beam also makes the home more earthquake resistant.

A twelve-inch adobe block wall has an R-value of about 4.0. Although adobe walls offer little resistance to heat loss, their ability to store heat helps offset the low R-value. Direct solar gain on walls also helps compensate for their low insulation value.

Bond Beams and Lintels

Bond beams are required in adobe construction, as in many other forms of natural building. As you may recall from previous chapters, bond beams provide a means of attaching the roof to the walls and the foundation; they also distribute the roof load over the entire wall. This is especially important in adobe structures, as a concentrated load could cause the wall beneath to give way.

Builders install concrete and wooden bond beams, although the latter are not approved in some areas. Concrete bond beams are poured between wooden forms (as in rammed earth construction) laid on top of the adobe wall. They are reinforced with rebar to increase the tensile strength of the structure, and anchor bolts are set in the concrete after it is poured to provide points of attachment for the top plate. The top plate serves as a point of attachment for the roof frame.

As shown in figure 5-7, rafters and vigas (peeled logs that span the ceiling and rest on opposing walls) may be laid directly on the bond beam. Vigas should never be placed directly on an adobe wall, because adobe walls lack compressive strength and will buckle under the point pressure.

Lintels are required in adobe building. As Paul McHenry reminds us, lintels are "bridges over openings in the walls where windows, doors, and passages occur." Builders use a wide assortment of materials to make lintels. Wooden timbers and steel-reinforced concrete are two popular options. Whatever your choice, lintels must be strong enough to support the weight of the wall and roof above it and must be anchored securely in the wall on either side of the opening. To cut corners, some builders make the bond beam do double duty over doors and windows by simply adding steel reinforcement over each opening. For more details on lintels, see *Creating Architecture from Earth* and consult with a building engineer or architect for advice on the thickness of lintels for various spans.

Finishing the Walls

One of the final stages in building an adobe home is plastering. Although some homeowners leave their walls unplastered, most choose to apply plaster to the inside and outside surfaces to protect the blocks, cover up rough work, hide mistakes, and, of course, to make the wall look nicer. Plaster and plastering techniques have been described in chapters 2, 3, and 4.

Earthen plaster is the most commonly used material worldwide. Nearly identical to the mud used as mortar, except that it contains a higher clay content, earthen plasters may be modified by adding a variety of sometimes bizarre materials thought to enhance the durability of this protective layer, among them straw, dung, rice fibers, animal blood, egg yolks, oil, ceramics, lime, cement, and asphalt emulsion.

Despite these efforts, mud plaster wears down over time. Wind and rain slowly eat away at the surface, causing it to erode at a rate of about one inch in twenty years in dry climates like the desert Southwest of the United States. In wetter or windier climates, exterior plasters may wear down even more rapidly. Because of the natural weathering, adobe homeowners must periodically inspect their homes for cracks. Many owners apply a thin coat of mud mortar every year or so. Those who want no part of this annual ritual may choose to apply a sealant or a final coat of lime plaster. Lime plasters adhere well to mud and are quite resistant to weather—certainly more so than earthen plasters.

Some books on adobe construction suggest the use of cement stucco to finish adobe walls. However, cement stucco requires stucco netting attached to the adobe walls by special nails—which adds to the cost of the project. More important, cement stucco restricts the wall's ability to breathe. Moisture that becomes trapped inside the wall can damage the adobe, even in dry climates, according to some sources, and therefore should be avoided at all costs. For more on plastering, I recommend Steve Berlant's book, *The Natural Builder*, volume 3: *Earth and Mineral Plasters*.

One final note: Some builders, especially those working in colder climates, install rigid foam insulation on the outside of adobe walls to isolate the thermal mass of the wall from the cold, external environment. The foam is then covered with felt paper, chicken wire, and stucco. Another method of creating an insulated wall with adobe is to build a 2 x 4 stud wall on the outside surface of the wall. The cavities in that wall are then filled with insulation and exterior sheathing is applied. Yet another way to produce an insulated wall out of adobe is to build a double adobe wall, that is, a wall of adobe blocks with an air space in the middle. The air space is filled with insulation.

On its own, adobe doesn't provide much insulation. However, adobe makes up for this deficiency in many climates by its thermal mass properties. If you are afraid of losing heat through the walls, you may want to build much thicker walls. The increased thickness will retard heat loss.

The sense of creating shelter out of mud and earth is as primal as what must have been the first potter's amazement in creating an earthen vase; as the clay is manipulated, the potter gives it life and shape, resulting in a natural beauty that no straight line can offer.

ORLANDO ROMERO and DAVID LARKIN, *Adobe: Building and Living with Earth*

ADOBE AS AN ART FORM

In *Adobe: Building and Living with Earth,* Orlando Romero and David Larkin write, "Adobe is to the mason what clay is to the sculptor." Although not all adobe homes can be classified as art—some are pretty drab looking—adobe construction offers the potential to create houses of extraordinary beauty. Curved walls, archways, buttresses, niches, and a host of other features make many an adobe home stand out among the cheap imitations (Figure 5-8). Adobe is a form of art to live in.

Paul G. McHenry Jr.

"After complying with formal construction principles," say Romero and Larkin, "the adobe builder enters the realm of the sculptor." They add: "It is no wonder that the senses, even of the novice, are fired up, enthralled, mystified when they experience the beauty of adobe for the first time." And the second and the third . . .

Many dedicated adobe builders are artists, eking out a living building homes. It doesn't take much convincing to get them to confer an individual identity on each room, something one can only attempt to achieve in modern wood-frame houses with their sterile rectilinear efficiency. Through wall hangings, paint, furniture, and other features we try to give our homes a personality, but such efforts are often a vain attempt to hide drywall.

FIGURE 5-8.
Like other natural building materials, adobe lends itself to curved walls and other more organic architectural features that appeal to many people.

PASSIVE SOLAR HEATING AND COOLING: WHAT'S THE POTENTIAL?

One of the many benefits of adobe construction is that it is ideally suited to passive solar heating and cooling. Although we will examine natural heating and cooling in detail in chapter 11, a few words here will be helpful.

Adobe walls have tons of mass, but very little insulative value. Even though they lack the insulating properties of other wall systems, adobe walls perform well in some climates, especially deserts, keeping occupants comfortable year-round despite widely fluctuating outside temperatures. To understand how adobe works, consider a hot summer day in Arizona.

During the day, sun beats down on the walls of an adobe home. Sunlight is converted to heat that slowly begins to migrate through the thick adobe walls. Hot outside air also transfers heat to the walls, which also migrates inward. But because the walls are so thick, it takes many hours for the heat to move inward. In fact, if the walls are thick enough, the heat will never reach the interior. You will stay cool despite temperatures over 100 degrees F. How can that be?

The reason is simple. Before heat can migrate into the interior, night falls and the desert air becomes quite cool—downright chilly. Following the laws of physics, the heat that has slowly moved into the walls does an abrupt turnabout and begins to move outward. As night progresses, the heat reaches the exterior surface of the wall and is then dissipated into the air. The wall cools down and is ready to repeat the cycle the next day.

In the winter, heat in an adobe house comes from within. Heat from occupants, ovens, woodstoves, furnaces, and, of course, sunlight that streams into a passive solar house through south-facing windows, is absorbed by floors and walls. At night, however, the flow reverses. The interior walls and floor release their stored heat into the room and to the outdoors. If the ambient temperature doesn't fall too low for too long and internal heat sources continue operating, an adobe home can stay comfortable during the winter, despite the loss of heat to the outside.

The thermal performance of an adobe home (or any other solar home for that matter) can be improved by building mass walls inside the house in the direct path of incoming solar radiation. With sunlight striking them, these walls warm up during the day. At night, they give off their heat to the interior of the home.

Intelligent solar design also dictates the installation of floors with lots of thermal mass, such as earthen floors. Floors should be dark, too, to ensure maximum solar absorption. Clay-colored floors are an excellent choice.

The thickness of adobe walls ensures a stable interior temperature in arid climates year-round. In hot, humid climates, however, where summer nighttime temperatures fall less dramatically than in the desert, thick adobe walls tend to heat up and make the interior of a house very uncomfortable. Thin adobe walls work better. Adobe also functions poorly in very cold climates, unless walls are made very thick or heavily insulated.

ROOFS

With a properly engineered bond beam, any type of roof can be built on an adobe structure. In the arid southwestern United States, the flat roof is extremely popular. But pitched roofs are extremely common as well, for example, in rural New Mexico. Visitors to downtown Santa Fe will discover that many of the historic buildings made of adobe have pitched gable or hip roofs. Pitched roofs are more reliable than flat roofs, which require frequent repair or reroofing to stop leakage. Pitched roofs also repel rain and snow, and protect the walls better than flat roofs.

Adobe homebuilders often use vigas as roof trusses. Vigas not only support the roof, they add a stunningly beautiful decorative element. Vigas can be gathered on site or in nearby forests, or purchased locally from specialty lumber dealers. Daunted by the work of selecting, cutting, trimming, peeling, and transporting viga roof beams, many builders prefer to buy them from local suppliers who deliver them to the building site ready for installation.

Although vigas are an excellent material, they do pose some problems. One of them is that they taper. The diameter of vigas may vary by as much as

BUILDING NOTE

For maximum solar gain, thermal mass should be placed so that it intercepts sunlight streaming in through south-facing windows. Dark-colored floors and interior partition walls are ideal, but it is often difficult to design a home with a mass wall close enough to the windows without creating an intolerably sunny living space. See chapter 11 for more ideas on how you can get the most out of solar energy and live comfortably as well.

FIGURE 5-9.

To ensure adequate insulation in homes with exposed ceilings, 2 x 6s are secured to the first layer of decking. As shown here, 2 x 4s are then laid down perpendicular to the 2 x 6s, creating a 10-inch cavity for insulation. Another layer of decking is secured to the 2 x 4s. Roof felt and roofing are then applied.

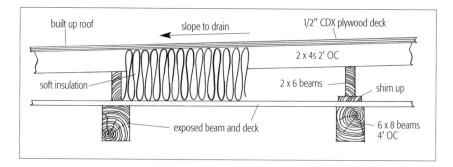

four inches. Although the taper can be used to provide roof pitch for proper drainage, some logs may taper more than others, making it difficult to build a roof. Vigas may contain irregularities, making it difficult to attach the roof decking. Although irregularities can be trimmed off with an axe or a hatchet to create a fairly level roof surface suitable for decking, many builders prefer to buy pretrimmed vigas with uniform taper.

Vigas or any other roof trusses are secured to the bond beam, either directly or via a top plate. Bond beams and top plates distribute the load over the entire wall. After the vigas are in place, wood planking or decking is nailed to them. As noted in chapter 4, be sure any planking you install is attractive, as it will be your ceiling. Many builders prefer tongue-and-groove planking for roofs because the interlocking joints reduce the unsightly cracks that develop when decking shrinks. Because tongue-and-groove planks lock together, they form a stronger surface than nonjointed planks.

Although wood in a ceiling provides little insulation, the thicker it is, the greater its insulative value. Even so, most of the R-value in a roof of this nature will come from rigid foam insulation secured to the outside surface of the planking (or decking), as described in chapter 4. Soft insulation, such as batting or loose fill, can also be installed in this type of roof and ceiling. As shown in figure 5-9, soft insulation is packed between 2 x 6s attached to wooden shims nailed to the decking. After the insulation is in place, roof decking is nailed to the 2 x 4s and roof felt is attached.

Roofs can also be made from 2 x 10 or 2 x 12 joists (not recommended because they are cut from old-growth trees), wooden trusses (a better option),

FIGURE 5-10.

I-joists are an excellent choice for closed ceiling designs. Here the I-joists supply the main framework of the roof and provide ample space for insulation.

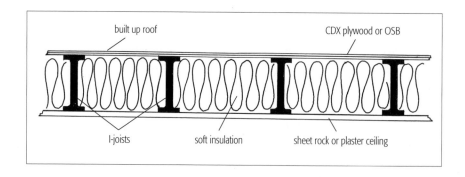

and I joists (the best option from an environmental standpoint), as shown in figure 5-10. Joist or truss roof systems are easy to pack with insulation, especially soft insulation materials.

FLOORS

A wide assortment of floors work well in adobe homes. Earthen floors, soil-cement, brick on sand, concrete, and wood are all perfectly suitable (see the "Floors" section in chapters 2, 3, and 4). Wood floors are very popular. They have a certain rustic appeal that many people like in adobe homes.

Wood floors are set on floor joists, attached to the stem wall with a nailing plate or ledger, as shown in figure 5-11. The floor joists or I-joists are placed 16 inches on center above grade. This not only creates a crawl space, but keeps the wood well off the surface of the ground and therefore reduces contact with moisture. Insulation should be placed between the floor joists to reduce heat loss and increase comfort. A vapor barrier should also be installed to reduce the flow of moisture into the house.

After the floor joists are in place, a decking material is nailed to them. Known as the subfloor, it usually consists of 5/8-inch plywood, OSB, or particle board. After the subfloor is in place, hardwood flooring is installed. Most builders use tongue-and-groove floorboards, which are nailed to the subfloor. Although laying a hardwood floor may look easy, the job requires a lot of experience and skill. Owner-builders may want to leave this part of the building project to the professionals.

Oak is a popular flooring material, but more and more green builders are using a hardwood substitute made out of bamboo. It's attractive and durable. (chapter 14 describes some environmentally sustainable hardwood flooring options. Check them out!) Carpeting and tile can also be installed over the subfloor.

ELECTRICITY AND PLUMBING

Locating plumbing lines in adobe walls is unwise because adobe, like rammed earth and straw bale, can be severely damaged by water that condenses on or leaks from water pipes. Over time, moisture will cause internal softening of an adobe wall, which could lead to major structural damage.

Plumbing is best run through interior wood-frame walls (partition walls), through the slab, or between the joists in the crawl spaces of a wooden floor. If you do run them under a floor, however, be sure to insulate the crawl space or insulate the pipes themselves and keep runs as short as possible to minimize heat-up time, heat loss, and freezing.

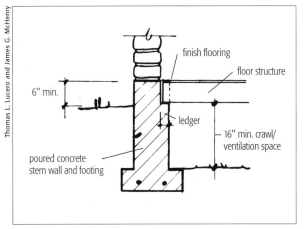

FIGURE 5-11.
Floor joists rest on a ledger in the stem wall. Be sure to insulate the floor and apply a vapor barrier to prevent moisture from seeping into the house.

Electrical wiring can be run through interior partition walls, floors, and ceilings. It can also be run along the surface of adobe walls, later to be covered with plaster. As in other natural homes, standard electrical wiring (Romex) is placed in metal conduit in channels chiseled in the surface of the wall with a sharp tool, such as the claw of a hammer. If you don't want to mess with conduit, use wire approved for underground applications.

Wiring can also be run through the attic or through the insulated space between roof trusses in a closed ceiling. If you have built with vigas and have an exposed ceiling, electrical wiring can be strung along the upper surface of the roof decking—between the decking and the rigid foam insulation. Drops are then run to wall outlets and switches. As in Earthships, special provisions must be made to permit the wiring to pass along concrete bond beams so it won't be left exposed. Channels are dug in concrete bond beams shortly after the forms are removed (while the concrete is still setting up) or knockout blocks are installed at strategic locations prior to the pour. The blocks will be knocked out after the forms are removed, leaving a channel to carry drops. L-shaped PVC pipe can also be installed prior to the pour, as discussed in chapter 4.

SHOULD YOU BUILD AN ADOBE HOME?

Over the years, millions of people have built and lived, quite happily, in adobe homes. But as with other natural, sustainable options, it is important to weigh the pros and cons of adobe construction.

Advantages of Adobe Homes

- Adobe is widely available and could provide housing for millions of people, especially in impoverished nations.
- Adobe codes have been adopted by many building departments. Even if yours doesn't have a code, there are model codes they can use.
- Adobe is a locally available resource. It is inexpensive and has low embodied energy.
- Adobe bricks can be made on site by hand or by machine. They are also commercially available.
- If made on site or locally, adobe bricks reduce transportation demand, cutting energy use and pollution.
- Making adobe bricks requires minimal skill. For the owner-builder, adobe is one of the easiest of all natural materials to work with.
- Adobe walls are fairly easy to build, requiring little skill and few tools.
- Adobe bricks are fairly inexpensive, about $2,000 to $3,000 for a 2,000-square-foot home.
- Local manufacture of adobe contributes to the local economy.
- Adobe's thermal properties make it an ideal building material for homes heated by passive solar energy.

- Adobe homes stay warm in winter and cool in summer in arid climates with cool nights.
- Adobe is a great sound insulator, and living in an adobe home is quiet and peaceful.
- Adobe is easily recyclable. Broken bricks can be wetted and mixed with mud. Intact bricks from demolished walls can be used to build new walls.
- Although very few people realize it, adobe can be used to build houses of a wide variety of architectural designs, even Victorian.
- Adobe building lends itself to artistic expression.
- As Paul McHenry writes in *The Adobe Story*, "The softly flowing walls and rounded corners create a gentle, comforting, and secure feeling space."
- Adobe walls are fireproof.

Disadvantages of Adobe Homes
- Building adobe bricks and walls is labor intensive and can be costly if you hire others to do it for you.
- Adobe wall construction is hard work.
- Traditional adobe brick making is limited by weather and climate. It is unsuitable to cold or wet areas.
- Adobe bricks cannot be made during freezing or wet weather.
- Squirrels and insects, notably termites, can burrow through adobe walls, weakening them. Plaster reduces this problem.

Adobe is a versatile building technology. It is relatively easy to learn and requires few tools. Its simplicity and the universal availability of building materials make adobe a superb candidate for housing the world's people, rich and poor alike.

But adobe architecture is on the decline. "Adobe seems to be a forsaken material in the eyes of the world," writes McHenry. After millennia of dependable use in almost every climate, he laments, we have turned our backs on adobe. Our love affair with modern building materials and the associations we make between adobe and poverty have spelled doom for this marvelously simple means of satisfying one of the most basic human needs. Even in developing nations, where people are exposed to Western ways through television and other avenues, adobe is looked on with disdain. Hopefully, the high cost and high embodied energy of most modern building materials, the pervasiveness of poverty, and the local abundance of adobe will bring the world back to its senses.

An adobe structure, properly built, is in that earthen comfort zone that tempers the harshest of climates; it is efficient shelter in the searing desert heat of Arizona or Saudi Arabia, the windy reaches of East Anglia, or the snowy mountain regions of Asia.

ORLANDO ROMERO and DAVID LARKIN, *Adobe: Building and Living with Earth*

6

COB HOMES

Y VARIOUS ESTIMATES, about one-third to one-half of the world's population live in homes made from earthen materials. Many of these structures are primitive dwellings fashioned from sticks and mud. However, many earthen buildings, like adobe and rammed earth homes, are elegant and graceful structures. Another candidate for this list of potentially stylish, highly sustainable, healthy natural structures is the cob home.

The word *cob* is an Old English root meaning a lump or rounded mass. It has nothing to do with corn cobs, as commonly surmised by those hearing the term *cob home* for the first time. Cob homes are mud-walled buildings, sometimes referred to as monolithic adobe. This term was coined by Steve Berlant, cob builder, educator, and author of *The Natural Builder*, volume 2, *Monolithic Adobe Known as English Cob*, a must-read for anyone interested in building a cob home. Cob is made by mixing soil, sand, straw, and water. To build a wall, the cob is then applied to the foundation, one handful or shovelful at a time. It is worked by hand—massaged, as it were—to create smooth-surfaced walls ranging in thickness from 4 to 24 inches.

Cob construction lends itself to sensuous curved walls, arches, and niches. Like adobe, cob homes can be as much an expression of one's artistry as a place to live. In the words of Becky Bee, cob builder and author of *The Cob Builders Handbook*, cob construction is "like hand-sculpting a giant pot to live in." But don't think that the walls of a cob structure are fragile like a clay pot. Nothing could be further from the truth. Most cob homes are built with thick walls, often two feet thick at the base. When the cob dries, it achieves a

FIGURE 6-1.
This cob house in Oregon is finished with lime plaster.

BUILDING NOTE

Cob is also known as puddled adobe, and coursed adobe, and direct molding. The term *monolithic adobe*, however, may be useful when dealing with building departments.

STEVE BERLANT,
The Natural Builder

hardness similar to stone. The walls are often whitewashed or lime plastered, creating a structure of extraordinary beauty (figure 6-1).

Cob homes are suitable in a wide range of climates, even harsh ones, such as the wind-battered, rain-drenched coasts of Great Britain and the soggy temperate rain forests of Oregon. Cob walls are strong, permitting builders to fashion two-story homes out of this unique building material. And cob homes can be built in earthquake-prone regions of the world (with special measures to protect them).

Cob is fireproof and environmentally friendly. The material has low embodied energy and is widely available. Cob construction is also simple to learn and forgiving of mistakes. "It requires dedication more than physical strength, and willingness to experiment more than skills," says Becky Bee. Even children enjoy building cob walls, and can do a pretty decent job (figure 6-2).

Cob is ideal for owner-builders. Most of the work is done by hand or with simple hand tools. There's no need for expensive machinery or noisy power tools. Cob also allows for a freedom of expression unrivaled in the building world. You can build a conventional-looking cob home or take advantage of its sculptability, creating a one-of-a-kind structure. A plethora of books, videos, and workshops provides the details you'll need to build a cob home; the resource guide lists many of the most valuable ones. Several of them merit mention. For basic coverage, I

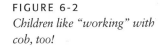

FIGURE 6-2
Children like "working" with cob, too!

Although cob is suitable for a wide variety of climates, it may not be suitable for places with extremely cold winters. Heat loss through cob walls can be substantial.

In addition, special precautions should be made to anchor roofs to walls in regions with high winds. This can achieved by installing roof plates, as explained in the text.

In seismically active areas, cob requires a good foundation with sound anchoring of the walls to the foundation.

STEVE BERLANT, *The Natural Builder*, vol. 2,
Monolithic Adobe Known as English Cob

recommend Michael Smith's *The Cobber's Companion* (it's clearly written and wonderfully illustrated) and Becky Bee's *The Cob Builders Handbook* (it is also well written and fun to read). If you want to learn more, especially the latest developments, I strongly urge you to read *The Natural Builder*, volume 2: *Monolithic Adobe Known as English Cob*, by Steve Berlant. Steve's book is especially useful. It presents a lot of information that will help you make wise building decisions; it will also help you deal with building departments that have never heard of or approved a cob structure. Attending a workshop is essential for the owner-builder. The Cob Cottage Company and Groundworks offer an assortment of hands-on experiences that are fun and extremely useful.

WHERE IT ALL BEGAN . . . NO ONE REALLY KNOWS

Where cob originated is a mystery. Cob structures have been built in the Middle East and Far East, but details of activities in these regions are sketchy. Some people believe that cob construction may have begun in Europe about 800 years ago. But Native Americans also built with cob, perhaps even earlier. Casa Grande, a ruin in Arizona, is a massive, multi-story earthen structure built over 1,000 years ago. It is made of puddled or "coursed" mud. The Pueblo Indians built five-story buildings of cob 800 years ago. Where these peoples learned about cob or what inspired them to begin experimenting with mud as a building material is anyone's guess. They may have gotten the idea from cliff swallows, elegant little birds that build precarious nests on the undersides of rocky ledges from little gobs of mud (figure 6-3).

Cob was a popular building material until very recently. In England, tens of thousands of cob homes were built in 1700s and 1800s. In the 19th century, Australia and New Zealand witnessed a massive upsurge in cob construction with thousands of structures being built.

Despite its initial popularity, cob construction plummeted with the advent of the fired brick. (This is getting to be a familiar story, isn't it?) Today, like other natural building technologies, cob is staging a comeback. Many advocates promote it as a sustainable alternative to modern housing—one that is accessible to ordinary folks.

Cob is a versatile, durable building material. Its durability is attested to by the fact that many cob buildings built in the 1500s and 1600s remain stand-

FIGURE 6-3.
Cliff swallows build small cob cottages of their own, one beakful at a time. Did humans get the idea from them?

ing today (figure 6-4). According to one estimate, approximately 20,000 cob homes are still in use in England; most of them were built in the 1700s and 1800s. Many of these beautiful structures command a high price on the open market. In Devon County in southwest England, an estimated 20,000 cob barns and fences dot the countryside as monuments to a craft that has all but faded from the scene.

Fortunately, the durability of cob homes compensates for the loss of builders with the knowledge and skills of this ancient craft. Many modern-day cob builders have learned their trade, in part, by studying these buildings. Developing techniques and passing them on to others, this inspiring group of natural builders is mending the break in the chain. They too are responding to the need for a new kind of housing, the depressing onslaught of environmental problems, and the growing desire of many to live sustainably.

FIGURE 6-4.
This beautiful old home in England is made from cob. Tens of thousands of similar structures exist in England.

Cob's versatility is witnessed by the fact that cob homes can be found in countries spanning a wide range of climates, many of them cold, windy, and wet. You can find cob homes in England, New Zealand, Australia, South Africa, Denmark, Canada, and parts of the United States. The Cob Cottage Company and Groundworks, listed in the resource guide, have built or helped build cob structures in most states west of the Rocky Mountains. Cob is also being used in combination with other natural building techniques such as straw bale, adobe, and tire homes as you'll discover in chapter 10.

FOUNDATIONS

Many kinds of foundation can be used in cob construction, but because cob walls are so heavy, you will definitely need a strong one. You will also need a foundation that keeps water away from the walls of the house. Standard building practices are all that's required to meet this goal. When building on a concrete or concrete block foundation, be sure the stem wall is built high enough to prevent contact with surface water. A grade beam on a rubble trench foundation serves the same purpose. (Standard concrete foundations and rubble trench foundations have been discussed in chapters 2, 3, and 5.) The foundation must prevent groundwater from being drawn into the wall, which would weaken it, perhaps even cause it to collapse. French drains and waterproofing serve this function well.

Stone foundations are an excellent option. When properly built, they provide sufficient support for heavy cob walls (figure 6-5). Mortarless stone foundations (dry stone foundations) also retard water migration into cob walls. In addition, stone foundations look pretty spectacular when combined with cob and can be built by first-timers—after a little instruction and practice.

Cob is ideal for those with sensitivities to modern construction materials, as well as those trying to live simply.

STEVE BERLANT, *Monolithic Adobe Known as English Cob*

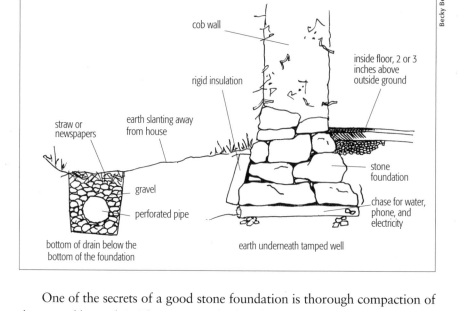

Becky Bee

One of the secrets of a good stone foundation is thorough compaction of the ground beneath it. This prevents the foundation from settling, which could cause walls to crack. Another secret is having a large selection of rocks on site. The best rocks are those having at least two parallel flat surfaces. You will want a variety of sizes, too. Yet another secret is patience. Stone masonry requires skill, of course, but patience is a vital to the success of any stone project. Build slowly and methodically. (For a general overview of stone foundations, consult *The Cob Builders Handbook* and *The Cobber's Companion*. A more detailed discussion of stone wall building appears in chapter 9, and there are many references listed in the resource guide).

As in all home construction, it is important to slope the ground away from the house to ensure that water drains away from the structure. To further reduce the risk of water damaging your foundation (and your cob walls), dig a drainage ditch around your house. Place porous drain pipe in the ditch and then fill the ditch with rock. This simple measure diverts water away from the foundation and protects your home from damage. It is especially useful in wetter climates (figure 6-5).

BUILDING A COB WALL

The first step in building a cob wall is to prepare the building material, the mud. As with other earthwalled homes, the best mud is made from subsoil. Topsoil is unacceptable because it contains organic matter that can decay and weaken earthen walls.

Creating Cob

Cob walls are built with a mud containing sand, clay, and water. Straw is often added to increase tensile strength and to bind the layers of cob together into a

BUILDING NOTE

Cob walls should be slightly wider than the foundation to prevent water from pooling at the top of the foundation—that is, at the juncture of the cob wall and foundation.

strong, monolithic cob wall. Unlike rammed earth building, the mud used in cob construction needn't be as precisely controlled. Sand content can be as low as 50 percent and as high as 85 percent, according to some sources. Sand should have a coarse or jagged texture. Coarse grains of sand interlock unlike the round grains of most beach sands, and therefore produce a wall with greater compressive strength.

The clay content in cob can be as low as 15 percent and as high as 50 percent. In *The Cobber's Companion,* Michael Smith recommends a 10 to 30 percent clay mixture. For an excellent summary of the components of cob, including the functions of each, see *The Natural Builder*, volume 2, *Monolithic Adobe Known as English Cob.*

Many people are able to use the soil (actually the subsoil) from their building site to make cob, sometimes augmenting it with sand or clay. The more cob components that can be dug from the actual site (footprint), the less disruption to the land. If you're one of the unlucky few whose subsoil isn't right, you will have to import a suitable soil mix from a commercial source or a willing neighbor. Whatever the source, be certain to subject your soil to careful testing before you start building.

The first test your soil mix must pass is the standard jar test (see pp. 44 and 128). This simple procedure allows you to determine the approximate ratio of sand and silt to clay. If your mix contains the right proportion of sand to clay, you will need to run the drop test. For details on this test and other soil tests, refer to *The Cobber's Companion.*

If your soil has passed so far, it is time to make some sample test bricks. Because the ratio of sand to clay is critical, your test bricks should initially contain only these components. Once you've determined the proper ratio of sand to clay, you can add varying amounts of straw and make more test bricks. Although you won't be using bricks to build your home, test bricks allow you to determine the strength and durability of your cob. Check out the books listed in the resource guide for details on sample brick tests.

If your soil fails the brick test, you will have to make some modifications to your soil mixture. The bricks will tell you what you need to do. If they crack, the mix probably has too much clay. Add a little sand and make some more test bricks. If your first test blocks fall apart, your mix probably has too much sand. Too much sand reduces the compressive strength of the walls, which means they can't support as much load. Add some clay and make some more test blocks. Be sure to keep records so you know how much you added. The feel of a test brick can also be used to determine your course of action. If sand rubs off the test brick, add some clay to the mix.

Sample cob bricks should also be drop tested to see how strong they are. Watch out for your toes and watch out for your pets! Blocks that break upon impact are no good. Michael Smith's book includes a trouble-shooting guide that lists problems with cob mixes and ways to rectify them.

What sets cob apart from every other building technique and draws so many artists and creative people is its extreme fluidity of form. Since cob requires no formwork and is not made of uniform blocks, it provides absolute liberation from the straight, flat walls and right angles which plague modern construction.

MICHAEL G. SMITH, *The Cobber's Companion*

BUILDING NOTE

If test bricks crack, add more sand.
If test bricks crumble, add more clay.
If test bricks break easily, add more straw.

MICHAEL SMITH, *The Cobber's Companion*

Mixing Your Cob

Cob must be thoroughly mixed in preparation for cobbing. In olden days, soil was dumped in a pit. Water and straw were added, and cattle tethered to a central post were driven around in circles until the mud was thoroughly mixed. Manure incidentally added to the cob was believed to add strength to the walls. If you don't have cattle, don't despair. Human feet work pretty well.

For small projects, some builders mix cob on tarps. The cob mix is shoveled or dumped onto the tarp. The worker then grasps the opposite corner and pulls it toward herself or himself. The process is repeated many times until the dry ingredients are blended to perfection. After they are thoroughly mixed, straw and water are added and mixed to create the consistency of cookie dough. One of the best ways of combining water with the straw and mud is by old-fashioned foot power (figure 6-6). This is a great job for children. Most adults also get a kick out of dancing barefoot in the mud, too, so you shouldn't suffer from a shortage of willing laborers.

Cob can also be mixed in wheelbarrows, plastic lined pits, and mortar mixers. Cement mixers are sometimes included in the list, but the folks with whom I've talked don't recommend them. The mud sticks to the walls of the mixer. Tractors can also be used to mix cob. Steve Berlant told me he once mixed 80,000 pounds of cob with a tractor in ten hours!

Many cob builders mix at the end of each work day, then let the mud sit overnight. This tactic produces a mix that is considerably more plastic and much easier to work with.

Sculpting a Wall

After proper mixing, cob is transported to the foundation in a wheelbarrow or by hand in loaf-sized units, cobs. Workers often form a "cob brigade" to transport cob to the wall. It's a great place to get dirty! Missed catches will soil a T-shirt in no time. The technique quickly alerts you to people who may be harboring a grudge toward you or humanity in general. So watch out for misanthropes!

Cob is placed on the wall in one of three ways, depending on the applications. They are pisé (flattened cob), cob loaves, and shoveled cob. Each offers advantages and disadvantages.

Pisé, or flattened cob, is probably the simplest method. In this technique, cob is first flattened into pancakes. Fully mixed cob is shoveled onto a hard surface such as plywood, then flattened by walking on it. The cob pancakes (measuring about 1½ to 3 inches thick) are then lifted off using a flat-tined garden fork or pitchfork. The flakes are placed on the wall, and tamped by foot once again. This procedure bonds the cob to the foundation and, as wall building proceeds, bonds one layer of cob to the next.

Pisé works well while the walls are wide and low to the ground. However, the higher the wall, the more difficult it is to "fork" cob into place and the

BUILDING NOTE

There are no mortar joints in a cob wall as in adobe buildings. This adds to the building's strength and integrity.

STEVE BERLANT, *Monolithic Adobe Known as English Cob*

more precarious the balancing act required to tamp the cob down. Because it requires so much work, this technique is not very popular among cob builders.

The second method, cob loaves, harkens back to early days in England when cob was applied in loaf-shaped masses. It is still very popular. After mixing, cob is kneaded into manageably sized loaves that are laid on the wall, often in special patterns, a topic discussed in more detail other books on the subject. The cob loaves are then "massaged" by hand to shape the wall.

The shoveled cob technique is something of a hybrid between pisé and cob loaves. In this procedure, cob is shoveled onto walls, then worked by hand. Because it bypasses the loaf and pancake stages of the previous methods, it saves a lot of labor and is very popular among some cob builders.

In all three systems, individual layers of cob must be skillfully melded together to enhance the structural stability of the wall. This is achieved by compression—either walking on the wall or pressing down on the cob as the wall is massaged and shaped—and by use of a tool known as a cobber's thumb. Looking very much like a large thumb, the wooden cobber's thumb is used to make depressions in the cob, extending through a newly applied layer of cob into the layer below. This pushes strands of straw in the upper layer of cob into the one below and effectively knits the laminae into a cohesive wall. The depressions also serve as keyways that lock adjacent layers of cob in place. Builders wet the surface of previous layers to counteract drying. This too makes for a tighter bond.

Exterior cob walls vary in thickness from 8 inches to 4 feet, depending on the climate, structural requirements, and level of ambition of the cobber. Most houses, however, have walls that range from 12 to 24 inches thick at their base. Walls are thicker at the bottom than the top because the upper portions of the walls support less weight. Most builders taper their walls about 2 inches for every 3 feet they gain in height. The folks at The Cob Cottage in Oregon recommend that load-bearing exterior walls be a minimum of 12 inches thick at the top and that interior, non-load-bearing walls be at least 4 inches thick.

When building, care must be taken to ensure that walls are plumb. A three-foot carpenter's level comes in handy. However, to compensate for wall taper the level must be fitted with a triangular piece of wood or rigid foam that mirrors the taper of the wall, as shown in figure 6-7. In cob building, as in adobe and straw bale construction, it is important is to check the wall for

FIGURE 6-6.
"Workers" mixing straw into the cob by foot.

BUILDING NOTE

Michael Smith recommends building a sacrificial wall before building your house, so you can get the hang of cobbing before you start your project.

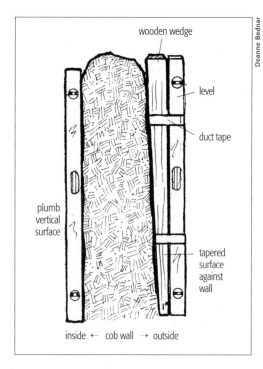

wooden wedge

level

duct tape

plumb
vertical
surface

tapered
surface
against
wall

inside ← cob wall → outside

FIGURE 6-7.
*A level with a wooden wedge
taped to it will help you ensure
proper taper of the outside
surface of the wall.*

BUILDING NOTE

The rate of building depends on the
weather and the size of your work-
force, but racing to build a structure
would be missing the point and halv-
ing the fun. Unlike conventional,
modern building practice with its fre-
netic pace, power tools, and scope
for errors and accidents, cob making
is a peaceful, meditative and rhyth-
mic exercise.

IANTO EVANS, cob builder,
writer, and teacher

plumb frequently. Although you may be inclined to "eyeball"
the wall, remember that your eyes can easily trick your brain
into thinking a wall is straight. A level never errs. It is an indispensable tool! Even if you don't care if your walls are plumb,
Michael Smith points out that "it is still advisable for structural reasons to keep them close to vertical."

As the wall gets higher, the weight of the cob in the upper
layers may cause the lower sections to bulge or, as it is known
in cob-builder parlance, *oog*. Ooging is a sign that you've built
enough cob wall for the day. It is the cob builder's time clock.
Take heed of the warning. If ooging occurs, stop immediately.
Start on a new section, or mix a batch of mud for the next day
and go home and relax.

After the wall dries a little, shave off the cob cellulite with
a machete or shovel or some other sharp implement. (Oh, if we
could only dispense with those extra pounds of fat so easily!)

According to one source, cob walls are built approximately three feet at a time. However, not all people I've talked
to agree. Steve Berlant says that experienced cobbers build no
more than 12 to 18 inches of wall per day. Beginners, he says, should consider
daily gains of about 6 inches.

As the walls of a cob house grow, workers continue to check them with
the level. They also rake a short piece of 2 x 4 over the vertical face of the wall
to ensure a flat surface. (If you don't, the walls will be lumpy.)

At the end of each work day, before the tools are put away and your hands
are washed, prepare the surface of the wall for the next layer of cob by poking
holes with the cobber's thumb 3 or 4 inches apart. These depressions will help
not only to secure the next layer, but also to accelerate drying by providing a
route for moisture inside the wall to escape. When work resumes, the top of
the wall is sprayed with water and new cob is laid on the wall. Pressure on the
loaves or shoveled cob pushes the new cob into the holes in the previous layer
so the newly applied cob now has a solid foothold. Further adhesion is
achieved by poking strands of straw from the new layer into the previous one
using the cobber's thumb.

Although it is best not to let the cob dry out between applications, it may
be necessary from time to time to attend to other matters. If this is the case,
cover the top of your cob wall with a layer of wet straw, then a layer of burlap
or newspaper, and finally with a sheet of plastic. If the cob dries, thoroughly
rehydrate the old cob before placing a new layer on top. Be sure to use the
cobber's thumb to wet the cob and sew the layers with additional bunches of
straw. As Michael Smith warns, once a wall has dried, "you will never achieve
quite as good a bond as you do between successive layers of wet cob." So, if at
all possible, don't let walls dry out.

Sculpting Shelves, Furniture, and Fireplaces

One of the neat things about cob is that it can be used to sculpt many interior features of a house. Cob bookshelves, niches, and benches add to the beauty and grace of the curved walls (figure 6-8). Sculpted window seats are particularly attractive and useful. They make a room look much larger without increasing the actual floor space. However, if you build in furniture, be sure it contours to your body for comfort. And remember, if you add a window seat or other furniture, you will need to support it with an extension of the foundation. Even wide shelving will require you to beef up the foundation and the cob wall under it.

Bookshelves, niches, and benches are typically constructed as the cob walls are made. Small sticks and straw secure them to the walls. Shelves and benches are also built out slowly. Adding too much cob at once may cause the structure to break away from the wall.

One technique to achieve this goal is known as corbelling (figure 6-9). As Michael Smith explains, "Corbelling allows the deliberate and controlled widening of a wall as it goes up, in such a way that the weight of the projection and anything that rests on it is carried into the wall." This technique

Robert Bolman

FIGURE 6-8.
Cob lends itself to organic forms, like this sculpted seat.

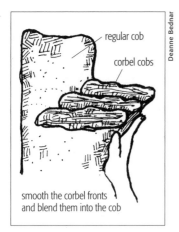

regular cob

corbel cobs

smooth the corbel fronts
and blend them into the cob

FIGURE 6-9.
Corbelling.

The sculptability of cob becomes apparent when you create niches, recessed spaces for bookshelves, candles, or sculptures. Niches can be carved into cob walls, the earlier the better. If the cob dries too much, it becomes harder to work with.

FIGURE 6-10.
Details of window bucks used in cob construction.

utilizes elongated loaves of cob (corbel cobs) containing long strands of straw (figure 6-9). When building a cob shelf, for example, the first row of corbel cobs is situated above a purposefully thickened region of the wall that serves as the base. After the first row has been secured, a second layer of corbel cobs is added, projecting a couple of inches out from the first. This process is repeated until the shelf is complete.

One of the marvelous things about cob is that if you don't like what you've done, just knock it down and start over. Rather than toss the dried mud in the trash bin, you simply rewet it and use it again.

Cob is fireproof. Because of this, cob can be used to build ovens, stoves, fireplaces, and chimneys. While nothing adds to the charm of a room better than a fireplace, most of them are terribly inefficient. They suck as much heat out of a room as they produce. Unless you can build a really efficient fireplace, like the Russian fireplaces, you're better off installing a woodstove for heat.

All structures, whether chairs or bookshelves or fireplaces, can be "shaped to fit curved walls precisely, seeming to grow out of them," in the words of Michael Smith. Cob even allows you to build a child's seat into the wall and permits you to accommodate annual growth by adding a little cob each year, until the child becomes an adult.

Windows and Doors

Installing windows and doors in cob walls requires a fair amount of patience, attention to detail, and skill. Let's consider windows first.

The most common practice is to install sturdy wooden window frames similar to those used in rammed earth, straw bale, and adobe homes. Two-by lumber—for example, 2 x 10s—is a good choice (figure 6-10). If you build

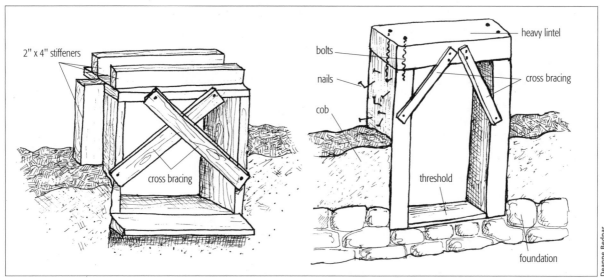

2" x 4" stiffeners

cross bracing

bolts

nails

cob

cross bracing

heavy lintel

threshold

foundation

window frames out of thinner wood, add 2 x 4 stiffeners to strengthen them. The stiffeners also serve as a keyway to help anchor the frame to the cob.

The top of window frames must be spanned by lintels. Embedded in the cob above each window, lintels extend at least 4 inches on either side of the opening (For the exact distance, check with a builder, architect, or building department official.) Because lintels may settle when cob dries, special precautions must be taken to prevent glass from cracking. See *The Cobber's Companion* for ways to avoid this.

To save time, money, and resources, however, some builders simply use thicker lumber in their window frames to support the weight of the cob and the weight of the roof above the window opening. The added thickness is necessary only in the top piece of the frame. As a general rule, a lintel built into a window frame should be one inch thick for every foot the lintel spans horizontally. A 3-foot-long lintel, for example, should be 3 inches thick.

Window frames are installed directly in the cob wall. When the cob wall reaches the height of the window sill, nails are driven into the frame so that it resembles a riled porcupine. The nails hold the frame in place. The frame is then set in the wall. To prevent the frame from being pushed out of square, cob is added equally to both sides of the window frame. Strong cross-bracing, shown in figure 6-10, may also be used to eliminate this potential problem. Other precautions may be required as well (see books listed in the resource guide).

Window frames can also be installed directly into the cob with gringo blocks, discussed in chapter 5. Gringo blocks are large pieces of wood inserted into the cob wall alongside window openings. They serve as nailers for window frames.

After a window has been cobbed in place and the cob begins to dry, shrinkage cracks may begin to appear between the wood frame and the wall. They can be filled with cob plaster to protect the wall and keep the wind out.

When positioning windows, it is important to remember that cob walls typically taper as they ascend. Therefore, make sure that the frame is recessed at the sill to take into account the taper in the wall (figure 6-11).

Windows can also be formed out of odd-shaped pieces of salvaged glass. Glass such as this is typically installed into cob walls without frames. To protect your hands and prevent the glass from breaking, tape the edges, then set the glass in place and build up the cob on either side. The glass should be embedded in the cob one quarter inch. If you really want to spice up your dwelling, consider embedding colored glass bottles in cob walls. At night, inside lights will make the wall shine like a gem. During the day, as sunlight shines on the walls, the light show switches indoors to the delight of the house's occupants (figure 6-12).

BUILDING NOTE

Shrinkage in cob walls can be reduced by adding sand or a small amount of lime.

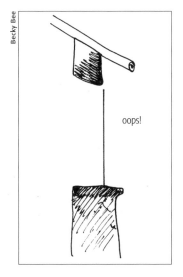

Becky Bee

oops!

FIGURE 6-11.
Make sure window glass is placed properly to account for the slope of the outside wall!

FIGURE 6-12.
Glass bottles in cob walls makes the cob home sparkle like a jewel.

Robert Bolman

Door frames are installed in much the same way as window frames. The frames are made of sturdy lumber and are attached to the foundation. Once the frames are in place, the cob walls on either side of them are built up. A keyway system with 2 x 4 stiffeners and nails, like those used to secure window frames to cob walls, ensures a snug fit. This is especially important for doors, which tend to be heavy and get a lot of use (and abuse). If you don't secure a door frame to the cob well, opening and closing of the door will cause the frame to work loose. Finally, the structural integrity of the door frame is also enhanced by installing a lintel over the door. Once the rough frame is in place, the door can be installed.

Plastering the Walls

Cob walls are durable and able to withstand the elements. However, most people prefer to apply a surface coat of plaster to protect the cob from wind, rain, sun, and snow and to enhance its beauty and longevity. Interior plaster also reduces dust.

Although you've already been treated to many details of plastering in previous chapters, there are a few things worth repeating and a few new facts you should know. First of all, cob builders recommend using a plaster that allows your walls to breathe. Breathable plasters permit moisture to escape, rather than build up in walls where it could do considerable damage. Lime and earthen plasters are ideal. Cement and synthetic stuccos, on the other hand, seal off a wall and may trap moisture inside. In addition, cement stuccos are brittle and tend to crack. Tiny cracks permit moisture to enter the wall. Steve Berlant strongly warns against the application of cement-based plasters. "A cement stucco finish will destroy a cob building in sixty years," he says. "There's ample proof of this point."

Earthen plaster is used by many cob builders. Earthen plaster contains the same ratio of straw, sand, and clay as the cob used to build your walls, but the straw used in the mix is typically shredded so you don't end up with the long strands of straw in the surface coat. The cob mix is passed through screens to remove small pebbles, large clay particles, and large pieces of straw. Some cob builders add fresh or finely grated dry cow or horse manure to their plaster. Other plaster enhancers include human or animal hair, cactus juice, ground psyllium seed husk, Elmer's glue, mica, vermiculite, pumice, flour paste, and colorizers. These additives serve a variety of functions. Some are believed to minimize cracking. Others enhance the strength of the plaster. Still others increase water resistance or improve the adhesion between the plaster and cob. Some supposedly make the plaster easier to apply. Others enhance its beauty. For details, check out the references listed in the resource guide.

Plaster is applied to cob walls after they have thoroughly dried. That is because as the walls dry, they shrink and crack a little. If you apply plaster to them before they dry, the plaster will crack and you'll need to go back over the

Cob buildings dry at different rates in different climates and different seasons. In England, it may take a year or two for a cob structure to dry sufficiently to plaster. In extremely dry climates, drying occurs so quickly that you may need to slow it down, by wetting and rewetting the walls, and by hanging wet burlap and plastic over them as they are being built.

FIGURE 6-13.
This tool, known as a hawk, is used to transport and hold plaster while plastering a wall.

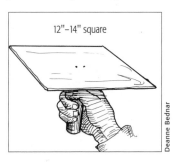

12"–14" square

Deanne Bednar

walls to repair them. If time is of the essence—for example, winter is bearing down and you need to finish the job so you can move in—plaster the interior walls and save exterior plastering for spring. Although you may be tempted to fire up a space heater or keep a fire burning in the woodstove to dry your walls, don't fall victim to your impatience. Accelerated drying causes the walls to shrink and crack, especially if the clay content of your mix is on the high end of the acceptable range. You'll make a lot of extra work for yourself. The safest route is to wait until the walls are completely dry before you begin plastering.

To plaster a wall successfully, first spray the surface with water. This will enhance the adhesion of the plaster to the cob. Then apply the plaster in two to three thin coats, either from the top of the wall down or from the bottom up. Earthen plaster is applied by hand, by trowel, or by brush. It can also be sprayed on walls with a machine. Most builders use their hands or trowels. (Some swear that pool trowels with rounded corners are the only way to put on a finish coat!) Plaster is transported to the wall on a hawk, shown in figure 6-13, then scraped onto the wall and smoothed out with upward sweeping motions. The first layer fills most of the indentations and surface cracks in the walls. The second coat completes the job, creating a nice, smooth surface.

Generally, the thinner the layer of plaster, the less it will crack. If you have never applied plaster, start in an out-of-the-way place in your house, for example, under countertops or in closets. It will take a while to learn the technique. Or sign up for a plaster workshop and get some experience first.

Some cob builders decorate their walls with stones, tiles, or shells embedded in the plaster. Or, you can "carve" designs in the wall. It is up to you and that wild imagination you've kept bottled up (figure 6-14).

The color coat is applied last. Many cob builders polish the color coat to create a smooth, flat surface. After the walls have partially dried, wipe them down with a damp sponge or rag to remove surface imperfections. Large areas can be smoothed by rubbing a short section of 2 x 4 on the wall in a circular motion. Then fine polish by rubbing the wall down with a small circular piece of plastic. Believe it or not, the lid of a yogurt container works extremely well. This aligns the clay particles and creates a smooth, durable surface.

Earthen color coats can be pigmented by adding colored clays. After drying, earthen plaster can also be painted with a lime wash. Lime wash is made by mixing builder's lime (not agricultural lime) in water to create a solution with the consistency of skim milk. Apply with a brush. Two to three thin coats are generally sufficient. The result will be stunning.

Perhaps the best way to finish the exterior of a cob home is to apply a lime-based plaster, either directly on the cob or over two layers of earthen

BUILDING NOTE

Earthen plaster is more finely sifted and wetter than cob used to build walls.

Earthen plasters must be watched vigilantly and repaired immediately in moist climates to avoid further damage due to moisture penetration. This is especially important in cob and straw bale homes.

FIGURE 6-14.
This marvelous dolphin was carved into the wall of a cob bus stop at Sundial Community in Rico, Colorado. A host of other creatures and visages adorn the wall of this funky structure.

plaster, if you want a smooth wall. Always use builder's lime (Type S hydrated lime), not agricultural lime.

Lime plaster is "fun to work with, breathable, and easy to repair," according to Berlant. It adheres well to earthen plaster and is extraordinarily durable. In most instances, two coats will suffice, the second of which contains the color pigment. Over time, lime plaster converts to a harder substance, limestone or calcium carbonate for the chemists among us. This natural process increases its water resistance.

In dry climates, lime plaster walls must be allowed to cure slowly—a period of ten days—to prevent cracking. In England, where humidity is amply provided by neighboring oceans, you won't have to worry about the lime plaster drying too quickly. In arid climates, like the desert Southwest or the outback of Australia, however, the walls must be sprayed at least three times a day for ten days to achieve the proper rate of drying. Some plasterers hang burlap or plastic curtain over the walls to retard evaporation.

Details of lime plastering can be found in Steve Berlant's book, *The Natural Builder*, volume 3, *Earth and Mineral Plasters*. Although lime is fun and easy to work with, it can cause serious burns, so wear goggles, long-sleeved shirts, and long pants when working with it. When handling dry lime, wear a respirator. Read up on the process before you begin working with the material. It's quite an art!

As beautiful as they are, exterior earthen plasters must be watched over and repaired immediately when cracks are discovered to prevent moisture from entering. If the walls are not sealed and not protected by overhangs, they may need to be resurfaced every year or so, according to Moab, Utah, builder Doni Kiffmeyer, one of the leaders in earthbag construction (an alternative building technique discussed in chapter 10). Sealed walls are more durable and involve considerably less repair. However, you will have to reapply the sealer every few years.

In closing, while exterior walls can be coated with lime, interior walls are often plastered with an earthen plaster finish. Because these walls are subject to a much milder climate, earthen plaster is quite suitable and generally easier to apply. However, a lime wash may be desired to reduce the dust from walls. Natural paints can also be used to add color to the interior walls.

ROOFS

A wide assortment of roofs are possible on cob homes. Like all other forms of alternative homes, roof loads must be distributed by a top plate, except perhaps in very small structures where the roof load is negligible.

Anchoring a roof to a cob wall requires some special techniques not yet encountered in our exploration of natural building. For example, rafters are often secured to cob walls by deadmen, wood embedded in the cob wall about 18 inches from the top of the wall (figure 6-15). In such instances, rafters rest

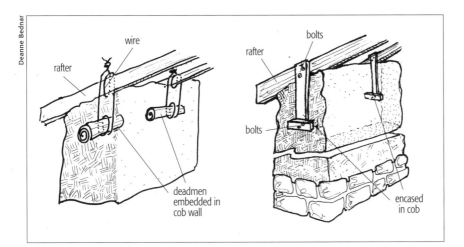

Deanne Bednar

wire

rafter

bolts

rafter

bolts

deadmen
embedded in
cob wall

encased
in cob

FIGURE 6-15.
*Rafters (shown here) and top
plates (not shown) are attached
to cob walls by using deadmen.*

directly on the cob wall. A top plate is not required, although as noted earlier
this is generally not advised except in smaller structures. Resting a rafter directly
on the cob wall could cause the way to give way. When top plates are used,
they're connected to the deadman—either by wire or long anchor bolts. Roof
trusses and rafters are attached to the top plate as in other types of construction.

Concrete bond beams provide the builder with another option. Poured
into notches carved in the top of the cob wall or made from concrete-filled
blocks, bond beams are generally so massive that they need not be anchored
into the underlying cob wall. The roof is secured to the bond beam via a top
plate attached to the bond beam by anchor bolts. Although rarely encountered
in cob structures, bond beams provide an additional measure of safety, especially in windy areas, and may be required by some building departments.
They also enhance the structural stability of a cob home and are vital in earthquake-prone regions.

Some builders have used all-thread (threaded rod, which is discussed in
chapter 3) to anchor top plates to cob walls. All-thread is first attached to the
foundation by anchor bolts, then buried in the cob walls as they are built.

Although the walls of many cob homes are load-bearing structures, that is,
they support the weight of the roof, some builders prefer to build non-load-
bearing walls using post-and-beam construction techniques described in the
straw bale chapter (chapter 3). Post-and-beam structures bear the roof load and
eliminate the need for top plates and bond beams. They also make it possible to
build a roof before beginning the cob work. In rainy climates, the roof provides
shelter for workers and protects the walls while the work is underway.

When building a post-and-beam cob structure, posts should be secured to
the cob wall using gringo blocks or deadman. Most folks find it advantageous
to align the wooden support beams with the inside surface of the wall (figure
6-16). This not only protects beams from weather, it adds to the aesthetics of
the interior walls. It also makes it possible to repair cracks that form as the cob
dries.

BUILDING NOTE

In post-and-beam cob structures, the
roof is secured to the beams, which
are secured to the posts, which are
secured to the foundation.

Ianto Evans

How long have you been building cob homes?

Ten years ago Linda and I built our first cob cottage, the first in North America in 150 years.

Worldwide, housing is in short supply. How could cob construction help alleviate this chronic and serious problem?

This is a rather complex question. I don't personally see cob as *the* solution to the world housing problem. I think a change in attitude is needed more than anything else. We need a shift in our thinking toward natural building, which will help us create more affordable housing here and abroad. We also need to deal with unbridled consumerism and the way we humans behave. What we at Cob Cottage are doing is chipping away at one corner of these concerns, knowing that other people are chipping away at others.

There is a specific application that I think is very promising. My colleagues and I have been talking with folks in the United Nations who want to provide the 20 million people currently living in refugee camps with useful skills for after they leave the camps. Unfortunately, host countries don't allow refugees to build houses, because they're afraid these displaced people will set up permanent settlements and never return to their homelands. The UN has suggested, however, that

cob might be suitable temporary housing for a number of reasons. In most places, the soil is suitable for cob construction. After the refugees leave, the roofs could be taken off the cob homes, and the walls would then return to the earth. And the refugees would have a skill that they could use in their home countries to make money and to rebuild villages.

What role do you see for cob construction in the coming years in the more developed countries?

The future of cob is tied up with industrial building cartels, the power of industry, and the profit motive. At the moment, cob isn't a threat to the building industry because it is seen as so marginal and ridiculous. If it gets to a point where it is a threat to those industries, I think that they will begin to put immense resources into preventing cob from happening. Ironically, that pattern has been commonplace in the history of industrialism and free enterprise.

One of the reasons I have opposed establishing building codes for cob is that the powers that be (building supply manufacturers like cement manufacturers) will see to it that the codes are written in a way that ensures the use of their products.

My personal strategy has been to get thousands of cob homes built and for these homes to be very obviously successful before anybody starts trying to pass

codes. Then, at least we're part of the way down the road. While a code can be thought of as legitimizing, it can actually be undermining. We have to be very careful.

What are the major changes in cob construction since you began working and experimenting with it?

The big advance we've been involved in is the custom mixing of cob. We have found that by modifying the mixture, we can custom create cob to perform different roles within any given structure. But it is not just making mix A for walls and mix B for floors. As you build walls, the structural strength and amount of insulation needed varies from foot to foot. If you start building with the thought in mind that the process is more like growing a tree than putting up a stick-frame building, then you are able to constantly modify the mix so it does everything it needs to do in every micro part of the building. It's the kind of thing that we're unable to do in conventional homes. One thing that carefully measured mixes enable us to do is to create sparingly constructed walls of incredible strength.

Will cob walls fall down in the rain?

Contrary to what many people fear, unstabilized cob walls are very resistant to weather. Cob walls can withstand long periods of rain and wind without weakening. Even so, cob homes should be equipped with generous roof overhangs (eaves) and good foundations. Plaster finishes add an additional measure of protection. [Author's note: For more details on weatherproofing a cob house, see Steve Berlant's book.]

What are the most common mistakes people make when building with cob?

I could write a book on this. In fact, I am. I think we've made most of the worst errors ourselves. Probably the biggest one is starting off with a building that is too large and can't be finished in a single season. Two additional mistakes are building in a bad location and not observing and understanding the ecology of your site. Yet another problem is rushing into it. Drainage and flooding can be a big issue. We helped on a cob house in Texas that was built in a floodplain. The owner would not be convinced of the dangers of building in a known flood zone, and his home washed away in a flood, along with many other conventional homes. If the house had been built ten feet higher up the slope, it would have been fine.

Poor solar orientation is also a major problem. Like building too close to neighbors who can block your access to the sun, say by planting a stand of Douglas fir that shade your home. Don't build on a west-facing slope or place big windows on the west to take in a view. In the summer, the west windows let in so much late afternoon sunlight that the house can overheat. Another common mistake is building a poor foundation. I always remind people that if your roof is inadequate, you can rebuild it. If your foundation is insubstantial, you have to take the house down to fix it.

Most of the problems have nothing to do with the cob itself. More than three-quarters of them have to do with poor siting, poor organization, and not spending enough time getting to know your bioregion, microclimate, and local ecology.

What does a cob cottage cost?

Cob is one of the cheapest building materials imaginable. Often, the soil removed during site work is enough to build the walls. The owner-builder can supply the labor, inviting friends to join in the excitement of hand-sculpting a house. With inventiveness and forethought, the costs of other components (doors, windows, roof, floors, etc.) can be extensively reduced.

Cob Cottage works primarily with recycled materials, hand-worked lumber, and local building materials such as poles, bamboo, stone, and cedar shakes. Beautiful, durable homes have been built for less than $5,000.

Are there any commercial builders who specialize in cob homes?

There are lots of commercial builders interested in cob. At the Cob Cottage Company we connect clients with commercial builders trained in cob building. Gradually, we're trying to focus on training builders. This year we will probably see four or five groups of commercial homebuilders begin building cob homes for clients in California, Texas, Washington, and British Columbia.

Does cob building require an expert?

Absolutely not! Once the basics are understood, cob building is amazingly simple. In a week-long workshop, you can learn how to select materials, prepare a mix, and build a wall. We have taught men, women, and children of all ages and abilities everything they needed to know to build: site selection, foundations, windows and doors, attachment of wood and other materials, detail work and finishing. Course graduates with no previous building experience leave feeling confident and enthusiastic about building their own cob cottage.

How can you learn more about cob? The only proven way to learn about cob building is to try it! The Cob Cottage Company offers hands-on cob construction workshops in most parts of North America. Contact us for our current schedule. We supply videos and slides of finished cob buildings.

IANTO EVANS is responsible for beginning the cob renaissance in North America and was the originator of the Natural Building Colloquium, an annual week-long national meeting of natural building practitioners. One of the world's experts on cob construction, Ianto is co-owner of the Cob Cottage Company, which offers training, workshops, and publications on cob and natural building.

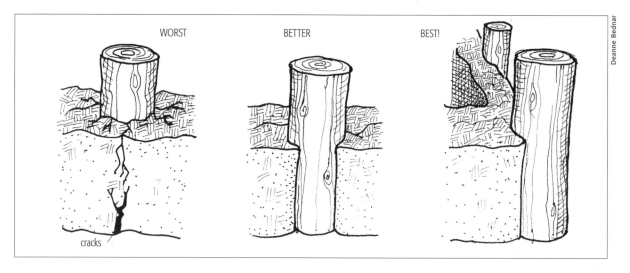

WORST   BETTER   BEST!

cracks

Deanne Bednar

Deanne Bednar

FIGURE 6-16.
Posts in post-and-beam cob structures can be placed in several locations. To avoid cracking, and to improve the aesthetics, placement along the inside of the wall is best.

FIGURE 6-17.
This cob home built by SunRay Kelly sports an organic roof made from fallen lumber.

Paul Lacinski

Because cob homes can be sculpted into a variety of forms, some builders prefer more organic roofs, meaning those with a free-flowing design, such as the one shown in figure 6-17. This innovative cob structure was designed and built by SunRay Kelly, who lives in the state of Washington. SunRay used peeled logs gathered from his hillside property to frame the walls and roof of this building. Taking advantage of natural features of the trees, he was able to create an architecturally unique roof design. SunRay produces some of the most intriguing, stunningly beautiful, and personable architecture of our times. If organic roofs appeal to you, this may be an approach you will want to explore.

Organic roofs require enormous creativity and special skills. Remember, a roof may be innovative and exciting, but it has to be structurally sound and waterproof, too. Otherwise, the artistry is for naught. Organic roof construction isn't a job for the uninitiated. As explained in chapter 4, special care must be taken when attaching the decking to log trusses. Planks are typically used.

Before you undertake a project of this nature, I'd suggest you start with a small structure, perhaps a shed or two to hone your skills. You may want to consult with people who have experience with organic roofs. Their assistance may be invaluable. Whatever type of roof you build, it is essential to build a wide eave or overhang and to install gutters to keep water off the cob walls.

One of the most common problems in cob structures results from poor roof design. The design shown in figure 6-18a may cause serious problems, for example, as the roof pushes out on the walls, the walls are forced to lean outward. Over time, they may give way. This problem can be solved by installing cross-braces, which transfer the weight of the roof vertically downward instead of outward at the top (figure 6-18b).

Ceiling insulation is essential to maintaining a comfortable interior temperature and to reducing fuel consumption. Ceiling insulation can be applied internally in an attic space or between rafters, or externally as rigid foam on top of the roof decking (see chapter 14).

While many types of roof and roofing materials can be used on cob homes, thatch roofs look particularly good. Thatch roofs are also fairly durable and reliable even in rainy climates. A well-built thatch roof can last up to 60 years. Thatch roofs, however, require special skills, proper materials, and a lot of time.

Thatch is still commonly used in Asia, Africa, and Latin America. In the United States however, it may be difficult to obtain approval from local building departments who fear they are a fire hazard. In Great Britain and other parts of Europe, thatch has a long and successful history, so building department officials may be more amenable to the idea. For a more detailed discussion of thatch roofs see *The Cobber's Companion, The Cob Builders Handbook,* and Barbara Jones's article on the subject in volume 26 of *The Last Straw*.

Sod or living roofs are also popular among cob builders. Details of the living roof are presented in chapter 3 and are discussed in *The Cobber's Companion* and *The Straw Bale House*. Read everything you can on alternative roofs, especially living roofs, before you commit to a particular roof type. The stakes are too high to enter into a project of this nature without thorough knowledge and skill.

Some builders in arid climates are experimenting with arched or domed roofs made from cob. This architectural feat is achieved by corbelling, a technique discussed earlier in this chapter. Corbelling takes great patience and skill. As shown in figure 6-19, each course in a corbelled arch or dome protrudes a couple inches beyond the course below it. The "base" of each corbel cob is anchored by cob. For optimum results, let each course dry a bit before the next one is applied. A rigid dome form can also be used, but the time required to build an arched structure out of plywood can be considerable. Cob domes and arches are best used in arid climates. But even in these regions, the cob should be waterproofed. If the cob becomes wet, it could collapse. (For this reason, some builders frown on cob roofs. Be very careful.) For more details on cob arches and domes, see *The Cobber's Companion*.

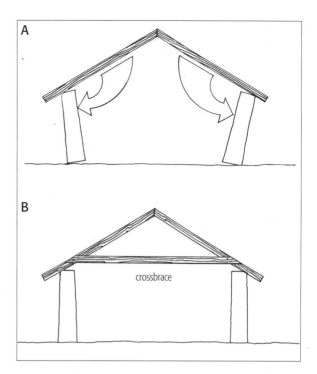

FIGURE 6-18 A, B.
Cross braces (bottom) transfer the weight of the roof downward instead of outward, preventing the walls of a cob home from pushing out and tumbling over (top).

FIGURE 6-19.
Corbel cobs can be used to build arched doorways and windows.

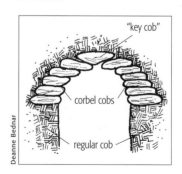

Cob is one of the cheapest building materials imaginable. The owner-builder can take great satisfaction in supplying the labor, building little by little in leisure time, or inviting friends to share the excitement of hand-sculpting a whole house.

IANTO EVANS

FLOORS

Floors are installed after the walls and roof have been finished. One of the most popular floors in cob homes is the earthen floor. The floor options discussed in previous chapters (soil-cement, brick on sand, concrete, and wood) are also viable choices. Remember that, like adobe, cob walls should start 2 to 6 inches above the finished floor to prevent the wall from getting wet if a pipe breaks, a toilet overflows, or a large aquarium springs a leak and floods the interior. This also protects the walls from outside standing water and splashback from rain dripping from the roof. Prolonged exposure to water at the base of the wall could cause it to weaken and collapse.

ELECTRICITY AND PLUMBING

Electrical wiring and plumbing are installed in cob homes in floors, between floor joists, and in interior partition walls and ceilings. If necessary, electrical wire can run in exterior cob walls, but it must be placed in conduit (or you must use wire rated for burial.) Conduit, in turn, is situated in grooves carved out of the walls—usually an inch or so beneath the surface. After the conduit is in place, a fresh layer of mud is applied to cover it up. Conduit can also be placed along the top of the foundation and mudded in with the first layer of cob.

Electrical boxes for switches and outlets are placed in interior walls or embedded directly in cob walls. Some builders attach electrical boxes to wooden plugs and insert the assembly into the walls where needed. The boxes are then cobbed in place. Other builders wait until the walls are finished, then carve out openings for boxes.

Plumbing, as in all other earthen homes, is run through slabs, between floor joists, in ceilings, and in interior partition walls. It should never be placed in cob walls, for the same reasons it is excluded from walls made of adobe, rammed earth, and straw bales. Small leaks or condensation on pipes can cause a cob wall to soften and weaken.

WHAT DOES IT COST TO BUILD A COB HOME?

Cob buildings are dirt cheap—if you provide the labor and use recycled materials. Many people who have built them brag about a cost of $10 per square foot. But as with any alternative form of homebuilding, the cost depends on many factors. The fancier the house, the more materials you have to buy, the more work you subcontract, the higher the cost.

Cob is ideally suited for the do-it-yourselfer. If you want to build inexpensively, live inexpensively, have some building skills or have skilled friends who will help you, and have modest desires, you can erect a structure at a fraction of the cost of a regular home. If you want to scrounge for materials, for example, used windows and doors, you can bring the cost down further. But don't forget to count your time. You can easily spend five to ten hours locating a high-quality used door, that you could pick up new at your local building supply store for $150. You may have to construct a door jamb as well, which could add a couple more hours of labor and will require a little more lumber. If you are earning $10 to $15 per hour at your regular job, and are staying home to work on the house, then you're really not saving much. But if you have plenty of time on your hands, building your own home is a kind of self-employment. You aren't paying yourself directly but you are avoiding having to pay someone else. (And as Ben Franklin reminds us, "A penny saved is a penny earned.") Building your own home isn't just about saving money. It is about fulfilling dreams and being self-sufficient, two values you can't assign a price to.

If you hire a contractor or subcontract a substantial portion of the labor, expect your cob home to cost more than low-end cob estimates. If you want a large structure with expensive tile, handcrafted cabinetry, and fancy lighting, expect the cost to be higher still. Please be careful, my friends. Altogether too many people are drawn to natural homebuilding by outlandish cost claims. Be realistic. Before you embark on a building project, make a list of what you will need, how much time it will take, subcontractors you will have to hire, and how much all of them will cost. Get reliable estimates. Then add up estimated costs and tag on another 20 to 50 percent. Although cost estimating sounds like a daunting task, some guidance is available. Chapter 16 outlines all of the steps involved in building a home and lists references that will help you make accurate cost estimates.

SHOULD YOU BUILD A COB HOME?

By now, you are aware that each form of natural building has its list of pros and cons. Let me hasten to add that disadvantages are not unique to alternative building. Conventional homebuilding has disadvantages, too. But because we are so familiar with conventional techniques, we forget the disadvantages, among them, their egregious depletion of the Earth's natural resources. So, don't damn alternative building for a few faults until you've compared it to conventional practices.

Advantages of Cob Homes

• Building with earth can be fun and relatively inexpensive, especially if you do the work yourself, scrounge for materials, build a simple structure, and have modest desires when it comes to finish materials.

AT THE HEART of this approach is a deep regard for nature and for the wisdom of traditional cultures. By working with natural materials close at hand . . . cob builders can provide their own housing at moderate cost, and their homes don't contribute to the cycles of deforestation, mining, and pollution driven by modern building practices.

ROBYN BUDD, book review of *The Cobber's Companion*

COB AS A MEDIUM is inherently artistic and allows the builder to creatively shape and sculpt spaces that feel good to live in and delight the senses. A builder, using cob, does not have to be an experienced tradesperson, since hands, shovels, and simple tools do most of the work.

STEVE BERLANT, *Monolithic Adobe Known as English Cob*

- Cob structures lend themselves to curved walls and other visually inviting features. It is easy and fun to fashion niches, furniture, and shelves out of cob.
- Cob construction encourages artistic expression. Walls can be designed as they go up, providing maximum creative freedom. A cob home can be as much art as it is a home.
- Curved, tapered walls appear to be highly resistant to earthquakes.
- Cob construction does not require forms as in rammed earth construction.
- Cob is applied directly to the wall, eliminating the need to make mud bricks, as in adobe construction.
- Cob homes are ideally suited for passive solar heating.
- In some areas, such as arid desert regions, cob stays warm in the winter and cool in the summer.
- Thick cob walls insulate the interior space from outside noise.
- Cob construction relies on locally available resources.
- Cob construction requires much less wood and fewer manufactured building materials than many other forms of building.
- Cob construction is relatively easy to learn and requires minimal building skills. Even kids can become skilled cobbers.
- Cob construction is labor intensive, but requires very little outside energy or power tools.
- Cob construction is forgiving. Mistakes can be easily rectified.
- Cob integrates well with other earth-friendly forms of architecture, including straw bale and adobe.
- Cob is easy to remodel, even when fully cured. For example, windows can be added and walls can be extended to create new living space.
- Cob can be used to fashion niches, furniture, and shelves.
- Cob homes are durable.
- Cob walls are fireproof.
- Building with cob poses little danger to workers on the site.
- Cob construction encourages community participation. Young and old, family and friends, can participate in meaningful ways.

Disadvantages of Cob Homes

- Getting approval for cob construction may be difficult in some areas. In the United States, only a few jurisdictions formally approve cob construction. Many cob builders therefore choose to avoid building departments. Although this is an option, it is a risky one. If you are caught, a building department can force you to tear down an unpermitted structure. Note: Some legally permitted cob buildings are now starting to appear. These may make it easier in the future for others to follow suit.

Steve Berlant's book *Monolithic Adobe Known as English Cob* contains a wealth of tips on engineering data about cob construction, and how to get a permit for a cob home.

- Cob construction, while relatively simple, is labor intensive. That can be a plus or a minus, depending on your personality and your attitudes about the value of hard work.
- Stone foundations, which are often used in cob construction, are also labor intensive. A small, 20-foot round building will require 8 tons of stone, which are usually collected by hand, stockpiled, and then set in place.

Cob is a remarkable housing alternative. To many people, its benefits clearly outweigh its negatives. But earth-friendly architecture is only half of the battle to living lightly on the land. To be sustainable, the home should incorporate passive solar design, solar electricity, recycled materials, rain catchment, graywater recycling, and other features discussed in Part 3.

7

CORDWOOD HOMES

JUST AS I WILL NEVER again view a truckload of used automobile tires headed for a landfill as a waste product or think of a bale of straw as neatly bundled animal bedding, I will never view a neatly stacked pile of wood awaiting its fiery demise in quite the same way again—that is, now that I've seen cordwood homes. Cordwood houses are low-cost, natural homes built from firewood laid widthwise in a wall one course at a time and then mortared in place with cement (figure 7-1). Popping up in Canada and the northeastern and midwestern United States at a surprising rate, cordwood homes in North America now number in the hundreds.

Cordwood construction has a long, mysterious history. Some of the oldest documented structures are found in Siberia and northern Greece and are estimated to be 1,000 years old. But where this building technology began is anyone's guess. Several experts on the subject think that cordwood construction may have had multiple origins.

In the United States, some of the first cordwood homes appeared in southern Wisconsin around 1850. The oldest known specimen, built by David Williams, a farmer originally from western New York, contained an estimated 20,000 fourteen-inch logs. Although the Williams house was torn down in 1950, a piece of the wall is preserved in the Webster House Historical Museum in Elkhorn, Wisconsin. Not too far from Williams's farm another cordwood structure, an old grain mill, was built around the same time. It was torn down about 60 years later.

FIGURE 7-1.
This lovely cordwood home built by Rob Roy in upstate New York is the home of Earthwood Building School. Notice the living roof and photovoltaic panels.

Cordwood construction has many names, among them, piled wood wall, stackwood wall, stackwall, and firewood wall.

In the late 1800s and early 1900s, numerous cordwood structures were built in Wisconsin's Door County Peninsula. Over forty houses, barns, and other structures are found in this "hotbed" of cordwood construction, according to Rob Roy, North America's preeminent expert on the subject and author of the *Complete Book of Cordwood Masonry Housebuilding.*

Cordwood construction was also popular in Canada's Ottawa Valley and along the St. Lawrence River in Quebec. Although many of the cordwood structures built in these regions are still standing, finding them can be difficult, because many owners have plastered over exterior walls or covered them with clapboard siding, either to halt deterioration or to hide what was largely considered to be—you guessed it—poor people's housing.

One of the best remaining examples of this unique architecture, now listed on the National Register of Historic Places, is a large cordwood structure in the hamlet of Jennings, Wisconsin. Located in the north central portion of the state, this structure served as a general store, saloon, bunkhouse, post office, and home to John Mecikalski and his family (figure 7-2).

Some historians speculate that cordwood construction may have arrived in the United States from Sweden. However, Jack Henstridge, cordwood builder and author of *Building the Cordwood Home,* thinks that cordwood construction may have been introduced to North America by the Vikings about 1,000 years ago. Henstridge bases this assertion on evidence of the remains of a round cordwood struc-

FIGURE 7-2.
The Mecikalski general store in Jennings, Wisconsin, was built from cordwood.

FIGURE 7-3.

In cordwood construction, insulation (I) is sandwiched between two mortar joints (M). Logs are embedded in the mortar.

BUILDING NOTE

Circular structures encompass the most floor space with the least building material.

One of the interesting things that happens to you when you step inside a dwelling that has curved instead of square corners is that it feels right, it feels comfortable; it feels secure, you feel part of it.

EDITH TAYLOR,
cordwood homeowner

---

ture built with a clay mortar in Newfoundland in an area where the Vikings settled. More likely, cordwood building "may have been spontaneously conceived by many different people at different times and at different places," says Roy. At this time, we can only speculate.

Cordwood homes began a modest revival in the 1970s thanks in part to a new set of pioneers, including Rob Roy, Jack Henstridge, Richard Flatau, and a host of unsung heroes who are now living in the fruits of their labor. Thanks to efforts of these pioneers, those interested in building a cordwood home now have access to numerous resources on the subject, including books, videos, and workshops.

A survey of these sources reveals three types of cordwood homes. The first type relies on post-and-beam construction, discussed in chapter 3. In this popular building technique, a post-and-beam framework is erected on a foundation. Cordwood is then stacked between the posts and mortared course by course.

The second architectural style is a load-bearing, curved wall structure. In this design, the cordwood is stacked and mortared *without* a wooden support structure. As in load-bearing straw bale homes, the roof load is supported entirely by the walls.

The third version is called stackwall construction. In this design, corners of logs or log ends are build first. Cordwood is then stacked between the built-up corners, which bear the load of the roof.

In all three types of construction, the length of the logs defines the thickness of the walls. Sixteen- to twenty-four-inch walls are commonplace. As shown in figure 7-3, the mortar in cordwood masonry is applied in two parallel sections (known as joints) with a cavity between them. This cavity, cradled by the inner and outer mortar joints, is filled with insulation.

FOUNDATIONS

Cordwood walls can be erected on virtually any foundation as long as it can support the load. Poured concrete foundations are a good choice from an architectural standpoint, but because they must be wide enough to support the cordwood masonry wall and because they must be dug deep into the ground to prevent frost heaving in cold climates, a concrete foundation can be a very expensive option. For these reasons, most people turn to other less costly and less resource-intensive alternatives. One option is the concrete block foundation.

Concrete block foundations are less expensive than poured concrete foundations and can easily be made wide enough to support the thick cordwood walls. For details on this foundation, consult Rob Roy's book, *Complete Book of Cordwood Masonry Housebuilding.* You may also want to check out other books on foundations.

Some innovative builders have used railroad ties or pressure-treated timbers, laid on a well-tamped layer of crushed stone, for foundations. Researchers at the University of Manitoba laid railroad ties on #2 stone in a shallow rock-filled excavation trench (figure 7-4). For more details, see *Stackwall: How to Build It* by A. Lansdown and K. Dick.

Although relatively inexpensive, railroad ties and pressure-treated lumber foundations are not as durable as stone or concrete foundations. Another concern is the toxic chemicals used to preserve the wood. Railroad ties are drenched in creosote, a toxic organic goo derived from crude oil. Pressure-treated lumber contains copper and arsenic. When used in raised-bed gardens, creosote and wood preservative may leach into surrounding soils. Some studies suggest that vegetables grown in contaminated soils could pose a risk to people. Although I doubt their use in foundations will have an impact on the occupants of the house, caution is still advisable.

Rubble trench foundations are another good option for cordwood homes. They're relatively easy to build, inexpensive, and very effective. Similar to the

FIGURE 7-4 A, B.
Details of railroad tie rubble trench foundations. Top view (A). Side view (B).

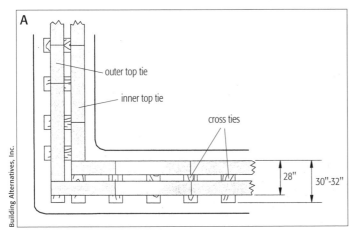

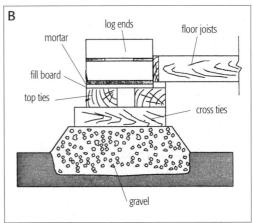

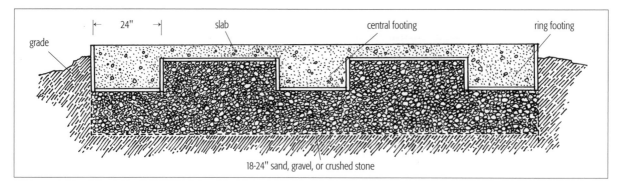

18-24" sand, gravel, or crushed stone

FIGURE 7-5.
The floating slab consists of a monolithic slab on a thick bed of sand, gravel, or crushed stone. The ring footing that extends around the perimeter of the structure is the stem wall and footing.

BUILDING NOTE

If sand is used in a floating slab, it should be laid down in six-inch layers, then compacted using a tamper. Application of thicker layers will dramatically reduce compaction.

---

rubble trench is the "floating slab." A favorite of Frank Lloyd Wright and many cordwood homebuilders, the floating slab foundation is ideally suited for cold climates and round cordwood houses. This unusual foundation consists of a concrete slab poured on a pad of coarse sand, gravel, or crushed stone—18 to 24 inches thick (figure 7-5). Like the rubble trench foundation, the porous base helps to keep moisture away from the house. Without water, there is no freezing. No freezing means no heaving. Without heaving, the foundation should not crack.

When building a floating slab foundation, surface vegetation and topsoil are first stripped from the site three feet beyond the perimeter of the house. Topsoil is stockpiled for later use—say for a garden or for a living roof. Next, coarse sand, crushed stone, or coarse gravel is laid down and compacted, about six inches at a time for best results. Porous drain pipe is installed in the base, and plywood forms are erected to form a ring footing, approximately 24 inches wide, depending on the width of the cordwood wall, and about 9 inches deep. The footing extends around the circumference of the building and is reinforced with rebar. After the forms are in place, electrical conduit, plumbing, foam insulation, and wire mesh (for floor reinforcement) are installed. Then concrete is poured.

Wooden floors can be laid on the slab. But to do so, you must secure furring strips to the cured concrete. This provides a means of securing wood flooring and also elevates the flooring to reduce moisture absorption. Another (I think better) option is to install joists and lay a wood floor on them. If you select this option, you won't need to pour a slab. You will only need to form and pour a ring footing and a central footing to support the weight of the floor joists that radiate from the center of the structure to the periphery like the spokes of a wagon wheel. (For more information on the floating slab foundation, see Rob Roy's book, *Complete Book of Cordwood Masonry Housebuilding.*)

BUILDING A CORDWOOD WALL

After a foundation has set, it is time to build your walls. For a post-and-beam structure, you must first erect the wooden frame. Many homebuilders use large timbers, which, though they provide a sturdy skeleton, come from

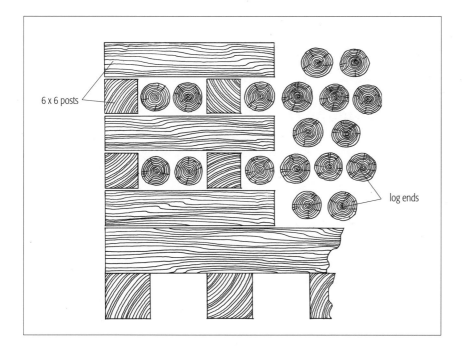

6 x 6 posts

log ends

FIGURE 7-6.
Stackwall corners can be built using posts or hewn timbers or a combination of round and split wood. The corners bear the roof load.

mature trees, often old-growth trees. To minimize your demand for lumber and reduce the destruction of our old-growth forests, consider using recycled or salvaged timbers (from old barns, for example) or some of the engineered wood products discussed in chapter 14. Although not perfect, they require less wood and are made from smaller-diameter trees. Another option worth considering is the use of wood panels akin to those built by progressive straw bale builders (see chapter 3).

One of the advantages of the post-and-beam technique is that after the frame is erected the roof can be built, creating a sheltered working space in which to complete the next phase of the project. In rainy climates, roof protection can be invaluable.

After the frame is constructed, walls are built between your posts. The technique for building walls is discussed shortly.

In the second method, stackwall construction, the walls are begun in the corners. Figure 7-6 shows a corner made from wooden 6 x 6 or 8 x 8 posts or timbers. Corners can also be built out of round wood, stone, and cement blocks, as shown in figure 7-6. After the corners are completed, the intervening walls are constructed.

Wall sections are built one course at a time, as explained shortly, with special care to be sure they always remain plumb (vertical). A string run from one corner to the next is used to ensure plumb. Those interested in learning more should read *Stackwall Construction: Double Wall Technique* by Cliff Shockey, and Lansdown and Dick's *Stackwall: How to Build It.*

The third type of cordwood construction, the round wall, is a load-bearing structure. Walls are built directly on the foundation without support as in

BUILDING NOTE

Post-and-beam construction is well accepted by building departments and provides an additional measure of security to occupants, for a post-and-beam frame can support a huge roof load and also serves as a solid anchor for the roof, so it won't be blowing away in the wind.

post-and-beam or stackwall corners. This part of the process is similar in all three types of cordwood home and will be discussed below. For details, read the books listed in the reference section, view the videos, and attend a workshop or two where you will get hands-on experience.

Starting with the Wood

Before we study wall construction, let's consider the wood that will be used to build walls. Cordwood is gathered and dried well before the foundation work begins. This phase of the project starts a few months to a full year before construction begins. When asked what kind of wood should be used for cordwood homes, Jack Henstridge responds, "Use what you've got. Almost any wood can be used, as long as it is sound."

Most builders work with wood newly cut from the forest. Venturing into nearby woods (with permission from the landowners if the property isn't yours, of course), select and fell a number of trees. Trim the branches off and cut the logs into manageable sections. Then strip the bark from the logs. This facilitates drying and also roots out insects hiding beneath the bark, where they feed on nutrient-rich tissues of the trunk.

Removing the bark from logs is tedious but you can reduce the amount of work by cutting trees in the spring. In spring, the sap rises—ascending from the roots where it was stored over the winter to the trunk and branches to nourish new growth. Rising sap loosens the bark from the underlying woody tissues. One cordwood builder found that soaking the logs in a stock tank for a week loosens the bark and makes the job a lot easier. Ken Campbell, who built a cordwood home near my house, says that "even under the best of conditions, barking a log is hard work."

Drying times vary considerably, depending on the climate, the season, and the weather. In summer, drying may require a couple months. If trees are cut in the fall, however, expect drying to take six to seven months. If you live in a moist climate and the drying season is wet, or if you happen to have an exceptionally wet year, it may take even longer.

Although most tree species are suitable for cordwood, some species are more resistant to rotting than others. Check out the *Complete Book on Cordwood Masonry Housebuilding* for a comprehensive list. If you've got a choice, use the most decay-resistant species available. If you don't, use what you have and don't sweat it.

No matter what wood you use, it is important to select trees that show no signs of rotting. Be on the lookout for heart rot fungus. This organism causes the wood in the center of the trunk to decay and renders the wood useless. It is found in standing trees, both dead and alive.

After trees are cut and set out to dry, be sure to protect them from moisture. Raise logs off the ground (for example, by placing them on pallets) and cover them with a tarp or makeshift roof. Never lay them directly on the ground to dry.

Another factor to consider is the degree of contraction and expansion various species undergo. Wood is like a sponge, absorbing moisture during rainy, wet periods, and giving it off during dry periods. The absorption and release of moisture in response to environmental conditions causes wood to expand and contract. However, some woods are more absorbent than others. The best species are least absorbent. They are subject to the least amount of expansion and contraction (shrinkage).

One of the problems with the expansion and contraction of logs is that it causes wood to crack. Cracks along the length of a log, known as checks, provide an avenue for air movement. If they're not sealed, your cordwood house will be drafty. A more troublesome problem arises from radial shrinkage, which occurs when the radius of a log decreases as it loses moisture. This phenomenon causes logs in a wall, known as log-ends, to pull away from the mortar, creating cracks around the perimeter of each log.

Radial expansion is another, more serious problem. As a log in a cordwood masonry wall expands, it compresses the mortar around it. This causes the mortar to crack, which weakens the joints. Rob Roy tells of having to dismantle entire walls whose mortar had crumbled to bits because of radial expansion.

Cordwood homebuilders have devised a number of strategies to prevent these problems—or treat them. One of the most important preventive measures is proper drying. In other words, make sure your logs are properly dried before using them to build walls. Proper drying brings out the cracks *before* log-ends are placed in a wall. Large longitudinal cracks can be stuffed with strips of newspaper to eliminate drafts. After the wall is completed, the checks can also be sealed with caulk.

Proper drying permits radial shrinkage before wall construction commences and thus reduces the size of the cracks around each log-end. Expect to caulk the cracks around the log-ends after a couple years. Silicone caulks work best, although newer products may be even better. One of them is Perma Chink, an acrylic latex co-polymer. Although it is expensive, it seems to last a long time. Log Jam, made by Sascho Industries, is another good choice.

Radial expansion of logs, although relatively rare, can be avoided by not overdrying dense hardwoods. If hardwood logs are too dry when they are placed in a wall, they will absorb moisture from the atmosphere and from rain. This moisture will cause the log-ends to swell and crack the mortar. Cracks in the mortar compromise the structural integrity of the wall and could cause it to collapse.

The ideal wood for cordwood construction, according to Rob Roy, is northern white cedar, which is a relatively porous softwood. It has good insu-

BUILDING NOTE

A year or two after you finish your cordwood home, expect to fill in the spaces that form around the logs. This is a time-consuming process.

lating properties, is relatively easy to work with, looks terrific, and weathers well. It is also resistant to rotting, and once dried undergoes little expansion or shrinkage.

Round Logs or Split Logs?

Cordwood homes are built out of round or split logs. Although round logs lay up very nicely, most people prefer split wood. It dries faster and allows for more uniform mortar joints (figure 7-7). Uniform mortar joints result in a stronger wall. (Logs should be separated from one another by at least ¾ inch of mortar.) Uniform joints are easy to point, the process of compacting and smoothing out the mortar between the ends of the log-ends after a wall is completed. Splitting logs also reduces or eliminates large cracks that develop in large "rounds." Although wood splitting adds many hours of hard labor to your project, it does reduce the time spent stuffing and caulking cracks.

Cordwood homes require a lot of wood. According to Rob Roy, a round cordwood home measuring 38 feet in diameter with 8-foot-high, 16-inch-thick walls requires 8 cords of wood. (A cord is a stack 4 feet wide by 4 feet high by 8 feet long.) A 24-inch-thick wall requires 12 cords of wood. These are the thicknesses recommended for colder climates. Although windows and doors reduce the amount of wood you will need, plan on cutting a lot of wood for your home.

FIGURE 7-7.
Cordwood homes can be built with round, peeled log ends or split logs. Some builders use a combination of both.

Mortaring Log-ends

The "glue" that holds the cordwood wall together and provides structural support is cement mortar. The mortar used in cordwood homes is a special mix of fine-grained sand, cement, hydrated lime, and sawdust. Although the addition of sawdust to cement may cause experienced bricklayers to cringe, it is effective in reducing the amount of cracking. The sawdust must be thoroughly wetted, usually soaked in water overnight, then added to the lime, cement, and sand, and mixed with additional water in a wheelbarrow or cement mixer.

Cement mortar is applied by hand to a clean foundation. Cement handled by bare hands dries skin unmercifully, causing cracking and considerable pain. A good pair of rubber mason's gloves will protect your hands from damage. Your loved ones will appreciate the precaution, and you will too.

As shown in figure 7-8b, mortar is placed along the foundation in two parallel courses separated by a small cavity into which insulation will be poured. The most widely used form of insulation is dry or slightly dampened sawdust, treated with lime, usually two shovelfuls of lime per wheelbarrow load, or about 1 part lime to 12 parts sawdust. If you don't have access to sawdust, other forms of loosefill insulation can be used, including vermiculite, Per-

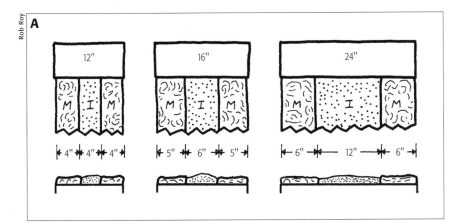

FIGURE 7-8 A, B, C.
The top figure illustrates the relative thickness of the mortar joints and insulation space in walls ranging from 12 to 24 inches wide. The bottom two figures show details of cordwood construction.

lite, treated cellulose, wool, and fiberglass. From an environmental standpoint, sawdust is probably the best insulation, and it is the easiest to work with.

After the mortar is laid, log-ends are pressed across the mortar joints using a slight vibrating motion to promote adhesion. As Rob Roy notes, "The superior thermal characteristics of the wall are to be found in the mortared portion, so don't skimp on the mud [mortar]."

Sawdust is poured into the cavity and then a second course is laid. When this is finished, a third course is laid, and so on until the wall is completed.

Cordwood walls can be laid up randomly or in interesting patterns. Although random stacking is preferred, some builders express a love for patterns. This approach, however, is not recommended for beginners. It is time consuming and produces embarrassing results if you make a mistake.

As in most forms of construction, care must be taken to ensure that walls are plumb. A four-foot level is very useful. After a wall has been completed, smooth out the cement mortar. This process, called pointing, increases the attractiveness of a home, strengthens the walls by tightening the joints, and makes the mortar joint more water repellent. Pointing is accomplished with a flat knife, for example, a kitchen knife with its tip bent up. The knife is drawn across the mortar while the concrete is slightly wet and therefore still "plastic." Sufficient pressure is applied to smooth out the surface, but not too much so as to push the mortar into the insulation space.

BUILDING NOTE

At no point should the inner and outer mortar joints meet. This would eliminate the insulation space and result in a loss of heat, an "energy nosebleed," to quote Rob Roy.

BUILDING NOTE

Round and split logs can be used in a cordwood wall to create a more interesting pattern.

FIGURE 7-9.
*Log ends can protrude from
the wall creating small shelves
(as shown here). Pegs can also
be driven into log ends to
support "normal" shelving.*

Pointing a wall offers an opportunity to let your
creative impulses run wild. Some builders etch patterns in the mortar. Others carve faces of people and
animals. Although this may sound tacky, I've seen
some very attractive art in the cement mortar of cordwood homes. Using different length logs provides
additional room for artistic expression. A slightly
longer log-end protruding into the interior of a home
may interrupt an otherwise visually monotonous wall
(figure 7-9). Two logs spaced a few feet apart serve as
a support for a bookshelf. If your creativity is still
straining for expression, you may want to place colored glass bottles in the wall. During the day, sunlight
striking the exterior walls causes the bottles to glow; at night, interior lighting
will cause the wall to glow like a jewel in the sun.

Mass and Insulation All in One

Cordwood walls provide thermal mass and insulation. Although the log-ends
have some mass, the bulk of the thermal mass in a wall is provided by the mortar. Insulation is supplied primarily by the sawdust in the mortar joints. But
once again, the log-ends also add to the total insulation value of the walls.
How much insulation can you get out of a cordwood wall?

The R-value of a wall made from cordwood varies, depending on its thickness, the type and amount of insulation, and the type of wood. R-values for
16-inch walls range from R-16 to R-20, the lower value for walls built with
log-ends made of dense wood such as oak and the higher value for softer
woods such as cedar. These figures are comparable to 2 x 6 stud walls with
either fiberglass (low end of range) or cellulose insulation (high end).

Although the R-value of cordwood walls is acceptable, it pales in comparison to straw bale walls. But the comfort of a home is not determined entirely
by R-values. As you may recall from chapter 5, thermal mass also plays an
important role in determining interior comfort levels. In a cordwood home
heated by a woodstove, for example, heat from the stove warms the indoor air.
Heat is then transferred to the walls, where it accumulates in the inner mortar
joints and to a lesser degree in the log-ends. Although some heat passes
through the log-ends and through the insulation, escaping to the outside
through the outer mortar joint, a great deal of heat is retained in the interior
mass. When the woodstove burns out during the wee hours of the night, the
heat radiates back into the room.

In the summer months, the situation is reversed. The main heat source, the
sun, warms the exterior walls. Although some heat migrates into the interior
of the structure, much of the heat is trapped in the outer mortar joints and the

BUILDING NOTE

If you want to maximize the R-value
of a cordwood wall, use the lightest,
airiest woods you can find.

outer portion of the log-ends. The result is a cool interior, despite oppressive outside heat. At night, energy stored in the outer mass of the shell is radiated into the atmosphere, preparing the house for another hot day.

Like many other forms of earth-friendly construction, the cordwood home stays warm in winter and cool in summer. Adequate ceiling insulation and passive solar orientation further add to the comfort levels and reduce utility bills.

Will Cordwood Walls Deteriorate over Time?

One of the biggest concerns people have about cordwood homes is that the logs will eventually rot, causing the walls to collapse. Rotting, caused by bacteria and other microorganisms, occurs very rapidly when wood is kept damp for long periods. Cordwood homebuilders employ several measures to avoid this problem. One of them is to select decay-resistant tree species. They also avoid trees that show any signs of decay and protect logs from rain and snow when they are drying. Good overhangs will protect walls. An adequate foundation is essential, too.

After the walls are built, there's not much one can do to prevent moisture from contacting the logs. Log-ends in a cordwood home absorb water from driving rains or atmospheric humidity. However, these sources of moisture are generally not a problem as long as the wood can periodically dry out. Just as a split rail fence endures inclement weather year after year without rotting, log-ends show little sign of deterioration despite heavy rains and high levels of humidity if they dry out from time to time. In most places they don't stay wet long enough to permit bacterial decay.

Like so many alternative forms of construction, the proof is in the pudding; many cordwood buildings in Canada, Sweden, and Wisconsin are over one hundred years old.

You can also treat log-ends with sealants to retard water penetration, but be prepared to treat your wood every couple of years. Sealants will darken the wood, too, which you may or may not like. Be careful what you use. Fumes will penetrate the log-ends and enter your home. Linseed oil is a nontoxic sealant used by some folks. Unfortunately, it is often mixed with turpentine, which is a natural product but really smells. Check out the resource guide for some of the more environmentally friendly sealants offered by companies such as Livos.

Windows and Doors

Windows are placed in boxes or window bucks similar to those used for houses of rammed earth, straw bale, adobe, and cob. The boxes are typically made out of 2 x 8s. A piece of lumber is tacked to the frame to create a key that will lock the window box in the wall (figure 7-10). The boxes are set on a level bed of mortar at the desired locations. After the rough frame is in place and leveled, wall construction continues around it.

BUILDING NOTE

Rob Roy advises against the use of sealants or wood preservatives. "Sealants reduce the breathability of the wood," he remarks, "and preservatives can contaminate the interior space through off-gassing."

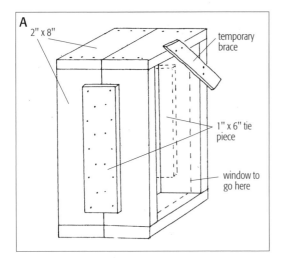

FIGURE 7-10 A, B.
Details of window buck construction for use in a cordwood home (A). The photograph (B) shows a cordwood wall under construction. Notice the symmetrical pattern created by large and small round logs.

Door frames are built the same way. However, if you are planning on hanging a heavy wooden door, use large posts to create a more durable frame. (One of the best descriptions for window and door frame construction and sill placement for windows is found in Lansdown and Dick's book, *Stackwall: How to Build It*. This book also contains details on positioning door and window frames.)

To prevent the window and door frames from buckling under the weight of the wall and roof above them, install lintels as discussed in chapters 2, 3, 5, and 6. As you may recall, lintels bridge the space above doors and windows, carrying the load from above and distributing it to the walls, rather than the window or door frames. Lintels come in many varieties, but the type preferred by many cordwood homebuilders consists of two parallel 6 x 6 or 4 x 8 posts. Lintels should extend 6 to 8 inches into masonry on either side of the door or window.

Wooden lintels not only perform a valuable function, but also add to the aesthetics of the structure. For more details, look at Rob Roy's *Complete Book of Cordwood Masonry Housebuilding* and Lansdown and Dick's *Stackwall: How to Build It*.

ROOFS

Cordwood construction lends itself to a variety of roof structures, including living roofs (see chapter 3). Post-and-beam structures offer the widest range of roof options. Roofs are attached to the beams, which are supported by the vertical posts. In load-bearing walls, roofs are attached to the wall via top plates. Failure to install a top plate is asking for trouble, except in cases of very small buildings, such as sheds and saunas. In larger structures, any concentrated load bearing down on one point in a wall can cause a cordwood wall to buckle, even collapse.

Many different types of top plate can be installed, the most common being two parallel 2 x 6s or 2 x 8s that run along the exterior and interior of the wall, as shown in figure 7-11. (See chapters 2 and 3 for further discussion of top plates.) The top plate for cordwall buildings can be attached to the underlying wall with 5-inch spikes spaced approximately 2 feet apart and anchored in the mortar. The tips of the spikes should be hooked to help hold them in place. Details of top plate construction, alternative ways of attaching top plates to cordwood walls, and roof options are covered in various books listed in the resource guide.

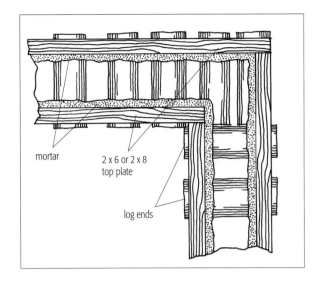

FIGURE 7-11.
The top plate spreads the roof load over the wall and prevents buckling and collapse of the wall.

FLOORS

The simplest floor for a cordwood home is an earthen floor. Soil-cement floors are a good option, as are brick or flagstone-on-sand. Or, you can pour a concrete slab, then stamp it to create realistic patterns that resemble tile or even flagstone. (The effect is impressive!) Concrete can also be tinted or colorized. Tile and carpeting are frequently laid over poured concrete floors. Be sure to check out the high-quality tile, carpet pads, and carpeting made from recycled materials. They're easy to obtain and cost competitive with conventional products. They are laid directly over a concrete floor. Or, you can lay a wooden floor. This is done by securing wood subflooring on floor joists, or on furring strips nailed to a concrete slab (see chapter 5). A number of environmentally sustainable wood flooring options are discussed in chapter 14.

ELECTRICITY AND PLUMBING

Electricity and plumbing pose no special problems in cordwood construction. Electrical wires are often run in interior partition walls, in the ceiling cavities, or in the floor—either beneath the slab or between floor joists. Although electrical lines can be run in exterior walls in conduit, this approach is pretty challenging. The irregular pattern of log-ends makes it difficult to run a straight line from ceiling to outlets and switches. Another possibility is surface-mounted raceways, moldings applied to the wall, usually at the floor level, to convey electrical wiring. (For more on wiring, see the article by Paul Mikalauskas in the 1999 proceedings of the Continental Cordwood Conference, edited by Rob Roy, which is listed in the resource guide.)

Plumbing runs along with electrical wiring in interior partition walls, in ceiling cavities, and in the floor. It is very difficult to run rigid pipe through cordwood walls.

Rob Roy

How long have you been involved in cordwood construction? How many homes and other structures have you built or helped build?

My wife Jaki and I have been building cordwood houses for 24 years, beginning with Log End Cottage in 1975. We have built four cordwood homes for ourselves, as well as numerous other saunas, sheds, other outbuildings, and guest houses at Earthwood Building School. We've worked on homes and other cordwood buildings for other people in Chile, British Columbia, and all over the U.S.

What sparked your interest in this type of building?

We needed a low-cost, easy-to-build system when we came to upstate New York from Scotland in 1975. We had tried horizontal log construction in Arkansas a year earlier and found it to be heavy going. Our first exposure to cordwood masonry was in the April 1974 *National Geographic*. We saw instantly that it was something we could do, and that it was beautiful. During a Sunday drive up to Ontario, we came across a farmer actually building a cordwood wall! What a break. We were on our way.

What is it like living in a cordwood home?

Log End Cottage was charming, but took a lot of firewood to heat up here near the Canadian border. The cordwood walls were only 8 inches thick, not nearly thick enough for this climate. (Okay for a sauna, though!) Earthwood, the round two-story home we have lived in since 1982, has 16-inch-thick white cedar cordwood walls. We heat the place, just over 2,000 square feet, with 3.25 cords of firewood per year (average for 10 years).

Aesthetically, we love the cordwood walls. They combine the warmth of wood with the pleasing textural relief of a good stone masonry wall. I can't imagine living in a home without at least some of the walls in cordwood masonry. By the way, round houses also have a great natural feel of coziness to them, sort of like a womb, I guess. That feeling is particularly pronounced in the 10-foot diameter cordwood sauna, just outside the lower story door.

What are some of the most common mistakes people make in cordwood masonry construction? And how can they be avoided?

There are lots of potential pitfalls in cordwood, and the best way to avoid them is to research thoroughly and, perhaps, attend a cordwood workshop somewhere with somebody who knows their stuff. Some of the common mistakes are:

a) not taking the bark off the wood (bugs love the space between the bark and the epidermal layers)

b) not seasoning softwoods long enough (the wood can shrink)

c) seasoning hardwoods too long (the wood can expand and actually break the wall up, potentially a more serious problem than shrinkage)

d) not spacing the log-ends correctly, that is, creating mortar joints that are too large or too small

e) not taking the time to point the mortar properly (there are lots of good reasons to point the mortar, including aesthetics, greater wall strength, waterproofness, ease of repair in case of shrinkage, and more)

f) not putting insulation in the inner mortar cavity (a "dead air" space just doesn't do it!)

g) not addressing mortar shrinkage with an appropriate mortar admixture

h) mortar that's too soupy (it shrinks a lot)

These are some, but not all, of the potential problems. But all can be overcome if care is taken. The smart man learns from his mistakes, but the wise man learns from the mistakes of others. Ancient proverb, that's why it's sexist. But it goes for women, too.

Are there any special hazards like fire or insect damage that people should know about?

To prevent insect problems, bark the wood and don't use any wood that already has active insect infestation. In termite-infested areas, incorporate a termite shield in the masonry foundation, and walk around the house once in a while. Termites don't particularly like boring into wood on end grain. Insects have not been a problem for us at all.

As for fire, it is very difficult to get a cordwood wall to burn. Jack Henstridge in New Brunswick has put a blowtorch to a 16-inch cordwood wall for an hour with no change in temperature on the other side of the wall. The mortar takes the heat out of the fire, and there are no direct air conduits inside the wall for oxygen to rush around in, as with a framed wall. Cordwood walls are probably the most fireproof of all wooden wall systems.

How difficult is it to obtain a permit to build a cordwood home?

I only know of one person, a lady in Utah, who has been flat out refused when she applied for a permit to build a cordwood home. The local code enforcement officer wasn't the most cooperative guy around. In other cases, I know of people who have had to jump through a few hoops, but eventually the project went through. Again, you've got to do your homework, as with any alternative building system, and allow plenty of lead time.

What are the major changes that have occurred in cordwood masonry since you built your first home? In other words, what are the major innovations in this building technology?

The addition of soaked sawdust to the mortar has all but eliminated mortar shrinkage, by retarding the set of the mortar, something we learned on the very last panel at Log End Cottage. However, the sawdust has to be absorbent. Some people have used hardwood sawdust like oak, with poor results. In experiments conducted during the summer of 1999 at Earthwood, we found that by introducing just three ounces of liquid cement retarder—we used Daratard-17 by W. R. Grace and Company—we were able to retard the set of the mortar just as well as with the wet softwood sawdust. Other cement retarders have been used with success in Sweden and Chile. However, don't mix the retarder and the sawdust in the same batch. One couple ended up with crumbly mortar that way. Remember, the slower the set, the less mortar shrinkage you will have.

Another exciting development in the past five years is that the round-house people are using a 16-sided post-and-beam frame. This enables the roof to go on before the cordwood work commences, so that the work can take place safe from both rain and sun. "Lomax corners," prebuilt corner quoin units for use with the stackwall system of cordwood masonry are also exciting.

We had the first Continental Cordwood Conference in 1994 here at Earthwood. The second was held in August of 1999 in Cambridge, New York. All the latest developments are shared at these conferences. The collected papers are available through Earthwood Building School.

What advice would you give someone who is thinking about building a cordwood home?

Research. Read books. Look at videos. Visit people who have built successful cordwood homes. Try to get to a workshop. Don't repeat other people's mistakes. Make friends with the code enforcement officer. Don't rush them. Oh, and allow plenty of time for your project. It will take twice as long as you expect. The best piece of advice, I think, is to build a small building, such as a sauna, and learn from that. If you can't build the small one, don't start the big project!

ROB ROY is the author of *Complete Book of Cordwood Masonry Housebuilding* (Sterling, 1994), *The Sauna* (Chelsea Green, 1996), which tells how to build a cordwood sauna, and *Stone Circles: A Modern Builder's Guide to the Megalithic Revival* (Chelsea Green, 1999). He is director of the Earthwood Building School, 366 Murtagh Hill Road, West Chazy, NY 12992, established in 1980 with his wife Jaki. Tel: 518-493-7744. You can visit the Earthwood Web site at: http://www.interlog.com/~ewood.

SHOULD YOU BUILD A CORDWOOD HOME?

Cordwood construction offers advantages and disadvantages that you should consider before embarking on this path.

Advantages of Cordwood Homes

- Perhaps one of the greatest advantages of cordwood construction is its potential cost. According to Rob Roy, the do-it-yourselfer can build a cordwood home dirt cheap, provided he or she obtains wood on site and does all of the work—or finds willing friends to donate their time. Estimates as low as $10 per square foot make this a perfect technology for those with little money, and a lot of time, and a free wood supply. According to this estimate, a 2,000-square-foot home would cost $20,000. This does not include the cost of your land, water, septic system, and permits. And, you won't be living in the lap of luxury, either. A heating system for a 2,000-square-foot home costs $5,000 to $15,000. Good-quality windows cost $5,000 to $10,000. To buy tiles and have them installed could run you another $8,000. Run a careful cost estimate before you commit, and read Rob Roy's book *Mortgage-Free!* for a lots of cost-saving ideas.
- Cordwood homes are relatively energy efficient. The walls provide mass and insulation, and cordwood homes tend to be warm in the winter and cool in the summer.
- Like other alternative building technologies, cordwood homes are easy to construct. Individual log-ends are lightweight and mortaring is pretty simple to learn.
- Advocates of this building technology love its aesthetics. Rob Roy writes, "A cordwood wall combines the warmth of wood with the pleasing relief and interest of stone masonry." For those who like a more rustic-looking home, cordwood construction is a good choice. The cordwood homes I've visited look a lot nicer in "person" than in picture. From a distance, they look like stone houses.
- Cordwood homes use wood, which is often locally available and a renewable resource, making this building technology a logical choice in many parts of the world. It's not a viable technique for people living in prairies or deserts, however.
- Cordwood homes can be built with high-quality waste wood from discarded wooden fencing, waste from sawmills or log home builders, and peeler cores from plywood companies.

Disadvantages of Cordwood Homes

- Cordwood construction is extremely labor intensive. Numerous trees must be felled, cut into logs, "barked," split, and dried prior to use. Laying up walls is also labor intensive.

- Cordwood homes require enormous amounts of wood, especially post-and-beam structures.
- Cordwood homes require an enormous amount of cement, which has slightly higher embodied energy than wood and earthen building materials.
- Enlarging or modifying a cordwood home requires special skills and training (but then so do all other forms of building). It is difficult, for example, to knock out a new doorway unless you have carefully planned in advance.
- Although wall building doesn't require advanced carpentry skills, completing a cordwood home does; if you have to sub out work, it will add substantially to your costs.
- Cordwood homes require a fair amount of maintenance—you usually have to seal expansion cracks a year or two after you get in.

8

LOG HOMES

LOG HOMES ARE ONE of the most popular natural buildings. Each year approximately 70,000 log homes are constructed in the United States alone. The popularity of log homes can be traced to many factors, perhaps most important being their aesthetics. Few structures are as visually appealing as a quality-crafted log home. Interior spaces are warm and inviting, especially when bathed in light from a blazing fire burning in a stone fireplace. Log homes possess a rustic, back-to-the-earth look that appeals to many modern warriors who have grown jaded by the rectilinear world of modern cities and suburbs (figure 8-1). What better way is there to break out of the mold? Another factor accounting for the popularity of log homes is their "accessibility" in the marketplace.

Each year, thousands of North Americans build their dream homes out of log. Relatively few of these homes are built by individuals from scratch with chain saw and axe, however. The vast majority of owner-built homes are made from log home kits. Reasonably priced and relatively easy to assemble, a log home kit makes the dream affordable and achievable. For those who want a log home but don't want to mess with assembling a kit, there are hundreds of builders who specialize in log building.

Modern log homes vary dramatically, from modest one- or two-room log cabins to sprawling spacious homes owned by doctors, lawyers, and retired software engineers. However, as appealing as log homes are, they rarely make a sustainable contribution to the housing stock. Gary Dunkin from the Smart Shelter Network in Montrose, Colorado, pointed out quite emphatically at a workshop on natural building in 1999 that the cedar log home kit comes with a huge environmental price tag—a one-acre clear-cut somewhere in Canada. In his view, kit homes and contractor-built homes made from imported logs are an environmental travesty and totally indefensible. Many people agree. The only environmentally sound log home is one built from local woodlots and forests, and then, only if tree stands are being sustainably managed and harvested. Creating a truly sustainable home, whether from logs or mud, requires many other choices, among them energy-efficient design, passive solar heating, renewable energy for electricity, catchwater and alternative wastewater treatment systems, environmental landscaping, and the use of nontoxic, green building materials.

BRIEF HISTORY OF LOG HOME CONSTRUCTION

Log home construction has a long history. In North America, the very first log cabins were built by the Swedes and Finns who settled the heavily forested Delaware Valley in the early 1600s. They imported the technology from Scandinavia, where it had been practiced for at least 800 years. Log homes came to the forests of Pennsylvania at a slightly later date when the Germans and Swiss settled the region. From these centers of early activity, log construction spread as pioneer families who adopted the techniques settled the uncharted American frontier. Soon, log barns, schools, churches, inns, and jails began to appear throughout the South and the old Northwest Territory. Even Native American tribes adopted log construction.

Early log structures in the United States consisted of two basic types: the round log home and hewn log home. Round logs were used primarily to build temporary structures—for example, remote cabins, barns, forts, lean-tos, and other hastily built shelters. Homes that people intended to live in day in and day out, however, were typically built from hewn logs—that is, logs that were

Although log homes are a form of natural building, most of them are far from sustainable. If a log home is in your dreams, your challenge will be to do it right—to make it a structure that contributes to rather than subtracts from the long-term health of the environment.

FIGURE 8-1.
Log structures enjoy a long history in Europe and the United States. Many of our ancestors lived in log homes built using hand tools and locally available lumber.

Clear-cutting is a way of harvesting trees that is almost universally frowned upon by environmentalists. It often leaves an ugly, nearly lifeless scar. Clear-cutting, however, is appropriate for some tree species. Seedlings of sun-loving species do best in clear-cuts. Other species, however, require shade for regrowth.

Clear-cuts vary in size and impact. In Austria, a nation that produces as much lumber as the U.S. Pacific Northwest, clear-cutting is commonly employed on private wood lots, but clear-cuts are small and skillfully blended with the natural topography so that they are barely visible. In the U.S. Pacific Northwest and southwestern Canada, huge, ugly clear-cuts are the norm. Check out your source of logs, in person, to see how logs are obtained.

FIGURE 8-2.
Corner notching of round and hewn logs is a functional art required to hold the structure together and prevent water from seeping into the joints.

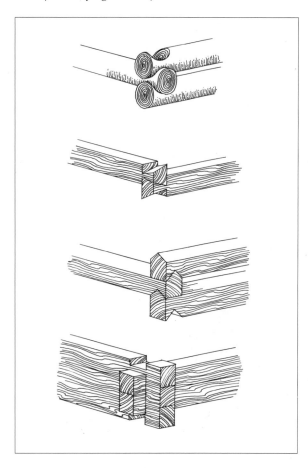

partially or completely squared off. Skill and craftsmanship marked the hewn log homebuilder, and the result of his labor was a home that survived the seasons and the wear and tear of many generations of early Americans.

The appeal of log cabins in the early days of American history was multifaceted. Log construction relied on an abundant, indigenous resource, trees, and required a few simple tools, an axe and a knife. If need be, a single man could build a cabin, although working with two or more people greatly facilitated the process and resulted in considerably less cussin'. Log cabin construction required less labor and fewer building materials (such as nails) than wood-frame houses. A well-built log structure provided shelter from wind and rain, and was virtually impenetrable to arrows and bullets.

Like many other natural and sustainable building technologies, the log home fell to the wayside over the next two hundred years. Several forces brought its near demise. One was the availability of lumber, brick, and stone—which were all commonly perceived to be better building products. Like early straw bale homes, log homes were perceived by many people to be a primitive structure best suited for temporary habitation, abandoned when money became available for a "real" house. Over the years, many log structures were converted into kitchens attached to a wood-framed house—or were converted to sheds or stables when a modern home could be built. Ironically, today's log structures are anything but primitive. They are rustic but elegant.

OVERVIEW OF LOG CONSTRUCTION

Exterior walls of a log structure consist of peeled logs that are laid horizontally and interlocked by a variety of ingenious corner notchings (figure 8-2). Logs vary in size from 6 inches to 10 inches or more, and may be round, square, or some combination of the two. Large logs are particularly attractive, but they come from older trees, often from old-growth forests, and should be avoided, given the horrendous indignities already visited upon the old-growth forests of the world. Especially large logs also require special equipment to hoist them into place.

In the early days of log home construction, the spaces between adjacent logs, the chinks, were packed with mud, moss, or a combination of the two, to prevent drafts and to deter small mammals, insects, and snakes. Saplings were sometimes jammed in the chinks, then mortared in place. Today, chinks are filled with a variety of materials such as cement mortar. A synthetic chinking material is also available.

FOUNDATIONS

Early log homes were often built directly on the ground or on simple stone foundations, piles of stones located in the corners and in a few strategic locations along the exterior walls. Both techniques failed the test of time. Wood exposed to ground moisture and snow deteriorates over time, causing walls to collapse. Inadequate foundations cause the walls to sag and may cause roof collapse.

Log homes must be built on a solid foundation that protects them from moisture. The simplest and least expensive option is a pier foundation, an option not previously discussed in this book. Piers are made from poured concrete or piles of flat stones, either mortared or not (figure 8-3). Wooden piers are also used, but should be made from decay-resistant wood. Even then, don't expect a wooden pier to hold up as well or as long as a stone or concrete pier.

Concrete piers are placed on footings that extend below the frost line to prevent frost heave. The tops of the piers should be at least 18 inches above the ground to protect the bottom logs from snow accumulation. (This arrangement also raises the floor off the ground.)

Another foundation, introduced in chapter 2, is the monolithic slab or slab-on-grade. The slab-on-grade consists of a thickened edge that serves as a

Part of the durability of hewn log homes lies in the fact that hewing removes the less dense sapwood on the outside of the log, leaving the more decay-resistant inner wood. Hewn logs also fit more tightly together, creating a more airtight and waterproof wall structure.

BUILDING NOTE

When raising a floor off the ground on piers, be sure to insulate the floor and install a vapor barrier. You should also attach a skirting to block the flow of air under the structure.

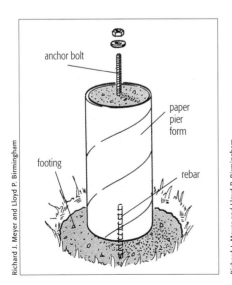

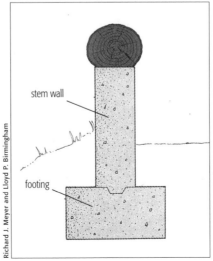

FIGURE 8-3.
Pier foundations made from poured concrete are inexpensive and easy to make.

FIGURE 8-4.
Log homes can also be built on the traditional poured concrete stem wall and footing.

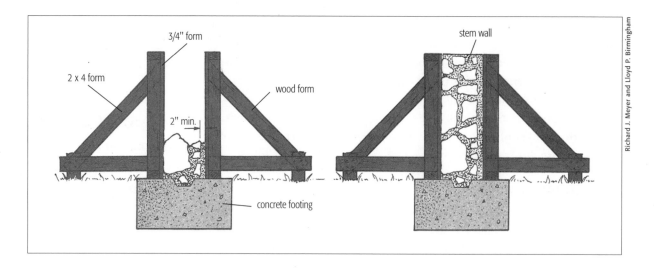

FIGURE 8-5.
Stone foundations can be made using the slip form method.

FIGURE 8-6.
Cross section through a poured concrete foundation, showing placement of peripheral drain, rigid foam insulation, and other details.

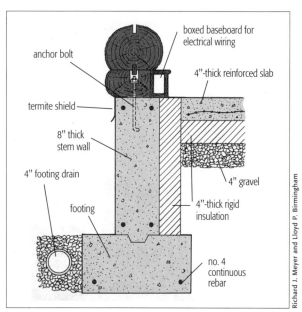

footing and stem wall, and a slab, upon which your finished floor will be laid. According to Monte Burch, author of *Complete Guide to Building Log Homes,* even though it requires more concrete than piers, the slab-on-grade foundation may cost less than a pier system because it eliminates the need for floor joists.

The poured concrete foundation, consisting of a footing and stem wall, is the next option (figure 8-4). Stone facing applied to the stem wall creates a more aesthetically appealing foundation and blends nicely with log walls.

Stem walls can also be constructed from stone (see chapters 6 and 9) or concrete blocks (see chapter 7). Figure 8-5 shows a relatively simple technique used to build stone foundations, known as the slip form method. In this technique, stones are placed along the outside of wooden forms. Concrete is then poured into the form. When set, the forms are removed and *voilà;* you have what appears to be a genuine stone foundation. (This technique is described in greater detail in chapter 9).

Foundations can also be made from fly ash concrete, insulated concrete forms, or Faswell blocks—all discussed in chapter 14. Several of these innovative approaches create a highly insulated foundation wall that reduces fuel consumption, makes your home more comfortable, and saves you money.

Waterproofing the external surface of a concrete foundation below grade and protecting a wall from termites are essential for the long-term future of a log home. A good foundation drain is also vital.

Even though log walls are extremely massive, they should be attached to the foundation. If you are building your home on concrete piers or are pouring a concrete stem wall, sound anchorage can be provided by L-shaped anchor bolts embedded in the top of the foundation stem wall (figure 8-6). If you are building a concrete block foundation, anchor bolts are embedded in concrete poured into the top row of blocks. Anchor bolts protrude through the first log of the wall, which is usually a smaller, flattened log known as a sill log. They are secured with a washer and nut.

For those interested in learning more about foundations for log homes, consult Monte Burch's book, *Complete Guide to Building Log Homes*.

LOG HOME OPTIONS

If you decide to build a log home, you have three options. First, you can hire a qualified builder to put up a house for you. Unlike all other natural building technologies, you should find many builders to choose from—unless you live in a desert.

Your second option is to purchase a log home kit. This popular approach offers you a wide variety of suppliers and kits. For kit homes, check out Jim Cooper's book, *Log Homes Made Easy*. Before you go this route, be sure to investigate the source of the logs. If they are being transported from a distant forest, consider another option. Check around to find a company that uses logs sustainably grown and harvested from local forests. You don't want your home to contribute to environmental decimation.

Log kits come in three price ranges. The least expensive are the partial shell kits. They consist of logs for the walls and heavy timbers for the roof and floors. You will have to purchase the rest of the building materials such as doors, windows, and roofing to complete the project.

In the medium price range are those suppliers who provide "dried-in" shell kits. These include the materials needed to dry in a house—that is, all of the materials required to build a structure that is protected from rain and snow. The owner is responsible for all the costs of interior work, including electrical wiring. If this option appeals to you, shop around and compare kits and prices so you can make fair comparisons. Some manufacturers go only as far as the roof decking and tar paper. Others include shingles.

The most expensive log kits supply virtually everything you will need for the project, from windows to doorknobs. They make your life as simple as possible and reduce the number of trips you need to take to the hardware store. Because many owner-builders are not experienced in home construction, this option is often the most desirable. And even though it may appear to be the most expensive, appearances can often be misleading. If you had to purchase all of the materials separately, it could cost you more. Kit home suppliers often receive volume discounts and pass the savings or part of them on to you. A complete kit will certainly mean a lot less time shopping for what you need. One of the disadvantages of this approach, however, is that it limits your options. For example, it may prevent you from using recycled materials. For

Choosing logs from a local forest reduces transportation costs, energy demand, and environmental damage caused by the production and consumption of fossil fuel energy.

BUILDING NOTE

Kit homes are delivered by truck to the job site, so you will need to find a way to protect the building materials from the elements and theft.

more on the subject, see Cooper's book. He provides numerous worksheets, including one for keeping track of what various kit manufacturers offer. Be sure to compare on the basis of quality as well. Then talk to others who have built homes and get their honest assessment. You may also want to view the Log House Builder's Association of North America's Web site. Not strong advocates of kit homes, they assert that it is "often easier to build a log house of your own [using "real" logs that you can purchase from the same sources as the "kit" manufacturers], than it is to assemble a kit log house."

The third option for getting a log home is to build it yourself from scratch. This can be an extremely arduous and sometimes dangerous undertaking, especially if you fell and process the logs yourself. To make the process a lot safer and faster, you can purchase peeled, dry logs that have been milled to eliminate taper. If possible, select trees from sustainably harvested forests, preferably forests near your home.

CUTTING AND PREPARING LOGS

If you decide to build your own home from scratch, you can cut your own logs from a nearby forest, obtain logs from abandoned log structures, or acquire them from a logger or a log supplier in your area; the last option saves you an enormous amount of time and energy. To cut trees and prepare the logs yourself, you will need a source of logs—either a forested portion of your own property or a woodlot on a neighbor's property. A friendly neighbor might be willing to let you remove some trees to thin his or her woodlot—and they may be crazy enough to help you, too! Well, maybe not. Another option is a nearby state or national forest. In this case obtain permission or you risk a fine.

Although most people cut trees with chain saws, a good sharp axe or a sharp buck saw is a more sustainable option. They're great exercise, exhaust free, and a lot quieter than chain saws. Whatever route you take, be sure to think safety. Wear safety glasses or other eye protection when cutting with an axe or chain saw. Ear plugs are also essential when using a chain saw. Cutting wood with a chain saw, axe, or hand saw is hard work. Get in shape before you start. A good book on the subject of chain saw safety is *The Good Wood-cutter's Guide* by Dave Johnson.

Before heading for the woods, it is wise to estimate the number of logs your house will require. Know what species of tree you are going to use and how to identify that species with and without its leaves. Also know the length of the logs needed for your home. Draw up plans before you start cutting trees.

Although you can build a log home out of any tree species, some are more resistant to insects and moisture damage (rotting) than others, as noted in chapter 7. For more information, contact the U.S. Forest Service or refer to Burch's *Complete Guide to Building Log Homes*.

Most log homebuilders use logs that are 6 to 12 inches in diameter. In general, the larger your building, the larger the diameter and the longer the logs

Logging is backbreaking, dangerous work. Unless you have an ideal spot with enough logs in the immediate area of your home site, you'll spend more time logging than you will building.

MONTE BURCH, *Complete Guide to Building Log Homes*

will need to be. Hands off the old growth, however. For best results, strive for uniformity among logs. This makes building easier and produces a more attractive finished wall.

Log homes require about 50 trees for a small home and 75 to 100 for a larger structure. If trees are selectively harvested and logs are carefully hauled out of the woods (being sure not to damage saplings and ground vegetation), your home can be built without causing irreparable harm. Selective harvesting can even enhance the health of the forest.

The best trees usually come from the heart of the forest, that is, the thickest patches where trees compete for sunlight and tend to grow straight and tall. When trees fall in a dense forest, however, they often damage or kill neighboring trees. Study tree cutting and practice your craft before you head into the woods. (Tree cutting is well beyond the scope of this book. If you want to learn more, check out the books listed in the resource guide. One thing I've noticed, however, is that although these books give sound advice on how to cut trees, they offer little information on doing it in an environmentally sensitive manner.)

Finding the right trees and the proper number can be a time-consuming task. If your knowledge of dendrology is limited, hire a state forester or a local logger to help out—or at least to provide you with some on-site training. You will be surprised at how many acres of forested land are required to find the number of trees you need. As Jim Cooper points out, "An area that at first glance seems to have enough trees to build a fort may only provide a handful of usable trees. The rest may be the wrong species or size, may have too many limbs, or may be crooked or twisted."

If you are inexperienced at tree felling, hire a professional to do the job or at least hire someone to teach you how to do it right. This is dangerous work. You'd be amazed how many things can go wrong. You'd be especially amazed how quickly a chain saw can slip or an axe can miss the target, cutting into your flesh. Take your time. Plan an escape route with each tree you cut—just in case. Practice on easy trees first. Wear safety equipment.

After trees are felled, remove the branches and haul the logs to the job site. To reduce damage, use a metal skidding pan, a device that keeps the leading edge from digging into the ground as a log is hauled behind a team of horses or a tractor (Figure 8-7). A team of horses or a single, well-trained horse has less environmental impact than a gas-guzzling tractor. Both have exhaust problems, but the horse's won't produce acid rain! Horses can also maneuver in places a tractor operator would never dream of entering. Hiring a team of horses or buying or renting a tractor, of course, will add to the cost of your project.

Logs are piled according to size, then cured (a log builder's term for dried). For proper curing, logs must be suspended on pallets or skids or on other logs at

BUILDING NOTE

If you are using a tractor to haul trees out, you may need to build a road, which causes more damage. Even if you don't need a logging road, driving a tractor over the same ground can do significant damage.

FIGURE 8-7.
A skidding pan prevents logs from digging into the earth and reduces damage to the environment.

Richard J. Meyer and Lloyd P. Birmingham

least six inches off the ground. Curing takes six months to two years, depending on climate and whether the bark is stripped off or left intact. Leaving bark on logs slows the curing process and tends to cause discoloration. It also provides shelter for insects. Because of this, most builders peel (bark) their logs. Even then, drying may cause trees to twist or warp, which makes them less desirable and harder to work with.

Peeling logs is time-consuming work, best done in spring when the sap is running and the cells of trees are saturated with moisture. High water content makes it easier to remove the bark. Bark is peeled using a draw knife or a peeling spud, a long, hand-held, chisel-like device. The process is described in some of the reference books listed in the resource guide.

Even though it is best to peel logs in springtime, it is often the worst time of year to cut trees in a forest, as the ground is usually muddy and difficult to navigate. Tractors, even horses, may bog down in the mud. Even if you can work, count on slow going and figure on creating a lot more damage to the land than at other times of the year.

To avoid the hassles of mud, cut and stack logs in the fall or winter, leaving the bark intact. Peel the bark off just prior to using them in the spring or early summer. Some builders prefer to build homes with unpeeled logs. Although this may create a more rustic look and may appeal to your sense of aesthetics, it is generally not recommended. Bark on interior walls will flake off after the home is built, creating a cleaning nightmare. Bark also harbors insects and may trap moisture that could cause logs to decay over time.

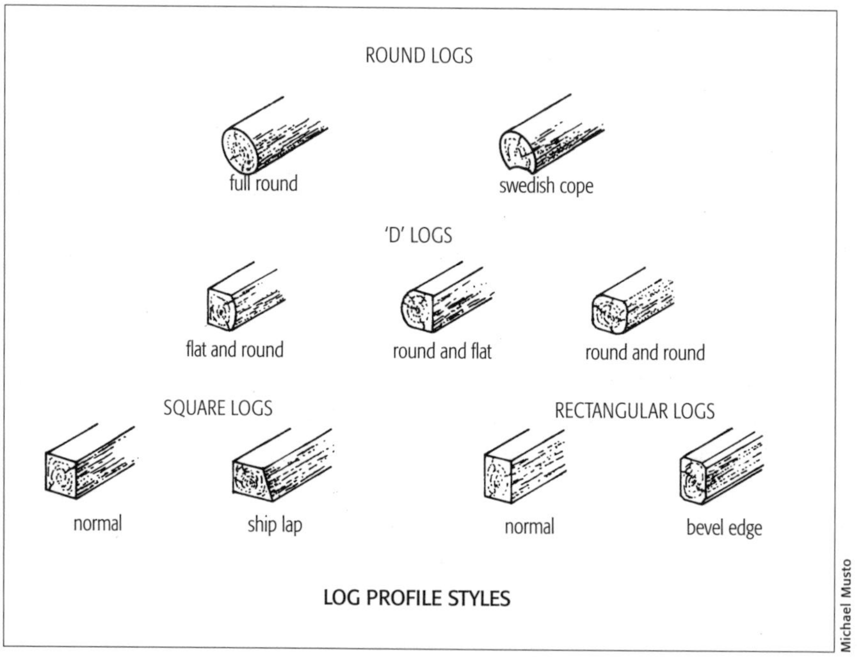

BUILDING NOTE

If you are building a hewed-log house, it isn't necessary to peel bark off the logs. Hewing removes the bark and part of the outer layers of wood.

FIGURE 8-8.

Logs are often hewn to create tighter joints and improve the performance and lifespan of a log home. This figure shows the many styles of round and hewn logs.

After the trees are cut, peeled or hewn, and cured, you can build your walls, the part of the project many people find the most exciting and satisfying. This is the phase in which the house begins to take shape, and all those images you've scratched on paper materialize.

BUILDING LOG WALLS

Log wall construction begins with the bottom logs, known as sill logs. These rest directly on the foundation and are attached by anchor bolts. Take your time when preparing and installing sill logs. Because they "set the tone" for the whole house, sill logs should be your straightest logs. They should be carefully hewn with an axe or saw to create a flat surface that rests on the foundation. Hewing isn't complicated, but it does take patience and practice. And it is best done while the wood is still green. After a log has cured, the process becomes much more difficult as the wood is typically much harder. Details, including the tools you will need, are outlined in several reference books.

Sill logs must be laid squarely on the foundation so that the entire structure will be square. If you make a mistake now, it will be magnified as the wall gets higher. If termites are a problem in your area or code requires it, install termite shields on the foundation beneath the sill log (figure 8-9). Consult with your building department, a local structural engineer, or an experienced builder to determine whether termites are prevalent in your area and to find acceptable ways of deterring these hungry, cellulose- munching critters from gaining entrance to your humble abode.

As shown in figure 8-10, each log must be notched to fit in place. Many different notching schemes have been devised over the years (figure 8-2). Some of them are extraordinarily complex and difficult. Whatever system you choose, be sure that it sheds water. Otherwise water will seep into joints and could cause the wood to rot.

One of the simplest notches is the saddle notch. Even though it is structurally simple, the saddle notch may take a first-time builder hundreds of attempts to master. Notches are marked with a tool called a scribe, then cut with axes, chain saws, and chisels. (For details on cutting notches, consult Robert Chambers's *Log Building Construction Manual* and Monte Burch's *Complete Guide to Building Log Homes*.

Once the sill log is in place and anchored to the foundation via anchor bolts, walls go up one log at a time. Building walls with round logs is probably the simplest method. As shown in figure 8-11, however, round logs leave large chinks that must be filled with cement mortar or some other chinking compound. To

In log construction, like other alternative building techniques discussed in this book, the log wall is pretty much the finished product. You won't have to install insulation, drywall, and siding. The log is the wall. It provides insulation and mass as well.

FIGURE 8-9.
The sill log is anchored to the foundation by an anchor bolt. The termite shield, shown here, prevents termites from burrowing into and eating the walls.

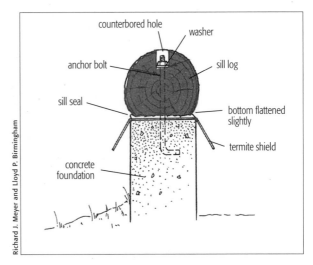

Richard J. Meyer and Lloyd P. Birmingham

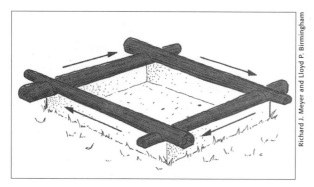

Richard J. Meyer and Lloyd P. Birmingham

FIGURE 8-10.
Sill logs are flattened on the bottom surface so they rest firmly on the foundation.

BUILDING NOTE

A bead of caulk is typically applied to the joint to reduce air infiltration in the chinkless design.

---

reduce cracking, log builders use a cement mortar containing lime. This mixture is less brittle than standard cement mixes and thus expands and contracts with changing temperature and as logs expand and contract in response to changes in atmospheric humidity. Another option is a synthetic chinking product made from a rubbery material that readily expands and contracts without cracking.

Another disadvantage of working with round logs is the requirement for periodic maintenance. Small cracks may form in the chinking, allowing water to seep into the walls, which could cause the wood to deteriorate. Round log walls are also structurally weaker than chinkless designs, and if the chinks are large, heat loss between logs may be extensive. For these reasons, round logs are not often used in building log homes, except in remote cabins.

New log homes are designed to eliminate chinks. Chinkless designs require more time up front, but greatly reduce maintenance; they also result in stronger, more airtight, water-repellent, energy-efficient walls. Figure 8-12 shows the Scandinavian or chinkless style that is popular in the northern United States. As illustrated, the lower surface of each log is cupped so that when logs are placed in the wall, they fit snugly. This joint is extremely strong and repels water very well.

Logs can be joined using splines, as shown in figure 8-13. Splines are made out of dimensional lumber inserted into grooves in top and bottom logs.

FIGURE 8-11.
A wall being made from round logs. Notice the large chinks that must be filled.

FIGURE 8-12.
The chinkless style, using round (shown here) and hewn logs, permits a much tighter fit, which reduces penetration by moisture and air infiltration.

Monte Burch

Monte Burch

Grooves are cut out with chain saws, circular saws with dado blades, or heavy duty routers. They make the joint even more airtight.

Chinkless round log construction is popular among owner-builders and professional builders in the northwestern U.S. and southwestern Canada. Chinkless square log construction enjoys its greatest popularity in the eastern U.S. Square log home construction requires the builder to hew each log, creating two to four flat surfaces (figure 8-8). Splines create even tighter joints.

Whatever option you choose, let the ends of the logs run well past the corners. Because log ends will absorb water, they will crack and slowly deteriorate. The further the logs extend past the joint, the longer the house will last.

How Are Logs Inserted in a Wall?

A 20-foot-long log can weigh more than 600 pounds. Stacking logs is therefore extremely difficult and dangerous, especially as wall height increases. A large log that gets away is potentially life threatening.

Fortunately, our forebears developed a number of handy techniques that will assist the contemporary log builder. With a few simple tools, logs can be lifted into place with relatively little personal effort and with very little danger to workers. The tools make the job so easy that even a person working alone can build a wall, although having at least one coworker is generally preferable. One of these techniques is the log ramp, shown in figure 8-14. It uses a couple of notched logs, a pulley, and ropes. (For more on this procedure and others, see Monte Burch's book.)

If you don't have an aversion or a moral objection to power machinery, you can rent a small crane or a large tractor with a bucket to lift logs in place. These options require more money and fossil fuel and are clearly less sustainable than the rope-and-pulley approach our predecessors used—quite successfully, too.

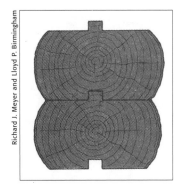

FIGURE 8-13.
Splines can be used for round or hewn logs to create a tight joint and a strong wall. This reduces air infiltration, too.

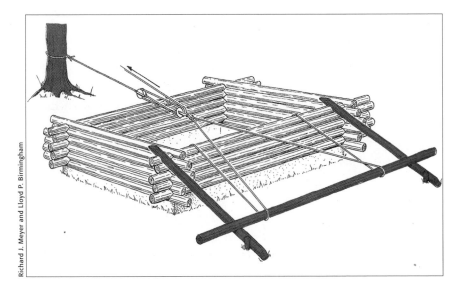

FIGURE 8-14.
This innovative device, known as a log ramp, was used by pioneers to build log homes. It is one of the simplest and cheapest ways of lifting logs onto a wall.

Shrinkage and settling can be minimized by artificially drying logs
before they are used. Many log kits
come with kiln-dried logs. Remember, though, this requires fossil fuels
and adds to the costs of the project,
both economic and environmental.
Also, kiln-dried logs will absorb a
substantial amount of moisture if
they are shipped to a moister climate.
This causes logs to expand, which
could cause damage to walls.

As walls go up, they must be kept plumb. A considerable amount of finagling may be needed to fit logs tightly together, as perfectly straight logs are rarely encountered unless you've had them milled.

Logs that interlock at corners and along their full length generally don't need to be secured any other way. However, some builders still prefer to join logs with long spikes that anchor one course to the next.

Log homes don't require top plates to attach the roof frame, as in many other natural building technologies. However, as shown in figure 8-15, builders typically leave a gap of 1 to 6 inches between the top log and the log below it. This gap, or settling space, is filled with a compressible form of insulation and is designed to accommodate settling and shrinkage. The size of the space varies with the type of wood, its moisture content, and the design of the home. (The best discussion of settling I've read is found in Robert Chambers's book, *Log Building Construction Manual*. If you are planning on building a log home, read the section on settling very carefully.)

Settling is not a possibility, but an inevitability, in log homes. It occurs as the wall increases in height and the weight of the upper logs begins to be transmitted to the lower logs. Weight bearing down on these logs causes them to flatten somewhat, which reduces the height of the wall over time. Another problem that reduces wall height is shrinkage. When a log dries, it shrinks, causing the wall to settle.

If you have not accommodated settling caused by shrinkage and compression, look out. You are in for some big surprises—and some costly repair work. Settling damages doors, windows, interior partition walls, cabinets, ceilings, and roofs.

The solutions to this problem are relatively simple. One of the easiest is to let the walls settle before installing windows, doors, and interior walls. If you don't have the luxury of time, leave a settling space at the top of the wall to accommodate this very natural phenomenon. Settling spaces are hidden by trim. As Jim Cooper points out in his book, *Log Homes Made Easy*, "In the past, log home manufacturers and builders who did not incorporate settling features have, after several years, faced windows and doors that were difficult to operate. In extreme cases, glass in windows has even cracked. A simple settling space over windows and doors during construction can avoid major headaches and expense later." By far the most comprehensive treatment on settling and ways to accommodate it is found in Robert Chambers's *Log Building Construction Manual*.

Preserving Logs

Peeled logs must be treated with preservatives to reduce rotting. This is especially important for the lowermost logs in exterior walls, which are more likely to get wet from snow and rain. Although preservatives greatly extend the lifetime of a log home, they contain highly toxic chemicals, such as pentachlorophenol, creosote, and various chemical compounds containing arsenic and copper. Because preservatives should be applied every two to three years, they can pose a health risk to applicators. (For details on log preservation, see Monte Burch's book.)

Although preservatives are not used on interior surfaces of log walls, which are not subjected to the elements, builders apply a finish coat consisting of boiled linseed oil containing three parts linseed oil to one part turpentine. Two coats are usually sufficient. Unfortunately, turpentine is a highly volatile solvent that may persist many months after the occupants move in. To protect yourself, your family, and your pets, use an environmentally friendly wood finish product. Although it may cost more, it is well worth the investment.

Before applying a finish coat, thoroughly clean walls to remove bark, dirt, excess caulk, and mortar. A steel-bristle brush works well to remove bark. Next, wash the walls with water containing a small amount of bleach.

If you want a more conventional-looking interior wall, or if you want to beef up the insulation in your log home, you can build a frame wall against the log walls or attach furring strips to the logs. Insulation can then be installed between furring strips or framing members. Drywall or artificial wood paneling is then attached. Because it pains me even to think about this process, you'll have to extract details from the reference books listed in the resource guide.

WINDOWS AND DOORS

In olden days, pioneer families made windows out of oiled paper or thin pieces of animal skin. Although neither method permitted an outside view, they did let light into the dingy, dark interior. Today, the log homebuilder has access to a wide range of doors and windows. Doors and windows are handled much like those of straw bale, cordwood, rammed earth, adobe, and cob homes. That is, they are placed in window and door bucks (figure 8-15).

Window and door frames are usually made from 2 x 6s. Four- and six-inch timbers also look very nice in log homes. Frame lumber can be milled from trees cut on your property or made from dimensional lumber purchased at a local building supply store. To prevent the window frame from being compressed as the logs settle, leave a small space between the top of the buck and the log that spans it, the header. To accommodate settling that occurs in the logs on either side of the rough opening, install splines in the window and door frames. These fit into a continuous vertical notch or groove in the ends of the logs (figure 8-16),

FIGURE 8-15.
Compression and shrinkage of logs as they dry causes walls to settle. Smart builders accommodate this in part by creating a settling space over doors and windows.

Monte Burch

Robert Chambers

How long have you been building with logs?

I graduated from the Mackie School of Logbuilding in British Columbia in 1983, and have been building with logs since then. I started teaching hands-on workshops three years later, in 1986.

What got you interested in log home construction?

I have loved the feel and look of log cabins since I was a child—my family canoed and camped in northern Minnesota, which is home to many old log buildings. The Civilian Conservation Corps buildings constructed in the 1930s are among my favorites—they have a scale and quiet dignity that is difficult to match.

I was in graduate school in Vancouver when I saw books and magazines about log construction. I didn't know that anyone was still building with the old, scribe-fitted, handcrafted techniques, and I was intrigued. Five years later, my best friend and I spent the summer in British Columbia learning to build.

Why build with logs in this day and age of inexpensive, convenient building materials?

Exactly. Why live in a home made of inexpensive, convenient materials? This certainly sets a very low standard for the place in which we live: it must be cheap and easy. Let's think about setting a higher standard for our homes.

Wouldn't most of us prefer to live surrounded by natural materials instead of synthetic materials? Why wear rayon and nylon when there is cotton and wool? People, often by necessity, spend a large part of each day in synthetic environments. Living in a log home is a tonic to this: It is natural, organic, full of unique shapes and soothing colors and textures. A handcrafted log home is a commitment to something unique, timeless, and even unusual that appeals to many people who desire something out of the ordinary.

Is log building environmentally sound? Does it have a negative impact on forests?

The devastating forest clear-cuts we all see are not for log homes—they are for toilet paper, newsprint, and 2 x 4s. Building with natural materials like logs can be environmentally sound. But it is obvious that natural materials must come from nature—not from a factory.

Logs certainly provide a very healthy environment for humans, which should be important to all of us. They do not off-gas formaldehyde (like OSB and plywood). They do not need to be painted. They are one of the best materials for people with hypersensitivities and environmental allergies.

Many log homebuilders use trees that have been killed naturally by fires or insects. Selective cutting from private lands is the biggest source of good trees where I live—in fact, few builders I know get trees from national forests. Many of the logs I use come from plantations. I recently built a home using only trees that were blown down in a windstorm. And probably most of the large timbers now used in log buildings are recycled from demolished buildings.

I think that clear-cutting old-growth forests is a very big mistake. There is too little of the primeval forest left, and it should be protected. Good log-building logs that are not harvested by devastating the land are available in most places.

It seems as if you use more wood building a log home than a conventional home. Is this true?

A 1,600-square-foot log home uses fewer than 90 trees. The trees we use are so beautiful that I cringe when I think of them being sawn up into 2 x 4s—they keep their beauty and dignity when they are left whole.

The size of the home is a much better indicator of the amount of timber used than whether the building is made of logs or 2 x 6s and plywood. Far too many new homes are being built that are 3,000 square feet or larger. The building materials they require, and the energy that it takes to heat them every year—this is where the waste of resources really adds up.

I built a small home for my parents that used just 24 logs, each 24 feet long. The total log waste from this cabin fit into a pickup truck without coming above the wheelwells. This was efficient building.

I am not certain of how much energy it takes to convert 24 trees into lumber, or how many gallons of phenol-resorcinal glue to convert 24 trees to plywood or OSB, or how much sawdust and planer-mill shavings are burned. But it must add up.

The average handcrafted log home also has a longer life expectancy than the average stick-built home. The longer a home is lived in, the more cost-effective it is. I have visited log homes in Russia that have been lived in continuously for 400 years—it seems unlikely that the 2 x 4 houses built today will last until the year 2400.

What are the most common mistakes made by owner-builders and how can they be avoided?

A common mistake is to design the home, build the foundation, have the logs delivered, and then commit to learning how to build with logs. This is too late in the project. Attend a hands-on workshop first, even before you decide on a design, and you will end up with a better home, and have a happier experience, with fewer surprises. Get the best-quality logs you can afford—for easy building they should be straight, have little taper, and be smooth.

What's your advice to people who want to build a log home but are considering a log home kit? Is building with a kit home any easier or cheaper than building with logs you harvest yourself?

I have not built a kit log home, but some of them look nice. There are hundreds of factories building kit homes, and I am certain the price and quality of materials varies dramatically. A kit home may be easier; certainly it requires fewer new skills to put together, and the logs are much smaller and lighter, but it is not necessarily cheaper.

Get references from people who have lived in their log homes (kit or handcrafted) for more than five or six years. Some problems do not show up in the first five years. Call these homeowners, and visit their homes.

What advice would you give to someone who wants to build their own handcrafted log home?

Handcrafting a log home requires skills that most people can learn with hands-on instruction and practice. Seek out the best in the field and learn from them.

ROBERT CHAMBERS builds handcrafted log homes in River Falls, Wisconsin. He travels the world each year to teach log construction techniques to beginners, owner-builders, and experienced logcrafters. He recently published a textbook: *The Log Building Construction Manual*.

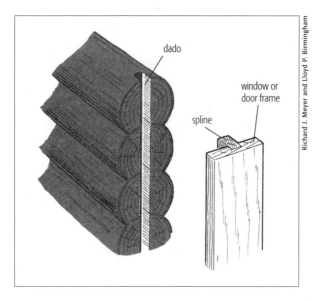

FIGURE 8-16.
Splines in window and door frames and corresponding grooves in the adjacent log walls allow settling to take place and protect windows and doors from costly damage.

FIGURE 8-17.
Interior walls also require settling spaces to accommodate the settling of the exterior walls. Adjustable support posts (jacks) allow owners to lower interior walls as the outside walls settle.

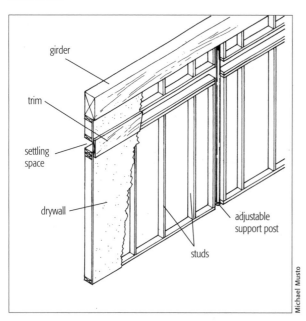

and allow the logs to settle without taking the window frame with them.

Once the window buck is in place, it must be squared and plumbed, then nailed in place. Drive nails through the lower board only, not along the sides where settling will occur. Then nail the windows to the frame and stuff insulation between the rough frame and the log-ends. Install trim to hide the evidence. Doors are handled in much the same way. If your log home is for seasonal use, consider installing lockable doors and shutters to provide extra security against people and hungry bears.

INTERIOR WALLS

Interior walls in log homes are often made from dimensional lumber, although logs can be used to enhance the beauty of a home. Conventionally framed walls will settle less than log walls; because of the different rates of settling, interior walls of dimension lumber should be designed to let the exterior log walls settle around them. If not, serious damage can occur. Interior walls will buckle, and drywall will crack.

To avoid major structural problems and costly repair, builders often construct free-floating interior walls. An example is shown in figure 8-17. The wall contains a settling space which is hidden by a piece of trim. A jack (an adjustable support post) is also situated in the wall. As the exterior walls of the house settle, the roof and ceiling drop. To accommodate this phenomenon, the interior wall is periodically lowered by adjusting the jack.

ROOFS

The roof for a log structure can be framed out of logs or dimensional lumber. Log roof-framing is more difficult than building with dimensional lumber and limits your choices. Because traditional wood framing offers more possibilities, especially when it comes to insulation, you may want to frame with dimensional lumber or have the lumber milled from your own logs, using a portable sawmill or at a local sawmill.

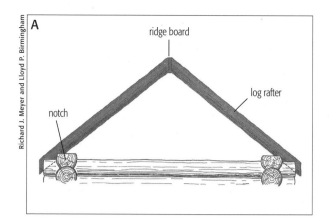

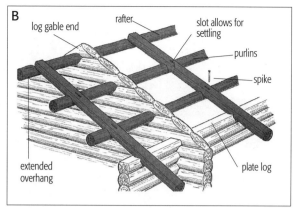

Richard J. Meyer and Lloyd P. Birmingham

The most common roofs found in log homes are gable, L-shaped gable, hip, L-shaped hip, and gambrel. In the United States, the most popular roof for a log home is framed with log rafters. As illustrated in figure 8-18a, log rafters rest in notches carved into the top logs of the exterior walls, sometimes called plate logs. Rafters are attached by wooden pegs, bolts, or spikes. Rafters join at the ridge line, the apex of the roof, where they are attached to a ridge pole or ridge board. Decking is then nailed or screwed to the upper surface of the rafters. Rigid foam insulation is attached to the decking and shingles are applied.

Although log rafters produce a beautiful ceiling, they don't allow much room for insulation. Unless you use a very thick layer of rigid foam insulation, heat loss through the ceiling will exceed recommended levels for sustainable housing.

In Canada, log builders use a combination of log rafters and purlins, as illustrated in figure 8-18b. Purlins run the length of the roof. Rafters are set in notches on the purlins and roof decking is secured to the upper surface of the logs. For aesthetic reasons, most builders use tongue-and-groove planking instead of decking. Rigid foam insulation is then put in place and the roofing materials are attached.

Dimensional lumber—either storebought or milled from your own logs—can also be used to frame roofs of log homes. Other options include engineered structural lumber such as I-joists, or prefabricated trusses, introduced in chapter 2. For details on building a roof for a log home, read Burch's *Complete Guide to Building Log Homes* and Chambers's *Log Building Construction Manual*.

FLOORS

Your choice of floor for a log home depends in large part on the type of foundation you choose. If you select a slab-on-grade, or monolithic slab, your options are limited to tile, brick, slate, flagstone, and carpet. Wood flooring

FIGURE 8-18 A, B.
Log rafters are a popular framing material for log homes. They join in the center and attach to the ridge board (A). Purlins and rafters are also popular. Purlins, shown on the right (B), run the length of the roof. Rafters fit into notches in the purlins.

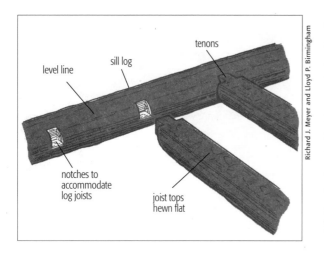

Richard J. Meyer and Lloyd P. Birmingham

FIGURE 8-19.

Floor joists made from logs fit into notches in the sill log. The top of the joists must be flattened to accommodate the subflooring.

can be installed on a slab, but special precautions must be taken to protect the wood from moisture—for example, by placing a vapor barrier underneath the slab and using furring strips to attach floorboards.

If you choose a pier or stem wall foundation, you must install floor joists. Some builders use logs fitted into notches in the sill log, as shown in figure 8-19. If the span is long, your floor may need the additional support of a log girder, running lengthwise from one end of the house to the other (figure 8-20). Consult an architect, structural engineer, or your building department on this matter.

Dimensional lumber can be used to frame a floor. Floor joists are either inserted into notches in sill logs or rest directly on the foundation on a ledger. Another floor framing option is the I-joist.

Once the floor joists are in place, subflooring is installed. Many builders use plywood, particle board, or OSB; however, there are a host of environmentally friendly substitutes for traditional floor decking—even a product made from straw. After the subflooring is in place, install your tile, carpet, tongue-and-groove wood, or bamboo flooring. (Check out these and other green flooring products in chapter 14.)

ELECTRICITY AND PLUMBING

Electrical wiring is typically run through interior partition walls, in floors, in ceilings, or in floor boxes (chases that run along the base of the floor). It can, however, be placed in exterior (log) walls. In order to accommodate wire, holes should be drilled in the logs to house the wire (figure 8-21). Recesses to accom-

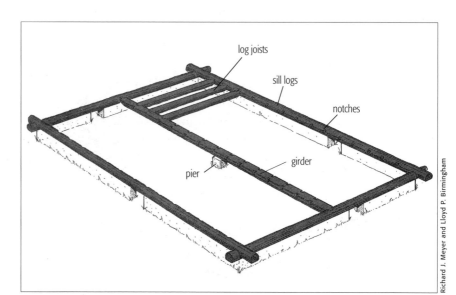

Richard J. Meyer and Lloyd P. Birmingham

FIGURE 8-20.

Details of a log floor showing log girders required to span long distances.

modate switch and outlet boxes must also be carved into the log walls. In log homes, plumbing is typically run through interior partition walls, in floors, or in ceilings.

HOW ENERGY-EFFICIENT ARE LOG HOMES?

If you are like me, you've heard conflicting stories about the energy efficiency of log homes. One source says wood is "not a particularly good insulator." Another source proclaims that wood is "nature's best insulator." As in many issues, the devil is in the detail. The claim that wood is one of nature's best insulators is true. Wood has an R-value four times greater than concrete, six times greater than brick, and fifteen times greater than stone. Comparing wood to concrete blocks, bricks, and stone, however, is useless. Most homes in industrial nations are made out of 2 x 6 or 2 x 4 studs with fiberglass or cellu-

lose insulation. When log walls are compared to stud walls using R-value as your sole means of distinction, log walls fail miserably. A 6-inch white pine log, for example, has an R-value of about 8 and a Western red cedar of the same diameter is a measly R-6.5. Compare that to a 2 x 6 stud wall with fiberglass insulation (R-19) or with cellulose (R-22), and you can see the concern many people have about log structures.

But don't close the book on the subject just yet. Wood also has substantial thermal mass and stores heat much like rammed earth, adobe, cob, and cordwood. When thermal mass is taken into account, researchers have found that log walls are equal to conventional insulated framed walls in terms of heat loss. In fact, when the U.S. National Bureau of Standards performed tests on log walls, they found that a 6-inch log wall equaled or exceeded the energy performance of any other type of wall during all seasons except the dead of winter. In the winter, insulated framed walls won by a small margin. The thermal mass made up for the difference in R-value.

Jim Cooper points out, however, that too many people see the tag "energy efficient" as a license to ignore sensible energy conservation measures. "They run amuck with energy-inefficient cathedral ceilings," says Cooper, put "glass in the wrong quantity and the wrong place, and disregard for the role of house siting in maintaining energy efficiency." Don't fall into that trap with this or any alternative building technology.

Building a Passive Solar Log Home

Log homes can be designed to capture solar energy for heat. The key elements of passive solar design are outlined in chapter 11. Follow them carefully and

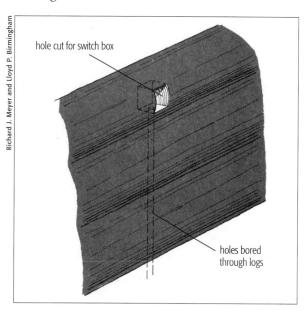

Richard J. Meyer and Lloyd P. Birmingham

hole cut for switch box

holes bored through logs

FIGURE 8-21.
If you must run electrical wire through log walls, you can. You will need to drill holes in the logs as they're placed and carve out notches for the electrical boxes.

you will find yourself in an enviable position—being free, or virtually free, of heating bills.

Unfortunately, most log homes fall far short of their solar potential. Many face the wrong direction, oriented for view rather than solar gain. Many are exceedingly large structures with dizzyingly high vaulted ceilings. Both of these mistakes increase overall heat demand and require a lot more building materials. The homes are natural, yes, and beautiful, but they do not contribute to the goal of a more sustainable human presence on the planet.

SHOULD YOU BUILD A LOG HOME?

Many writers and builders wax poetic about log homes. Kit home manufacturers, for example, have touted their homes as being "maintenance free." And some manufacturers still persist in boasting about the relative ease with which one can assemble a log home. For people with lots of experience in log home construction, this might be so. But for the vast majority of people whose experience with a hammer is limited to pounding in nails to hold pictures on a wall, a good sober look at the pros and cons is necessary before deciding whether this technology is appropriate.

Advantages of Log Homes
- Log homes are beautiful and comfortable to live in.
- Wood is a renewable resource.
- Cutting trees for a log home can promote a healthier forest.
- Walls of log homes have excellent insulation value, on a par with wood-frame homes.
- According to some sources, log homes actually use less lumber than a standard wood-frame house, because there is less waste. I have never been able to find data to support or refute this idea. (Changes in industry practices that put more parts of a tree to use, however, may take this advantage away.)
- Logs harvested from local forests reduce the energy required to transport building materials.
- The use of logs, rather than milled lumber, reduces the amount of energy required to build a house.
- Kits are available from many suppliers, and kits may make the task of building a log home much easier. (There's some disagreement on this benefit, as pointed out in the chapter.)
- Many builders are qualified to build good log homes.
- Log homes can be built in many styles.
- Log homes can be designed to capture solar energy for heating.

Disadvantages of Log Homes
- Log home construction is time-consuming, sometimes noisy, and potentially dangerous work.

- Log homebuilding requires a lot of strength and physical stamina.
- Log cutting can scar the landscape and damage forests.
- Preserving wood uses highly toxic chemicals. Interior finishes used in log homes may be toxic.
- Building a log home from scratch requires numerous fairly costly tools.
- Log home construction requires special skills.
- If special precautions are not taken to accommodate the settling of exterior walls, serious structural problems can occur.
- Predicting shrinkage and settling is not easy. Different tree species behave differently. Logs from the same species may also vary. The amount of shrinkage and hence settling also depends on such factors as when the wood was cut, how it was dried, the age of the tree, and local weather conditions.
- Log homes require a fair amount of upkeep (periodic applications of preservatives) to protect logs from the UV light, moisture, insects, and to maintain the color of the logs.

Log homes are one of the most popular of all natural building techniques, but before you decide on this building option, examine the pros and cons. If you are considering a kit home, check out the log source. If the logs come from clear-cutting, decide if you want this on your conscience. Also consider the relative amounts of wood needed to make a log home versus other natural buildings such as cob, adobe, rammed earth, or tire homes. If you are undeterred by all of this, build slowly and carefully. In his book, *Building and Restoring the Hewn Log House*, Charles McRaven advises, "Do not hurry. Almost all the problems modern owners of hewn log houses encounter stem directly from their own frenzy to get inside." The same can be said about any log home.

9

STONE HOMES

HUMAN BEINGS have been living in stone structures for thousands of years. The first stone shelters were caves in rocky terrain and cliffs. In Northern Europe, our ancient ancestors sought shelter against cold winter winds and dreary rains inside numerous caves where they often hovered around fires and slept under thick animal skins. In the Middle East, people carved their own caves in cliffs of soft stone. As Karl and Sue Schwenke point out in their book, *Build Your Own Stone House,* these caves represented the first stone home construction. "They were, in fact, stone houses by subtraction."

Over time, the Schwenkes write, our ancestors found that they "could reverse the process, and build with stones by addition." Using flat rocks, these early stone builders constructed simple, yet effective walls. Except for the addition of mortar, this building technology has changed very little since early times.

Over the years, humans have used stone to build a wide assortment of structures, including houses, cottages, castles, churches, barns, aqueducts, roads, walls, and walkways (figure 9-1). For centuries, the craft of stone building passed from generation to generation. Thus, even though the Stone Age ended nearly 4,700 years ago, stone remained a predominant building material many years later. Looking around my little community in the foothills of the Rocky Mountains, I've discovered dozens upon dozens of stone walls, steps, garages, and houses. In the farm country where I grew up, northwest of Rochester, New York, there is a similar abundance of stone structures. Many of these homes were built from glacier-polished stones, called cobblestones.

Paul G. McHenry Jr

FIGURE 9-1.
Paul McHenry Jr., author of numerous books on adobe construction, standing next to an incredible stone wall in Peru. Imagine the work required to shape these stones to create such incredibly tight joints—and then imagine how hard it was to hoist them in place!

Stone, along with wood, was a popular building material well into the 20th century. But in a now all-too-familiar story, stone construction began to wane with the advent of newer products that were easier to work with, more consistent, and structurally more predictable than stone. Nevertheless, stone remains popular in many countries for building foundations, decorative walls, retaining walls, walkways, patios, and steps. Stone homes also continue to be built for their rustic elegance. Few buildings blend as well with the native landscape as a sensibly designed home built out of native stone. Like so many other natural homes, stone can be fashioned into simple, sustainable living spaces or elaborate, costly buildings that require an extraordinary expenditure in energy and resources.

Charles McRaven writes in his book *Building with Stone*, "In the course of your building projects you will sooner or later come to the use of stone, the oldest, most durable, and certainly one of the most beautiful materials." Like other natural and potentially sustainable building technologies, stone homes offer many benefits to people and the planet. They also have some significant drawbacks, ones you should consider carefully before embarking on a full-scale building project.

FOUNDATIONS

Stone houses are built on a mortared stone foundation or a poured concrete foundation. Mortared stone foundations must be tightly sealed to prevent water from penetrating the mortar joints where it could freeze, cracking the foundation. Stone foundations are typically laid on a poured concrete footing reinforced with steel, usually rebar (figure 9-2). Stone foundation walls can also be laid on crushed stone. These rubble trench foundations allow water to drain away from the foundation wall and percolate into the subsoil below the

Stone buildings seem a natural outcropping of the Earth. They blend into the landscape and are a part of it. Stone houses are poised, dignified, and solid. . . .

HELEN and SCOTT NEARING,
Living the Good Life

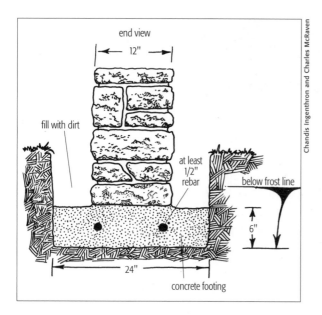

end view

12"

fill with dirt

at least
1/2"
rebar

below frost line

6"

24"

concrete footing

Chandis Ingenthron and Charles McRaven

FIGURE 9-2.
Details of a stone foundation.

> *A brick is a brick is a brick, but one stone isn't like any other stone in the world; each is its own challenge. Putting them up in useful, permanent, and beautiful structures despite their shapes will take a lot out of you. And give a lot of satisfaction in return.*
>
> CHARLES McRAVEN,
> *Building with Stone*

frost line, thus protecting the house from frost heave (see chapter 2).

Building a stone foundation requires the same skills needed to build any wall. The only trouble with a stone foundation is that you are building it in a ditch you've excavated, which makes the work a lot more difficult. Hunched over in the trench, you will find that it is much harder to build than above ground. Because the foundation is so important to the integrity of the house, you may want to hire a professional to handle this part of the project, or build a foundation out of poured concrete.

Poured concrete foundations consist of a steel-reinforced footing and stem wide enough and strong enough to support the massive stone walls of your home. A more detailed description of concrete foundations is presented in Charles Long's book *The Stone Builder's Primer*.

STONE WALLS

Building a stone wall is relatively simple, provided you have a ready supply of high-quality stones, a lot of time and patience, a modicum of skill and experience, some muscle power, a good eye for straight and plumb, and the proper tools. The first step in building a stone home, or any stone structure for that matter, is to gather rocks.

Gathering Stones

Stone walls for homes and other purposes can be made from different types of stone. A knowledge of geology is helpful, but as in so many things, a bit of common sense will serve you just as well. There's a simple rule you must follow to build a solid stone wall: Use strong, durable stones. Flaky, soft, easily crushed rocks won't hold up. In my region, the ground is strewn with decomposed granite that crumbles in your hands. This rock makes lousy stone walls. Stone structures in the area are made from a much more solid form of granite. It doesn't take a degree in geology to tell the two apart. If you want to read a bit of geology that will help you understand the nature of various stones, check out John Vivian's *Building Stone Walls*. His coverage is easy to understand and practical.

In climates where stone is subject to freezing and where snow and rain are common, good stones must not only be strong and durable, they must be dense and impervious to moisture. If not, water will seep into them through fissures and cracks or through loose-grained surfaces, such as those found in less dense sandstones. When water that has penetrated the interior of a rock freezes, it expands and causes cracks to form. More water can now enter and, upon freezing, widen the crack. This natural process destroys the exposed surface of rocks in a wall and can, if severe enough, cause a rock to crumble—not a desirable outcome for stones in the wall of a house. Most books on stone building provide tables that list the features of various stones. Find out what rocks you have on site, and then consult the charts to see if they are appropriate.

Although many different types of stone will work, some of the best are those made of hard shale or schist, according to veteran stonemason and author John Vivian. These stones are desirable because they split nicely thanks to natural flat cleavage planes. This, in turn, produces a nice flat top and bottom, ideal for laying up walls. Sedimentary rock such as sandstone is relatively soft and easy to work with, but it wears faster than other types of rock. Limestone, a popular building material in the midwestern United States, is also easy to work with, but it wears faster than harder stones. A sandstone or limestone wall, for instance, will only hold up for ten or twelve thousand years. The most difficult stones to build with are igneous rocks, such as granite and basalt, because they are difficult to split and shape.

Stones can be gathered from your own land, nearby fields, old stone fences, abandoned quarries, stone homes that are being demolished, and the foundations of dilapidated log homes or homes that have been razed by fire. As a boy, I worked for a farmer who paid me to remove rocks from his fields to protect his farm equipment. Don't assume that farmers or landowners will look kindly on your stone gathering. People can become pretty persnickety about such things. Be especially careful about cannibalizing stone fences. Always ask permission.

Acquiring stone from previous structures such as foundations or buildings that are being demolished is a great way to collect a lot of material—and very quickly. It also eliminates stone selection work—careful screening of stones to be sure they meet your criteria—so long as the original builder knew what he or she was doing.

Stones can be purchased from local quarries or even from overseas suppliers. While imported stone sounds appealing, it greatly adds to the embodied energy and the cost of the structure. If the stone should have an undesirable aesthetic effect, you could regret it for the rest of your life. "Stone houses using cut stones or those far removed in origin from their resting place, appear to us to have a transient out-of-place look," remark the Schwenkes. If you want a house that blends with the environment, your best bet is to use native stone.

Whatever your source, house building requires fairly large stones—each weighing up to fifty pounds or so. You'll also need a lot of flat stones, to fashion strong corners.

If you gather your own stones, you will need a wheelbarrow or a stone sled. A team of horses, a tractor and trailer, or a pickup truck can be used to haul stones from the field to your building site. Figure 9-3 shows a simple sled, known as a stoneboat, that can be pulled by people or horses.

BUILDING NOTE

When working with stone, a good pair of steel-toed boots and leather work gloves are essential.

Don't make the mistake of thinking that stone is free. Just transporting it can mean anything from the cost of a wheelbarrow to the cost of a hernia.

CHARLES McRAVEN,
Building with Stone

BUILDING NOTE

There's nothing worse than importing huge amounts of stone from another area only to find that it doesn't match—that the house sticks out like a sore thumb because the color of the stone simply doesn't match the soil or local rock outcroppings.

FIGURE 9-3.

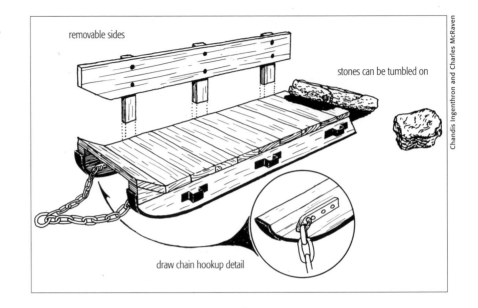

removable sides

stones can be tumbled on

draw chain hookup detail

Chandis Ingenthron and Charles McRaven

Transporting stones from field to the building site can be a major undertaking. Some builders use this device, a stoneboat, to drag stones. Horses can also be employed for this grueling work. Wheel barrows and trucks work well, too.

Stone and mortar weigh about 150 pounds per cubic foot.

Scouring fields for rocks, bending over to pry them loose and inspect them, lifting stones, and hauling them to your sled or truck is strenuous work that is hard on the back. Once you haul your rocks to the building site, you will need to unload them. And later, when you build the walls, you will need to lift the stones again to put them in place. Proper lifting techniques mean using the knees rather than the back to reduce back strain. Be sure you are in good shape before you start.

Lifting large stones may require two or more people, or some mechanical assistance. A tractor with a bucket is handy and reduces the chance of hernias and strained backs. Charles McRaven rigs up a boom on the front of his truck to lift stones onto a trailer. These devices add to the cost of building, but if they prevent you from getting injured, the investment is well worth the extra cost.

If you are building a stone home, be prepared to move a lot of stones. A small decorative wall 3 feet high, 2 feet wide, and 20 feet long requires a thousand stones or more, depending on the size of the stones, and weighs five tons or more, according to John Vivian. By the time you have gathered and loaded them, hauled them to the worksite, unloaded them, then placed them in the wall, you will have moved twenty tons or more of dead weight. A full-size house will require many more stones. "Taken one stone at a time, though," building a house requires no more work "than any other middle-aged, out-of-shape weekend putterer could handle, or indeed, could benefit from," notes Charles Long in *The Stonebuilder's Primer*. So go for it!

Well, not so quickly. Vivian and others recommend that those contemplating building a stone home get a physician's okay *before* they embark on a project of this magnitude, especially desk workers whose sedentary life would make a sea anemone appear active. "In any event," Vivian notes, "plan to take your time and use carts, ramps, barrows and levers to move larger stones.

There's little point in hurrying to complete a wall that will likely endure into the next millennium. And no point at all in busting a gusset doing it."

Building a Stone Wall

According to Charles McRaven, the best way to learn stonemasonry is to begin by laying up stone dry—that is, a stone wall without mortar. He goes on to say, "I've found that one basic rule will keep you out of trouble with stonework better than any other: Each stone should be laid so that it will stay in place dry, whether you intend to mortar it or not."

So, before embarking on an ambitious homebuilding project, hone your skills by building a 10-foot stone fence or two. First clear the ground. Remove sod and tree roots that could interfere with your wall. For best results, use stones that are flattened on at least two surfaces, preferably the top and bottom. If your rocks are round and too hard to break, forget it. To build a dry-stone wall out of round rocks is very difficult. If your stones have one flattened surface, use a stonemason's hammer and chisel to shape the other surface. Wear goggles or shatterproof eyewear to protect your eyes from flying stone chips. Another option is to use a sledgehammer to crack open large rocks. This will usually yield some usable pieces, but it is very hard work.

Stone walls are built two stones wide. Stones should slope toward the middle for best results. To begin, lay a single course directly on the ground. Next, lay your second course with the stones overlapping joints in the first course, as you would if you were laying brick. This results in a running bond. Single stones should occasionally span the width of the wall to provide additional

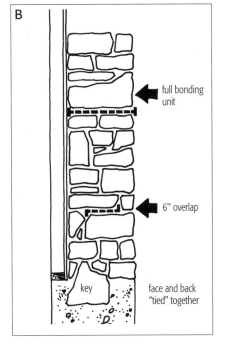

full bonding unit

6" overlap

face and back "tied" together

face could split off

key

key

Ian S.R. Grainge

FIGURE 9-4 A, B.
One of the secrets of building a strong stone wall is to prevent continuous vertical joints (A), which create a weak joint that could give way, causing the face of a wall to split off. To prevent this unfortunate event, stone masons bridge the wall with longer stones (B). This ties the face and back of the wall together very nicely.

Linda M. and Charles McRaven

FIGURE 9-5.
Another key to successful stone masonry is the one over two, two over one pattern shown here. This creates a running bond that eliminates vertical joints.

strength (figure 9-4). Continue to build the wall, one stone at a time. Place heavier sloped stones toward the outside.

The layout of stones in a stone wall is determined largely by aesthetics and structural requirements—primarily the need to overlap stones to prevent the formation of long vertical joints (figure 9-4). This is accomplished by following the rule "one over two and two over one." Translated, this means that when building a wall, the joint between two stones in the lower course should be spanned by one in the stone course above. In the third course, two stones should butt in the middle of the stone below (figure 9-5).

Because stones won't always fit nicely in place, you will need to insert smaller rocks here and there on the inside of the wall (figure 9-4). Some of these stones serve as wedges, holding the larger rocks in place. Wedge solidly, but don't permit the wedges to carry much weight. If a wedge slips out of place or is crushed, the wall could crumble.

While all of this sounds fast and efficient, it is not. Plan on spending a great deal of time sorting through your rock pile, trying to find the perfect stone, at least at first. Once you become more experienced, you will become quite adept at selecting rocks and laying them in the wall. You will also become adept at using your rock hammer to shape rocks to fit.

Like some stonemasons, you may want to plan your pattern carefully before laying rocks in a wall. Or, like others, you may develop an intuitive sense, a more seat-of-the-pants approach, to laying up stone according to texture and color. Once you get going, you will find that the process moves along pretty swiftly. "It usually takes only a couple feet of wall for the beginning stonemason to get into the rhythm and magic of laying stone," says Magnus Berglund, author of *Stone, Log, and Earth Houses.*

Once you have mastered drystone wall construction, it is time to learn how to work with mortar. Mortar is not a glue that holds misshapen rocks in odd spaces. Rather, it helps to hold well-fitted stones in place. It also makes a stone wall airtight—hence the necessity of becoming a good drystone mason before you tackle a mortared wall. To ensure a good fit, many experienced stonemasons first test stones before applying mortar.

Stonemasons use masonry cement for mortar or, more commonly, a mixture of Portland cement, sand, and lime, which is cheaper. The lime permits some expansion and contraction of the cement mortar after it sets up and thus reduces cracking. You can mix

FIELDSTONE is an ideal medium for amateur builders. The massive strength in solid walls of stone comes more from the material than from skill. And if the lines are not exactly straight, no critic's eye can tell which wiggles are the builder's and which of them came from the stone. Little mistakes can disappear in the rough-hewn texture of the rock.

CHARLES LONG, *The Stone Builder's Primer*

your own using formulas given in most stone building books. Or, ask local masons for the best mixture for your area.

Mortar is mixed in a cement mixer, wheelbarrow, or a homemade mortar trough. If it is too thick, it is difficult to mix and hard to work with. If it is too thin and watery, it will run and make a mess. Before applying mortar, you must clean the upper surface of the foundation. A stiff wire brush will sweep off loose, crumbly concrete. Dirt should be scrubbed off the foundation as well for better adhesion. After the surface is cleaned, it must be moistened to prevent the mortar from drying too quickly. (If it dries too quickly, it will crack.) Apply water using a hose, a bucket of water, and a brush, or a watering can.

With the surface cleaned and moistened, begin mortaring. Trowel on a thin layer of mortar a few feet at a time. The thickness of the mortar depends on the shape of the stones. The more irregular they are, the more mortar you will need.

Stones are embedded in the mortar by "rocking" them back and forth. This ensures that all surfaces of the stone are covered with mortar. Rocking also works out air bubbles that could weaken a wall. Some mortar will squish out. Scrape it off and apply it to the end joints, that is, the vertical joint spaces between adjacent stones in a wall.

After the first layer of stones is in place, start the next course. Clean and moisten the surface before you begin. Slowly, your wall begins to grow one stone at a time.

Some stone builders like to lay up stone in regular layers. Many like to keep courses level, as shown in figure 9-6. This is best achieved when you have access to stones with two parallel flat surfaces. Having a level surface makes it easier to set the next course, and it also makes it easier to bridge the inner and outer rows of stones.

At the end of a day's work, or sooner if you're working on a hot, dry day, you will have to go back over the surface of the wall to remove excess mortar or to apply additional mortar to fill the face joints—that is, the joints between the stones. This process is known as pointing. To fill gaps, apply additional mortar with a trowel, then run a wet finger over the mortar to smooth it out.

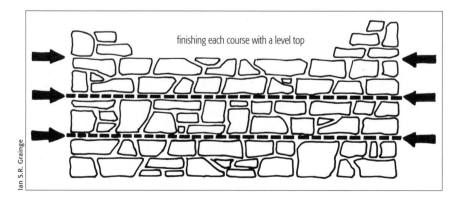

finishing each course with a level top

Ian S.R. Grainge

FIGURE 9-6.
Stones that are flat on two parallel sides permit stonemasons to lay level courses. This also makes it easier to bridge the inner and outer faces.

FIGURE 9-7.
Elaborate pointing and the use of cobblestone (rounded river rocks) results in a stunningly artistic pattern.

BUILDING NOTE

The crucial thing to remember about mortaring is that drying needs to be controlled so that it does not occur too quickly. If the water evaporates too quickly, the chemical reaction occurring in the mortar stops and some of the cement and sand will not bond and the mortar will be weaker.

This produces a solid, airtight mortar joint. Many masons believe that walls look nicer if the mortar joints are recessed a bit as opposed to running mortar flush to the surface of the wall. Others apply mortar so it protrudes beyond the stone. Some of the most interesting mortar work I've seen is on the cobblestone houses of western New York State (figure 9-7).

To avoid contact with mortar, use waterproof kitchen gloves when pointing. As you may recall from chapter 7, the lime in mortar dries the skin and causes painful cracking after only a few days. If your hands come in contact with mortar, wash them at the end of each day with water and vinegar. Vinegar contains acetic acid, which neutralizes lime and thus prevents your skin from dying out.

As your wall grows taller, you may find that some stones stick out further than the rest. That's okay. In fact, it gives stone walls a more rustic look. However, be sure not to stray off course. In other words, be sure your walls are as straight and plumb as possible. String lines and plumb bobs are used to accomplish this task. Both techniques are outlined in McRaven's *Building with Stone* and other reference books listed in the resource guide at the end of the book.

Stone walls are built about three feet at a time to give the lower section of the wall time to cure. Curing involves complex chemical changes in the mortar that increase the bonding between cement particles and sand. To optimize curing, let the wall dry slowly by periodically spraying it with water over the next week. This is especially important in hot, dry climates. If you don't, the mortar will dry too quickly and the bonding between cement, sand, and stone in the wall will be impaired, causing the mortar to crumble. If the problem is serious enough, the wall could collapse.

Stone and mortar walls should be kept covered in between work sessions, especially if you are building in dry climates. Moistened burlap bags or burlap and plastic are often used to wet the walls after each course and provide some shade for them as well. Cold weather reduces the rate of curing. Consequently, mortar should not drop below 40 degrees Fahrenheit for at least two days after application. If you are working in cold weather, cover the wall with insulated blankets at night to keep the heat in and promote proper curing.

A second three-foot-high segment of the wall can be built two days or so after the previous section has been laid. The mortar will be rigid enough to permit you to work without damaging previous courses. When the mortar is fairly dry, begin brushing the excess off using a stiff wire brush.

Stone walls can support a lot of weight, but like concrete have little tensile strength. This weakness can be serious in seismically active regions of the world. Masons address this problem by reinforcing walls with steel rods set in the horizontal mortar joints between stones (see figure 9-12). Some masons use steel rods manufactured specifically for this purpose. Fortunately, most stone homes don't require steel reinforcement. Check local codes to be sure. (Code books usually include stone homes with other forms of masonry, such as brick, concrete, and adobe.)

In stone homes, wood used for door and window frames and top plates must be protected from mortar. Because cement is hygroscopic, it absorbs moisture after it has cured. Wood placed directly against cured concrete will absorb moisture that the concrete has absorbed from the atmosphere. If the wood remains damp for long periods of time, it may begin to rot. To prevent wood from decaying, most builders treat wood with a chemical preservative. Another, healthier option is to apply roofing felt to wood surfaces that will come in contact with mortar.

Stone walls can also be built with earthen mortar. In fact, I ran across an old stagecoach stop just north of Moab, Utah, that had been built in the early 1800s with stone and mud mortar and looked as good as new—thanks in large part to the dry desert climate. Mud mortar may be appropriate in damper climates, as long as you take adequate measures to protect walls from water. After all, cob and adobe homes hold up well in wet climates if protected by eaves and solid, waterproof foundations adequately elevated above grade. You might consider adding some lime or a little cement to your earthen mortar. Both materials enhance the durability and water resistance of mud mortar.

Building Stone Walls with Slip Forms. If all of this seems like too much work, you may want to consider the slip form technique, introduced in chapter 8. A slip form is a wooden structure used to lay up stone foundations and stone walls. The name comes from the fact that the forms are used to build one section at a time, typically about a foot and a half. After a section is completed, the form is raised—or "slipped" up—and work continues, as in rammed earth wall construction (chapter 2) and light clay-straw wall building (chapter 10).

The slip form method is discussed in detail in Karl and Sue Schwenke's book. Although this technique makes wall building easier, it does require construction of wooden forms—eight at the very minimum, but usually more. This, in turn, means more wood, more labor, and higher costs—all of which may be justifiable if it helps you build better walls.

In the slip form method, forms are first placed on the foundation, then secured with braces and wires as shown in figure 9-8. After the forms are in place, squared, and plumbed, stones are selected. Many builders put them in place to see how they fit before mortar is added. Stones are typically laid along the outer face of the form. Concrete fills the rest of the cavity. As a general rule, the largest stones are used for the lower portions of the wall—both for strength and appearance.

It's best when you start a stone building to realize that it is going to take longer and require more patience than any single project you've ever undertaken, aside from raising your children or writing an encyclopedia.

MAGNUS BERGLUND,
Stone, Log, and Earth Houses

BUILDING NOTE

Be certain walls are plumb and straight when using slip forms, because you will be laying down large sections of wall. Unlike the drystone or mortar technique, a mistake here can affect a large portion of wall and can't be fixed.

FIGURE 9-8.
Details of a slip form used to make stone walls.

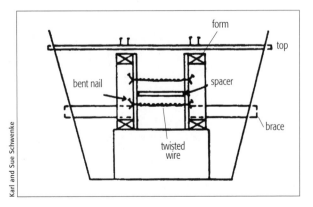

FIGURE 9-9 A, B.
The modified slip form method relies on a frame interior wall (B) that is later filled with insulation and finished with drywall or some other material. Sheathing and a waterproof laminate are laid against the stud wall and the stone and mortar wall is built against them.

When you are satisfied with the placement of the stones, pull the stones out, shovel a little mortar into the form, then begin replacing stones, paying special attention to preserving the pattern. After the first course of stones (filling the form) is in place and has cured a day or two, the forms are slid upward, and so on until you reach the top of the wall.

In *Build Your Own Stone House*, the Schwenkes note that the biggest mistake people make when using the slip form method is placing the stones too close together. This not only detracts from the beauty of the finished wall, it may make it weaker. They recommend a 2-inch joint on the outer face of the stones as optimum.

While the slip form method sounds pretty straightforward, it is still hard work. Bending over a form to lay stones strains the back and legs. This technique also uses a lot more mortar than a standard stone and mortar wall—in fact, one-half to two-thirds of the volume of the wall will be mortar. As Charles Long writes, "Slip form building is virtually a poured concrete wall with stones set in the outer face." And while slip forms can reduce the time required to lay up stone because you are building only one stone face, form work can be extremely time consuming. Thus, much of the time you save compared to laying stone using conventional mortar-and-stone wall-building techniques may be lost in form work. Another problem is that the interior surface of the walls of your house will be cement and must be finished with drywall or stucco to cover this raw, rather unsightly building material.

Another similar method, something of a mix between slip form and standard mortar and stone construction, has been pioneered by Charles Long. He builds a standard stud wall, using either 2 x 4s or 2 x 6s, on the inside of his foundation stem wall (figure 9-9). He then secures a half a sheet of sheathing (1½ feet by 8 feet long) such as particle board or chipboard to the outside surface of the stud wall. A thin layer of waterproof laminate (or sheet metal) is placed on the outside surface of the sheathing to prevent moisture from ruining the sheathing and to make it easier to slip the sheathing up. Any rigid

Dan Maruska

waterproof material will do. Next, he builds a wall of stone and mortar against the single-sided form, making sure the outer surface is plumb and straight. The next day, after the mortar has dried a bit, the sheathing and waterproof laminate are detached, slid up, and resecured so that another section of wall can be built.

This technique offers several advantages over the conventional slip form method. It allows the stone builder to see what the wall looks like as he or she works. It also provides a plumb building surface to work against. This, in turn, makes it easier to build a stone wall. Another benefit is that the wall appears more like a conventional stone and mortar wall, rather than a stone veneer wall, one complaint about the slip form method. Finally, the stud wall creates an insulation space. But be ready for a lot of framing. If you're not a skilled carpenter, you may want to get some help on this part of the project or have an experienced carpenter teach or assist you.

Stone walls come in a variety of styles. According to Magnus Berglund, "The nature of the stone determines the pattern to some extent. Square or rectangular stones lend themselves to geometric patterns, whereas random sizes and shapes lead to flowing, organic patterns." Round stones from riverbeds (cobblestones) have been used with stunning results in New York State. Although cobblestone construction sometimes results in a less orderly pattern, some creative stonemasons in the Rochester area have devised fascinating patterns in their homes (figure 9-10). Extremely large stones can also be used when building a wall. Some builders use variety of sizes. Before you get started, study stone walls and stone buildings in your area, see what you like, then acquire the stones you need.

Strengthening Stone Walls: Buttresses, Corners, and Intersecting Walls. The strength of a stone wall varies considerably with thickness. Generally, the thicker the wall, the stronger it is. In one-story houses, most builders construct 12-inch exterior walls. In two-story houses, the lower wall is typically 16 inches thick and the second-story wall is 12 inches thick. The type of roof you install also affects the thickness of the stone wall, as different roofs put different loads on walls. Check with local codes and local masons to determine how thick your walls should be.

Walls can also be strengthened by buttressing. Buttresses are external supports built perpendicular to the exterior wall at strategic locations along their length (figure 9-11a and b). The stones of a buttress interlock with the stones of the exterior wall. A footing is located beneath each buttress for support.

Corners and intersecting walls are far more common ways of strengthening the exterior walls of stone homes than buttresses (figure 9-11c). Corners require skill and a good supply of long, rectangular stones to create a tight, interlocking pattern to withstand the loads placed on them. Some builders

FIGURE 9-10.
The herringbone pattern in this cobblestone masonry wall illustrates the artistry of western New York's stonemasons.

For a long, long time, ordinary farmers and untrained folks have been laying up fieldstone houses and barns. The techniques and the styles have been as individual as the people with the trowels.

CHARLES LONG,
The Stone Builder's Primer

Stone Homes / **219**

FIGURE 9-11 A, B, C.
Buttresses may be required to provide additional lateral support for garden walls (A). Stones in the buttress are interlocked with those in the wall (B). Corners (C) and intersecting walls (not shown) in stone foundations and walls are typically used to provide structural support.

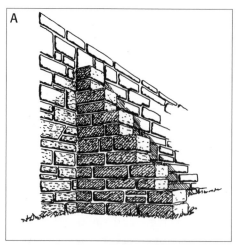
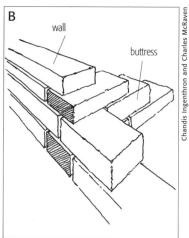

BUILDING NOTE

As a rule, buttresses are not required for most stone houses, unless they have extremely long walls—over 24 feet in length. Rather, buttresses are most often needed for tall stone walls—for example, garden walls or privacy fences.

BUILDING NOTE

Vapor barriers are placed on the warm side of the wall. In cold climates, this is the inside. The vapor barrier intercepts warm moist air, blocking its movement through the wall.

insert metal ties between the stonework in corners for additional rigidity and strength. See Ian Cramb's *The Complete Guide to the Art of the Stonemason* and John Vivian's *Building Stone Walls* for advice on building durable corners.

Insulating Stone Walls. One of the major drawbacks of stone walls is that they provide very little insulation. As a result, heat moves through them very quickly and interior spaces can become quite cold and uncomfortable. In the winter, moisture may condense on interior wall surfaces. The wall is said to sweat, even though this process is a form of condensation—like water condensing on the outside of a pitcher of ice water. Mold may form on the wall, creating an unsightly and possibly unhealthy mess.

To reduce heat loss and condensation, many builders insulate stone walls. One common technique is to build a double stone wall, as shown in figure 9-12. Insulation is stuffed in the space between the inner and outer walls and a plastic vapor barrier is installed to halt the flow of moisture from inside the house.

(If not, water will soak the insulation, robbing it of its insulating qualities.) Steel rods are typically used to bridge the inner and outer segments of the wall to provide additional strength.

This double-wall technique is a simple, elegant solution to the problem of condensation and heat loss. It does have a drawback, however: It increases the number of rocks and the amount of labor required to build a wall. To avoid these problems, some stone homebuilders erect conventional stone walls, then construct stud walls against the stone. Insulation is placed in the stud cavities. Stud walls consist of either 2 x 4s or 2 x 6s, and are attached to stone walls by masonry ties embedded in the mortar on the inside surface of the stone wall. Because a stone face is irregular, however, shimming is required to ensure a flat, plumb interior frame. Furring strips can also be secured to the stone wall. Insulation is applied and drywall or paneling is then nailed or screwed to the 2 x 4 frame or to the furring strips. The easiest way to provide insulation space is by building a stud wall on the foundation, as in the method pioneered by Charles Long (described earlier).

A 2 x 4 stud wall with fiberglass insulation has an R-value of about 12 and a 2 x 6 stud wall with fiberglass insulation has an R-value of about 20. Cellulose and a host of other natural insulation materials described in chapters 14 can also be used. Some of them, like cellulose, increase the insulative value of the wall with little, if any, additional cost. One popular insulation material used by stone builders consists of a bubble plastic sandwiched between aluminum (figure 9-13). So effective is this material that a ½-inch-thick layer has an R-value of just under 10. Although it costs more than most common insulation materials, bubble insulation can save money by reducing lumber costs. (You can get by with furring strips rather than 2 x 4s or 2 x 6s.) Bubble insulation holds up well in moisture. Unlike virtually all other forms of insulation, its R-value does not plummet when it is wet.

While the stone wall/wood frame method cuts down on stonework, it adds a significant amount of framing, shimming, wall board installation, and finish work. How much time you save over the double-wall system is debatable. From an aesthetic standpoint, the finished wall may even be less desirable than an exposed stone surface. Nonetheless, too much stone in the interior of a home can become visually tiresome, even overbearing, while a mixture of stone and finished wall may add variety to the home.

DOORS AND WINDOWS

Doors and windows are installed in rough frames (window and door bucks) made from wood, as in

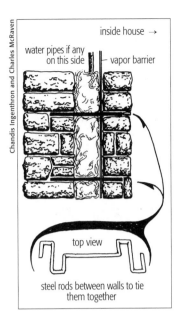

FIGURE 9-12.
Steel rods may be used to tie inner and outer walls of a double stone wall.

FIGURE 9-13.
Bubble plastic sandwiched between thin layers of aluminum provides incredible insulation.

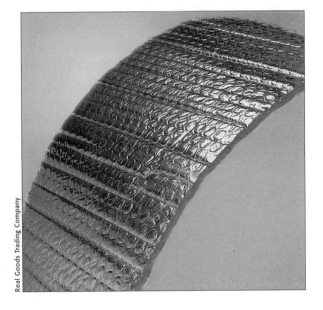

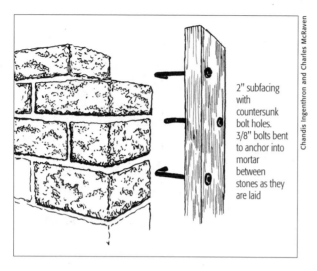

Chandis Ingenthron and Charles McRaven

2" subfacing with countersunk bolt holes. 3/8" bolts bent to anchor into mortar between stones as they are laid

FIGURE 9-14.
Anchor bolts from door and window bucks are embedded in the horizontal mortar joints.

BUILDING NOTE

Mortar and wood expand and contract as temperatures change. During the coldest weather, there is a tendency for the two to pull apart, thus creating a crack that may permit air to flow through. A key system helps prevent this unfortunate circumstance.

other natural homes. Window and door frames are made from two-by lumber, typically 2 x 6s or larger. To secure the rough frame to a stone wall, builders often use 10-inch anchor bolts set in the mortar between the rocks abutting the frame, as shown in figure 9-14. The mortar-embedded ends of the anchor bolts are L-shaped, a feature that locks them more securely in the mortar. After the rough frames have been installed and bolts have been tightened, set the windows in place and secure them to the frame.

Creating a solid anchorage for doors and windows is important, particularly for doors. Opening and closing a door, especially a heavy one, places a lot of stress on anchor bolts. Over time, they may work loose from the mortar. To spread the load and to reduce the chances of a door coming loose, builders typically use a large number of bolts—a dozen or so for every 7 feet of vertical frame. Set each bolt deeply in a bed of solid mortar.

In addition to anchoring the window and door bucks to the stone wall, you should create a key system. The key ensures a tight fit between rough frames and the surrounding mortar. It also eliminates leakage, uncomfortable drafts, and energy loss. For ideas on building effective keyways, see Charles Long's book.

A word to the wise: Plan your openings carefully. Unlike log or cob homes, in which openings for doors and windows can be cut out or enlarged with relative ease, in a stone home a small mistake is uncorrectable. It's a good idea to build or purchase your windows and doors long before you begin laying up walls.

Lintels must be installed over doors and windows to transfer the load from above door and window openings to the walls beside them. The walls, of course, must be strong enough to bear the additional load.

The stone homebuilder has the unique advantage of using a material that is ideally suited to the task (figure 9-15). The simplest yet weakest option is the single-stone lintel. Another, stronger option is the inverted **V** made from stones. The strongest is the semicircular arch. Perfected by the Romans, the arch has had such a significant impact on building that the word we use to describe the art, practice, and science of building, *architecture*, is derived from this simple yet highly effective design.

Arches are built using temporary wooden forms. Wedge-shaped stones are mortared in place, one at a time, starting at the sides and working toward the middle. The center stone, the keystone, is positioned last. The form is removed after the mortar has cured, although many stone masons wait to remove it

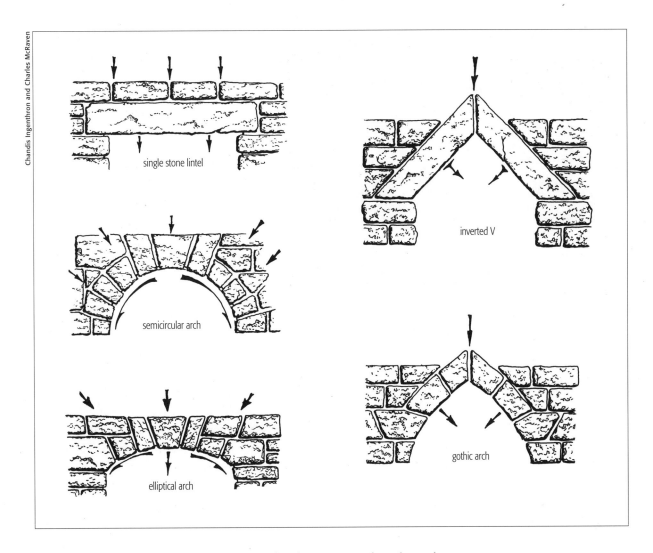

single stone lintel

semicircular arch

elliptical arch

inverted V

gothic arch

until after the wall above the arch is completed. Care must be taken when building an arch. The stones must fit tightly and the mortar joints must be of excellent quality. For more details on lintels and arch building, see the books of McRaven and Long listed in the resource guide.

ROOFS

The roof on a stone home must be securely anchored to a wooden top plate which, in turn, is attached to the rock wall by **L**-shaped anchor bolts (figure 9-16). Anchor bolts are set in the mortar between the stones. Consult with a structural engineer or your local building department for the number, length, and spacing of anchor bolts. Rafters or trusses are then attached to the top plate, as shown in figure 9-17. If you've opted for the stone and stud wall system of Charles Long, the rafters are also attached to the stud wall, although the bulk of the load will be borne by the stone wall.

FIGURE 9-15.
Stone can be used to form lintels and archways. The results are strong and beautiful!

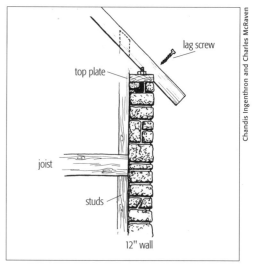

FIGURE 9-16.
Anchor bolts embedded in the mortar secure the top plate to the stone wall.

FIGURE 9-17.
Roof rafters rest on the top plate.

FLOORS

Like roofs, floor options are many and varied. For wooden floors in a stone house, joists are typically set on a 2 x 4 sill plate supported by a ledger in the foundation wall, as shown in figure 9-18. Sill plates are secured by L-shaped anchor bolts set in the mortar. The wood used for the sill plate should be treated to prevent rotting as well as insect damage. A sill gasket, a thin strip of foam inserted between the ledger and the sill plate, prevents moisture from contacting the wood. Be sure to insulate between the floor joists to keep your house warm in the winter.

FIGURE 9-18.
Floor joists rest on a sill plate anchor bolted into the stone foundation. Pressure-treated lumber or a sill gasket should be used to prevent the sill plate from rotting.

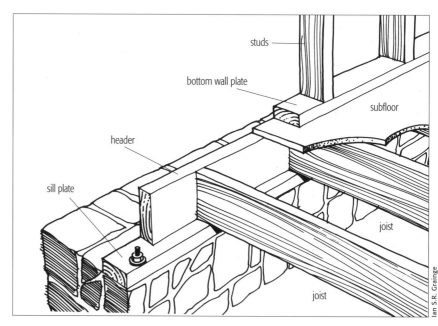

ELECTRICITY AND PLUMBING

Plumbing and electrical lines in a stone house typically run through interior walls, in the floor, and in ceiling spaces because of the massive, impervious nature of the exterior walls. If you build a double stone wall, wiring and plumbing can be run in the insulation space. To service light switches or outlets on exterior walls, run wiring through conduit embedded in the mortar between stones. If you are building a stud wall on the inner surface of the stone wall, the insulation space provides a space for plumbing and electrical. If you are using furring strips to build out from the stone wall, however, there won't be enough room to accommodate electrical conduit. You'll have to run it elsewhere.

SHOULD YOU BUILD A STONE HOME?

Why would anyone in their right mind want to build a house out of stone? Let me count the reasons.

Advantages of Stone Homes

- Stone is a natural building material.
- Stone is ubiquitous.
- Stone is durable and strong.
- Stone homes are very attractive and blend well with the natural environment.
- Stone building is relatively easy to learn, although becoming an accomplished stonemason can take a lifetime.
- Stone is inexpensive, if gathered locally.
- Stone offers thermal mass needed for passive solar heating.
- Stone walls can be airtight, reducing heat loss.
- Stone walls don't burn.

Disadvantages of Stone Homes

- Stone is a heavy building material and requires great physical stamina.
- Stone homes require lots of stone, which, even if gathered on or near your site, is extremely labor-intensive and grueling work.
- Stone building can be dangerous. It is also hard on the body. Strained backs, hernias, and more severe injuries are all possible.
- Stone poses certain design limitations and is inappropriate for some building applications.
- Stone is a poor insulator. Special design features must be incorporated to create a well-insulated stone wall, adding to the cost.
- Without insulation, stone walls may experience condensation during cold months, which may promote the growth of potentially harmful molds.
- Because stone is so heavy, stone walls in houses may require more costly foundations to support them.

Real stonemasons are a rare and talented breed. Their skills take years to learn and develop. A simple wall in a simple house, however, is the sort of thing that untrained yeomen and illiterate peasants have built for themselves for millennia past. It is only in this recent era of the specialist that we have begun to doubt our ability to do such simple things for ourselves.

CHARLES LONG,
The Stone Builder's Primer

- Stone walls can be built two or more stories high, but lifting stones to a second story height can be extremely difficult.
- In temperate and cold climates, stone work must cease over winter months because freezing adversely affects mortar and could weaken the wall.

Building with stone is an exercise in patience, strength, and quality craftsmanship. It probably requires more labor than any other natural building technique discussed in this book and will cause more than its share of back strain, tired muscles, and frustration. If you can overcome these obstacles, you will have a safe, comfortable, and healthy home that lasts for centuries. Even if you are not planning on building a home with stone, the techniques are useful to know for building retaining walls, garden walls, patios, stone fences, or even erosion barriers for streams. Stone structures lend themselves to a feeling of harmony—something many of us strive to achieve.

EMERGING NATURAL BUILDING TECHNIQUES

R AMMED EARTH, STRAW BALE, Earthships, cob, cordwood, and the rest of the natural homes we've been exploring may soon be joined by some rather brazen newcomers. Now in their early stages of development, these techniques appear to offer many of the advantages of other natural homes—among them, comfort, health, and the potential to enhance humankind's prospects for a sustainable future. Like all building techniques, they will undoubtedly have a few drawbacks. It goes with the territory.

This chapter examines six "new" or emerging building technologies: earthbags or superadobes, papercrete, cast earth, light straw-clay, bamboo, and hybrid homes.

BUILDING WITH EARTHBAGS OR SUPERADOBES

In Moab, Utah, near the world-renowned natural stone arches and magnificent rock formations that have converted this old uranium mining and milling town into a mecca for hikers, backpackers, and, most recently, mountain bicyclists, live Kaki Hunter and Doni Kiffmeyer. Their house is across the street from a local baseball field. To protect themselves from stray softballs, Kaki and Doni erected a privacy wall. But it is not an ordinary fortification (figure 10-1). It is built out of earthbags, polypropylene bags filled with soil and stacked like bricks to form a wall (figure 10-2). When completed, the walls were covered with earthen plaster.

Knit your eyebrows in puzzlement at this seemingly odd idea. I know I did. But before you dismiss it as lunacy, remember that people have been using

I consider every attempt to do something new to be a good example, so long as we are receptive to the lessons we learn and are willing to share those lessons with others.

MICHAEL G. SMITH

FIGURE 10-1.

These funky-looking walls are made of earthbags covered with a generous supply of adobe mud.

Earthbag building technology, clearly still in its infancy, has been tested by Gernot Minke of Germany and Nader Khalili of the California Institute of Earth Art and Architecture (Cal Earth).

FIGURE 10-2.

Earthbag wall under construction.

FIGURE 10-3.

Kaki Hunter fills an earthbag on the wall in a specially built stand. After it is filled, the bag will be laid down on the wall.

sandbags for decades to hold back flood waters and to build bunkers to protect soldiers in times of war. It didn't require much of a stretch for folks looking for alternative building materials that were accessible to people, affordable, and kind to the Earth to begin experimenting with sandbags as the building blocks of shelter. What could be simpler than filling bags with soil, piling them up to build a wall, then laying on a thick coating of plaster to hide the evidence?

The soil used to fill earthbags is similar to that used to make rammed earth walls. Containing about 5 to 25 percent clay, the soil is moistened slightly, then transferred to bags using a coffee can or a shovel (figure 10-3). Doni and Kaki fill the bags, which can weigh up to 100 to 200 pounds each, in place using a special bag stand of their own invention (figure 10-3). Once the bag is full, the top is folded over and the bag is laid on the foundation. Because it is filled in place, there's no need for lifting.

Earthbags are staggered like bricks. Overlapping the bags results in a running bond that strengthens the wall. After an entire course is laid, workers

FIGURE 10-4.
*Kaki tamps earthbags before
starting on the next course.*

FIGURE 10-5.
*Barbed wire laid down on the
earthbags serves as a mortar,
knitting courses together.*

BUILDING NOTE

Bags at the ends of rows are precompressed using a tamper before being set in place to create a solid end. Nails are used on some bags to pin the corners to help hold the soil in place.

tamp the bags using a hand tamper (figure 10-4). Tamping compresses the soil, which hardens upon drying. After the bags are tamped, two parallel strands of four-point barbed wire are laid down on the first course of bags as a kind of mortar for the next layer (figure 10-5). The next course is then laid and so on.

After the soil in the bags is compacted and dries, it sets up. Each earthbag becomes a solid block as hard as the walls of a rammed earth home. Because the process doesn't require extensive and costly form work, earthbags are considered to be a user-friendly form of rammed earth construction that is ideally suited for owner-builders. In earthbag construction, the bags act as the forms. Once the blocks have hardened, however, the bags are superfluous. They have done their job.

Earthbags have been used to construct garden walls, sheds, small shelters, and modest-sized homes. Nader Khalili, who pioneered the ceramic house described in chapter 5, built a code-approved museum out of earthbags for the city of Hesperia, California. The building is a showcase that could popularize this novel form of natural home building.

The polypropylene sacks are relatively inexpensive and readily accessible. Bag manufacturers will gladly sell you rejects, typically misprints, at a generous discount, especially if you are buying in volume.

Earthbag homes can be built using a variety of foundations, from the ever-popular rubble trench to rammed earth tires to that old standby, the poured concrete foundation. Some natural builders are experimenting with earthbag foundations for straw bale and cob homes. In Mexico and Nova Scotia, gravel-filled bags are being used as foundations to reduce water penetration in straw bale buildings.

BUILDING NOTE

Burlap bags also work well for earthbag construction. Some builders are also experimenting with long tubes of bag material that are filled with earth and stacked on top of one another like a coiled ceramic pot.

The thickness of the walls of an earthbag home depends on the size of the bag. A 50-pound rice bag tamps out at about 15 inches wide, 5 inches high, and 20 inches long—and weighs 100 pounds when full. A 100-pound bag tamps out at 18 to 22 inches wide and weighs 200 pounds when full.

To create window and door openings, Kiffmeyer and Hunter use temporary wooden forms that are placed in the walls during construction. Bag walls are built around the forms. When walls are completed, the forms are removed. Windows also may be plastered into the wall with or without a frame. Because they need more secure anchorage, door jambs are attached to wooden blocks, called strip anchors, similar to those used in adobe construction. Strip anchors are laid in the wall between earthbags. Glass can also be mudded into window openings as in cob buildings—that is, without a frame (see chapter 6).

After the walls are completed, the roof is added. Domed roofs are very popular. They do not require a top plate. Kaki Hunter writes, "A dome is the strongest, best use for earthbags. Why compromise a near-perfect design?" However, in wet climates, domed roofs must be extremely well protected from water or they risk collapse. Standard roof structures described in previous chapters are also possible. Attaching the roof to the wall does pose special problems, except in post-and-beam buildings. In load-bearing structures where the earthbag wall supports the roof, top plates must be installed to provide a point of attachment for the roof framing. They must also be secured to the wall beneath. Rebar can be driven through the top plate into the underlying bags. An L-shaped rod would be most effective. Vigas that rest directly on the bag wall must also be attached by rebar. When one wants to attach the roof to the foundation, say in a hurricane zone, strapping can be used, as it is in straw bale construction (chapter 3).

After the roof has been constructed, the walls are covered with plaster. Earthen plaster is ideally suited and can be applied directly to the bags without metal lath or chicken wire. Cement-based stuccos, on the other hand, require a roughened surface; lath or chicken wire provides sufficient adhesion. Electrical conduit is run between bags of adjacent courses. Plumbing is best run through the ceiling or floors.

Earthbag construction is truly in its infancy, but holds great promise. Although it may seem odd, the work is fast, fun, and satisfying. It only took me a couple of hours to "master" the intricacies of the procedure and to develop the skills to build a retaining wall. Workshops are available (see the resource guide) for those who want to build relatively inexpensive homes, sheds, offices, privacy walls, or barricades to deter stray softballs (figure 10-6).

OKOKOK Productions

FIGURE 10-6.
Made entirely from earthbags and earthen plaster, this neat little house built by Doni Kiffmeyer and Kaki Hunter is one of a handful of earthbag structures in the United States.

Earthbags can be used in a variety of climates. There's even an earthbag home in the Bahamas built by Hunter and Kiffmeyer. Getting approval from a local building department, however, may pose a challenge, especially when proposing a structure with load-bearing walls. As in other forms of natural building, post-and-beam dwellings with earthbags as infill may be the easiest to get approved. Smaller structures such as sheds can be built in some areas without the need for building permits.

Advantages and Disadvantages of Earthbags

On the positive side, earthbag building is simple, easy to learn, and relatively inexpensive. It requires few tools and no power tools. Earthbag construction is suitable for many different regions of the world, especially in less developed nations where labor is abundant and building capital is in short supply. It doesn't require as precise a soil mixture as most other earth-building techniques. Moreover, earthbag walls are fairly thick and thus lend themselves to passive solar design. And, lest we forget, this technique also permits construction of sensuous, curved walls.

On the negative side, this is a very new building technology. The name *earthbag* is bound to raise eyebrows at your local building department. Getting approval could be difficult. And because it is a new building method, experience is limited. Little is known about the performance of earthbag walls under various conditions, including earthquakes. If you choose this method, you will essentially become a guinea pig.

My advice is to start slowly. Take a workshop first. Then obtain additional experience by building a garden or patio wall. A small shed might be next. If you are satisfied with the results and your structures last, build yourself a house using logic and as much engineering expertise as you can muster.

PAPERCRETE: BUILDING A HOME WITH WASTE PAPER

There must be something about the West that stimulates innovative thinking, especially in home construction. New Mexico appears to be especially conducive to the pioneer builders. That's where you'll find Michael Reynolds, the Earthship builder. That's where much of the straw bale homebuliding work began. That's where Eric Patterson lives. A printer by trade, Eric grew tired of the mountains of waste paper discarded by his business every year. He decided to do something about it and began experimenting with mixtures of cement and paper with hopes of making inexpensive building blocks. Today, Eric is cranking out blocks and building all kinds of new structures.

Colorado is another state where innovation and experimentation in alternative building seem to be taking off. Mike McCain lives in Alamosa, another hotbed of natural building. He's building with papercrete, too.

Gordon Solberg

FIGURE 10-7.
Sean Sands holds a lightweight papercrete block for all to see.

BUILDING NOTE

Papercrete blocks can be made in a variety of forms. My sons and I made an experimental block in a cardboard box lined with aluminum foil. That was just for fun. Some builders pour the material in sandbags, then drop the sandbags in wooden forms. Others pour papercrete directly into wooden block forms. Still others make brownie forms, large wooden block forms. The papercrete is then cut into pieces like brownies.

Papercrete is typically used to make building blocks, like the adobe blocks described in chapter 5. When the cement hardens, papercrete is transformed into a lightweight block that holds its shape, even when wet (figure 10-7). Papercrete and paper adobe could also be poured into forms like those used in rammed earth, cast earth, or light straw-clay construction. Although there is much to be learned about this new building technique, it appears to offer many unique benefits. It is lightweight and easy to work with. Finished blocks can be cut with a saw. Papercrete is filled with tiny air pockets and thus has a high R-value, according to proponents. Estimates are R-2.8 per inch. Papercrete blocks have incredible compressive and tensile strength. Papercrete is inexpensive. Moreover, you can build an entire house out of blocks, including a domed roof (so long as you waterproof it), to eliminate wood use.

Like adobes, papercrete blocks are mortared in place, using a material of the same composition. They are then often covered with a papercrete stucco. This is typically sealed with a silicone material to prevent water from penetrating. On dome roofs, some builders have applied a layer of tar, which soaks into the papercrete, creating a waterproof layer.

Papercrete is made in large vats. Newspaper, cardboard, and magazines (even straw or sawdust, if paper is not available) are tossed in, soaked, then stirred. In short order, with good stirring, the paper is converted into a pulpy mass. Sand and cement are added and the slurry is poured into block forms. (The recommended mix consists of 60 percent paper, 30 percent sand, and 10 percent cement.) Bill Knauss of Tucson, Arizona, has pioneered the use of adobe mud rather than sand and cement to produce paper adobe. It hardens into blocks that have a higher R-value than plain adobe. Being lighter than adobe, it is also easier to work with. It is also relatively inexpensive. Sean Sands, who builds houses out of recycled newspapers and magazines in the City of the Sun near Columbus, New Mexico, estimates that material costs for walls built this way come to about 16 cents per square foot. (That's just materials cost; it doesn't include the labor to build walls.)

Walls are erected on waterproof foundations to prevent water from wicking up into the papercrete blocks. But because the walls are so lightweight, you don't need the massive foundations required for rammed earth, straw bale, adobe, stone, and other comparable natural building techniques. This cuts costs considerably.

FIGURE 10-8.
Sean Sands's papercrete house in Columbus, New Mexico. Papercrete blocks are lightweight and easy to work with. They put waste paper to good use.

Finished walls have a high R-value, according to the advocates. A 12-inch wall, they claim, has an R-value of 33. 6. Supporters also claim that papercrete walls have a substantial amount of thermal mass and thus act like adobe walls. They take on heat during the day and release it at night. If you build a papercrete home, you'll get thermal mass and insulation all in one—as you would in cordwood and log homes.

Because the cement and sand content are relatively low, compared to rammed earth and adobe, window and door openings can be cut out of finished walls. Some builders even key their blocks so they lock together.

Like adobe, papercrete lends itself to creative expression. Domes and vaults can be made using blocks or steel wires and mesh, which are then coated with papercrete and waterproofed to prevent collapse. But not all homes need to look as funky as the one shown in figure 10-8. Papercrete is suitable to a wide range of architectural designs.

Papercrete can be used to pour floors, too. A 4-inch layer under a 2-inch slab, says one source, makes a great floor. It's insulated and softer than plain concrete. Papercrete appears to be resistant to termites, although more testing is needed and precautions should be taken to prevent termites from penetrating a structure.

According to Gordon Solberg, author of *Building with Papercrete and Paper Adobe*, papercrete absorbs water like a sponge. Throw a glass of water at a paper wall and the water will be instantly absorbed. Uncoated domes and walls may become totally soaked after heavy rains. So, he protects walls by proper drainage, overhangs, and a high-and-dry foundation. It would also be wise to seal exterior walls with papercrete stucco that is sealed with a silicone-based product. Homestar waterproofing sealer seems to work well. Water beads up on walls sealed with the product, then runs off. For further protection, Eric Patterson paints the outside of domes with a sealant containing an elastomeric product. The acrylic materials expand and contract with changing temperatures and

BUILDING NOTE

Papercrete can be used to stucco earthbag structures.

thus protect the surface. Although water will penetrate the acrylic, the underlying silicone sealer will prevent it from penetrating any further.

As Gordon Solberg reminds his readers, "Papercrete and paper adobe are experimental. Mistakes have been made, and will continue to be made. Because they are so new, there is no proof that papercrete and paper adobe will last indefinitely." Then there's the issue of fire safety. Although these materials burn much less readily than wood, they do burn.

I like the idea of building with papercrete. It puts waste to good use, requires very little cement, and is easy to learn. Papercrete blocks could even be mass-produced and used for large building projects. I suspect papercrete will rise in popularity in the coming years, especially if preliminary claims about its compressive and tensile strength and cost prove to be true. Its resistance to water will also be crucial in determining its long-term future in the alternative building market. If you want to learn more, I strongly recommend *Building with Papercrete and Paper Adobe* by Gordon Solberg. You may also want to subscribe to the *Earth Quarterly*.

CAST EARTH HOMES

In many states, for example Florida, homebuilders construct dwellings out of poured concrete, creating a shell that can withstand hurricane-force winds. Concrete home construction requires an elaborate system of forms into which the concrete is poured. After the concrete sets up, forms are removed. *Voilà!* You have the exterior walls of a house. Stucco is typically applied on the exterior wall for aesthetics. Drywall is mounted on a modest wood framework (furring strips) secured to the concrete wall to create an appealing interior surface.

Imagine if you could do the same with earthen materials. That is, imagine if you could "pour" an earthen house rather than pack it.

Well, you won't have to imagine too hard. Builders are now experimenting with a technique called cast earth (figure 10-9). Cast earth homes are made of a

FIGURE 10-9.
This lovely home was made from earth, cast earth, to be exact.

Cast Earth Affiliates

slurry of soil mixed with 10 to 15 percent calcined (gently heated) gypsum that is poured into forms (figure 10-10). The mixture sets up quickly, and the forms are removed—usually on the day of the pour.

The outer shell of a cast earth building requires less labor (no packing) to build and can be erected in less time than a rammed earth home. The presence of gypsum counteracts the tendency of many clays to expand and contract as moisture levels change. This reduces cracking. Cast earth walls also require no rebar, even in earthquake-prone regions, according to the inventors of this building technique, because reinforcing steel has a much different frequency of vibration than earthen walls and would be detrimental to their structural integrity. The absence of steel reinforcement reduces labor and material costs. Cast earth construction requires the same tools and equipment needed to build concrete homes, but special training is required. Like rammed earth, cast earth building is for professional builders.

Wall strength of cast earth homes ranges between 600 and 700 pounds per square inch, about the same as high-quality adobe and rammed earth. But proponents note that the costs of cast earth are usually significantly lower than adobe or rammed earth because of the reduction in labor costs. If conditions are optimum for building a cast earth home, they assert, the costs could be as much as one-half to three-quarters of the cost of an adobe or rammed earth home. For larger projects, such as a subdivision or an apartment complex, where an economy of scale could be achieved, savings accrued by using cast earth construction can be considerable. Developers of this new building technology say that, under certain conditions, cast earth homes could even compete favorably with frame wall construction, across the full spectrum of building from high-end custom homes to mass housing. Like proponents of rammed earth, cast earth homebuilders point out that the lifetime costs for maintenance and energy (for heating and cooling) are substantially less than in frame construction—so savings can be had in the long term, too.

Cast earth homebuilding offers other advantages as well. Because the slurry is free-flowing and quick to set up, forms can be lightweight and flexible. Radius corners, undulating walls, curves, and angles are easy to make.

Another benefit of cast earth is that it is more resistant to rain than rammed earth and adobe. If unprotected, cast earth walls absorb rain but retain their original wet strength. Waterproofing can be added to the slurry or applied to the outer surface to enhance water resistance. Exterior plaster, stucco, or siding in extreme climates will add further protection.

Like rammed earth, cast earth walls can be insulated. Rigid insulation can be inserted in the forms prior to the pour.

Cast Earth Affiliates

FIGURE 10-10.
Unlike rammed earth homes, walls of cast earth homes are poured into forms all at once. This cuts down on labor and time and may help make cast earth one of the most viable forms of construction for the mass market.

Cast Earth has been demonstrated successfully in actual residences. It is a true "breakthrough technology," providing a product with all of the properties of traditional earth construction, augmented by superior aesthetics, rapid construction, and affordable cost.

Cast Earth Affiliates

Because the technology is machinery intensive, cast earth building lends itself to mass production. Concrete homebuilders might find the transition to cast earth simple and painless—not to mention profitable and good for the environment. If you are interested in having a cast earth house built, contact Cast Earth Affiliates for the name of a qualified builder in your area (their Web site, listed in the resource guide, provides a searchable database). Unfortunately, the do-it-yourselfer is out of luck. The secrets of the process are highly guarded and available only to members of Cast Earth Affiliates who undergo special training. As noted on their Web page, the years spent in developing this technology "have resulted in a body of knowledge without which a new aspirant could simply not make the process function." The bottom line: If you want a home built of cast earth you will have to hire a certified professional.

Advantages and Disadvantages of Cast Earth Homes

On the bright side, cast earth appears to be less labor intensive and less expensive than other natural building techniques. It produces thick mass walls, ideal for passive solar heating and passive cooling. This technique is suitable for virtually all climates, except extremely cold and perhaps hot, humid climates.

On the down side, cast earth is not for the owner-builder. It requires extensive form work and costly machines. And the process is proprietary, which means that the inventor is holding on to the secrets that make it successful.

BUILDING LIGHT STRAW-CLAY HOMES

Over 400 years ago, German builders began filling the walls of homes with a mixture of loose straw coated with clay, known as light straw-clay. Consisting of a post-and-beam frame and loose packed straw, the walls of these houses provide extraordinary insulation.

BUILDING NOTE

In straw-clay mixture, the clay retards mildew and also acts as a natural flame retardant.

FIGURE 10-11.
Light straw-clay homes may be made to resemble any conventional home.

Paula Baker-Laporte

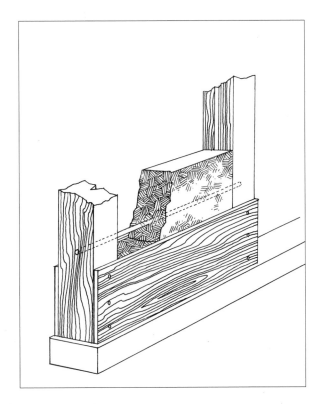

Straw-clay homes can be made to look like any conventional house, even fitting in with the neighbors in suburbs (figure 10-11). To understand why an earth-friendly form of architecture can straddle both worlds, consider a typical cross-section through the wall of a straw-clay home. As shown in figure 10-12, the walls consist of a timber frame. Siding can be attached to the outer surface. Drywall or plaster can be applied to the inner surface. Between the siding and the interior plaster surface lies a thick layer of straw coated with clay.

To build a wall, two 2-foot-wide sheets of plywood are attached to the wooden frame (inner post and the outer 2 x 4s). The space between the ply-wood sheets is then packed with loose straw previously coated with a clay slurry.

The clay slurry (or clay slip) is a watery mixture of clay with the consis-tency of thick cream. It is sprayed on a half a bale of straw spread out on an 8 x 12-foot working platform (plywood on level ground). The straw-clay mix-ture is then turned with a pitchfork or by hand to achieve a light, even coating of clay on the straw.

According to experts, the thickness of the coating varies according to end use. In north-facing walls (in the northern hemisphere) where insulation is the primary goal, the clay is applied in a very thin coat. On south-facing walls, a higher ratio of clay to straw is used to increase the thermal mass. The straw-clay mixture used for walls is placed in the forms and then worked into the corners and the edges by hand. Foot tamping comes next, followed by compaction with

FIGURE 10-12.
Cross section through a wall made from light straw-clay. Notice the forms secured to the timber frame that are used to contain straw-clay during packing.

FIGURE 10-13.
After the first course has been packed, the forms are moved upward and another course is created.

BUILDING NOTE

Some builders insert horizontal rods, such as hardwood saplings, in a straw-clay wall every two feet or so to help stabilize the wall.

wooden tampers. Straw-clay can also be used as ceiling insulation. In this case, a very thin coating of clay is best. It increases the insulative value of the straw yet weighs less, which, in turn, exerts less load on the ceiling.

After the first course of straw-clay has been tamped, a second set of 2-foot-wide plywood forms is nailed in place. After this section of form has been filled and tamped, the bottom form is removed and readied for the third course (figure 10-13). This process, similar to some methods of rammed earth and stone construction, continues until the wall is completed. As Bill and Athena Steen point out, the walls are essentially "rammed straw."

When the wall is completed, the forms are removed to facilitate drying. Any shrinkage that occurs is accommodated by stuffing cracks with handfuls of wet straw-clay. When the wall is completely dry, loose strands are trimmed off and the surface is wetted in preparation for plastering. (Wetting the surface helps the plaster stick and prevents rapid drying, which can cause plaster to crack.) Natural plasters are then applied. The final product looks like any straw bale or adobe home with its thick walls and plaster finish.

Advantages and Disadvantages of Straw-Clay Homes

Like other forms of alternative building, straw-clay has its pros and cons. Straw-clay construction produces thick walls containing plenty of insulation. This, of course, saves homeowners money on fuel bills and makes for a more comfortable interior. Straw is also an abundant and renewable resource. As noted earlier, like some other earth-friendly building technologies, straw-clay homes can look like conventional homes, making them potentially marketable to a larger audience (figure 10-11).

Straw-clay can also be used to build interior walls. Straw bale builders often turn to straw-clay when they need in-fill material for gaps between bales. Straw-clay has been successfully used as an insulating layer beneath earthen floors. Frank Anderson, a builder with extensive natural building experience, is using straw-clay bricks in place of adobe bricks. These lightweight blocks are much easier to work with than heavy adobes and they provide more wall insulation.

The main disadvantage is that straw-clay is fairly labor intensive and the mixture requires a long time to dry. If you're on a tight schedule, this can be a real problem. Another problem is that this technology requires the skills to build a post-and-beam frame. Fortunately, as in virtually all other forms of construction, workshops are available for those who want to learn the technology.

BUILDING WITH BAMBOO

In tropical countries, bamboo is the material of choice for building homes, furniture, musical instruments, and even for making paper (figure 10-14). Widely available and relatively easy to work with, this renewable resource has a long history in building. Visitors returning from the Far East often remark on the

Super-strong and durable, bamboo may one day replace wood products in American homes—and save the forest for the trees.

CAROL STEINFELD,
"A Bamboo Future"

FIGURE 10-14.
Bamboo is a remarkable building material with a long history of use. It can be incorporated into many natural building techniques described in this book.

extensive bamboo scaffolding used to construct office buildings and other structures.

Today, bamboo is finding its way into buildings elsewhere—for example, in Costa Rica, Mexico, Nicaragua, Australia, France, and, to a lesser extent, the United States.

In Costa Rica, the government has invested heavily in bamboo plantations and has built large earthquake-proof housing projects using this renewable resource. Bamboo is used to make a substitute for particle board, plywood, and flooring (chapter 14). Plyboo America in Kirkville, New York, for example, sells a tongue-and-groove bamboo flooring board that consists of three layers of bamboo veneer glued together under high pressure. According to the company, plyboo is as easy to work with as any floorboard and appreciably harder than most oak flooring. It does not shrink and swell as most hardwoods do, according to the company's literature. This makes it a very stable flooring product. The company is planning to add bamboo panel boards and moldings to its product line. Other companies in the U.S., Australia, and elsewhere also sell bamboo flooring, and many offer a variety of other bamboo products for use in cob and straw bale homes. One product of interest is bamboo structural members (columns, beams, and rafters) for new homes and renovations. Innovators in the building materials industry are now producing composite beams made from strips of high-strength bamboo.

Bamboo has made a modest appearance in modern homebuilding thanks to its many advocates, among them Jules Janssen and Victor Cusack, whose books explain how to grow and harvest bamboo and offer details on how bamboo is used to build homes. But there are other reasons as well. Bamboo is abundant. According to one estimate, approximately 35 million acres (14 million hectares) of the Earth's surface is covered with one of the world's 1,200 to 1,500 species of this hardy plant. Surprisingly, bamboo is widely distributed around the world.

In California and the Southwest, straw-bale buildings with bamboo superstructure and pinning have been built and could be prototypes of earthquake-proof homes of the future.

CAROL STEINFELD,
"A Bamboo Future"

Special care must be taken when harvesting and preparing bamboo, because it doesn't hold up to the elements very well. By harvesting culms (stems) during the dry season, one can avoid the powder post beetle, a pest that eats the wood. Use of culms in the interior of a house, away from the weather, will ensure a long useful life. Bamboo that is used for exterior applications, however, must not only be dry-harvested, it must be treated with preservatives. Another key to success is to use mature culms. Not only are immature culms weak and likely to shrink, crack, and distort, they are also vulnerable to powder post beetles. Cusack's book and Janssen's book offer advice on ways to treat bamboo to enhance its useful lifespan.

FIGURE 10-15.
Bamboo can be used to build elaborate roof frames as shown here.

Simon Velez, courtesy Darrell DeBoer

Far from being restricted to the tropics, as many people think, bamboo is found on several continents. Australia, for example, boasts at least three to five indigenous species of bamboo. Some species prefer cold climates. Moreover, many tropical species can survive temperatures as low as 25 degrees Fahrenheit, making them suitable for growth in temperate climates. Cold-climate species can tolerate even colder temperatures, down to 10 to 16degrees Fahrenheit.

Yet another reason for this plant's promise as a building material is its rapid rate of growth. According to Cusack, author of *Bamboo Rediscovered*, bamboo is the fastest-growing, most versatile "woody" plant in the world. It reaches harvestable size in a fraction of the time of timber. In the tropics and subtropics, many bamboo species grow to a height and thickness suitable for building in four to five years, compared to trees that take decades.

Bamboo species are typically divided into clumping and running varieties. Although many of the latter are highly invasive and consequently a bit troublesome, there are dozens of noninvasive, clumping species that can be grown without fear of spread. Although most clumping varieties are tropical or subtropical species, many of them are cold-tolerant and therefore suitable for planting in temperate climates. Besides being noninvasive, clumping varieties grow quickly and require minimal maintenance. They also grow well among trees, meaning they grow taller and straighter, provided there is sufficient water.

Bamboo is not really a woody species, but rather a member of the grass family. Although most species are hollow, some of the strongest ones are solid for at least half their height. For information on species that might be appropriate for your area and for your expected use, see Cusack's book. It lists many common species of bamboo, notes their size and temperature tolerance, and provides information on growing and harvesting them. It also describes their usefulness as structural components of buildings or furniture.

Building a House from Bamboo

In Japan, bamboo is used to build expensive homes for wealthy clients, but in most locations, bamboo is used to build small domiciles for the less fortunate.

As a building material, bamboo has many applications. It can be used to make columns, rafters, studs, flooring, and mats to cover walls and ceilings (figure 10-15). As a general rule, the thicker portions of the culms (stems) are used to make framing members, including floor joists, columns, beams, and roof rafters. The smaller-diameter portions are used for non-load-bearing structural components of walls and ceilings or for furniture.

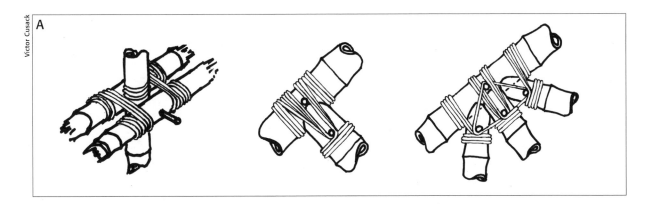

A

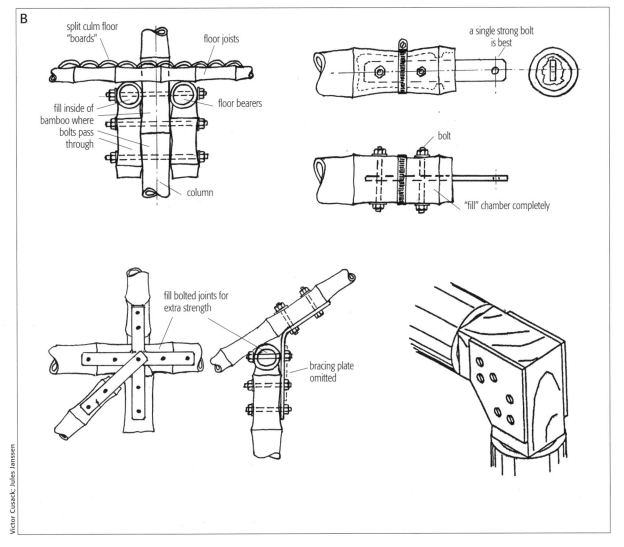

B

split culm floor
"boards"

floor joists

fill inside of
bamboo where
bolts pass
through

floor bearers

column

a single strong bolt
is best

bolt

"fill" chamber completely

fill bolted joints for
extra strength

bracing plate
omitted

Victor Cusack

Victor Cusack; Jules Janssen

FIGURE 10-16 A, B.

*(A) Traditional bamboo joints consisted of pins and lashing. (B) Modern bamboo
joints made from steel plates, braces, bolts, and glue are reportedly stronger and may
be required by local building codes for approval.*

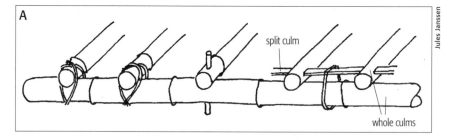

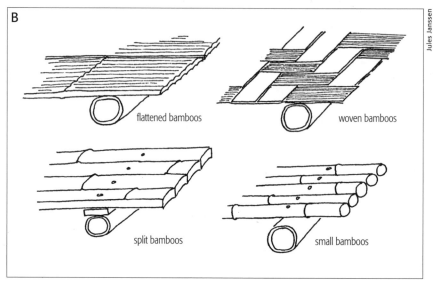

FIGURE 10-17 A, B.
(A) Bamboo floor joists may be secured in a variety of ways as shown here. (B) Bamboo can be fashioned into flooring that is laid over joists. Notice the many different options.

Bamboo frames hold up remarkably well, even in earthquake zones. However, because most species of bamboo split when nailed, special techniques must be employed to hold framing members together. Traditional builders joined bamboo frame members by pins inserted into holes drilled in bamboo culms. Pins were supplemented by lashing (figure 10-16a). Modern builders have successfully introduced joints made from steel plates, braces, bolts, and glue (figure 10-16b). Although less attractive, I'm told that modern joints are stronger than traditional joint systems.

Many bamboo homes in Asia have earthen floors made with packed dirt from the site. Because bamboo deteriorates when exposed to moisture, most bamboo homes are built on simple foundations that raise the columns off the ground. Foundations are made from concrete, stone, or brick. Innovative methods of embedding the bamboo in the foundation are described in *Building with Bamboo* by Jules Janssen.

Elevated floors are also desirable for comfort and cleanliness. Figure 10-17a shows ways in which the floor joists are secured. Flooring is made from small bamboos (canes) or split, flattened, or woven bamboo, and then applied to the joists, as shown in figure 10-17b. Walls are made from "bamboo studs" and covered with bamboo mats or vertical bamboo, either canes or split culms.

Bamboo is also used to make roofs. Elaborate trusses from large culms are lashed together or joined by pins and plates. Thatch can then be secured to the frame. Some builders use halved bamboos (split culms) that run from the ridge to the roof plate as shown in figure 10-18c. Roof culms are nested together to create a series of channels that allow water to flow down the roof. In some regions of the world, local builders make shingles out of bamboo.

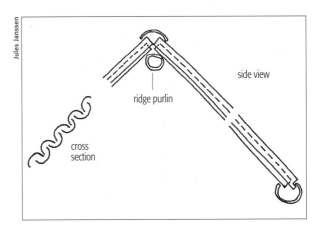

side view

ridge purlin

cross section

Jules Janssen

FIGURE 10-18.
Bamboo can also be used to make roofs.

Bamboo as Food Source and Landscaping

In addition to providing building materials, bamboo is a great source of food. Edible shoots can be harvested from plants without damaging them. Bamboo is also a marvelous decorative plant for outdoor landscaping. Many species have beautifully colored shoots and distinctive leaves. In addition, bamboo can be planted to enhance privacy, block the east or west sun, or divert wind away from a house (chapter 15). Some species are effective barriers to livestock. Fast-growing species can even be used to control soil erosion or to protect river banks. Be careful, however, when introducing foreign species into established ecosystems. A strong, invasive species could become a real pest, overtaking native species and changing the ecological relationships of the area.

Should You Build a Bamboo Home?

Bamboo is a remarkable building material, with many unique properties. It has many advantages over wood, notably faster growth. In fact, some species reportedly grow over three feet per day.

But bamboo is by no means a panacea. Sobriety is the order of the day for those who wish to experiment with this up-and-coming building product. To adequately consider bamboo, familiarize yourself with this long list of pros and cons, adapted from *Building with Bamboo* by Jules Janssen.

Advantages of Bamboo Homes

• Bamboo is relatively strong and stiff.
• Bamboo contains no knots or rays, radiating fibers like those found in trees. This feature helps bamboo distribute stresses more evenly throughout its length compared to dimensional lumber.
• Bamboo can be cut and split with simple tools.
• The outer skin of the bamboo culm contains a high concentration of silica, which makes it hard and durable.
• Bamboo can be grown in many climates and harvested within a few years of planting.

- Bamboo buildings hold up well against storms and tornadoes.
- Bamboo is hardy and easy to grow and harvest.
- Bamboo plants serve many purposes.

Disadvantages of Bamboo Homes

- Bamboo is hollow and round, which makes it more difficult to join than dimensional lumber.
- The outer skin of the culm contains a lot of silica, which dulls tools and makes it difficult to glue to other pieces.
- Demand has been so high in some areas of the Far East and management so abysmally short-sighted that some species are endangered.
- Bamboo is not very resistant to weather and requires chemical treatment for protecting it.
- Bamboo deteriorates rapidly when it comes in contact with wet or damp soil.
- Bamboo is highly combustible.
- Bamboo tapers, making it more difficult to build with than dimensional lumber.
- Bamboo varies in size from one piece to the next, making it difficult to build with.

HYBRID HOMES: COMBINING EARTH AND FIBER

Once upon a time a straw bale home was a straw bale home and an Earthship was an Earthship. One wouldn't think about marrying the two building technologies—or any others, for that matter. Proponents of the individual technologies, while definitely thinking outside the box of contemporary architecture, were stuck in their own little boxes.

Today, however, many of the innovators who brought us straw bale homes, cob cottages, packed tire homes, and other alternative building technologies are wandering outside the conventional limits of their unconventional fields. They are mixing and matching building techniques to produce a new generation of natural homes. A home like mine, for example, combines packed tires and straw bales. The bedrooms, kitchen, and dining room are made from tire U's. The living room is straw bale. Other builders use packed tires and earthbags to build foundations for straw bale homes. The list goes on. The possibilities are only limited by the imagination.

Bill and Athena Steen, best known for their work in straw bale construction, are pioneers in hybrid building. More and more, the Steens find themselves incorporating traditional clay building systems with straw bale. Rather than referring to themselves as "straw bale builders," this dedicated couple prefers to define themselves as "clay/fiber builders who happen to be using straw bales in the exterior walls." The Steens use clay for floors, walls, and

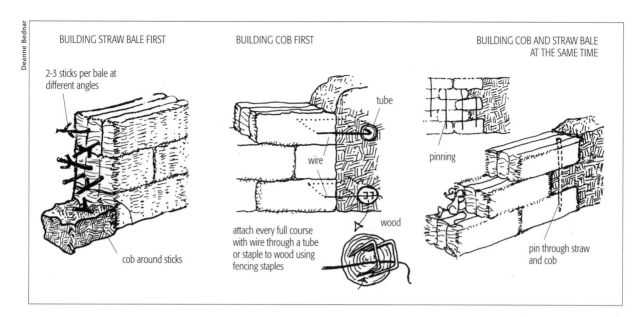

BUILDING STRAW BALE FIRST

BUILDING COB FIRST

BUILDING COB AND STRAW BALE
AT THE SAME TIME

2-3 sticks per bale at
different angles

tube

wire

pinning

cob around sticks

attach every full course
with wire through a tube
or staple to wood using
fencing staples

wood

pin through straw
and cob

plaster. Once the straw bales have been placed in the wall, they often use traditional English cob to fill odd spaces and to sculpt window and door openings, as well as shelves, bancos for seating, and free-form fireplaces. Larger
spaces in the wall that are too small for bales are packed with loose straw
coated with clay slip. The Steens have found that although straw-clay provides
less insulation than straw bales, it offers greater flexibility—for example, it is
more adaptable to building curved walls and thinner walls. They often use
straw-clay to fill smaller framed wall sections where bales would not fit—for
example, the gabled ends of a home.

As the Steens write on their Web page, "Straw bales can be used to create
super-efficient passive solar buildings, but they need the addition of adequate
thermal mass to store the heat gained during the day from the sun." Accordingly, they are now using adobe blocks to build the interior mass walls that
come in direct contact with sunlight pouring in through south-facing windows
during the winter months. Rammed earth and cob walls could also be used in
such cases. Figure 10-19 shows ways to join a cob wall with a straw bale wall.

As Michael Smith writes in *The Cobber's Companion,* "Cob and bales are
thoroughly complementary. . . . Because bales are large, rectangular blocks,
straw bale buildings tend to be boxy. Using cob for parts of these structures
softens corners" and "offers more sculptural options."

Earthen plaster is used in many natural forms of building. It adheres wonderfully to straw without stucco netting, which is required for cement stucco.
Another advantage of earthen plaster is that it allows the walls to breathe.
Moisture won't build up inside a wall and cause the insulation to mildew or
decay. Furthermore, earthen plasters are relatively easy to repair, especially
compared to cement stucco.

FIGURE 10-19.
*Three ways to tie straw bale
and cob walls.*

*Cob is the duct tape of
natural building.*

DONI KIFFMEYER and
KAKI HUNTER

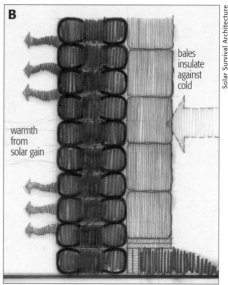

bales
insulate
against
cold

warmth
from
solar gain

FIGURE 10-20 A, B.
The strawbale earthship awaiting plastering (A). Straw bales are located on a foundation external to the tire walls, providing superior insulation for the mass walls (B).

FIGURE 10-21.
In my home we used concrete buttressing to support tire walls where they joined straw bale walls.

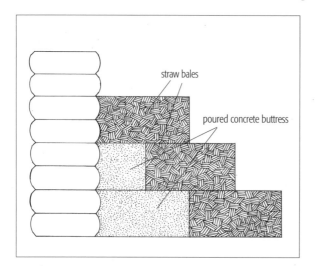

straw bales

poured concrete buttress

"By viewing building as a process of combining different, but complementary materials rather than adhering to a particular building system," write the Steens, "we have given ourselves the freedom to create structures that respond to a wide variety of contexts and circumstances. They can be elegant or simple, quick or detailed, inexpensive or costly, and probably most importantly, they can be built from predominantly local materials in whatever combination best matches the local climate."

Another example of a hybrid structure is the straw bale Earthship (my name for it). Shown in figures 10-20a and 10-20b, this building is an above-ground Earthship with a skin made of straw bales. The rammed-earth tire walls of the Earthship provide thermal mass; the straw bales provide much-needed insulation (figure 10-20b). In this hybrid design, the bale walls are covered with a layer of plaster to protect them against the elements and to add beauty.

Hybrid homes have their own challenges. As noted earlier, I built a straw bale living room for my otherwise all-tire home. To "marry" the two walls, my builders created a special concrete buttressing structure that supports the lateral load the tire wall exerts on the straw bale wall. The internal buttress consists of a block of concrete poured in conjunction with the foundation of the straw bale section, as sketched in figure 10-21.

Straw bale additions can be built on to conventional wood-frame houses with a little ingenuity. If your home has conventional siding, don't be dis-

FIGURE 10-22 A, B.
Outsulation anyone? This uninsulated metal building (A) is getting a straw bale facelift that will greatly improve its thermal performance (B).

mayed. As noted in chapter 3, you don't have to cover a straw bale wall with plaster. Any siding can be adapted to a post-and-beam structure. Even the Department of Housing and Urban Development's special loan program for renovating houses includes straw bale additions. Cob additions can also be added to conventional homes.

To retrofit existing structures, a skin of straw bales can be added as shown in figure 10-22. Cement block, adobe, concrete, and wood-frame homes, even mobile homes, are all excellent candidates for a straw bale face-lift. Matts Myhrman and S. O. MacDonald us the term "outsulation" to describe the exterior bale wrap. As the Steens point out, "The high insulation [or should we call it outsulation] value of the bales can bring those structures from thermal horrors to super performers." Many inefficient homes, such as those built for low-income residents and military personnel, could benefit from a straw bale wrap.

Hybrid earth-friendly homes are a fascinating outgrowth of efforts by many natural builders. Through trial and error they are finding ways that these remarkably different, yet highly compatible, building techniques can be used together. The result has been the development of new building systems that combine efficiency, ease of construction, durability, and even artistry.

Hybrid homebuilding is truly experimental. Information on hybrid structures is scarce, and when it does appear, it comes in dribbles far too small to be satisfying. *Build It with Bales, The Straw Bale House, The Cobber's Companion,* and *The Last Straw* will partially quench your thirst for knowledge. Workshops also provide some information.

Matts Myhrman and S. O. MacDonald advise, "By its nature, a hybrid structure often requires extra thought during the design process. Draw it, model it, get a 'second opinion,' and still expect to have to think on your feet once you get started." If sound building practices are followed, your experiments could prove successful.

Emerging Natural Building Techniques / **247**

Michael G. Smith

How long have you been building with natural materials? What sparked your interest in this type of architecture?

I've been around building design and construction since I was born. My father was an architecture professor and I grew up surrounded by architects. I learned to build by helping put an addition onto our house, which was built almost entirely of salvaged materials, reclaimed from the dump and from demolition sites. One of my Dad's main interests is vernacular architecture. As a kid, I spent a year traveling with him in France, Italy, Spain, and Portugal. He was studying what I would now call natural building, which of course describes almost everything before the last 200 years and a lot of recent buildings, especially in "less developed" countries. Although I never intended to become an architect or a builder, some of that early exposure must have worn off on me.

Later, I went to Central America to work on sustainable forest management and grassroots community development. I worked for two years with a small association of subsistence farmers in the Costa Rican jungle. They asked me to help them design and build a little eco-lodge on a site that was completely surrounded by swamp and accessible only by foot over more than a mile of narrow boardwalk. There was little choice but to use local materials from the site. We got a portable chain saw mill and sawed up trees that were already down on the site. That really opened my eyes to the possibility of building only with what was available within walking distance.

When I returned to the United States I wanted to educate Americans about the global consequences of our consumption choices: how what we buy impacts communities and ecosystems all over the planet, usually negatively. By using local materials, we minimize our environmental impact. We also reduce our contribution to pollution from transportation and manufacturing and to the neo-liberal trade colonization of the world, which may be the single greatest force threatening traditional societies today. I see natural building as a small but essential part of what is necessary for a sustainable future.

As I understand it, you began building with cob and now are experimenting with a variety of other materials that you use in conjunction with it. How many cob homes have you helped to build?

Yes, I joined Ianto Evans and Linda Smiley in Oregon in 1993 and helped them start up the Cob Cottage Company. I've been researching and teaching natural building techniques ever since. Until the last couple of years, I was mostly focused on cob. I haven't kept a very careful count, but I would guess I've been involved with thirty or more cob and hybrid buildings, including houses, cottages, studios, greenhouses, outbuildings, and playhouses. My involvement has ranged from design only to being the primary builder. Most typically, I provide consultation on siting and design for owner-builders and then lead one or more construction workshops to help get the structure built.

What stimulated your interest in using combinations of natural materials?

When we started the Cob Cottage Company, we didn't know very much about different kinds of natural building taking place in other parts of the country. Pretty quickly, we started hearing about straw bale, Earthships, and so on, and decided to get as many of the innovators together as possible so we could learn from one another. We sponsored the first Alternative Building Colloquium in 1994 in Oregon. It was attended by seventy-five natural builders from all over North America. We had a full week of hands-on demonstrations, slide shows, and stimulating conversations. We learned a lot about straw bale, light straw-clay, earthen plasters, adobe floors, and timber framing, all of which we started to incorporate in our buildings, along with cob. It was the first time that many of the people present recognized that we were not just isolated individuals, but actually part of a large group working on similar issues, and began to think of ourselves as a movement.

That first colloquium was so successful that it has been repeated six more times, in New Mexico, California, Maryland, and Oregon. These gatherings, now called Natural Building Colloquia, are the main networking event of the North American natural building community (with international participants as well). They have led to a lot of collaborations and have been a major force toward hybridization of different natural building techniques.

What advantages do you see in hybridization?

Almost all buildings, whether natural or not, are actually hybrids. You use different materials for the foundation than for the walls, and something else for the roof. You choose the materials that are most appropriate for the job. Straw and earth make great walls, but they make lousy foundations in rainy climates.

The way we use the term "hybrid" implies an even further breakdown of function. For example, the north wall of the building should behave differently from the south wall because of thermal differences and wind and rain directions. The interior partitions have a different job from the exterior envelope due to structural and thermal differences. Kitchen walls and bedroom walls have different needs for openness, sound, and view privacy. Each wall presents a slightly different problem, and the solutions vary. As a result, you may end up with several different wall systems in the same building.

Hybridization is fun. It gives the builder opportunity to play around with different materials and systems in the same building. It provides lots of room for creative problem solving. And it creates uniquely beautiful buildings, with lots of different facets and qualities. I'm currently working on a small cottage that combines rubble trench and earthbag foundations, cob, straw bale, wattle-and-daub, and light straw-clay walls, adobe floors, earth and lime plasters, and bamboo roof trusses—all in a couple of hundred square feet. The project has enabled us to train dozens of people in nearly every aspect of natural construction.

Hybridization does present challenges. It can make both design and construction more complicated. You have to figure out how the wall systems are going to connect structurally, and whether they're likely to react differently to different kinds of forces and influences, like earthquakes and moisture. When I'm using multiple wall systems, I emphasize the continuity of the foundation and often use a bond beam to hold the tops of the walls together, which I wouldn't necessarily do in an all-cob building.

What are some of the best examples of hybridization?

Cob and straw bale have already proven to be a very successful marriage. Cob is very simple, very available, very sculptural. It has good thermal mass, but not much insulation, and it goes up slowly. Straw bale, in contrast, goes up faster and is a better insulator than cob, but lacks the thermal mass and sculptability. They're natural complements to each other. At this point I've worked on six or eight buildings that combined straw bale on the north side and cob on the south. In many temperate climates this combination can perform better thermally than either system alone. Lacking insulation, an all-cob building tends to lose a lot of heat through the north wall in the winter. But cob on the south side can catch and hold the heat of the winter sun. With less thermal mass, the temperature in an all-straw building tends to fluctuate more quickly. The addition of cob holds more warmth in winter and more "coolth" in summer. Another good combination is straw bale exterior walls with cob partitions. This produces a very stable interior temperature, and the cob walls can be made much narrower than bales, which take up lots of space.

Cob works well in combination with lots of other natural materials. It is great for stuffing cracks between rocks, or between straw bales, or between infill bales and a frame. Even people who consider it too laborious to use as a major wall system still find it invaluable as a general smoother and crack-stuffer, and for sculptural elements.

Another combination I've been playing with lately is a hybridization of wattle-and-daub and light straw-clay. In a sense, each of these systems is already a hybrid. Wattle-and-daub marries a lath of woven sticks with a clay-based plaster. Light clay is a combination of straw and clay slip, usually rammed while still moist into a slip form. Both are traditional European infill techniques, relying on a post-and-beam frame to carry the load. I've used double wattle panels to make a permanent form into which dry straw-clay is lightly packed. This provides better insulation than either system alone. The daub on the inside of the wall can be made as thick as desired to increase the thermal mass. I've been very pleased with my initial experiments, but it only makes sense in a forest where you have lots of small sticks for the wattle. Or you could use split bamboo, wild cane, probably even old wire fencing.

What mistakes have been made and what can we learn from them?
Sometimes a structure fails, or a combination of systems turns out to be impractical. That's to be expected at this stage, and in fact encouraged. I consider every attempt to do something new to be a good example, as long as we are receptive to the lessons we learn and are willing to share those lessons with others. In general, our understanding of each of these systems individually is still very incomplete. When we combine multiple systems in the same building, the learning curve is even steeper. However, the potential for beauty and functionality and site appropriateness is correspondingly higher. The natural building resurgence is so new that I expect it will be quite some time before the dust clears and we are able to sort out what really works best in a particular situation.

Have you been watching the development of other new natural building "technologies" and, if so, what are some of the most exciting and promising ones? What makes them so promising?
I divide alternative building systems into two categories. The first includes everything that has been used over the millennia by traditional cultures. These include cob, adobe, rammed earth, wattle-and-daub, stone, timber framing, thatch, lime plasters and mortars, lashed and pegged bamboo, and many more. These systems, even if they are no longer popular or common or even respected, have proven that they work. Our job today is to relearn what our ancestors knew, and figure out whether it needs to be adapted or hybridized in order to meet modern building needs. We can learn a lot from historians and preservationists and by investigating surviving old buildings.

The second category includes building systems that fundamentally depend on some material or technology that has been invented recently. Baling machines are only about 100 years old, so straw bale construction falls in this class. Woven polypropylene bags, used for rammed earthbags, are quite new. Likewise rubber tires. There are a number of alternative building techniques that depend on Portland cement. These include papercrete and cordwood masonry as now practiced in North America. There's a lot of excitement about these techniques because they are perceived, rightly, as something new.

In my view, the jury remains out on everything in this second group. Ten or fifty or even a hundred years just isn't enough time to explore all the impacts and consequences of a new technology. How long will plaster protect earthbags from photo-degradation? How will straw bales survive fifty years of moisture cycles in humid climates? We can make guesses, we can even make computer models to back up our hypotheses, but ultimately only time will tell. The same is true of an old technology used in a new way or under nontraditional circumstances. And that blurs the distinction between the two categories because almost nothing we build today is really being done in a traditional way.

I can't predict which of the systems we are currently experimenting with will eventually turn out to be the most effective. I believe it will be different in every bioregion, reflecting the climate, the culture, and the availability of materials. It may even be different for individual builders, depending on their skill level, time and budget constraints, and personal preferences.

When you come right down to it, there are only a few basic natural materials that builders have been playing around with forever—stone, gravel, sand, clay, wood, and other plant fibers including bamboo and straw. Although I'm drawn to the simplest, most time-tested solutions, I'm intrigued by the idea of combining low-tech natural materials in a more scientific way. Lance Durand has been experimenting with what he calls "natural composites," combining fibers of different magnitudes with a binder under special conditions. Using just straw and clay, he has produced a material with a bending strength approaching that of wood. It's basically engineered cob. Another variation on the same theme is papercrete or fibrous cement, a mixture of paper pulp and cement. Papercrete's advocates claim that it is very light and insulative, yet has excellent

strength and water resistance. These are both very new technologies that deserve to be watched.

Another of my favorite up-and-coming building systems is rammed earthbags, which Nader Khalili calls "superadobes." The bags allow you to use a much wider range of soils than you can with most earth building techniques. Khalili and others have been making earthbag domes in dry climates. I've used them a number of times as foundations for cob and straw bale. I use gravel in the lowest courses of bags to prevent rising damp. These foundations are quicker, cheaper, and easier than other kinds.

The durability of the bags is the biggest unknown. And of course you can't really call polypropylene a natural material.

What advice do you have for someone who wants to construct a natural home, but wants to blend different natural building techniques?

I would say, "Go for it." First, learn as much as you can about natural building and about as many individual techniques as possible. The more information you have, the better decisions you will make.

The other advice I can offer is, "Think it through." Figure out in advance how the different materials will be connected. Make a model of the building. Think about the different natures of the various materials and how they will change over time or under dramatic circumstances like floods, fires, hurricanes, and earthquakes. A little structural redundancy probably does no harm.

How can someone best learn about these new techniques?

A combination of written information and hands-on training is ideal. It's fairly easy to get good information on many of these techniques individually. There is a good selection of books and workshops on straw bale, cob, and timber framing.

Journals and newsletters are one of the best ways to keep abreast of the art of natural building. We are learning so quickly that all the books are partly out of date by the time they are printed. *The Last Straw* has increasingly been covering other innovative building techniques besides just straw bale. Another good one is *Joiners Quarterly*. It covers not only timber framing but all the natural infill and roofing techniques that are compatible with it. *Adobe Builder* is a great reference for serious earth builders, particularly in arid climates. And the newsletter *Building with Nature* does a good job at framing the big picture of what is natural building and how to promote it. It's a great idea to organize a little natural building club or discussion group in your town, get a bunch of shared journal subscriptions, send people off to different workshops and have them report back, start out with some small hands-on projects like an oven or a playhouse, and see where it goes.

What is your idea of an ideal natural home?

An ideal home is one that grows organically out of the needs and resources of the site, the builders, and the inhabitants. Every one would be unique. Its construction would contribute to the health of the site and to the sustainability of the planet. By living in it, its inhabitants would become happier, healthier, and more motivated to take an active part in the restoration of the Earth.

MICHAEL SMITH teaches hands-on workshops, researches, writes, and consults about natural building. He is the author of *The Cobber's Companion: How to Build Your Own Earthen Home* as well as several book chapters and numerous articles about cob, natural building, and permaculture. He is a contributing editor to *The Last Straw* and *Earth Quarterly*. He lives at an intentional community in Northern California where he enjoys growing and eating food and taking long walks in the woods.

To contact him write or call: Michael G. Smith, P.O. Box 764, Boonville, CA 95415. Tel: 707-895-3302. E-mail: lorax@ap.net

THE NATURAL HOUSE / PART

Sustainable Systems

ENERGY INDEPENDENCE
Passive Solar Heating and Cooling

CREATING SUSTAINABLE SHELTER entails more than simply building a natural home. Much more! It takes a giant step. A leap from conventionality to new architecture. Actually, an enormous leap, for we can't bridge the chasm between our currently unsustainable way of life and a sustainable society by leaping half the distance. Creating a sustainable home requires careful site selection, reliance on renewable energy, low-impact water and waste systems, use of green building materials, and environmental landscaping.

In this portion of the book, we will turn our attention to these requirements, beginning with the use of solar energy, a surprisingly simple and effective means for heating a home. We will also examine ways to cool homes naturally. The chapter is not meant to provide you with all of the information you will need to build a passively heated and cooled home. Rather, it will provide enough background material for you to make sound decisions. And as always, you can use the resource guide to explore this subject in more depth.

SOLAR ENERGY

The potential for solar in most parts of the world is enormous. Even in those cold, cloudy winter regions like the northeastern United States, solar energy can support an efficient household very nicely. Unfortunately, solar energy is still thought to be the sole domain of homeowners in sun-drenched regions, such as the southwestern United States, Central America, northern South America, Africa, India, and Australia. But nothing could be further from the truth. Solar energy is just as viable a solution in Ottumwa, Iowa, and

Both people and planet pay dearly for the production and delivery of energy to maintain comfort in shelter. Shelter of the future must be inherently capable of both collecting and storing appropriate temperature for human comfort. . . . Human shelter must be redefined to become more integral with natural processes of the Earth to achieve this.

Solar Survival Architecture
Earthship Chronicles,
May 1998

Rochester, New York, as it is in West Palm Beach, Florida, and Tucson, Arizona. In areas with long, cold winters, capitalizing on the sun's free heat can result in major savings in fuel bills and a substantial decrease in human impact on the life support systems of the planet.

The first rule of sustainable building, then, is: Don't exclude solar from your plans simply because your home site is not as sunny as the residence of a friend who lives in Phoenix or the outback of Australia. Think solar. Go solar. Plan to get as much solar energy as you can. You can take advantage of solar energy for heat with little, if any, extra cost to your building project. All you have to do is design smart. As James Kachadorian, an engineer, solar builder in the Northeast, and author of *The Passive Solar House*, notes: By properly orienting your house and rearranging items, notably windows, that you were going to buy anyway, you can tap into the generous potential of the sun at little or no extra cost. Yes, that's true. It isn't going to cost you an arm and a leg to go solar. And over the lifetime of your home, it could save you thousands of dollars in fuel bills.

Solar has become an economically viable option in part because building specifications have changed. In the United States, materials and techniques that once added to the cost of building a solar home are now considered standard building practice. For example, double-pane, high-performance glass is found in virtually all new windows and patio doors. You don't have much choice. Insulation standards have also greatly improved. Vapor barriers, which prevent insulation from getting wet and thereby becoming less effective, and exterior house wraps, which reduce air infiltration, are now commonplace in new construction in the U.S. Solar homes aren't much different from non-solar homes anymore. Because of these and other changes, you won't be spending more to build a passive solar home.

Bottom line: Most of your heat (and electricity) can be supplied by nature, free of charge, by that massive ball of nuclear fusion, burning 93 million miles above our homes. So go for it!

Solar Systems: Consider Full Costs First

Solar systems come in many shapes and forms, some more complicated and more expensive than others. But several solar energy technologies are quite economical, especially if you take a long-term view and do a full-cost analysis.

The usual way of analyzing the cost of passive solar energy systems is the most short-sighted. It compares the cost of a passive solar system and the savings on your energy bills to the cost of installing a conventional heating system and the fuel bills paid over the lifetime of the house. Although passive solar energy does well in this comparison, other solar technologies such as photovoltaics and hot water systems don't fare well.

If you go a step further and compare the environmental costs, all solar systems start looking much better. Passive solar energy shows even greater promise than it did in conventional economic analysis. Unfortunately, this kind of full-cost analysis is rare. Environmental costs of conventional energy systems are rarely considered. In most countries, citizens and businesses burn fossil fuels with abandon, causing billions of dollars of damage to human health, ecosystems, and crops from air pollution. Mining, oil spills, and the rising sea level and violent storms caused by global warming further escalate the cost of this dangerous dependence on fossil fuels. But environmental damage goes largely unrecognized.

Subsidies further skew the economic picture. Many governments spend billions of dollars each year to subsidize fossil fuels while giving pennies to the solar industry. Each year, for instance, the U.S. government spends billions of dollars to protect oil tankers in the Persian Gulf. Subsidies camouflage the real cost of fossil fuels, making them seem a lot cheaper than they are. If we paid the true cost of a barrel of oil to cover the subsidies, we couldn't afford to drive cars. More important, were we to pay the full cost of conventional fuels directly and not through taxes, renewable energy resources would appear downright cheap!

Despite the terribly uneven playing field on which renewable energy and fossil fuel energy "compete," passive solar homes fare well in economic comparisons. Even if you spend a little extra to build a passive solar home, the added comfort, freedom from fuel bills, and environmental benefits may make the expenditure well worth it.

Types of Solar Systems

Of the many different types of solar energy systems, solar heating and solar electricity are the most prevalent. The first type provides space heating or hot water for the house, as the name implies; the second provides electricity to power televisions, computers, microwaves, fans, lights, and much more.

Solar heat may be provided by an active or passive system. Active solar systems rely on collectors, usually located on rooftops, to gather sunlight and convert it to heat. Pipes and pumps transport the heat to storage tanks, hence the term *active solar*. The stored energy is used to heat water for showers, baths, and washing dishes, or to heat interior spaces. Figure 11-1 shows how these systems operate.

Passive solar systems are a radically different, remarkably simple form of solar energy, consisting of buildings that collect sunlight and store its energy as heat to warm interior spaces. (Water for domestic use can also be heated passively.) They require no pumps or pipes or storage tanks—hence the term *passive*.

FIGURE 11-1.
Solar panels in an active system shown here absorb sunlight energy and convert it to heat. The heat is drawn off, usually by a liquid flowing through pipes in the panel. Heat is then transferred to water for storage and use.

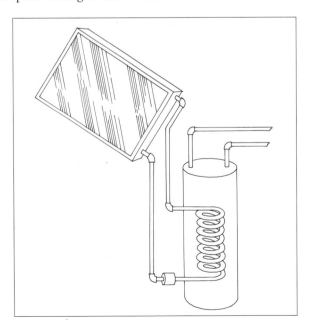

FIGURE 11-2.

Passive solar energy is a simple way to heat buildings, but care must be taken to design the structure to optimize solar gain and heat storage. Floors and walls serve as thermal mass but work best when placed in direct contact with incoming solar radiation. Care must also be taken to minimize summertime solar gain through overhangs (see eaves and cantilever).

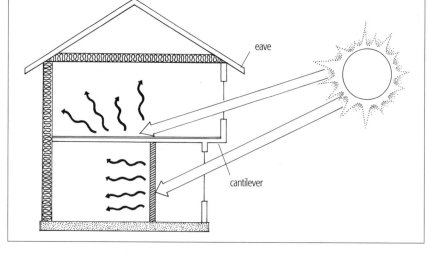

A dose of intelligence, forethought, and a basic understanding of solar cycles and ways to capture solar energy in a building are vital to your success. The idea behind passive solar homes is to design with nature, not against it.

In a passive solar home, sunlight streams through south-facing windows (in the Northern Hemisphere; if you live in the Southern Hemisphere, change "south" to "north" as you read this chapter). The sun's rays are absorbed by the floors and interior walls, then converted into heat. Some of the heat radiates into the room, warming the interior space; the rest is absorbed by the thermal mass of the floors and walls. It is stored there until the room temperature drops, usually at night or during long, cloudy spells, and then released to the interior. Figure 11-2 shows how a passive solar home operates.

A solar electric system, the subject of the next chapter, is a unique species in this diverse family of technologies. The only common feature that it shares with its solar heating relatives is that it is powered by sunlight. In a solar electric system, sunlight strikes special devices known as photovoltaic modules, also referred to as solar panels or PVs. Each module contains dozens of thin wafers made primarily of silicon (figure 12-2). The energy carried in the sun's rays generates an electrical current in the PVs, which is captured and stored in batteries, as shown in figure 12-2. The electricity generated and stored by such a system is used to power lights and home appliances.

Principles of Passive Solar Energy

Passive solar energy systems are mechanically simple; they contain only one moving part, the Earth, whose rotation ensures that the sun faithfully appears in the east each morning, crosses the sky above our homes, and then sets quietly but often dramatically in the west each evening.

In passive solar designs, sunlight is used to heat the interiors of our homes. However, for us to benefit from the sun's generous radiation, the house must be sited and oriented to ensure the sun's entrance, and designed to provide a means of storing the sun's energy for later use. That storage medium is called thermal mass.

Passive solar design is not rocket science. You don't need a degree in solar engineering to figure it out. In fact, the Anasazi Indians of the American Southwest built their mud and stone dwellings in protected canyons facing the south to keep themselves warm in the winter, and that was over 1,000 years ago. Long before them, the Greeks tapped into the sun's energy. So advanced were the Greeks in their understanding of the importance of sunlight as a source of heat that they treated solar access as a legal right. The Greek city of Olynthus was even designed so that all houses would have access to the sun—and this was in the 5th century B.C.! What is required to design a home for passive solar energy? (The following principles are adapted from *The Passive Solar House*.)

Principle 1: Choose a location with good solar exposure. To make solar energy work for you, choose a site with unobstructed access to the sun from 10 A.M. to 3 P.M. as many days out of the year as possible. As a rule, the more sun you receive, the greater the solar potential. If you live in the Northern Hemisphere, don't build your house on north-facing slopes (just the opposite for folks in the Southern Hemisphere). Avoid heavily wooded lots unless you plan to cut down a lot of trees on the sunny side of your house. That doesn't mean you

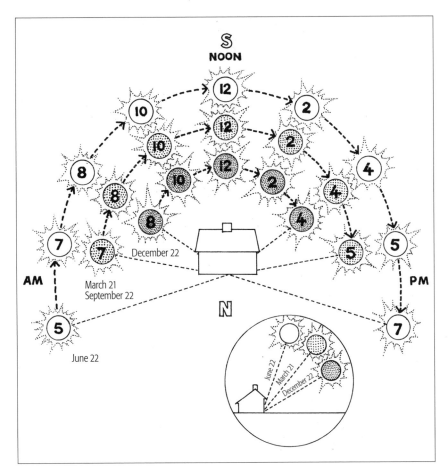

FIGURE 11-3.
This diagram shows the position of the sun in relation to true north, that is, the bearing angle, at sunrise and sunset on June 22, March 22, September 22, and December 22. It also tracks the sun across the sky, giving the position in relation to true north during the day. Note that south is shown at the top of the drawing. The insert at the bottom shows the altitude angle at 12 noon on each of these days.

can't have trees on your property. A few deciduous trees are beneficial, as they will help shade the house in the summer, but trees on the south side of the house may block the low-angled winter sun. A related rule of thumb is to watch out for obstructions. Avoid building areas where mountains, hills, cliffs, or other buildings will block your access to the sun.

To choose a good site, a subject discussed in greater detail in chapter 15, you need to assess its solar potential. You have to determine east and west on the lot (a compass is essential!) and know how the sun "moves" across the sky at different times of the year. As you may know, over the course of the seasons, the sun rises in different places. During the winter, for instance, the sun rises in the southeast and sets in the southwest. As days get longer, sunrise occurs further and further to the north. (This is due to the tilt of the Earth and its orbit around the sun.) By the time summer rolls around, the sun is rising in the northeast and setting in the northwest. Figure 11-3 illustrates the position of the sun at various times of the year and at various times during the day. Notice where the sun rises and sets on the following days: December 22 (winter solstice), March 21 (spring equinox), June 22 (summer solstice), and September 21 (fall equinox). Figure 11-3 shows the angle of the sun—that is, how high it is in the sky at these times of year. By studying these drawings you should be able to answer one of the most commonly asked questions regarding passive solar energy: "How do you keep from overheating in the summer with all the sun coming in?"

The answer is that during the summer the sun is so high in the sky that it doesn't enter south-facing windows. Although east- and west-facing windows allow sunlight to penetrate your house in the early morning and late afternoon, respectively, during the summer very little sunlight comes in through the south-facing windows the rest of the day. Contrary to popular belief, a solar home does not bake in the summer. If properly insulated, ventilated, landscaped, and perhaps earth-sheltered, your solar home should stay nice and cool during the summer. If it doesn't, you've done something wrong.

On any potential building site, first determine the approximate path of the sun at various times of the year. Then look for obstructions that could eclipse the sun during part or all of its journey. Be especially attentive to obstructions that will block the sun in the late fall, winter, and early spring when the sun is low in the sky—and when you need heat.

If all of this sounds too complicated, you can call on a solar designer or solar builder for assistance.

FIGURE 11-4.
The bearing angle is the location of the sun in relation to the points on a compass at any moment in time from sunrise to sunset. The altitude angle is the angle of the sun in the sky at a particular time.

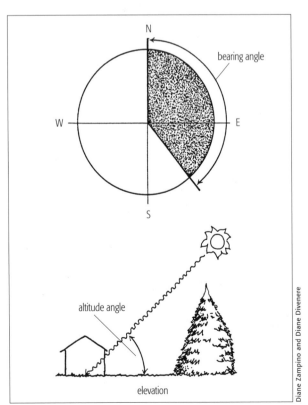

Diane Zampino and Diane Divenere

There are also some mechanical forms of assistance. One of them is the Solar Pathfinder, which will evaluate the solar potential of any site. This handy device, available from Jade Mountain, for purchase or rent, is set up on a site on a stand. According to the manufacturer, one quick siting gives accurate solar information for the entire year and shows energy loss due to shadows.

Charts and maps are also available. To use these tools, you need to be familiar with two solar measurements. The first measurement is the height of the sun in the sky, or altitude angle. Technically, altitude angle is the angle of the sun from the ground (figure 11-4). Go outside and take a look at the sun. Raise your right arm and point it at the sun. This forms a line from the ground to the sun. The angle between the ground and this line is the altitude angle.

The second measurement is the bearing angle. This is the position of the sun in relation to the points on a compass—either north or south. In this exercise, we will assign zero to north on the compass. On June 22, the bearing angle of the sun in my region at sunrise is about 63 degrees—that means it can be found 63 degrees from north. Go back outdoors with a compass. Determine where north is, then draw a line in the ground or put a stick on the ground extending from north to south. Locate the sun, then draw a line pointing to it. This is the bearing angle at that particular moment. If you go out again in two hours, the angle will have changed.

To determine the sun's course at various times of the year, first locate the latitude of your home on a map of the state or country you live in (figure 11.5). Next, refer to a chart that shows the bearing angle and the altitude angle at different latitudes at different times of the year and at different times of the day. This will allow you to plot the arc the sun makes through the sky at equinoxes and solstices, the four critical times of the year.

Figure 11-5 shows data for latitude 40 degrees June 22, the summer solstice, or longest day of the year. On the summer solstice, the sun cuts its highest arc through the sky. To form an accurate assessment of the solar potential of a building lot, you will also need to determine the path of the sun during the shortest day of the year, the winter solstice, when the sun carves its lowest arc across the sky. The bearing angle and altitude angles for the spring and fall equinoxes (about March 21 and September 22) can also be determined for these days from figure 11-5.

For proper solar design in a cooler climate, sunlight should freely enter a building from September 22 to March 21—a period of about six months. Sunlight is especially helpful in the spring, which is cool in many places. In the fall, however, while the Earth is still warm, you will probably need overhangs or window shades to reduce the amount of sunlight entering your home.

While you are studying the path of the sun, take note of the wind direction. You will need to know this so you can take precautions to protect your house from this notorious thief that shamelessly robs heat as it sails along exterior walls and windows of your home. Ask people in the area which direction

BUILDING NOTE

Think carefully about the number and size of east- and (especially) west-facing windows in a solar home. West-facing windows in hot, sunny climates can permit a lot of sunlight to enter, causing summertime overheating.

BUILDING NOTE

Whatever system you choose to collect solar energy for heat, it is extremely important to calculate the amount of sunlight your house will receive through your windows at your latitude and in your climate. You must next calculate heat losses through walls and windows, and losses due to air infiltration. These numbers are used to determine the amount of mass you will need. Many books on passive solar energy will help you in this endeavor. One book that provides worksheets to walk you through the process is James Kachadorian's *The Passive Solar House*. You still may want to hire a solar engineer to check your calculations. The price of his or her services will be well worth it.

FIGURE 11-5.

This map can be used to determine the latitude of your home or home site. Charts like this are then used to determine the corresponding bearing and altitude angles at different times of day on June 22, March 21, September 22, and December 22.

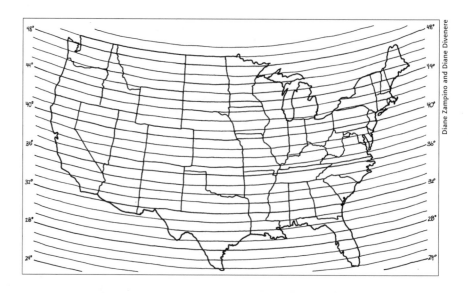

ALTITUDE ANGLE AND BEARING ANGLE AT 40° N. LATITUDE

	JUNE 22			MARCH 21, SEPTEMBER 22			DECEMBER 21	
TIME	ALTITUDE ANGLE	BEARING ANGLE	TIME	ALTITUDE ANGLE	BEARING ANGLE	TIME	ALTITUDE ANGLE	BEARING ANGLE
5:00 am	4.23	62.69	6:00 am	0	90.00	8:00 am	5.48	127.04
6:00 am	14.82	71.62	7:00 am	11.17	100.08	9:00 am	13.95	138.05
7:00 am	25.95	80.25	8:00 am	22.24	110.67	10:00 am	20.66	150.64
8:00 am	37.38	89.28	9:00 am	32.48	123.04	11:00 am	25.03	164.80
9:00 am	48.82	99.81	10:00 am	41.21	138.34	12:00 pm	26.55	180.00
10:00am	59.81	114.17	11:00am	47.34	157.54	1:00pm	25.03	195.19
11:00am	69.16	138.11	12:00pm	49.59	180.00	2:00pm	20.66	209.35
12:00pm	73.44	180.00	1:00pm	47.34	202.45	3:00pm	13.95	221.94
1:00pm	69.16	221.88	2:00pm	41.21	221.65	4:00pm	5.48	232.95
2:00 pm	59.81	245.82	3:00 pm	32.48	236.95			
3:00pm	48.82	260.19	4:00pm	22.24	249.32			
4:00 pm	37.38	270.71	5:00pm	11.17	259.91			
5:00pm	25.95	279.74	6:00 pm	0	270.00			
6:00 pm	14.82	288.37						
7:00 pm	4.23	297.30						

the wind comes from. It will usually be the direction that most storms come from as they move through your area, although wind patterns can change from one season to the next and odd wind patterns develop in valleys and other locations. (You'll learn more about wind in chapters 12 and 15.)

Principle 2: Orient your house to true south, plus or minus 10 to 20 degrees. Proper orientation is so important that the simple act of orienting a house—any house—toward the south (in the Northern Hemisphere), without a single thought given to solar design, will cut energy bills for heating by 30 percent!

Placing most of the windows on the south side of the house will give your home a big boost, raising the sun's share of your daily heat demand to about 50 percent. Adding insulation, thermal mass to store heat, and a few other energy measures can increase your reliance on the sun to 100 percent!

Figure 11-6 shows the efficiency of a solar home with respect to its orientation. If a house is oriented directly south, it gains 100 percent of the energy that it can absorb. Only a modest loss occurs up to 22.5 degrees either side of true south. However, as the angle from true south increases, the amount of sunlight entering a house falls dramatically. Therefore, to capture the maximum amount of sunlight, orient your house within 22.5 degrees of true south. If the dominant view occurs to the east, orient the long axis of your house to the south but make some design adjustment—an added window here or there—to take in the view.

True south (or solar south) is not the same as magnetic south. The compass gives a reading of magnetic south, but for solar design, you want to determine true south. True south is a line running directly from the North Pole to the South Pole. It is a geographic term. Magnetic south, on the other hand, is determined by magnetic fields. Although magnetic fields run more or less north and south, they do not run from pole to pole—that is from true north and true south. The map on page 206 of Kachadorian's *The Passive Solar House* shows the magnetic compass variations for your area. In El Paso, Texas, for example, true south is actually 12 degrees east of magnetic south. If you live there, you would locate magnetic south, draw a line pointing 12 degrees east of this line, and that would give you true south.

Orienting a house to the south means orienting its long axis east and west, so that the house presents the largest possible amount of exterior wall space to the south. As a general rule, a rectangular floor plan works best for passive solar design. When oriented along an east-west axis, the house gathers the maximum amount of sunlight from south-facing windows (in Northern Hemisphere) and north-facing windows (in the Southern Hemisphere). In desert climates, a rectangular floor plan minimizes the exposure of east and west walls to summer sun, which can cause a house to overheat. If you'd like to have a few west-facing windows to take advantage of a view, shade them with trees or an arbor planted with deciduous vines (figure 11-7). In cooler climates, solar gain from east- and west-facing windows may not be a serious matter.

The ideal solar design for most climates is one room deep, as Peter van Dresser points out in his book, *Passive Solar House Basics*, because incoming

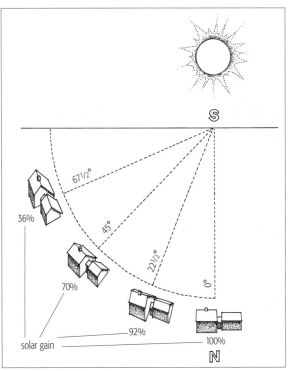

FIGURE 11-6.
A passive solar home should be oriented with its long axis perpendicular to true south and within 22.5 degrees east or west of true south for optimal solar gain.

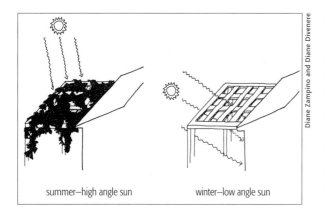

summer—high angle sun   winter—low angle sun

FIGURE 11-7.
Shade for walls can be provided by building an arbor (right) and planting vines that will spread over the structure (left). As an added bonus, you'll be graced with greenery, flowers, sweet fragrances, and the sound of birds!

sunlight will reach well into the interior of each room. In essence, each room becomes its own self-contained solar unit. There is no need to collect heat in one location and transport it to another to maintain comfort. Unfortunately, the linear or string arrangement is not always convenient for internal traffic flow. To address this problem, van Dresser recommends making buildings squarish with corridors, utility rooms, laundry rooms, pantries, workshops, and other less frequently used rooms along the north side. These rooms do not need to be as warm as dining rooms and living rooms.

Some builders argue that these constraints make for some rather drab-looking buildings. I beg to differ. With a little creative design, the basic rectangular or square-shaped home can be quite attractive. A clerestory roof, for example, adds dramatically to the aesthetics of a house, while providing a way for sunlight to penetrate back walls. Open floor plans like those popularized by Frank Lloyd Wright also add immensely to the warmth and comfort of a home.

Principle 3: Place most of the windows—your solar collectors—on the south side of the house in the Northern Hemisphere or on the north side in the Southern Hemisphere. Windows and patio doors are solar collectors in a passive solar system. When the sun is low in the sky during the winter months, sunlight will penetrate south-facing windows and ensure maximum solar gain. How much? When the sun is at its lowest point of the year, sunlight can penetrate more than twenty feet into a house.

In most instances, vertical glass suffices. Tilted glass, shown in figure 11-8, is also used, but is generally not advised. The logic behind tilted glass, explained in chapter 4 on Earthships, is that it permits maximum sunlight penetration during the winter months because sunlight strikes perpendicular to the glass at this time of the year (that is, when the sun is at the lowest point in the sky). Sunlight striking perpendicularly minimizes reflection and maximizes transmission, meaning more sunlight enters the house. The more sun that enters a house, the more heat you will gain.

Unfortunately, tilted glass also permits a lot of sun to enter in the spring, summer, and fall, which can turn a solar home into a furnace. Tilted glass is also more prone to water leakage than vertical glass. Skylights present the same problems—that is, they let in too much sun during the summer and can leak, if improperly designed or installed, causing considerable damage to the roof and ceiling. Skylights also provide an avenue of escape for winter heat. As Peter van Dresser points out, heat loss to the night sky from skylights usually exceeds what they gain during sunny days unless insulating shades are installed. One exception is a device called a Solartube skylight (figure 11-9). This small tube takes up a fraction of the space of a conventional skylight, and

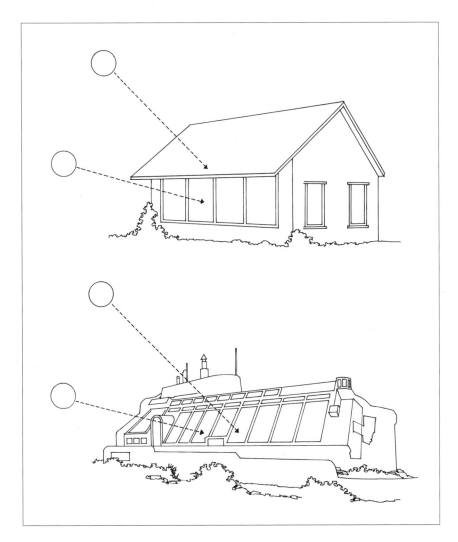

FIGURE 11-8.
Tilted glass (bottom) allows 12-month-per-year sunlight penetration and may cause overheating during the summer. Vertical glass with overhangs (top) prevents sunlight from penetrating into a home during the hot summer months.

FIGURE 11-9.
The Solartube skylight brings tons of diffuse light into a room without the massive loss of energy during the winter months.

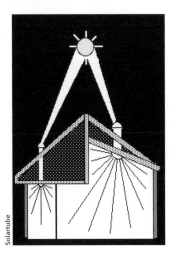

looks a bit like a ceiling light. It transmits huge amounts of light, thanks to an internal system of mirrors, and permits very little heat to escape.

As James Kachadorian notes, "Incorporate enough windows to provide plenty of daylight, but do not make the mistake of assuming that solar heating requires extraordinary allocations of wall space to glass. . . . A highly insulated, well-constructed home with a proper number and distribution of high-quality windows does not need much energy to maintain comfortable temperatures year-round." Consult with a knowledgeable solar engineer or solar designer to determine the amount of window space you will need—and the suitability of east- and west-facing windows. Or do the calculations yourself. Kachadorian's book and other passive solar energy books listed in the resource guide will prove helpful.

Principle 4: Provide overhangs to protect against excess sunlight penetration during the spring and the fall. Overhangs or soffits protect walls from driving rain,

FIGURE 11-10.
The success of a passive solar home depends in large part on the placement of mass inside the house. Mass placed in direct line of incoming solar radiation, either the floor or interior walls, as in this illustration, absorbs the most sunlight.

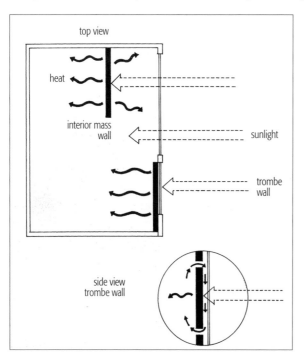

top view

heat

interior mass wall

sunlight

trombe wall

side view trombe wall

and are essential to the success of many natural homes, such as straw bale, cob, and adobe. Overhangs also shade the walls, reducing the amount of sunlight entering a home during times when excess solar energy would cause overheating.

I have found that designing the proper overhang is a matter of compromise. In the solar homes I've lived in, I've found that I need more sunlight in the spring and less in the fall. In my hometown, springs tend to be cold and falls tend to be very warm. Because of this, I often wish someone would invent a retractable overhang that could be pulled back in the spring to let more sunlight into the house and then extended in the late summer and early fall to block excess sun.

Principle 5: Provide sufficient, properly situated mass to absorb sunlight. Thermal mass is a material strategically placed in the house to absorb sunlight (figure 11-10). Thermal mass absorbs sunlight, converts it to heat, and then releases the heat into the room when room temperature drops, either during the night or during cloudy weather. The most efficient way to capture the sun's energy is to place the thermal mass directly in the path of incoming solar radiation. Walls made of cob, adobe, rammed earth, and packed tire are ideal. Stone and concrete walls also provide excellent thermal mass. Even the wood of a cordwood or a log home provides thermal mass, but not as much as earthen or concrete walls. The best place to locate thermal mass is in the interior of a house. Interior mass walls are optimum. Mass floors work well, too.

One of the easiest and most effective ways of locating mass in the path of the sun is by installing an earthen floor or pouring a concrete slab and finishing it with dark-colored tile or flagstone. Perimeter insulation—situated along the inside of the foundation—will help to retain heat absorbed by the floor. Subslab insulation may also help, although builders disagree. Some think it is vital; others think it is a waste of money. Van Dresser falls into the latter category. He argues that the heat that "travels a few feet down below the house is not lost to useful purpose and still contributes to the reservoir of warmth underneath the building." I'm not so sure. I insulated the thick concrete slab of an attached greenhouse with great success. The structure only dropped below freezing twice over one bitterly cold winter, and one of those times I had accidentally left the sliding glass door cracked open.

One of the most ingenious methods for capturing solar energy in floors was devised by James Kachadorian. This innovative designer and engineer creates thermal mass in the floors by installing concrete blocks under a concrete slab. The blocks are oriented in rows with the interior cavities pointing north and south to create continuous pathways for the flow of air (figure 11-10). Over the top of this block work, he pours a 3- to 6-inch slab, leaving vents in the floor along the north and south sides of the house, so room air can circulate through the labyrinth he has created. In this design, sunlight enters the house through south-facing windows, then strikes the interior surfaces where it is converted to heat. Hot air flows by convection into and through the subslab storage system, entering the vents at the back of the house, then re-emerging at the front along the south wall (figure 11-11). As it passes through the floor, the warm air gives off its heat, which is stored in the floor. At night, the heat radiates from the blocks and slab into the room, keeping the house warm year-round, even in the cold Vermont climate!

All thermal mass performs optimally when it is dark colored. So, plaster your interior walls with a dark material. And choose a dark floor tile, for example, a terra cotta (clay-colored) tile.

In passive solar houses containing inadequate amounts of thermal mass, heat from sunlight causes the temperature to soar, easily climbing into the mid to high 80s. However, because there is no mass to store the heat and distribute it during colder periods, heat quickly dissipates as the sun begins to set. At night, interior temperature drops, often 20 degrees or more.

The basic strategy is to design the house so that its own masses—mainly walls and floors—are so placed, proportioned, and surfaced that they will receive and store a large measure of incoming solar energy during the daylight hours and will gently release this stored heat to the house interior during the succeeding night hours or cloudy days.

PETER VAN DRESSER,
Passive Solar House Basics

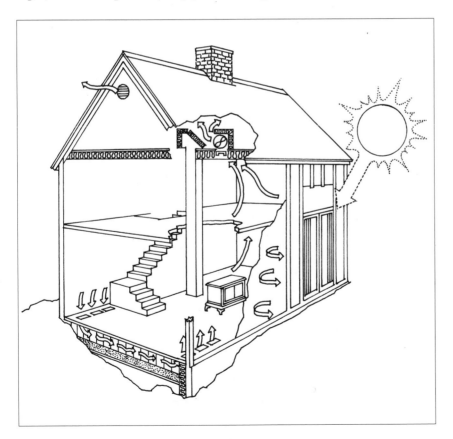

FIGURE 11-11.
Sunlight entering the home warms the inside surfaces and the room air. Warm air flows by convection currents under the floor, depositing its heat in the concrete blocks and slab. At night or during cloudy periods when the room temperature drops, the heat dissipates into the room, maintaining a comfortable interior temperature.

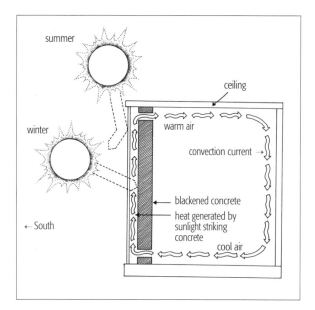

summer

ceiling

warm air

convection current →

winter

blackened concrete

heat generated by sunlight striking concrete

← South

cool air

FIGURE 11-12.

The Trombe wall is one of my favorite designs. It places mass in direct line of incoming solar radiation. To be successful, however, vents in the wall must be closed at night, or warm air will be siphoned out of the room after sunset.

Thermal mass greatly reduces temperature fluctuations in a house (if properly sized and placed). Because it absorbs heat, thermal mass prevents a solar home from getting too hot during the day. Because it releases heat into the room at night, it prevents the temperature from plummeting. Although temperature will rise and fall, the amplitude of the fluctuation will be diminished. Too much mass can even work against you. If it takes forever to heat up, it may prevent room temperature from climbing into the comfort zone.

Interior walls (mass partition walls) provide the best thermal mass, but to incorporate thermal mass inside a structure many solar designers often rob valuable floor space or design in sunrooms that serve little purpose other than heat collection.

Another highly effective means of marrying sunlight and mass is the Trombe wall. A cross-section of a Trombe wall is shown in figure 11-12. Located on the south side of the house, a Trombe wall consists of a mass wall made from adobe, rammed earth, or cinderblocks filled with concrete. On the outside surface, usually three to six inches from the mass wall, is a layer of double-pane glass. The surface of the wall is painted a dark color, usually black. Sunlight streaming in through the glass strikes the black surface, where it is converted into heat. Heat then slowly migrates into the house through the mass wall. By the time night falls, the heat has made its way to the interior surface, where it radiates into the room, keeping it warm and cozy.

Air vents located in the Trombe wall siphon heat out of the space between the glass and the mass wall. Cool room air enters the space through the lower vent, and is pulled upward between the glass and mass wall by convection, where it picks up heat. Heated air then flows through the top vent into the room.

The vent system adds to the efficiency of a Trombe wall, but to be fully effective, the vents must be closed at night. If not, air will flow in the reverse direction in the evening. Warm room air will flow through the top vent, through the air space, cooling off as it goes, then through the bottom vent, back into the room. This will strip warm air from a room, negating any gains and creating unpleasant drafts.

Principle 6: Insulate well to maximize heat retention. The sun can easily heat a home and create a comfortable living space, but only if three conditions are met. First, the structure must have sufficient and properly situated glass. Second, it must have sufficient and properly situated mass to store the sun's energy. Third, it must have adequate insulation to retain the sun's heat.

Most builders and solar designers prefer wall insulation ranging from R-22 to R-30 and ceiling insulation of R-40 to R-50. (R-value is a standard measure of heat retention). The lower end of the range is generally recommended for warmer climates and the upper end of these ranges is recommended for colder climates.

I prefer to overinsulate a house, because the research I've seen suggests that the most commonly used forms of insulation, cellulose and fiberglass, decrease in R-value over time as a result of settling or moisture accumulation. Additional insulation compensates for the reduction in R-value. The cost of increasing ceiling insulation from R-38 to R-65 in my 2,700-square-foot tire and straw bale home was under $1,000, an insignificant addition for greater comfort and reduced energy bills.

I also think you should insulate to the hilt even in hot climates. In the southern and the southwestern United States, for instance, summer cooling is vital to comfort. It is also costly. Very costly. A $200 to $300 monthly electrical bill is not uncommon. If you are living in a poorly insulated home with an energy-guzzling air conditioner humming alongside the house day in and day out, you are paying huge amounts of money unnecessarily to stay cool. The money you are spending could have financed a college education!

Unfortunately, most people think that insulation is meant primarily for cold climates where it is used to retain much-needed heat. Yet insulation works just as well to retain coolness (coolth) in a house. I like to think of a house as a big thermos bottle with windows and doors. Put hot liquids in the thermos and it stays hot. Put cold liquids in and it stays cold.

Insulation is important, indeed essential to the performance and comfort of a house. However, insulation is only as good as the people installing it. As Mark Freeman aptly points out in his book, *The Solar Home*, "Insulation that is crammed in too tightly, has gaps, or becomes wet loses most of its insulating value."

Many environmentally friendly homes described in this book have their own insulative properties. Straw bale offers the greatest insulation, as do straw-clay buildings. Earthen walls (adobe, cob, and others) have far lower R-values, but the thermal mass of these walls makes up for the lower R-values. However, the south-facing walls in many natural homes are built from conventional lumber—either a 2 x 4 or 2 x 6 frame—which makes it easy to install a large array of windows for solar gain. And many natural homes have conventional roofs. If your design calls for a conventional roof and a wood-frame exterior wall, you will need to insulate them. For walls, consider using 2 x 4s for framing, rather than 2 x 6s. The extra width of a 2 x 6 isn't structurally necessary, and a 2 x 4 frame with 1 inch of rigid foam exterior insulation on the outside of the sheathing has the same R-value as a 2 x 6 wall. But even more important, a 2 x 4 framed wall with a layer of exterior foam insulation suffers less bridging loss than a 2 x 6 wall with insulation packed between the studs. Bridging loss is the

BUILDING NOTE

The El Paso Solar Energy Association in Texas recommends that insulation in local homes in this desert climate be R-30 in the walls and R-45 in the ceilings.

conduction of heat from a building through the framing members in the walls and roof. Although you may not know this term, you've probably noticed the phenomenon on snow-covered roofs where the snow melts away in lines mirroring the rafters. Exterior rigid foam insulation blocks this heat loss. Because the foam is laid so that it overlaps the joints between sheets of exterior sheathing, it also reduces air infiltration. For more on insulation, consult Edward Harland's book, *Eco-Renovation*, Harry Yost's *Home Insulation*, or Alex Wilson and John Morrill's book *Consumer Guide to Home Energy Savings*.

Another important consideration in all homes, but especially in passive solar homes, is window insulation. Many solar homeowners make the egregious error of leaving south-facing (and other) windows uncovered at night. This is especially common in houses that have huge glass "walls" that are nearly impossible to cover with window shades. Glass is an extremely poor insulator. It conducts enormous amounts of heat to the outside. Because of this, windows can become a major source of heat loss even in a highly insulated, airtight structure. Don't be persuaded to leave your windows uncovered by the fact that you have installed energy-efficient windows. A high-quality double-pane windows has an R-value of about 4. While more efficient than a single pane glass, R-4 glass still loses tremendous amounts of heat, costing you much more than necessary to heat your home.

Cover windows with insulated drapes or shades. Good ones will add another R-3 or R-4 to windows. Better yet, consider insulated window shades that seal around the edge of the window with magnetic strips. They not only augment the R-value, they help block the flow of cold air from windows and are well worth the time required to make them yourself.

Insulated exterior or interior shutters can reduce heat even more. James Kachadorian installs rigid foam "thermo-shutters" on the inside of his homes. Interior shutters can be quite attractive and, if fitted tightly, are highly effective insulators. And they're a lot more convenient than the exterior shutters. (See Kachadorian's book for design ideas. Also read Peter van Dresser's *Passive Solar House Basics* for a discussion of the pros and cons of exterior and interior shutters before you make up your mind.)

Principle 7: Install a vapor barrier to protect your insulation from moisture. Moisture from kitchens, washing machines, bathrooms, plants, pets, and people in tightly sealed passive solar homes can raise havoc with most forms of insulation—not to mention window sills. Moisture moves from areas of high concentration to low concentration. In most climates, there is more moisture inside the house than outside. Water vapor, therefore, tends to migrate from the interior of a house to the exterior, passing through ceilings and walls to escape. When moisture strikes a cold surface, typically the exterior sheathing of the walls or the decking of the roof, it condenses, dampening or even soaking the insulation.

The R-value of virtually all forms of insulation deteriorates dramatically when it is wet. (The only exceptions are wool, rigid foam, and bubble insulation.) When wet, other forms of insulation become useless. Their R-value plummets. Besides greatly decreasing the R-value of your insulation, moisture escaping through ceilings and walls can collect on roof decking and wall sheathing, causing them to warp and rot.

Fortunately, several simple steps can stop the flow of moisture through walls and ceilings. First, reduce interior moisture sources. Cover aquariums, move indoor clotheslines outside, and make sure dryers are vented to the outside. Second, and more important, install a vapor barrier in walls and ceilings. Drape six-mil polyethylene along the framing members of exterior walls and ceilings, overlapping all seams, and then staple onto joists and studs. Vapor barriers should always be placed on the warm side of a wall or ceiling. In temperate and cool climates, vapor barriers are located just beneath drywall, as shown in figure 11-13. In hot, humid climates like the southern United States, vapor barriers are placed on the outside of the wall, to prevent moisture from penetrating in from the outdoors.

In most natural homes, vapor barriers aren't necessary. Walls of adobe, cob, rammed earth, packed tire, and earthbag homes, for example, have no insulation in them. They are all mass and therefore don't require a vapor barrier. Straw bale and straw-clay walls are packed with insulation, but the use of breathable plasters allows moisture to move in and out, so it doesn't accumulate in the wall. However, if a natural home has any conventional framing, for example, in south-facing walls and in the roofs, vapor barriers should be installed in these locations to protect the insulation. It may cost $100 to $300 to do an entire house, depending on the size of the job, but the expense is worthwhile.

In closed ceilings (that is, a ceiling space between rafters), a small air space should be left between the top of the insulation and the decking, as shown in figure 11-13. Ridge line and soffit vents are also needed to permit air to circulate through this space and carry off moist air. Don't build without them!

Vapor barriers are not required in attics if there is plenty of room for moisture to escape. Consult with a competent designer. If you feel the need to get rid of moisture that enters your attic, you can install a solar-powered roof vent, a small fan that vents the moisture from an attic space and is powered by a small PV panel. It will not only reduce moisture buildup in the winter, but will also cool the attic space in the summer, keeping the whole house cooler. Fans are listed in the Real Goods and Jade Mountain catalogues.

Principle 8: Seal your house against air infiltration. Unbeknownst to most people, our homes are a virtual Swiss cheese. There are so many air leaks in an average American home that if added up they would be equivalent to an open window measuring 3 feet by 3 feet. This leakage around windows, doors, and

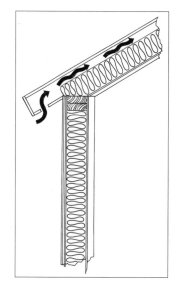

FIGURE 11-13.
In a temperate climate, the vapor barrier is located between warm, moist room air and the insulation. This prevents moisture from soaking the insulation. The success of insulation in a closed ceiling design depends on proper placement of the soffit and ridge vent and the presence of an air space between the decking and the insulation, which allows air to move over the insulation and draw off any moisture that may accumulate there.

BUILDING NOTE

Vapor barriers are harmful to earthen and straw bale walls. If moisture accumulates in a straw bale wall, it could cause the straw to rot. If it accumulates in an earthen wall, it could weaken it.

foundations wastes enormous amounts of energy and makes our homes expensive to heat and cool and, lest we forget, very uncomfortable. (Air leaks do, however, provide fresh air, which we all need to thrive.)

A solar home should be relatively airtight. James Kachadorian recommends that two-thirds of the air in a house should be replaced every hour. To investigate your air infiltration, hire a professional energy auditor or your power company to perform a blower door test. With all exterior doors and windows shut, a technician installs a large fan that fits snugly in the front door. The fan is then turned on and air is blown into or out of the house, forcing it through cracks and crevices. By measuring the volume of air blown into or out of a house over a certain period, the auditor can calculate the number of air changes, and locate infiltration leaks.

Adequate air exchange creates a good, healthy interior. It can be achieved by not sealing the house too tightly to begin with. Although this is an inexpensive option, it may result in unnecessary energy loss—that is, if you leave too many cracks. Proper air exchange can be achieved in a more controlled way by installing an air-to-air heat exchanger, a device that sucks new air into the house while venting indoor air. Much of the heat from the indoor air is transferred to the incoming (outdoor) air to conserve energy. Although convenient, this device consumes a lot of electricity and is inadvisable for off-grid solar electric homes. If you choose the air-to-air-heat exchanger route, seal all cracks in the building first. Use expanding foam insulation and caulk to seal all penetrations in the walls or foundations created by your electrical and plumbing lines. Seal around doors and windows, and around the top plate and sill plate.

Principle 9: Provide backup heat. Building departments are likely to require some sort of backup heating system, even if the house is designed to provide all of its heat from the sun. Because your house will be superinsulated, thermally stable, and heated by the sun, you won't need a large backup heating system. Whatever you do, don't let a heating contractor decide the size of the heating system you should install—they invariably oversize heating systems. Many heating contractors are unfamiliar with the full potential of passive solar design and are generally inclined to oversize systems to provide a large measure of safety and protect themselves from angry customers who complain about an inadequate system. Oversizing a system adds unnecessarily to construction costs and increases operating costs. An oversized system will operate inefficiently, shutting on and off more frequently than a smaller system. Essentially, you will pay extra to give your heating contractor some peace of mind.

A properly vented natural gas wall unit or two, a woodstove, or a pellet stove may be all that you need to supplement solar energy. Check with your building department first. Think twice about installing a radiant floor system. While nifty, it is an extremely expensive form of backup heat. Radiant floor systems installed in concrete slabs or earthen floors take a long time to bring a house up to a comfortable temperature. By the time they've got your house warmed up, the sun is out again. Radiant floor systems also require electricity to operate motors. A radiant floor system in a properly insulated passive solar home with adequate mass is like owning a Cadillac that you take out for a spin five or six times a year.

Principle 10: Protect your home from winds by landscaping and earth-sheltering. Earth-sheltering protects a home from heat loss. By nestling the north side of a house into a hillside, and perhaps burying the roof as well, a homeowner can prevent winter winds from stripping heat away from a structure. Dave Johnson of Vanguard Homes, who built my house with me, thinks of a house as a person wrapped in a blanket facing the sun. The blanket protects against the wind, while being open slightly in front to let the heat in.

Earth-sheltering also allows one to benefit from thermal energy stored in the earth. Below the frost line, the earth stays at a fairly constant 50 degrees Fahrenheit. An earth-sheltered house takes advantage of this constant temperature and itself remains at a fairly steady temperature year-round because of it. In winter, a little heat will bring the house up into the comfortable range. Solar energy can do that. Compare this to a wood-framed house that is exposed to the elements. If unheated, temperatures inside a wood-frame house fall pretty close to ambient temperature. Heating it to the comfort zone is much more difficult than heating a home that is tucked inside a 50-degree Fahrenheit earthen blanket. Earth-sheltering keeps a house cooler in summer. It may be 100 degrees outside, but the house stays a comfortable 70 degrees thanks to its cool earthen wrap. In summer, temperatures inside a wood-frame house can skyrocket.

Earth-sheltering involves partial to nearly complete "burial" of a house. Partial sheltering, known as earth berming, is achieved by burying the lower parts of the exterior walls of a house, creating a berm typically three to four feet high. This is a valuable means of protecting a house for those who live on the plains or in otherwise flat terrain. Houses can also be built almost entirely underground (figure 11-14). Proper design ensures ample light and generous views of the outdoor world—you won't be living in a dark cave, as some people fear. My office and bedrooms are underground (the walls and roof are covered with dirt; the south-face is exposed), yet the rooms are amazingly bright.

In his book *The Earth-Sheltered House*, Malcolm Wells sums up the virtues of earth-sheltered architecture. "It works," he says. "It really works.

FOR THE LAST several decades, architectural and engineering education related to the climate has been limited to choosing heating and cooling systems from one catalogue or another, usually supplied by equipment manufacturers. Thus architects and engineers have received the best training from the manufacturers of the products and unwittingly become their salesmen. They will go through their normal professional life without questioning the validity of such a route.

The basic questions—Why use mechanical coolers or heaters? Why use air conditioners?—have just begun to bother the public's mind.

NADER KHALILI, *Ceramic Houses and Earth Architecture*

FIGURE 11-14.

Earth sheltering a home is a great way to save energy and increase comfort. You can go all the way, placing a home almost entirely underground (shown here), or you can partially earth shelter a home with a berm. Most people who build underground homes design them to permit ample light to enter so they don't have the cavelike appearance of this home.

And, in addition to having a green 'footprint,' every underground (earth-sheltered) building is also silent, bright, dry, sunny, long-lasting, easy to maintain, easy to heat and cool, and fire safe." This approach, however, is not without problems—troubles largely wrought by improper design or construction (or both). If improperly designed and built, roofs can leak or, worse yet, cave in. Water can accumulate in the backfill around a house and seep through the walls if proper precautions are not taken. Backfill around the house can settle. As it sinks, it may strip rigid insulation board or waterproofing off the wall and rupture buried pipes. The moral of the story, says Wells, is not that earth-sheltering should be avoided, but that it should be done right. If this idea interests you, exorcise all images of living in a cave and get a copy of *The Earth-Sheltered House* to explore the potential of this technique.

Another effective way to protect your house from wind is to situate it in the protection of a grove of trees or some other natural windbreak, such as a small hill. Chapter 15 describes ways to protect a home from the wind in great detail.

Principle 11: Synchronize daily room use patterns with solar patterns. In passive solar homes, room use should be coordinated with solar patterns. Consider an example. Suppose you live in a two-story rammed earth home. You like to rise early, perhaps with the sun, and want to wake up to a warm room. By placing your bedroom on the southeast corner of the second story, you can take advantage of early morning sunlight to heat your room. The sun also provides you with a wake-up call. If you then shower and head to the kitchen for breakfast, place the kitchen on the first floor in the southeast corner, so that it is warm by the time you arrive.

Coordinating your daily activities with the layout of your home and the sun's predictable path across the sky not only provides comfort, it saves money and resources, as it reduces your need for backup heat—and artificial lighting. The sun serves you as you move through your day. *The Rammed Earth House* has an excellent description of home design in relation to solar energy and use patterns.

For a list of common mistakes, look at the sidebar on this page. One common error is the failure to provide sun-free zones. Too much sunlight can create terrible glare on computer screens. This causes eye strain and headaches. Your house needs a room or two where you can escape the sun to work at your computer or watch television. Curtains or shades may provide sufficient darkness.

When many people think about solar, they think about modern solar homes, such as that shown in figure 11-15. However, solar design is compatible with all forms of architecture, even a log cabin or a Victorian-style home.

PASSIVE COOLING

Many of the features that make passive solar heating work also keep a house cool during the long, hot summer months. As Paul Huddy of Eco-Engineering Associates in Tucson notes, "Good passive solar design can . . . give you a house that requires substantially less cooling . . ." than a standard home. "Besides slashing your heating and cooling bills . . . it also provides a house that is more comfortable, and naturally so, year-round."

James Kachadorian

FIGURE 11-15.
Although we tend to think of passive solar homes appearing much like this one, virtually any style can be made solar.

One of the most overlooked passive strategies for cooling a house is a reduction in internal heat sources such as light bulbs. Incandescent light bulbs, the kind most people use, produce a lot of heat. In fact, 95 percent of the energy consumed by a standard incandescent light bulb is converted to heat. Only 5 percent is converted to electricity. (Because they produce more heat than light they should really be called *heat bulbs*.) To eliminate the need to operate lights during daylight hours, many builders rely on a practice called daylighting. Daylighting provides light naturally, for example through clerestory windows, to illuminate the interior of a building. Because daylighting eliminates the need to run light bulbs during the day, it saves money and reduces cooling costs.

If you need direct light on a work area, try task lighting—that is, locate lights directly over specific work areas. Rather than wiring your home so that all the lights come on when you enter a room, set it up so that switches turn on a single light or a couple of lights near work spaces. Task lighting is easily designed into a house, but many electricians and builders overilluminate rooms with extensive track lighting or recessed ceiling lights, often with high-wattage light bulbs, providing far more light than you need to work at your desk in the corner.

Another strategy to reduce the heat and expense of incandescent light bulbs is to use compact fluorescent light bulbs. Compact fluorescents use about one-fourth the energy of a standard light bulb to produce the same level of illumination. A 125-watt incandescent light bulb, for example, produces the same amount of light as a 22-watt compact fluorescent bulb. Because compact fluorescent light bulbs are so much more efficient, they produce much less heat. Compact fluorescents are color adjusted too, so the lighting is warm and inviting, not cold and eerie like a typical old-style fluorescent bulb. Although they cost more than a standard incandescent light, they also outlast them—operating for 10,000 hours versus 1,000 hours. Because they outlast incandescent bulbs and use less electricity, they are actually much cheaper to own and operate.

Microwave ovens produce a lot less heat than standard ovens, although they draw an enormous amount of electricity, which can be troublesome in solar electric homes. Some people barbeque outside during the summer or use solar cookers to prepare food. And, lest we forget, air-drying clothes eliminates the internal heat production of clothes dryers.

Reducing internal heat sources won't cool a house by itself. You will also need passive cooling strategies such as thermal mass, shading, proper orientation, the use of light colors, nighttime radiation, vents, windscoops, and cool towers. This section will examine each option, providing an overview of innovative, environmentally friendly ways to meet your summertime cooling needs. Additional information can be obtained from references listed in the resource guide.

Task lighting addresses the overillumination craze now running rampant through the American building industry.

Proper solar design and orientation can reduce summertime heat gain substantially. Because the sun is high in the sky during summer months, little sunlight penetrates the south-facing windows. Eaves or overhangs further assist in keeping the sun out. Care must also be taken to limit east- and west-facing glass. Remember, east- and west-facing windows may provide wonderful views, but they can let a tremendous amount of sunlight into your house, causing overheating. This is especially true of west-facing windows, sliding glass doors, and patio doors.

If your plans call for a modest allotment of east and west windows, install shutters, curtains, or shades to block out the sun. Or, better yet, consider planting trees and shrubs that will intercept the sunlight. They'll make much better use of the sun's energy than you will. Another solution is to build patios or decks on the east and west sides of your house and cover them with arbors draped with deciduous vines. The vegetation shades you and provides additional coolness through transpiration (the loss of water from leaf surfaces). You may also be rewarded by the fragrance of flowers and sounds of birds chirping in the leafy sanctuary you have provided.

Thermal mass, which you integrated into your design to absorb sunlight and store heat during winter months, doubles as a cold-storage medium in summer months, especially in arid climates where evenings are cool—that is, where the nighttime temperatures drop surprisingly low because of the lack of moisture in the atmosphere. Adobe, cob, rammed earth, Earthbags, packed tires, stone, and cordwood homes all contain substantial mass.

Mass inside a house absorbs heat entering through windows or from internal sources, such as body heat, stoves, lights, or appliance motors. At night, windows are opened to permit cool air to flow into the house. These cool breezes strip the heat from the walls. When morning arrives, the windows are shut and the walls, cooled by the flow of night air, begin to siphon heat from the room, thus keeping the interior pleasantly comfortable.

In cob, adobe, and other earthen homes, thick exterior walls also help to keep the interior space cool during hot summer days. Exterior walls absorb heat from sunlight and warm air surrounding a house. Although heat migrates inward during the daytime, it reverses direction, moving outward, as night falls. The heat escapes into the cool evening air. The exterior walls, therefore, buffer the inside from temperature changes.

Some architects and builders recommend adding insulation to earthen walls to retard the movement of heat into a building during the summer months. As explained for rammed earth construction in chapter 2, rigid insulation can be attached to the exterior of

BUILDING NOTE

If ceiling fans aren't enough, and you've tried all other options, you may want to install one of the new, high-efficiency swamp coolers. They can cool a house for surprisingly little money, some say about $10 per month, which is cheaper than an air conditioner. One model that impresses some solar advocates is the Solar Chill, a solar-powered evaporative cooler that operates on one or two 50-watt solar panels.

PRINCIPLES OF PASSIVE COOLING

1. Reduce heat gain from heat sources inside your houses, such as lights and ovens, during summer months.
2. Provide adequate ventilation via windows to help cool the house at night.
3. Provide air movement, if needed in warmer climates, via ceiling fans.
4. Provide sufficient mass to store coolness.
5. Decrease solar gain in the summer through overhangs, shading, and insulation.

the wall or embedded in the wall. As a rule, though, if a wall is thick enough and protected from the sun by vegetation and overhangs, insulation is probably not required, at least in desert climates.

With a sufficient amount of thermal mass, good insulation, the right design, and good nighttime ventilation, your house will stay cool day after day in hot, arid climates as well as in temperate climates. In hot, muggy climates such as Indiana and Florida, however, evening temperatures do not provide much relief. Because nighttime temperatures in the 70s and 80s aren't sufficient to cool a mass wall, thick mass walls are not advised. In such instances, thinner mass walls, earth-sheltering, shade trees, and a light-colored exterior are your main allies in your quest to maintain a cool interior. Ceiling fans may also be needed. (See chapter 15 for more details on effective siting.)

Ceiling fans do not change the interior temperature, they simply move air around. Air movement across your skin, in turn, removes heat, which makes you feel a little cooler. This effect is especially important in hot, muggy climates where skin evaporation is reduced by high moisture levels in the air that surrounds our bodies.

Another ingenious idea for cooling homes, which has been used with great success in the Middle East and northern Africa, and recently in the United States, is the windscoop, or wind catcher. As their name implies, wind catchers are towers that capture breezes and funnel them into a house. They loom above a house, gathering up the cooler air. (Hot ground-level winds don't work as well.) As illustrated in figure 11-16, air enters the windscoop at the top, travels down through the shaft, and then passes through an underground tunnel where it is cooled by the earth. The cooled air enters the house, providing a comforting breath of cool, fresh air.

Wind catchers cool houses even when winds stop blowing. For example, when the winds cease during daylight hours, sunlight warms the upper por-

FIGURE 11-16.
The windcatcher swoops up air and funnels it to the house, providing natural cooling. The windcatcher also operates when winds are not blowing, as explained in the text.

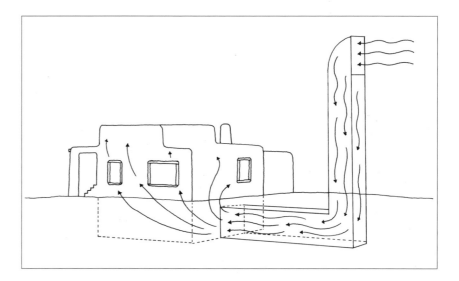

tions of the wind catcher. This warms the air inside the structure, causing it to rise. This creates a thermal siphon effect that draws air up and out of the house. During the evening, heat gained by the wind catcher during the day continues to create an upward draft that draws warm air out of the interior of the building. Wind catchers have doors that can be opened and closed to prevent wind from entering the home when it is not needed or to prevent heat loss during winter months.

Wind catchers can be integrated into the design of a house. To learn more about them, see Nader Khalili's book, *Ceramic Houses and Earthen Architecture*.

A well-designed solar home should not need air conditioning. If you have sufficient mass and proper insulation, you won't need to invest in air conditioning. If you do, you've probably made a design mistake. As Nader Khalili points out, "many environmental problems were a chain reaction caused by the movement of philosophy away from unity with nature. Climatic control is one example of this shift in human thinking and its effects." Design with nature, and you'll be rewarded many times over.

ACTIVE SOLAR ENERGY: SOLAR THERMAL SYSTEMS

Passive solar design can warm your home in the winter and help to keep it cool in the summer, all with little, if any, outside power. But it won't supply the hot water you need for showers and baths, washing dishes, and doing laundry. For these household needs, two basic solutions are an active solar water system that will heat your water with sunlight, or on-demand water heating.

An active solar water heating system is relatively simple. As illustrated in figure 11-17, it consists of a set of panels, usually mounted on the roof of a house, and a water storage tank located inside the house, typically in the basement or utility room. Pipes and pumps are required to move a heat transport fluid between the two components.

Although active solar systems vary considerably, and have become much more sophisticated and reliable in recent years (figure 11-8), let me explain how the older models worked. This will help you when researching new systems.

The most commonly used solar panels are flat rectangular boxes, the solar collectors. With an interior black surface and a pane of glass over the front surface to let light in, the solar panel gathers up sunlight and converts it to heat. Interior temperatures can easily climb to well over 200 degrees Fahrenheit. Heat created inside the panel is drawn off by fluid transported through pipes located in the interior of the box. When the sun warms the interior of the solar collector in the morning, a pump is switched on and fluid begins coursing through the pipes in the panels.

The heated liquid is carried into the house, where it is used to heat water held in a storage tank (figure 11-1). This is accomplished by a device known as a heat exchanger. In most systems, the heat exchanger is a series of pipes in the wall or the base of the water tank. Solar-heated liquid traveling through the

Solar hot water system designers have learned a lot from the mistakes of the past and the new generation of solar water heaters is much more reliable than those units that came out in the United States in the 1970s.

FIGURE 11-17.
The Thermomax Solar hot water system is one of the most innovative and promising new technologies for heating domestic hot water.

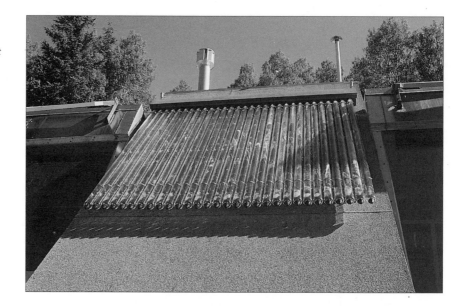

pipes in the heat exchanger gives up its heat to the water in the tank. Never is there any mixing of water and heat transport fluid. After giving off its heat, the fluid is pumped back to the roof to be reheated.

In most solar hot water systems, two water tanks are used; one is heated by the sun and one by gas or electricity. The solar tank feeds preheated water to the conventional water heater when the hot water in the house is turned on. On hot sunny days, solar heat may provide all of energy needed to keep water at a comfortable 120 degrees Fahrenheit. In fact, solar-heated water is often hotter than needed and must be cooled with water from the cold water line.

One of the many innovations in the field of solar hot water heating is the use of photovoltaic modules (PVs) to power the pumps required to move the heat-collection fluid back and forth between the solar panels and the heat exchanger. This simple change reduces electrical demand and represents a nice marriage of the two technologies.

Active solar hot water systems have become extremely sophisticated, using vacuum tubes filled with alcohol in solar panels (figure 11-17). Some operate without pumps, relying on convection currents and gravity to move fluids, rather than electricity. To learn more about these advanced designs, look at the ads in *Solar Today* magazine and contact the companies for more information. You may also want to attend a solar conference to see what the vendors have to offer or attend a solar hot water workshop put on by the folks at Real Goods. Be sure to check out John Schaeffer's *Solar Living Source Book* for a more complete description of the various options.

Active solar systems are wonderful devices, but they have some drawbacks. Some systems have temperature sensors, controllers, and moving parts. Experience has shown that sensors and controllers are the most common sources of breakdown. Moving parts in motors require periodic maintenance, and maintenance means fixing things yourself or hiring someone to perform the work for you. A single repair bill can easily erase the economic benefit in a given year. Unfortunately, solar thermal systems are also costly. For a conventional home, they don't make a lot of sense economically, unless you are heating water with electricity, which is an extremely expensive option in most places. For an off-grid home, solar thermal systems are more expensive than a conventional water heater fueled by natural gas or propane, although you can get some great buys on used systems. Check out Alternative Choices or call local solar dealers to see if they can find you a used system. Check the want ads, too. There's always someone trying to dump an old system. But be careful what you buy. The 50 percent solar tax credit offered in the late 1970s and early 1980s in the United States "gave rise to some pretty despicable, high-pressure door-to-door solar salesmen, and an abundance of poorly designed and sloppily installed solar hot water systems, giving a lingering black eye to the whole solar industry for many folks," according to John Schaeffer of Real Goods. You may be far happier buying a new unit, one that has benefited from the design mistakes of the past. Newer systems are simpler and therefore less likely to break down.

In recent years solar manufacturers have introduced new designs to simplify the systems, reduce maintenance, and lower costs. One simple system, which is used in many parts of the world, consists of a black tank mounted on the roof of a house, called a solar batch collector. Water is pumped into the tank by the water pump, which you have to have anyway, or by line pressure if you are linked to a municipal water system. The water is then heated by the sun and drawn off as needed.

Batch systems are particularly useful in warm or hot climates, for example, Mexico and Central America, where freezing is not a problem. In colder climates, tanks cool down too much at night. Solar batch collectors are simple, easy to install, and require little maintenance. Because water-filled tanks are heavy, your roof must be strong enough to support the additional weight. If you live in a fairly warm climate, give this option some serious thought. Batch water heaters are available from Real Goods or Jade Mountain for under $1,000.

A more technologically advanced system is the Thermomax solar water heater. This system consists of a series of parallel glass vacuum tubes as shown in figure 11-17. Each tube contains a black pipe containing alcohol. When sun strikes the pipe, the alcohol is heated and rises to the top of the tube by convection, where its heat is removed by a heat exchanger located along the top of the unit. The system is incredibly efficient, even in cloudy and cold weather. According to the company, it converts twice as much solar energy to heat as a conventional flat plate collector. The manufacturer claims that you can obtain 70 percent of your domestic hot water in worst-case climates. In good solar areas, they claim you can count on meeting 100 percent of your hot water demand.

BUILDING NOTE

Do-it yourself kits are available from Refrigeration Research in Brighton, Michigan. There are also books on building your own solar hot water system.

REBATES

One of the major drawbacks of solar thermal systems is cost. If you are lucky, you may be able to obtain a rebate from your power company or from your state or national government. Check it out before you buy. They may have some restrictions on the solar hot water models that are eligible for rebate.

FIGURE 11-18.
An on-demand water heater will provide all of the hot water you need instantaneously with much less energy than a conventional tank heater.

Real Goods Trading Company

There are two reasons why their system works so well. First, the vacuum tube is an excellent insulator. The absence of air molecules surrounding the alcohol-filled pipe virtually stops the escape of heat. In a laboratory test on another similar model, a single 50 degree Fahrenheit water-filled tube placed in a freezer at minus 13 degrees took six and one-half days to freeze. The second reason it works so well is that the round design of the tubes is ideal for capturing diffuse radiant energy. During cloudy weather, these collectors will capture 80 percent of the available radiant energy.

Thermomax is not the only high-tech solar water heater on the block, although it is one of my favorites. Check out the Sun Family water heaters, progressive tube collectors, and Heliodyne PV-powered systems offered by Real Goods and other distributors. Heliodyne, although more costly up front than many other units, offers the best value and the most trouble-free service in the long run, according to the technical staff at Real Goods. Moreover, this system is a favorite of many solar installers and the closed looped plumbing can't freeze, even in the coldest climates. Many of the components are pre-assembled, making installation much easier.

Solar hot water can also provide radiant floor or baseboard heat. But these systems require many more panels than a domestic solar hot water system designed to meet day-to-day hot water demands.

On-demand water heaters, also called instantaneous water heaters, are another option. They deliver heated water only when you need it (figure 11-18). In an instantaneous water heater, water begins to flow through the unit when the hot water is turned on. This ignites the burner, which heats the water coursing through the pipes. A few seconds after turning on the faucet, you have hot water.

Although most instantaneous water heaters have pilot lights and relatively low efficiencies, about 70 to 80 percent, they don't require a large tank of heated water awaiting your demand (and losing energy all the time). As a result, on-demand heaters use a minimum of 20 percent less energy than standard water heaters. Instantaneous water heaters are popular in parts of the world where efficiency is held in high esteem (that pretty much eliminates the United States).

Instantaneous water heaters are best placed close to the point of use. Because they deliver heat quickly, you waste very little water waiting for it to be delivered from a water tank. But as Michael Potts notes in *The New Independent Home*, even if a unit is installed some distance from its destination, on-demand water systems usually cut energy demand, because keeping a standby quantity of hot water is very costly.

Another advantage is that they never run out of hot water unless you run out of gas or water. As long as your unit is properly sized to keep up with demand, you will have hot water. Several manufacturers offer ten-year warranties on their heat exchangers, and because they are easy to repair and parts are readily available, a tankless water heater could last as long as your house, compared to the ten- to fifteen-year life expectancy for a conventional water heater. Tankless water heaters can also be used in conjunction with solar hot water systems.

Instantaneous water heaters are available from Alternative Choices, Real Goods, and Jade Mountain. Check out their catalogues for prices and specifications. Be sure to look into the newest instantaneous water heater of the bunch, the Targa demand water heaters. According to the manufacturer, they are 95 percent efficient, have no pilot light, and provide instantaneous hot water at a rate of 5 gallons per minute.

If you live in a region with lots of wood, you may want to consider purchasing a wood-fired water heater, a proven design that has been in use for more than forty years. Wood-fired water heaters are more efficient than conventional water heaters and require no natural gas or electricity. They're ideal for those striving for total independence from utilities.

In this chapter, I've presented some ideas on how to heat your house and your water using the sun. Study your options carefully. Talk to the experts and to others who are experimenting with these techniques and technologies. Learn as much as you can, and use your knowledge to design a home that fosters an enduring human presence. It is slow, painstaking work, taking place one household at a time.

BUILDING NOTE

If you are using a conventional water heater, you may want to consider installing a heat exchanger in your wood stove to capture excess heat to supplement your main source of hot water.

12

ENERGY INDEPENDENCE
Generating Electricity from the Sun, Wind, and Water

The age of photovoltaics is here. Almost every one has used, or at least seen, solar-powered calculators and watches. Almost everyone has seen photos of the satellites powered by solar panels. Now people are discovering that PV can be used "earthside." From tiny modules that replace throwaway batteries to acres of modules that replace obsolete electric power plants, PV is a fact of modern life.

JOEL DAVIDSON
The New Solar Electric Home

HUNDREDS OF SATELLITES orbit the Earth, soaring across the starlit sky like tiny shooting stars. Powered by sunlight captured by photovoltaic panels, this costly link in the global communications network performs all kinds of functions. Invented in 1954 specifically to provide power to satellites, the solar cell or PV has found many earthly applications. They are used to power road sign lights, calculators, and the occasional watch. They are even used by remote jungle communities of South America and isolated villages in Africa where they provide electricity to power light bulbs and TVs. According to one estimate, in 1999 approximately 400,000 families in the less-developed nations relied on small solar electric systems to provide power for fluorescent lights, radios, small appliances, and televisions. In the United States, over 40,000 homes and cabins are powered entirely by PVs. Many of these structures are situated in remote regions where connecting to the electrical power grid is not possible or is prohibitively expensive. More and more, however, photovoltaics are popping up in areas where electrical lines run nearby. Homeowners are by choosing to disconnect from the electrical power grid fed by coal-fired power plants, nuclear power stations, and hydroelectric dams. Utility companies are also getting into the act, eager to cash in on the solar revolution.

Wind power is also catching on. An increasing number of U.S. homes are now being supplied by small-scale wind power. After many failures in the 1970s and 1980s, small-scale wind power is making a strong recovery. Homeowners are installing hybrid systems consisting of photovoltaics and wind generators; two technologies that marvelously complement each other. In my area, I find that as storm clouds move in to block the sun, the winds often pick up.

Many power companies are adding wind power to their generating mix. My local utility, not known for its sympathy to environmental issues, has already installed a dozen large wind generators. California has been a leader in wind power for more than a decade, and at a surprisingly low cost. Combined with power companies in other states, these efforts could increase U.S. reliance on renewable energy. According to the Worldwatch Institute, new wind turbines built in 1998 pushed global wind generation to an all-time high. Worldwatch estimates that wind turbines generated roughly 21 billion kilowatts of electricity in 1999—enough for 3.5 million homes.

Wind power has become the world's fastest-growing energy source. The 2,100 megawatts of new wind energy added in 1998 was 35 percent more than the amount added in 1997. Although wind and solar are a long way from becoming mainstream energy supplies, they are well on their way. Wind and solar electricity are certain to grow even faster over the next decade as environmental problems such as global climate change and acid rain caused by the combustion of coal and oil continue to mount. Within most of our lifetimes, wind and solar could become the dominant sources of electrical energy, pushing coal and nuclear power aside. Hydropower will also continue to contribute energy, although a shift to small-scale operations may be a significant aspect of our quest for a sustainable energy regime.

This chapter explores ways to generate electricity for home use from three clean, renewable energy-producing sources: the sun, wind, and small streams. My goal is to introduce you to these energy-generating technologies and point you toward more detailed resources.

PRODUCING ELECTRICITY FROM SUNLIGHT: THE ANATOMY OF A PHOTOVOLTAIC SYSTEM

Sunlight can generate enormous amounts of electricity—certainly enough to power just about any home on the planet. This task is achieved with the aid of a solar module, consisting of many photovoltaic cells (figure 12-1). PV cells are commonly made of silicon and small amounts of boron and phosphorus. A typical PV solar module has thirty-six square or rectangular cells connected. Most photovoltaic systems (*photo* for sun and *voltaic* for electricity) consist of two or more modules mounted on the roof, the ground, a pole, or a special tracking device that follows the sun across the sky from sunrise to sunset.

When sunlight bombards a PV module, it causes the production of tiny electrical currents in each solar cell. The electrical currents produced by the cells in the modules combine to produce a much larger electrical current. Electricity produced by solar modules is typically transmitted to a bank of batteries where it is stored for later use. The most common batteries are deep-cycle lead acid.

Solar power is now the world's second fastest growing energy source —at an average growth rate of 16 percent per year since 1990. World solar markets are growing at ten times the rate of the oil industry. . . . Solar energy may now join computers and telecommunications as a leading growth industry in the twenty-first century.

CHRISTOPHER FLAVIN and MOLLY O'MEARA, Worldwatch Institute

Many solar modules come with ten- to twenty-year warranties, a fact that speaks highly of their reliability, but also comes as a shock to many people. What other product comes with a warranty that even comes close to this?

FIGURE 12-1.
The PV module consists of numerous PV or solar cells made mostly of silicon. The PV cells convert sunlight to electricity, cleanly, quietly, and reliably.

Silicon is the main ingredient of the three most popular and widely used PV cells. Silicon is made from silica derived from high-grade sand or quartz rock. Sand is the second most abundant substance on the Earth, next to water.

KEVIN JEFFREY,
Independent Energy Guide

Solar electricity is not like ordinary household electrical current. You can't use it to power household appliances, lights, or electronic devices. That's because PVs produce direct current (DC) electricity.

Direct current electricity is a steady unidirectional flow of electrons in the wires. Batteries in a flashlight produce DC power. However, most appliances and electronic devices that we use in our homes consume another type of electricity, known as alternating current (AC) electricity. Alternating current is produced by power plants and generators. It is delivered to our homes, offices, and places of business via electrical wires that are part of a nationally interconnected system, known as the utility grid.

Alternating current electricity changes directions extremely rapidly—120 times per second (or 60 cycles per second). That is to say, the electrons oscillate (alternate) back and forth in the wire. The energy of oscillating electrical currents is used to power appliances, lights, and a wide assortment of electronic devices. Why does the world rely primarily on AC electricity?

The answer lies in history and simple physics. When electricity was discovered and first put to use, it was the DC variety. Much to the dismay of its major proponent, Thomas Edison, DC electricity failed miserably in early experimentation with power distribution. It simply can't be transmitted efficiently over long distances. In order for DC electricity to travel any distance, thick and costly electrical wires are required; even then, current drops rapidly over distance.

A brilliant Yugoslav-American inventor named Nikola Tesla invented alternating current to rectify the situation. AC electricity provided the first practical means for generating large quantities of electricity and transmitting it economically over long distances. Although power is lost in the transmission of AC electricity, the losses are much lower than in DC. Thus, AC has become the dominant form of electricity worldwide. Most electronic devices are powered by it. To use conventional appliances, a solar homeowner must install a device that converts the DC electricity produced by PVs into AC. The device, known as an inverter, also converts the 12- or 24-volt DC electricity produced by solar systems into 110-volt current, which is required by household electronic appliances, stereos, televisions, and other electronic gadgetry.

Although most solar homes in the United States rely primarily on 110-volt AC electricity, it is not unusual for them to contain a few DC circuits for refrigerators, water pumps, ceiling fans, and even specially built televisions.

A typical solar electric system contains various meters to monitor how much electricity is being produced by the PVs and how much is being consumed in the house at any one moment. These meters permit the homeowner to monitor the battery voltage, which indicates their state of charge, or more practically, how much electricity they contain. (The higher the voltage, the more electricity they contain.)

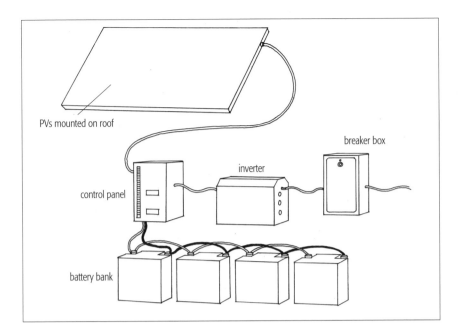

FIGURE 12-2.
A typical PV system consists of a PV array, a control panel, batteries, and an inverter, which feeds alternating current into the main service panel or breaker box.

One of the best ways to understand electricity is to think of it as water flowing through a garden hose. Water flows through a hose at a certain rate. As you open the spigot wider, it flows faster. As you close it down, it flows slower. Electricity also flows at different rates, depending on the load, that is, how much you are drawing at any one time. The flow of electricity through a wire is measured in amperes or amps. The higher the amperage the higher the flow.

Homeowners must monitor battery voltage to determine the state of charge of their storage batteries. Voltage is an electrical term that refers to the potential difference across the electrodes. In the water analogy, voltage is equivalent to the water pressure in the hose. In a battery system, high voltage means you have a lot of electricity stored in the batteries.

As a solar homeowner, you will need to know these basic terms, even if you hire someone else to design and install your system. Once your system is in place, you will be required to monitor its performance and determine the amount of power you have. You will be the system operator and in order to do the job right, you must become familiar with amperage and voltage. For a more thorough introduction, I urge you to read the basic electricity sections in Kevin Jeffrey's *Independent Energy Guide* or Joel Davidson's *The New Solar Electric Home.*

The meters in most solar electric systems are housed in a consolidated unit known as a power cen-

FIGURE 12-3.
The power center contains all of the circuit breakers, controls, and monitors needed for home-generated power. It makes the job of installing solar electricity much simpler and less costly.

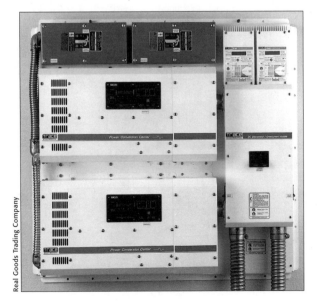

Real Goods Trading Company

Charge controllers are also called voltage regulators. The term voltage regulator *is a misnomer. These devices actually control the charging current by sensing battery voltage.*

BUILDING NOTE

In the United States, the National Electric Code calls for a disconnect between the PV modules and the first charge controller, and a second disconnect between the charge controller and the battery bank. A third disconnect must be placed between the battery bank and inverter. Details on disconnects and other important aspects of PV systems can be found in the reference books listed in the resource guide.

FIGURE 12-4.

This schematic shows a simple solar electric system consisting of a PV array, a voltage regulator, an inverter, AC load, and batteries. The regulator (actually a charge controller) is an important safety measure that monitors battery voltage and terminates the flow of electricity to the batteries when they are full of electricity. This, in turn, prevents overcharging, which can damage batteries.

ter (figure 12-3). In addition to the meters that monitor the performance of your system, power centers contain various safety devices. For example, power centers typically contain two charge controllers to protect the batteries. These monitor voltage and control the flow of electricity in a solar system to prevent the batteries from being overcharged or from being too heavily discharged. One charge controller is located between the PV array and the batteries. Its job is to keep the batteries from being overcharged, for example, while you and your family are on an extended trip and the house is using very little electricity but the solar system is producing excess power. Overcharging batteries can cause serious damage.

A second charge controller is located between the batteries and the wires leading to the main service panel (the circuit breakers). Also known as a low-voltage cut-out, it prevents batteries from being too deeply discharged. Like overcharge, deep discharge is extremely harmful to batteries. When the voltage drops below a preset level, indicating that the batteries are very low, the charge controller cuts out. This terminates the flow of electrical current to the main service panel, making it impossible to use any power until the batteries are recharged.

Power centers also contain safety disconnects (circuit breakers) that protect electricians servicing the system (see sidebar). Because power centers come prewired with all of the meters and safety equipment you will need, they are generally easy to install. This saves you a lot of time and frustration if you are wiring your own home, and also makes servicing easier for a professional electrician.

Power centers are approved by the Underwriter's Laboratory (UL) and are very popular among electrical inspectors, who typically know very little about solar electricity. When inspectors encounter a UL-approved power center, they breathe a sigh of relief because they know that the proper controls are in place for safe system operation. (For more on power centers, refer to at the catalogues of Jade Mountain, Alternative Choices, and Real Goods. If you can't decide, their technical people will help you select the power center for your home.)

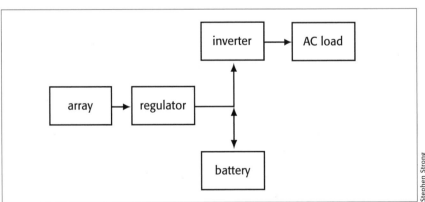

Most home solar electric systems are more complex than the one illustrated in figure 12-4. Many systems have a DC circuit or two and a generator for backup power and battery maintenance. In my house, for example, the water pump that delivers water from the cistern to the house is wired to a DC circuit. I also wired the ceiling fans for DC. Were I to build again, I'd also connect my refrigerator to a DC circuit. Direct current circuits feed DC electricity from the modules and batteries *directly* to ceiling fans, refrigerators, lights, appliances, water pumps, or any other DC electronic devices bypassing the inverter. Why bother with DC circuits and detour your inverter?

The inverter requires energy to operate. In fact, most inverters consume about 5 percent of the electrical energy passing through them. Bypassing the inverter, therefore, saves a lot of energy, especially when you are supplying appliances that require lots of electricity, such as refrigerators and water pumps. If this strategy saves so much energy, why not go entirely DC?

Although a lot of appliances and many types of electronic equipment use DC current, most were designed for use in RVs and therefore are rather small. A 9-cubic-foot refrigerator may hold enough food for a single person, but it won't suffice for a family. Because the market is limited, they are also more expensive than mass-produced AC appliances and electronic equipment.

Stand-Alone versus Utility-Intertie Solar Systems

Two major types of solar electric systems exist: stand-alone and utility-intertied.

A **stand-alone** PV system is totally independent of utility power. It consists of PVs, a battery bank, an inverter, and a power center. Most systems include a generator powered by gasoline, diesel, or propane for backup power and battery maintenance. Grid-dependent or utility-intertied systems are connected to the power grid so that they can receive varying amounts of electricity from the local utility.

In stand-alone systems, generators serve a variety of purposes. They can provide supplemental electricity if your household requires more electricity than your panels produce. If you undersized your system, for example, a generator will provide added power from time to time to keep things running smoothly. Or, if the weather takes a turn for the worse and clouds block the sun for an unusually long period, the generator can be fired up to keep you from running out of power. Generators are also used to supplement power when homeowners run energy-intensive equipment, for example, well pumps, washing machines, and power tools. Although a larger system could accommodate these loads, the need may be so infrequent that the extra expense is not worth it. It is cheaper and easier to install a generator to meet unusual loads. The generator also prevents deep discharging of batteries under such circumstances and therefore boosts their useful lifespan. As Steven Strong and William Scheller note in *The Solar Electric House*, adding a generator is "simpler and less expensive than increasing the size of the PV array and battery

Many people often say they do not want a generator in their system. I have not known anyone with has a generator who has not been happy to have it once they lived with their system for a while.

LAURIE CAMPBELL,
Alternative Choices

Although the thought of having a fossil fuel-operated generator may seem hideous, they're vital to stand-alone PV systems. The only recompense is that you don't need to use them very often. To learn more about batteries and generators, I strongly recommend Kevin Jeffrey's book, Independent Energy Guide.

storage to cover every extreme in weather and load conditions that potentially may occur." Generators produce AC power and can be used to run appliances and electronic devices in emergencies, for example, if your inverter breaks down. (By the way, if any component of your system is going to give you trouble, the inverter is the one that will!) Finally, generators are required to maintain your batteries in optimal condition, which ensures many years of trouble-free service.

One of the most important things you can do to keep your batteries in good working order is called *equalization*. Equalization is a controlled overcharge required for lead-acid batteries, the most common type in use today. It is effected by running the generator with the inverter set on the equalize mode. During this process, the generator feeds electricity to the batteries first very rapidly, then more slowly, and finally in a trickle, until they can absorb no more. Sensors in the inverter stop the process automatically.

Equalization drives deposits of lead sulfate off the thick lead plates inside each battery. The more deep discharging a battery sustains, the more lead sulfate will amass on the plates. Periodic equalization scours the plates, allowing them to accept charge more readily. It also extends the life of batteries. If not removed by periodic overcharge, lead sulfate will eventually ruin your batteries, reducing their lifespan from five to twelve years to a couple of years. Because batteries cost a couple hundred dollars each, you can't afford to mistreat them.

Generators also assist in keeping batteries highly charged. In between equalizations, I periodically run my generator just to keep the voltage high. It may only require a couple hours, but the extra effort is worthwhile in the long run. According to Laurie Campbell of Alternative Choices, a company that supplies PV systems and other renewable energy products, "Most battery banks that last only a few years are killed because they are never properly charged."

Besides equalizing your batteries and periodically boosting their voltage to keep them well charged, you must periodically check the water levels in your batteries. The object is to never let lead plates be exposed to air. If water levels have dropped, top them off with distilled water (not de-ionized water!). Some books on the subject say that this needs to be done every two to three weeks—although this may be a bit overzealous. Solar suppliers recommend a monthly checkup. Most PV installers would be tickled to death if you did it every two months! Don't cut corners. Get in the habit of equalizing and filling your batteries regularly.

In order for a generator to charge your batteries, you must have a battery charger, as shown in figure 12-5. Battery chargers convert the alternating current produced by generators to direct current that batteries can accept. Battery chargers are included in many inverters or may be purchased separately, then mounted near the control panel.

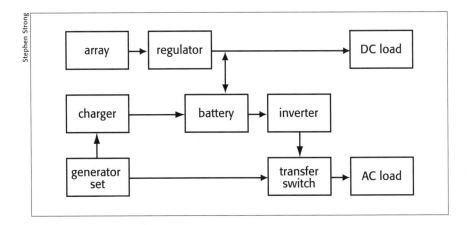

Stephen Strong

FIGURE 12-5.
In this stand-alone solar system, the generator supplies electricity to the system when loads are high or batteries are low or during periodic maintenance of the batteries (equalization). DC loads can be supplied directly from the batteries without passing through the inverter.

The **utility-intertie** or utility-dependent system is the second option for solar electric power. It is a hybrid system that relies on the local power company to provide you with electricity when your system comes up short. Utility-intertie systems may be designed to draw varying amounts of electricity from the electrical grid, from zero to nearly all of your electricity, depending on the size of your PV array and your demand.

Utility-tied systems are similar to stand-alone systems in many respects. They contain a PV array, an inverter, a power center, and various meters and controllers. Many have batteries for backup in case of a black out. Some don't.

In a utility-intertied system that contains batteries, utility power feeds directly into the battery charger (see figure 12-6). There's no need for a generator. The battery charger, in turn, converts AC utility power to DC electricity. DC electricity is then stored in the batteries and fed into household circuits via AC and DC circuits.

In this type of system, the PVs supply electrical power to household circuits when the sun is shining. Excess electricity is stored in the batteries. When the sun sets or clouds block the sun, electrical power demands are met by drawing electricity from the batteries. When voltage in the batteries drops below a preset level, current is drawn from the grid.

Most batteries do not die, they are murdered!

LAURIE CAMPBELL,
Alternative Choices

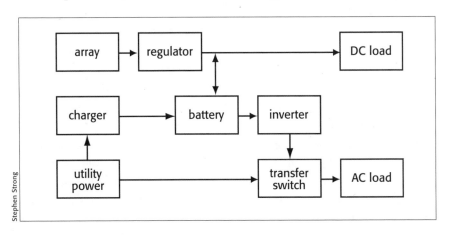

Stephen Strong

FIGURE 12-6.
Utility-intertie system. If you don't feel comfortable being disconnected from the electrical grid, you can use utility power in place of a backup generator, to supplement batteries or to meet excessive loads.

When excess solar power is being produced, it is shunted to the grid. The electrical meter runs backwards. When supplemental power is required, electricity flows from the grid to the house and the meter runs forward. In this system, the grid acts as a generator, supplementing power when PV output is reduced, and as a battery bank, "storing" excess power when PV production exceeds load.

The advantages of tying into the utility grid are very real and quite substantial. Despite assurances of experts and the growing success of solar electricity, some folks remain wary of solar electricity. For these people, the security of the electrical grid outweighs the disadvantages of being hooked up to the utility power lifeline. Others tie into the power grid to avoid use of a generator. They rely on grid power to keep their batteries charged. Still others tie into the grid to reduce or eliminate the need for batteries altogether. Utility-connected systems are cheaper than stand-alone systems; savings can be substantial. A generator costs $1,000 to $2,000. A battery bank for a house costs $2,000 to $5,000.

Utility tie-ins provide the level of convenience to which most people have grown accustomed. In addition, in utility-dependent systems containing batteries, there is no need to equalize your batteries. You won't be deep discharging them and utility power will maintain battery voltage at optimum levels to ensure a long life.

A utility tie-in requires little, if any, change in behavior. You can run power tools, the washing machine, a well pump, and anything else you want, all simultaneously, and all with impunity. You need not ration power on cloudy days or adjust habits to coincide with solar radiation. Stand-alone systems may require both.

Another advantage of a utility tie-in is that it permits one to incrementally expand a PV system, gradually reducing one's dependence on utility power and increasing one's dependence on solar or wind energy. This benefit may be attractive to those who want to ease into solar power or for whom cost is a barrier.

And, of course, a key advantage of the utility tie-in is that it provides you with the opportunity to shunt excess electricity back onto the grid. In 1978, the U.S. Congress passed the Public Utility Regulatory Policy Act, commonly known as PURPA. The law requires the nation's power companies to purchase excess electricity from small electrical generators, including individual households. By law, utility companies are only required to pay you what it costs them to generate power or buy it from other sources, usually about one-third of the price the consumer pays. However, because this fiduciary arrangement would require two meters on solar houses, one to monitor use and the other to monitor electricity you're delivering to the grid, most local electric utilities install a single electrical meter, one that runs forward when you're drawing electricity from the grid, then runs backwards when the house is producing

Hooking into commercial power adds a level of assurance to those who are venturing into solar electricity for the first time and can't see their way clear of that good, old, usually reliable power company electricity.

more electricity than it needs. In essence, the utility is reimbursing solar producers at the same rate they charge customers. (The gods must be crazy!) If, for example, a solar electric house connected to the grid consumes 100 kilowatt-hours (kWh) of electricity from the grid during a month, and produces an excess of 50 kilowatt-hours, the homeowner is only charged for the balance, 50 kWh. The term used for such an arrangement is *net metering*.

Which Option is Best for You? With so many options, potential solar buyers complain that it is difficult to choose a system. Before making a choice, carefully consider your available time, space, skills, money, and commitment to energy independence.

Stand-alone systems are pretty easy to use and quite reliable. The generator makes up for shortfalls in power production or periods of excess demand. Although generators burn fossil fuel and pollute a bit, your contributions to fossil fuel depletion and air pollution are much less than if you had stayed on the grid. Stand-alone systems do require periodic battery maintenance and care. You'll need to equalize batteries every three months or so, check battery fluid levels every month, and maintain voltage for optimum performance of your system. Running your generator may bother your neighbors, too! If you are independent-minded and want to get off the grid and out of the grips of large corporations, the stand-alone system is as independent as you can get. The system will cost more than a utility tie-in if you are close to the electric wires because of the need to purchase batteries and a generator.

To avoid having to maintain batteries, some homeowners install sealed gel batteries. Although they don't need equalizing and can't lose fluid (thus making your life a little easier) sealed gel batteries cost two to three times more than lead-acid batteries and don't last as long. In the long run, it is cheaper to spring for a good generator and to spend the time pampering your lead-acid batteries.

The economics of stand-alone systems change dramatically if your home is more than a half a mile from the power company's wires. Hooking up to the grid—either by running utility poles or burying the line—is extremely expensive, costing $50,000 or more for a mile run! That's enough money to purchase a couple of good-sized PV systems! In such instances, a stand-alone system is truly the only affordable way to provide electricity.

Stand-alone systems require a battery room, a safe and relatively warm location to house your batteries. Because lead-acid batteries release small amounts of hydrogen gas (which is combustible) when being charged, batteries must be placed in a vented space to avoid any chance of explosion. If you live in a hot or very cold climate, however, don't relegate your batteries to the garage. Batteries can overheat in the former and freeze in the latter. The ideal battery room should be kept between 50 and 80 degrees Fahrenheit. Lead-acid batteries operate best at about 70 degrees Fahrenheit.

BUILDING NOTE

Although utilities are required to purchase electricity from individuals, they may throw serious roadblocks in your way. They may, for instance, require the installation of extra meters, and some charge a monthly fee to read your meter. Because of this, some folks choose to install solar electric systems and hook to the electrical grid without permission. All that's needed is an appropriate inverter—one that's suitable for utility intertie. For more on this option, check out the Guerrilla Solar section of *Home Power* magazine's Web page at www.homepower.com.

BUILDING NOTE

Batteries are often stored in basements or utility rooms, but should never be placed directly on cement floors or against cement walls. Although some people will tell you that this drains the power, the real reason is that it cools the batteries, creating unequal temperature in the battery cells (colder on the bottom). This, in turn, causes them to function less than optimally.

NO HOOK-UP EQUALS MORE PANELS

The local electric provider wanted $1,000 to $2,000 to hook my house to the grid and $20 per month minimum, whether I used any power or not. By not hooking up to the grid, I saved enough money to buy a generator, and by avoiding the monthly minimum, I pay for an additonal module every 15 months!

If you locate your batteries in an unheated garage or shed, you will need to insulate the batteries against cold, and very likely even heat the battery storage space. If you can't maintain optimum temperature, your batteries will perform poorly and will provide far fewer years of service. In fact, many people who have placed batteries in garages or unheated outbuildings have found that their batteries mysteriously fail after a couple years. Some people have added gas heaters to their battery rooms, which adds to the cost and the overall energy demand of the system.

A specially vented battery vault in your house is one of the best options for housing batteries. Many people place batteries in sealed, insulated plywood boxes that are vented to the outside. This enclosure serves three functions. It keeps the batteries out of harm's way; it insulates them, buffering them from temperature extremes; and it prevents hydrogen gas from escaping into the room.

If you want the assurance of grid power, consider the utility tie-in. This system is inexpensive and simple to maintain. You won't have to buy a generator or mess with batteries. Not buying a generator will save you $1,000 to $2,000. Avoiding batteries will save you even more. This option sets up the grid as your storage medium and ensures an unlimited supply of power, unless of course . . . well, you know what I was going to say.

Although the utility tie-in is one of the least expensive solar electric options, it leaves you dependent on the power companies. However, one of the great things about solar electricity is that you can expand the system fairly easily, adding more panels or batteries over time, weaning yourself from utility power gradually as you feel more comfortable with this technology. If you are contemplating an expansion of your system, a word to the wise: plan for changes. Be sure to leave room in your design for additional panels, batteries, and a generator. If you think you may want to include some DC circuits, lay the conduit in the walls when you are building, so you can pull the wires when the time comes. It is costly and difficult to add them after the walls have been finished.

The same adaptability holds true for stand-alone system. If you have a stand-alone system but want to tie into the grid at a later date, you can do so fairly easily. If you can only afford a small PV system to provide DC electricity to a few circuits, that's fine. Install it and add an inverter and more panels and batteries later. You can add a wind generator, too, if you'd like.

For further information, read sections on solar electric systems in *The Solar Electric House* and *The New Solar Electric Home*. Another good book is Kevin Jeffrey's *Independent Energy Guide*. The folks at Real Goods, Jade Mountain, and Alternative Choices are very knowledgeable and accommodating. Local suppliers may be helpful, too. Solaronsale.com is also a great supplier. Most suppliers will design a system to meet your needs free of charge and will either deliver the parts for you to install, or install the system for you. Solar Survival Architecture also provides a complete PV package with minimal installation requirements.

Bear in mind that you can size a PV system so you never have to worry about energy efficiency or load management, but most people don't. They size their systems to save money. Or they end up with an undersized system because they use a lot more power than they estimated. In such instances, homeowners become acutely aware of power demands and often find creative ways of saving power.

A Word about Solar Modules

Photovoltaic modules come in many shapes and sizes. For the newcomer, the options can be confusing and overwhelming. Don't despair. If you choose a prepackaged system or enlist the aid of a professional to design a system for you, you really don't have to make a choice. Suppliers recommend modules and other components that they prefer.

On the other hand, if you are going to design your own system and choose the components, you are in for some work. The good news, however, is that the PV modules on the market are generally well made and very reliable. Many PVs come with a shocking-in-this-world-of-planned-obsolescence ten-year warranty, although I've seen some with warranties as low as five years and as high as twenty years. "PV panels are subject to rigorous laboratory testing, including repetitive extreme temperature cycling, temperature cycling at high humidity levels, wind loading exceeding 125 mph, and surface impact equivalent to one-inch hail traveling at terminal velocity of 52 miles per hour," writes Kevin Jeffrey in the *Independent Energy Guide*.

Catalog descriptions of the various photovoltaic modules typically list at least three parameters: watts, amps, and volts. Wattage is a measure of power (see sidebar). Module wattages range from 30 to 120, while amperage (the amount of current a module produces) varies from 2.3 to nearly 5 amps. The voltage of the various modules is pretty similar, around 17.

This information is useful to those who want to size their own systems. It can also be used to determine which option provides the most electrical power at the lowest cost. To determine which module provides the best value, calculate the price per watt of each module by dividing the wattage of a given module by the cost of the module. New PVs vary from $5.00 to nearly $8.00 per watt.

Many of the books listed in the resource guide at the end of the book walk you through the other calculations required to size your modules and batteries. Or have a solar expert work through the math for you, and design a system that will meet your needs.

A Word about Batteries

Batteries are a key component of most solar electric systems. According to Laurie Campbell at Alternative Choices, "The heart of any alternative energy system is the battery bank. It doesn't matter how much power you are making,

Photovoltaic panels are rated according to their peak power in watts at 25 degrees Celsius, as well as their voltage and current corresponding to that rated power.

KEVIN JEFFREY,
Independent Energy Guide

Wattage is a measure of power. It is determined by multiplying the amps a panel produces by the volts ($W = a \times v$). Electronic equipment, including appliances, televisions, and the like, are all rated by wattage.

or how fancy your inverter is, if your battery bank is bad, your whole system is bad." Joel Davidson, author of *The New Solar Electric Home*, notes that "any discussion of batteries becomes complicated because there are so many factors involved. To make things even more difficult, myths and misinformation abound. But the more you know about batteries, the happier you will be."

For most home systems, six-volt, deep-cycle batteries are sufficient. Deep-cycle batteries contain thick lead plates designed to handle deep discharges. In fact, the Trojan L16 deep-cycle batteries that are commonly used in solar electric systems can handle as many as 750 deep discharges over their lifetime—or so says the manufacturer. Nevertheless, the rule of thumb is to avoid deep discharging your batteries as much as possible. Joel Davidson argues that it may be wiser to buy twice as many batteries as you need so they are discharged at a shallower depth. This, he says, can double battery life more often than not. He notes that if you oversize your battery bank and discharge it to only 20 percent of its storage capacity, a battery bank that would last six years with deep discharge would last ten to fifteen years. You can also periodically fire up your generator to keep the charge level higher. Another strategy is to install more modules to keep the batteries at a higher charge level. Personally, I like this strategy. Even though a module costs more than a battery, it lasts much, much longer.

Batteries last longest if the voltage is routinely maintained near the upper end of the range. Your batteries just won't last as long if they're routinely maintained at low voltages. Find ways to use electricity more efficiently, or install an extra module or two, or fire up your generator to maintain voltage at the higher end of the scale and to extend the useful life of your batteries.

When you install a stand-alone solar electric system, you are the power plant manager and electrical consumer. To ensure maximum performance, you must pamper your batteries and you must manage your loads very carefully. For those who install stand-alone PV systems, the first six months can be nerve wracking as you get used to living with solar electricity. You may even have to fine tune your lifestyle and your PV system. But in time, you will figure it out. You will learn to hold your laundry until it is sunny and you have plenty of electrical power, or you will run the generator when the washing machine is on. If you use a microwave oven, you may find yourself taking food out and letting it thaw naturally first, then microwaving it. This can dramatically cut down on electrical demand and preserve your batteries, too.

A Word about Inverters

The inverter is a key component of all solar electric systems. Inverters are extremely sophisticated electronic devices. Some even contain microchips programmed to turn generators on when battery voltage drops below a certain set point.

Inverters are sized by determining the largest electrical load placed on them. That is usually when the washing machine, dryer, refrigerator, and well pump are on simultaneously. Most solar homes require a 2,400-watt inverter. However, if you have a deep-well pump over ¾ horsepower, you will probably need a larger unit.

You will have the choice of a sine wave or a modified sine wave inverter. Sine wave inverters produce a cleaner form of electricity. They are better suited to feeding electricity onto the grid and work better with many electronic devices, such as stereos and even some compact fluorescent light bulbs. Although they cost more, sine wave inverters are well worth the investment. If you can't afford one, however, install a modified sine wave inverter plus a small pocket sine wave inverter for the circuits servicing sensitive electronics.

Some inverters give off an irritating buzz. Check with the manufacturer or supplier before buying a unit, and install the inverter in a location that mutes any sound it produces. Otherwise it will drive you nuts while you are trying to read.

Sizing Your System

The size of a solar electric system depends on many variables, including solar insolation (the amount of sunlight you receive), the efficiency of your system, electrical demand, and temperature. Quantifying these variables determines how many PV modules and how many batteries you will need.

Solar Insolation. Solar radiation varies from one region of the world to another. Most places, however, experience enough sunlight to make a solar electric system feasible. Ultimately, the more sunlight the more electricity you will be able to generate.

To determine if your site has sufficient solar insolation, consult local experts or solar maps and charts. They will tell you the average amount of sunlight in your location. Keep in mind that weather patterns in your area may be quirky, causing more or less sunlight than the average for your region.

Assuming that your region offers adequate sunlight and your lot is open and properly oriented toward the sun, the next step is to determine the size of the system to meet electrical demands. Electrical demand is calculated by examining utility bills, then making adjustments for new appliances and other anticipated uses. If your usage is high, implement a household conservation strategy and base your demand on reasonable reductions over the current levels of usage. Another way to estimate demand is to list the power consumption (wattage) of all appliances, electronic devices, and lights, then estimate the daily usage of each item on the list—that is, how many hours each one will be used. Multiply the wattage by the hours of usage, and add the results to obtain total household consumption. The size of the system is then calculated based on estimated demand, to which you should add some efficiency losses (losses occurring during storage and inverter use). Most suppliers base battery bank size on these factors but add three to five days of storage capacity to tide you

STEPS IN SIZING YOUR PV SYSTEM

1. Determine your average daily electrical requirements.
2. Determine PV electrical output in your area and determine the size of the PV array needed to match your electrical demand.
3. Determine battery storage requirements to meet your demand over a three-day low-to-no-sun period.

BUILDING NOTE

Solar Energy International and the Center for Renewable Energy and Sustainable Technologies sell a CD-ROM program that will size and design a complete residential PV system. The program comes with solar data from over 300 sites around the world. This program works on Macintosh and PCs.

over during bad weather. (Formulas for calculating these figures are included in most books on PVs.)

Efficiency of Your System: Optimizing Solar Gain. The size of a system also depends on solar gain, or how much sunlight the system captures and converts to electricity. Solar gain varies within any given region. One variable is the orientation of the modules. For optimum solar gain, PVs should point as close to true (not magnetic) south as possible. Avoid shady spots, even if shade covers a small portion of the array. Partial shade can dramatically reduce the efficiency of a PV. Snow accumulation on a small portion of my array reduces efficiency dramatically, from 2 amps to 20 amps.

Tracking systems, such as the one shown in figure 12-7, increase the output of PV solar modules by about 40 to 50 percent in the summer and 15 to 20 percent in the winter, according to one source. In latitudes above 45 degrees, however, enhanced wintertime performance is negligible.

Trackers come in two basic varieties: active or passive models. Active trackers contain small electrical motors that rotate the mounted modules during the day, causing them to follow the course of the sun. Although they consume small amounts of electricity drawn directly from the PVs, motorized

FIGURE 12-7.
Tracking devices like this one allow your modules to "follow" the sun from sunrise to sunset, greatly increasing their efficiency in most areas. The extra cost, however, may not always make this a wise investment.

Michael Potts

units are still slightly (10 to 15 percent) more efficient than passive trackers. (They do a better job of following the sun.) Passive trackers contain a refrigerant gas located in the tube-steel frame. When heated, the gas shifts position, keeping the PV modules oriented toward the sun.

Tracking devices increase the efficiency of a PV array in most instances and therefore reduce the number of modules required to produce a given amount of electricity. However, trackers cost more than standard roof and pole mounts, so much more that it is usually cheaper to mount more PVs on fixed poles than to install a smaller number on a tracker. Tracking devices can get knocked out of kilter in strong winds. Moreover, moving parts require maintenance. Maintenance can be costly. So why would anyone purchase a tracking system?

People choose trackers because they don't want modules on their roofs, or they have aesthetic objections to standard roof mounts, or heavy snow accumulation makes it difficult to keep the modules free of snow. Trackers are ideal for solar water-pumping applications. As Laurie Campbell of Alternative Choices points out, some people install PVs on trackers that supply electricity to DC well pumps. Pumps siphon water from the ground, delivering it to a holding tank. From here it flows into the pressure tank inside the house. Because DC pumps have limited capacity, it is often necessary to pump all day long. Trackers ensure a steady supply of DC power from sunrise to sunset, which is sufficient time to fill storage tanks.

Roof and pole mounts, although they don't allow the modules to track the sun, can be adjusted to optimize solar gain during the annual solar cycle. By simply changing the angle of the module to the sun, keeping the surface of the module as close to 90 degrees to the incoming solar radiation during each season, a homeowner can significantly boost a module's power output. Adjustments are required at least twice a year—at the beginning of summer and at the onset of winter.

Roof and pole mounts keep PV modules cooler, which enhances their efficiency. Although you may be inclined to mount modules flat on a roof for aesthetic reasons, this is not recommended. Surface mounting limits air circulation around modules, causing temperatures to rise and reducing heat output. For more on mounting, see the *Independent Energy Guide* and *The Solar Living Source Book* by the folks at Real Goods.

Being Efficient Upfront. Proper orientation and the use of tracking devices increase efficiency and reduce the size of the system you will require. Yet another way to reduce the size of your system and cut costs is by careful and judicious load management—that is, using only what you need and using it efficiently. Simple measures such as turning off lights when you leave a room and switching off stereos and televisions when you leave the house are vital.

It cannot be stressed too often—PV systems work well if you are energy-conscious. You must be willing to keep an open mind and find creative ways to use and not use energy.

JOEL DAVIDSON,
The New Solar Electric Home

Next, ensure that all lights, electronic equipment, and appliances are as efficient as possible. Replace incandescent lights that are on for any length of time each day—for example reading lights or office lights—with compact fluorescent bulbs. (Lights that are not on for extended periods can be incandescent, as it is hardly worth the investment to place a $12 compact fluorescent light bulb in a closet that you only open when you are looking for the Scrabble game.)

If your appliances are inefficient, get rid of them if you can afford new ones. Recycle your discarded machines or sell them to someone looking for a used model. Washing machines are especially troublesome. Most top-loading machines use enormous amounts of energy and water. Replace a top-loader with a front-loading (horizontal axis) machine. They use one-third as much electricity as the upright models sold in the U.S.—and they have the added benefit of using about half as much water. Some models spin dry clothes extremely well, leaving very little residual moisture. This cuts down on drying time and cost.

If you have to have a dishwasher, shop around. Energy- and water-efficient models are available; they're also reasonably priced. Better yet, wash your dishes by hand.

If you have an electric stove, leave it in your current house. The large burners on an electric stove use about 1,200 watts of power when on high. The smaller burners consume about 600 watts of power. Although you may prefer to cook with electricity, an electric stove will sabotage your efforts to live on solar electricity. Use natural gas or propane to cook. You may also want to build or buy a solar cooker.

Next to electric stoves, refrigerators are the most energy-consuming device in our homes. Although new models are considerably more efficient than those manufactured just a few years ago, they are still not efficient enough for an affordable off-grid or stand-alone solar home. One of the best electric refrigerators is made by SunFrost. Refrigerator/freezer units are available in AC and DC models and in a variety of capacities from 4 to 19 cubic feet. SunFrost refrigerators are available from Alternative Choices, Real Goods, and Jade Mountain. Although a super-efficient refrigerator will cost a small fortune, around $2,700 for a 19-cubic-foot model, they are still a better choice than installing a less efficient unit. For the off-gridder, a less efficient refrigerator will require additional modules and batteries that will more than offset any savings gained. According to my sources, a conventional refrigerator requires twelve to twenty-one solar modules. SunFrost refrigerators require only three to eight modules.

How much difference can efficiency make? A well-designed solar electric home uses about one-fourth to one-third of the electrical energy of a conventional home. This doesn't mean that you are required to sit in the dark twiddling your thumbs while your neighbors watch TV. It simply means meeting your demands more efficiently. To illustrate how significant the difference is, consider a personal example. In the early 1990s, I lived in a passive solar

BUILDING NOTE

While front-loading washing machines are a great way to save energy and water, be careful which one you buy if you have a modified sine wave inverter. I found out the hard way that the Frigidaire front-loader won't operate on modified sine wave electricity. Nor will the Maytag Neptune. The only unit I know of that will operate on modified sine wave electricity is the Staber. Unfortunately, it's pricey (about $1,100 to $1,200) and not as efficient as either of the previously mentioned models—although it is still much better than a vertical axis machine.

home. Although we received most of our heat from the sun, used energy-efficient appliances, and had energy-efficient lighting, electrical demand was still pretty high. Sure, we had purchased an energy-efficient refrigerator, but it couldn't hold a candle to a SunFrost. And then there was the electric stove. According to my calculations, a stand-alone photovoltaic system for that residence would have cost $40,000 to $50,000. In contrast, the system on my current house, which relies on a gas stove and was designed with efficiency in mind, cost slightly less than $13,000 (that included $9,300 for the system and $3,400 to install it). Need I say more?

Let a Professional Size Your System. Sizing a system is a fairly complex task with lots of room for error. One possibility is to hire a qualified professional. A PV designer will examine solar potential and electrical consumption, based on historical records or projected demand for a new house, then select modules, batteries, and the other system components that you will need. To determine projected demand, some installers provide customers with detailed worksheets to list all of the appliances, electronic devices, and lights that they anticipate using. You then list the wattages of each appliance. After this, you estimate how many hours you will be using each one on an average day. This gives an average daily power consumption.

Consider Buying a Package. Although you can size and design your own system, it requires a lot of knowledge, time, and effort. You will have to study books on solar electricity, scrutinize the specifications of the PV modules on the market, determine the best inverter for your needs, and on and on. It is no easy task. If that sounds like too much work, consider purchasing a package. It will save you a lot of work and, if you're buying from a reputable dealer, should provide you with components that work well together.

Many suppliers such as Alternative Choices, Jade Mountain, and Real Goods offer package deals, that is, complete systems ready for installation. Alternative Choices, for example, sells several cabin systems that include two to four PV modules, a couple of batteries, an inverter, and other essential equipment (notably battery cables, meters, and controllers). They also sell five residential system packages, ranging from those suitable for small homes to those designed to meet the needs of larger, custom homes.

Working Backwards from Your Budget. For some people, budget rather than demand may determine the size of the system. If, after making the calculations or hiring a professional to do them for you, you discover that you can't afford a system large enough to meet your daily needs, install a smaller system and cut back on demand or tap into the grid for part of your electricity. (Be sure to include the cost of the utility tie-in.) If you plan to convert to a stand-alone system, be sure to buy an inverter that is large enough for your future needs and

BUILDING NOTE

Whether you design and install the system yourself or hire someone to do it for you, I strongly recommend adding capacity. People invariably underestimate their electrical demands. Oversizing your system will cover uses you have underestimated or missed. It will also reduce deep discharging of your battery.

be sure to include a designated battery room in your plans. You can expand your system as money becomes available.

Economics of Solar Electricity

Solar electric systems cost more than electricity from the grid. There's no way around that fact—unless you live in a location with no easy access to grid power. However, even in locations near grid power, the economics of solar electric can be pretty attractive provided your house is efficient and provided you take a long-term view. Consider a somewhat simplified and highly personal example: I paid $9,300 for my solar electric system (modules, batteries, control panel, and roof mounts) and $3,400 to install it. According to various sources, PV modules should last forty to fifty years. Batteries should last five to twelve years given the level of care I take to maintain voltage and fluid levels. Using 50 years for the lifetime of the modules and ten years as the lifetime of the batteries, and amortizing the cost of installation over a ten-year period, my annual cost for electricity is $730 or about $60 per month!

Because I paid the full cost up front, rather than over time, you could argue that I'm losing 5 to 8 percent per year on my $13,000 investment in solar electricity. If you assume that I could have invested the money at 5 percent, I'd be losing about $54 per month. If I add that to my solar electric bill, my monthly payments shoot up to about $115 per month.

To be accurate, however, I have to take into account some avoided costs, for example, installing an electric line from the road to my house, which would have cost about $2,000 or about $16 per month over a ten-year period. If this is subtracted from the cost of the solar system, my monthly bill is about $99. The power company also wanted a $20-per-month minimum just for the privilege of being hooked to their power supply. I doubt that I would have used much power, so that figure can be subtracted from the amortized cost, giving me a $79-per-month electric bill.

Although this is a little steep, especially because my house is so efficient, I'm never without power. When neighbors' homes are darkened by a power outage, I go about my business unaffected. And, I'm blessed with the ability to live on clean, renewable energy. Had I installed the systems myself, using one of the preassembled solar electric systems, say from Solar Survival Architecture, I could have saved at least $3,400, or about $28 per month, bringing my bill down to around $50 per month. Rebates and solar tax credits, not available to me at the time I installed the system, could have lowered the price even more.

Making the Decision to Go Solar

To create a sustainable presence, we must end our deadly dependence on fossil fuels by converting to clean, renewable technologies such as solar. But going solar is not always a clear-cut decision. To help you sort through the maze of issues, I've listed some of the pros and cons for you to mull over.

Advantages of Solar Electricity

- Solar energy is an abundant, renewable resource available worldwide.
- PV systems reduce our dependence on power companies and reduce our impact on the environment.
- The cost of PV systems may be offset by government tax credits or rebates from the utility companies.
- PV systems buffer homeowners from rising utility costs.
- PV modules are reliable, safe, and sturdy—they could outlast you!
- PVs can be used in virtually any climate, the sunnier the better.
- Solar electric systems have few moving parts and, except for the batteries, require little or no maintenance.
- PV systems can be expanded as one's needs change.
- PV systems produce AC electricity for conventional appliances, electronic equipment, and lighting, as well as DC electricity for a host of other applications.
- PV systems can be used in conjunction with commercial electricity, wind generators, microhydroelectric systems, and fossil fuel–powered generators.
- PV systems free homeowners from power outages.
- PV suppliers and installers are found in many locations worldwide.
- Excess electricity generated by PV systems can be sold to utilities, often at full market value.
- Installing a PV system is cheaper than supplying grid power to homes a half a mile or more from the closest electric lines.
- PV systems can be sized to meet any demand.

Disadvantages of Solar Electricity

- Solar electricity is costly. In most cases, it will cost you more than conventional grid power.
- Purchase and installation of a PV system requires a large cash outlay. In effect, you'll be paying for years of electricity all at once, unless you finance the system.
- Renewable energy systems require some form of storage, such as a battery bank.
- Batteries for PV systems are expensive and require periodic maintenance.
- Batteries require storage space, which may add to the cost of construction.
- Batteries must be replaced every five to twelve years, depending on how well you care for them.
- Batteries are expensive and can cause injury. (They contain sulfuric acid!)
- Batteries release hydrogen gas when charging, which can ignite and explode.
- PV systems require an inverter to convert DC to AC electricity, which is used in most appliances, lights, and electronic devices. A good sine wave inverter for a house is expensive.
- Stand-alone PV systems require costly backup generators that consume fossil fuel and generate air pollution.

- PV modules must be periodically adjusted to maximize solar gain. This may pose a risk to you, if your modules are mounted on a roof. Be sure to secure your ladder and don't get on a roof if it is wet or snowy!

WIND ENERGY

Wind power is the world's fastest-growing source of energy, according to Christopher Flavin of the Worldwatch Institute. This clean, renewable energy source has become the shining star of the renewable energy movement.

Wind has a long history of dependable service in the United States and other countries such as Holland. According to the National Renewable Energy Lab, from the late 1800s to the early 1900s, more than 8 million wind turbines were installed in the U.S. Early machines were used to pump water for livestock, grind grain, and produce electricity, largely in rural areas that had no other source of electricity. But America was a nation with a mission to bring electrical power to all rural areas through an interconnected network of wires, the grid. When rural electrification was completed, wires had been dutifully strung from town to town at exorbitant expense. Grid power produced by large centralized power plants was available to all who wanted it. Windmills, like so many other environmentally beneficial technologies, began to go the way of the dinosaur. Progress had once again conquered intelligent design.

In the 1970s, wind energy made a resurgence, not just in the U.S. but in many other countries as concern over oil shortages grew. But early experiments with residential and commercial wind energy machines were largely unsuccessful. Costs were high and wind turbines proved unreliable. Since that time, however, dramatic improvements have occurred in wind generator technology. Along with the technological improvements have come stunning decreases in cost. As a result, wind has skyrocketed to the top of the energy chart, fulfilling its promoters' early promises of providing a clean, reliable, and cost-competitive alternative to fossil fuels.

In 1999, the United States had approximately 2,500 megawatts of installed capacity—equal to nearly two and a half large nuclear power plants. Although this is not much and only produces enough electricity for 2 to 2.5 million people, wind energy is projected to rise dramatically in coming years. Wind farms currently generate electricity for as little as 4 to 7 cents per kilowatt-hour, clearly cost competitive with coal, the nation's other cheap source of electricity. Making conditions even more favorable for this renewable resource, wind-generated electricity comes without the huge environmental cost of coal-fired power plants.

Wind energy has enjoyed enormous popularity in other countries. In Germany, a world leader in wind

WHITE HOUSE SETS GOAL OF 5 PERCENT WIND POWER BY 2020

The Clinton administration sponsored a new initiative, "Wind Powering America," to increase the use of wind energy in the United States. The initiative sets a goal of providing 5 percent of the nation's electricity from wind power by 2020, with the federal government leading the way by buying 5 percent of its electricity from wind power by 2010. The initiative also aims to expand the number of states in which wind power is being generated.

energy, over 2,800 megawatts of wind-generating capacity were in place in 1998. In the northern state of Schleswig Holstein, wind power provides 15 percent of the total electrical demand. Spain, Denmark, and India are becoming major players as well. As Christopher Flavin notes, "Wind power is a far larger potential energy source than most people realize. In the United States, the states of North Dakota, South Dakota, and Texas have sufficient wind capacity to provide electricity for the entire nation."

Wind turbines or generators consists of four basic parts: a set of blades, a power shaft, a set of gears, and a generator (figure 12-8). When wind passes over the blades (technically known as the rotor), it causes them to spin. This turns the shaft, which drives the generator, which produces electricity.

The amount of electricity a wind generator produces depends on the amount of wind, the size of the blades, and the generator. The amount of wind an area receives, in turn, depends on local weather patterns, proximity to water bodies, and geography.

Wind generators fall into three categories: (1) very large units used by power companies on monstrous wind farms or by remote villages where electricity is too expensive to import via wires; (2) medium-sized units large enough to provide all of the electrical needs of a household; and (3) small units used to supplement other power sources, for example, a PV array or grid power.

In this chapter, we will be concerned with small- and medium-sized wind generators suitable for homes and farms. These systems may be stand-alone or grid-connected, as in PV systems discussed earlier in the chapter.

What Do You Need to Know about Wind Power?

If you are contemplating the use of wind power or the addition of wind power to a solar electric system, you should research a few things before getting out your check book. According to Laurie Campbell at Alternative Choices, "The most important thing to determine . . . is average wind speed" at your site. The average daily wind speed is measured by a device known as an anemometer. Anemometers may be purchased at scientific supply outlets such as Edmund Scientifics, or at alternative energy suppliers such as Jade Mountain or Real Goods. Jade Mountain will even rent you a unit.

The anemometer is a relatively simple device mounted on a pole at the projected height of your wind generator. Ideally, it should be left in place to record wind speed for a year or more. If you don't have the luxury of a year's time to collect data, you

OUTPUT OF WIND GENERATORS	
Large:	100,000 to 750,000 watts
Medium:	1,000 to 5,000 watts
Small:	300 to 600 watts

FIGURE 12-8.
The anatomy of a wind generator, also known as a wind machine or wind turbine. Power output is directly proportional to the size of the blades.

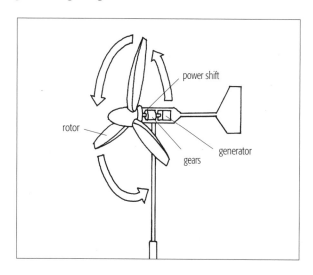

FIGURE 12-9.
This wind generator (Air 403) is one of the smallest on the market. It is used for small applications or to supplement home electrical power provided by other sources such as PVs (shown here).

can shorten the test period by taking a limited sample, a couple months' worth of data, then comparing them to data from a local weather station. This also allows you to estimate wind speed at your site during the rest of the year. (This technique is explained more fully in *Independent Energy Guide* by Kevin Jeffrey.)

Another useful measuring device is a wind totalizer. A totalizer displays wind speed and total wind occurring over long periods. That is, it adds up the wind to give you a total. This is much more valuable than periodic measurements of wind speed and will help you determine whether your site receives enough wind to make a sound investment in a wind generator. (Wind totalizers can be rented from Jade Mountain.)

Experts differ on the average daily wind they think is needed to qualify a site for wind power. Most estimates range from 7 to 12 miles per hour—that is an average of 7 to 12 miles per hour over a full year. This includes windy periods and those placid days when there's no wind at all. According to the American Wind Energy Association (AWEA), average speeds of 10 to 12 miles per hour pertain to larger commercial systems, and 7 to 9 miles per hour average wind speed is suitable for the smaller units, the kind used by homeowners.

Having the right amount of wind is important because the blades of most small wind machines don't begin to rotate until the wind speed reaches a speed of about 5 miles per hour. To produce maximum power, however, most machines require winds of 20 to 30 miles per hour (figure 12-10).

Measuring wind speed takes a lot of time. If you're not inclined to spend the time and energy required to determine average wind speed, consult the *Wind Energy Resource Atlas of the United States*. This book contains detailed colored maps that depict annual and seasonal average wind speeds throughout the United States. The National Wind Technology Center, also listed in the resource guide, sponsors an on-line data base. A local airport or television station may also be able to provide you with average wind speed data. For overseas readers, wind associations in other countries are listed in the resource guide.

Data from maps, local airports, or television stations should suffice, but be careful. If your site is surrounded by hills, large trees, or some other obstruction, local wind speeds may be much lower than average wind speeds at a nearby airport. They could also be greater!

Paul Gipe

Strong, steady winds are the key to producing maximum power from a wind generator. The more wind the better because the energy available from wind increases dramatically as wind speed increases. For the mathematically inclined reader, the equation says that energy output increases with the cube of wind speed. For example, if the wind speed doubles, the amount of energy you can get out of it increases by 2^3, which is eight times.

As the AWEA points out, wind is "seldom a steady, consistent flow. It varies with the time of day, season, height above ground, and type of terrain." However, even though a wind generator does not operate all of the time, it can produce a significant amount of energy. Because of the exponential increase of power output as wind (speed) increases, sites that have short periods of high-velocity winds can sometimes produce more power than sites that have steadier, more dependable, but less intensive winds. However, as Laurie Campbell notes, "Of course you need to make sure your high winds blow often enough to keep your system charged."

What Type of Wind Generator Should You Buy?

The type of wind generator you should buy depends on how much wind your site receives, how much electricity you want to garner from the wind, and how much money you have in your checking account. Consider wind first.

If wind at your site tends to gust—that is, you have short periods with high-velocity winds—you will want a system that operates most efficiently at high wind speeds. If your wind speeds are constant but blow at lower velocity, select a system that functions optimally at lower wind speeds. The power output graphs in figure 12-10 show the instantaneous and monthly output of a popular wind machine.

The second factor that influences your choice of wind machine is the amount of electricity you need. If your goal is to supplement your home PV system or even grid power, a small $600 system such as the Air 403 from Southwest Wind Power, or the $1,000 Whisper 600 or Windseeker 503 are among

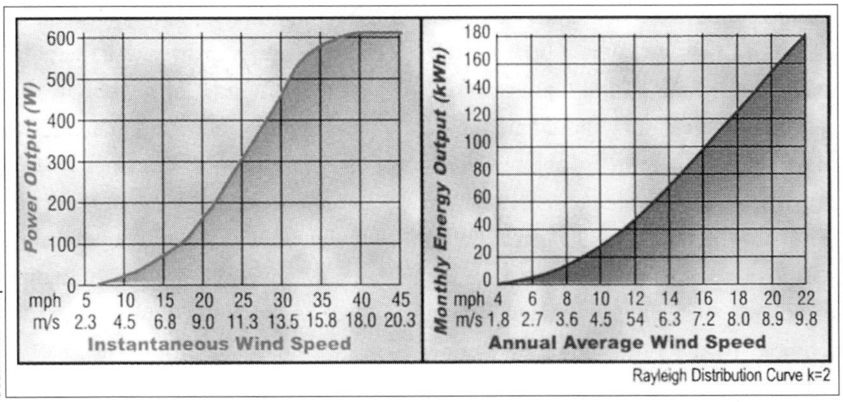

> *The U.S. contains enough usable wind resource to produce more electricity than the nation currently uses. The majority of this usable resource is in the Great Plains region. North Dakota alone has enough suitable wind resource to supply 36 percent of the electricity consumed in the United States.*
>
> The National Renewable Energy Lab

FIGURE 12-10.
Graph of wind speed vs. output measured in watts (left) for an Air 403. Note that wind speeds over 25 to 30 miles per hour are needed for optimal power output. The graph on the right shows monthly energy output related to average wind speed. As a rule, the greater the average speed, the greater the output.

WIND POWER DENSITY

As you determine whether you will be able to tap into wind in your area to produce enough power to make the investment worth it, you will encounter a measurement called *wind power density*. This engineering term is simply the watts (a measure of power) of electrical energy per square meter of air space. It is given at two heights above the ground, 10 meters (33 feet), and 50 meters (164 feet). The wind tables and maps classify areas by their wind power density on a scale of 1 to 8, 1 being the lowest and least likely candidate and 8 being the best. The American Wind Energy Association recommends a rating of 4 or more for large-scale wind facilities. Small-scale users like you should follow the same recommendation.

FIGURE 12-11.

Wind generators or wind machines can be mounted on poles or elaborate lattice towers (shown here). Although towers add considerably to the cost, they're sturdy and reliable and relatively easy to climb should service be necessary.

Paul Gipe

your best choices. (The numbers after the brand names refer to the approximate wattage the units produce at peak power output.) If you want to generate lots of power, consider installing larger models such as the Whisper H1500, 3000, or 4500. They sell in a price range of $2,000 to $6,000. Most units come with two- or three-year warranties. For more details, be sure to check out the manufacturers listed in the resource guide; the Small Wind Turbine Manufacturers Web site is particularly helpful. You can also learn a lot by reading Paul Gipes's *Wind Energy Basics* and by studying the material on wind power in the *Solar Living Source Book* published by Real Goods.

Installing a Wind Generator

As a rule, the higher your wind generator is off the ground, the more electricity it produces. That is because wind speed increases with altitude, and power increases as wind speed increases. The reason wind speed increases with altitude is that ground obstructions often interrupt air flow. Trees, bridges, and buildings all block the flow, slowing the wind down and creating weird eddies. Flowing across the surface of the ground also slows wind. Therefore, the higher your generator, the less likely that ground-level obstructions will interfere.

For optimal performance, wind generators should be mounted at least 20 to 30 feet higher than any obstruction within 500 feet of the unit. Therefore, if there is a large tree within 500 feet of your wind generator, the unit should be mounted on a tower that is at least twenty to thirty feet higher than the tree. Remember to leave room for future growth when dealing with trees. A 20-foot cottonwood may be 40 feet tall ten years from now.

The most economical type of tower is the lattice tower, which is secured by guy wires (figure 12-11). Repair work and maintenance require a careful climb to the top of the tower. Hinged towers are also available. They are easier to install than lattice towers. Because they permit the operator to lower the wind machine for repairs and maintenance, they are safer than fixed towers.

Wind generators should be located close to the battery bank, usually no more than 100 feet from the house to minimize the distance the DC electricity must travel. Remember, DC electricity doesn't travel well; line voltage drops significantly over long distances. If you must place your machine further away, you may need to install a thicker (lower-gauge) electrical wire. It is not only more expensive than standard wiring, it is harder to find. Or, you may be required to install a higher-voltage wind machine. This will also add to the costs.

One manufacturer, Southwest Windpower, manufactures a model (Air 403) that is designed to be mounted on a roof on the side of the house so that it protrudes above the roof. This model is best mounted on a pole (1½-inch schedule-40 pipe) securely attached to the house wall. For optimal performance, it must project at least 5 to 9 feet above the roof line. Although the wind generator is fairly quiet, it can set up vibrations in the pole that are transmitted into the house. Think twice about this installation! Some wind energy experts recommend against it unless you can dampen the vibration. I have been fussing with mine for three years to get the unit to work. When it does run, it produces an annoying vibration.

Economics of Wind Power

The favorable economics of wind power that thrill so many of its advocates pertain to the very large turbines engaged in commercial energy production. Medium- and small-sized turbines, the kind you and I would buy, are not as economically advantageous. But don't dismiss them, they are just not quite as favorable from an economic standpoint as their larger cousins. AWEA says that "as a rule of thumb, if economics are a concern, a turbine owner should have at least a 10 mph average wind speed and be paying at least 10 cents per kilowatt hour for electricity" to be economical.

Wind generators vary considerably in cost. The smaller units of 500 to 1,000 watts are about $500 to $1,000, give or take a little, plus installation. Larger units run from $6,000 all the way to $22,000 installed.

According to the American Wind Energy Association, wind turbines typically lower electricity bills by 50 to 90 percent when installed in suitable locations. AWEA also notes that it is not uncommon for wind turbine owners with totally electric homes to have monthly utility bills of only $8 to $15 during the low-demand months. Be wary of such claims. Do your homework. The contribution a wind generator makes depends on many factors, including the size of the unit, the average wind speed, whether or not it has been installed correctly (high enough and away from obstructions), and the amount of electricity you use.

Two remarkable characteristics of most wind machines on the market today are that they are reliable and require very little maintenance. Commercial-grade wind machines, for example, have a down time of only about 2 percent, which is pretty spectacular in the electrical generation industry. Most modern small turbines have few moving parts. This minimizes breakdown.

Today's wind machines are designed for a long life, up to twenty years according to AWEA, and to make matters better, they operate automatically! Some small models such as the Whisper series and the Air 403 model are a brushless design that minimizes generator wear. (The brushes are electrical contacts in the generator. If anything will wear out in a generator, it will be the

Most experts agree that wind power can be more cost effective than PV solar systems . . . if the average wind speed is 10 miles per hour or greater.

KEVIN JEFFREY,
Independent Energy Guide

brushes.) Moreover, all wind generators automatically disengage when wind speed becomes excessive—usually over 120 miles per hour. They do this by turning their rotors away from the wind. Models in the Whisper series, for instance, tilt back or lie down if the wind blows too hard. This feature, called tilt-up governing, prevents damage to the machine in extremely high winds. The Bergey wind turbines have a trademarked feature called autofurl. It relies on aerodynamics and gravity to steer the wind generator out of the wind during intense gusts, shutting down the generator. Some units come equipped with an air brake that controls rotor speed in high winds.

Most wind generators come with a charge controller (or voltage regulator). Like those found in PV systems, its job is to shut down the wind generator when batteries are in danger of being overcharged. If a generator doesn't come with a voltage regulator, you will need to install one in the circuit.

Another advantage of wind generators is their rapid energy payback. An energy payback period is the length of time an alternative source of power such as a wind machine requires to produce the amount of energy that was required to manufacture the device. For a wind machine, the payback period may be as short as three months! So not only does a wind machine produce clean power, each unit "reimburses" society for the power consumed by manufacturers. (Don't expect that from conventional power sources!) Wind machines also have a rapid economic payback. Three separate studies have shown that utility-sized wind turbines situated in regions with ample amounts of wind will pay for themselves within a year. Not a bad investment for a power company. Residential wind machines also have a rather rapid economic payback. But they require six to fifteen years, depending on the amount of wind in your area and other factors. Like PVs, wind generators can be used in conjunction with utility power, feeding excess power into the grid. In such instances, the payback period may be shorter.

How Large Should Your Generator Be?

The size of your generator depends on your electrical demand and conditions at the site. According to AWEA, the average U.S. home consumes an amazing 9,400 kWh of electricity per year—or about 780 kWh per month. This would require a 5,000- to 15,000-watt wind generator. (The range results from variation in the average wind speed in different locations.) An environmentally friendly home, consuming one-fourth to one-third as much power as the average energy gluttons most people live in, would require a much smaller generator. In such instances, a 5,000-watt unit might be the upper end of the range. An even smaller unit would suffice if it were used in conjunction with a PV system.

Pros and Cons of Residential Wind Power

Wind is one of those ideas whose time has come . . . again. But like everything else in this world, wind power offers benefits and drawbacks.

Advantages of Residential Wind Power

- Wind energy is clean, safe, and renewable.
- Wind energy is abundant in many locations worldwide and is especially abundant in many U.S. coastal and midwestern plains states.
- Wind energy systems reduce water and air pollution. Over its lifetime, a small residential wind turbine will offset approximately 200 tons of carbon dioxide, a greenhouse gas.
- Wind energy systems can be economical.
- Wind machine technology is quite advanced. Most machines on the market today are very reliable and require very little maintenance.
- Like PV systems, wind systems are modular—that is, they can be expanded as demand increases.
- Wind generators can be used in conjunction with photovoltaics or other power sources.
- Electricity from wind generators can be fed into the grid in utility-dependent systems.
- The use of a windmachine to generate electricity protects homeowners from potential price hikes.
- Wind generators do not interfere with television reception.

Disadvantages of Residential Wind Generators

- Wind generators produce DC electricity that must be converted to AC for most household uses. This requires an inverter.
- To install a wind system requires a good understanding of the wind resources at your site.
- Like PVs, wind systems, especially medium-sized ones, require a substantial financial outlay.
- Wind machines are relatively quiet, but they do make some noise.

SMALL-SCALE HYDROELECTRIC

Flowing water is an enormous source of energy, primarily electrical energy, supplying 24 percent of the world's electricity, according to the National Renewable Energy Lab. However, most hydropower facilities are large-scale projects involving huge dams, reservoirs, and extensive distribution systems. Because hydroelectric facilities don't produce clouds of pollution, most people view them as a clean, renewable source of electricity. Unfortunately, this view is greatly mistaken. Hydroelectric plants are economically and environmentally costly. Concrete used to build dams, for instance, is an energy-intensive material. The production of cement and aggregate used to make concrete causes substantial amounts of air pollution. Dams flood river valleys and ruin wildlife habitat, displace people and farms, and destroy many environmentally benign forms of recreation such as rafting, canoeing, and kayaking. Dams change the character of rivers in profound ways. Cold water from the bottom

FIGURE 12-12 A, B.
A microhydroelectric system consists of four parts: a forebay, pipe (penstock), a turbine, and a generator (A). The photograph (B) shows the blades and nozzles inside a turbine.

of a reservoir, for instance, may adversely affect native fish populations acclimated to warmer waters of the stream. However, one form of hydroelectric power available to homeowners is infinitely more earth-friendly. It is known as microhydroelectric or small-scale hydroelectric. A microhydroelectric system is installed on small streams and rivers.

Microhydroelectric systems consist of four parts. The first is the forebay, or intake source, a small box that sits in the stream as shown in figure 12-12. The forebay diverts stream water into a pipeline or penstock, the second component of the system. The water is then transported by gravity to rejoin the

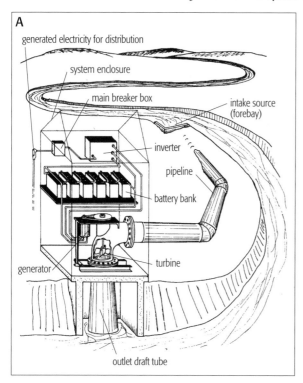

stream some distance below the intake. Before being returned to the stream, however, water is propelled through one or more nozzles, striking the blades of a turbine, the third component. As in a wind generator, the blades are connected by a shaft to an electrical generator, the fourth part of the system. When the turbine blades spin, the generator produces DC electricity, which is then delivered to the house.

The microhydroelectric system can run 24 hours a day, 365 days a year if you live near a stream that runs year-round. If your system generates enough electricity to meet all of your needs, you may not need batteries. Or, you may only need a small battery bank to store energy for times when loads are large—for example, when running power tools that exceed the system's electrical output.

If you don't want to install batteries, you can use the grid to "store" your excess electricity. Some homeowners, however, use excess electricity to heat water or provide space heat. Electric space heating and electric water heat are inefficient and wasteful when using grid power generated by coal, natural gas, or nuclear power plants. When excesses arise from microhydroelectric systems, however, electric space heating and electric hot water make more sense. If you have lots of extra electricity, why not use it to warm rooms or heat water for showers? Michael Potts suggests that an even better use might be to share electricity with your neighbors, in a sense creating a local neighborhood utility. You could also divert less water from the stream.

The success of a microhydroelectric system can be assessed for two features: first, its ability to provide you with electricity and, second, its ability to do so in a manner that is not harmful to the environment. The success of a system from the perspective of power production depends on many factors, including the size of the system, the amount of water available to you, the gradient or amount of fall, and year-round availability. The larger the generator, the more water you can siphon out of a stream (without damaging aquatic life); the steeper the gradient and the greater your access to year-round flows, the better your system will perform. In many cases, however, microhydro systems are seasonal. In colder climates, stream flows diminish during winter months. In warmer climates, stream flows may decrease during summer and fall months when precipitation is lowest.

"Every hydro project presents a unique combination of water source and topography and requires more careful surveying and planning than an equally productive wind or photovoltaic installation," notes Michael Potts. As a general rule, to install a small turbine you will need a fall of at least twenty feet, provided you have enough water flowing through the stream. A site with a fall of 100 feet would be ideal.

Fall and water volume are complementary variables. The less fall, the more water you need. The greater the fall, the less water will be required.

As Potts writes in *The New Independent Home*, "Falling water is the simplest and most cost-effective alternative energy source." Provided, of course, you have a good year-round source of running water in a nearby stream. If you are located near a stream, it may be well worth your effort to explore installing a small hydroelectric facility. As John Schaeffer notes in the *Solar Living Source Book*, "Small hydro systems are well worth developing, even if used only a few months out of the year." Moreover, current designs are simple, well constructed, reliable, and inexpensive. The average cost for a good, reliable home system is around $750 to $1,500.

If you are interested in learning more, obtain a copy of Real Good's *Solar Living Source Book*. It has an excellent section on microhydroelectric, including a worksheet that will assist you in analyzing the potential of your site. This book also provides charts for determining the amount of water you will need flowing through your system at different amounts of fall. The technical staff at Real Goods will run the calculations needed to size your system, including the size of the wire and pipe. They will also provide you with data on costs and electrical output to help you decide whether the investment makes economic sense.

Diverting water from a stream affects the flow in the section of stream between the forebay and the generator. This may disrupt stream life. Even small changes in stream flow can damage aquatic life. To ensure that your system won't damage the stream, an analysis of the environmental impact should be performed.

One thing that won't be affected by a microhydroelectric project are downstream water rights, an issue of great concern in the West, where water's

If you don't want to worry about a conservation-based lifestyle; always nagging the kids to turn off the lights, watching the voltmeter, basing every appliance decision on energy efficiency; then you had better settle down next to a nice year-round stream! Hydropower users are often able to run energy-consumptive appliances that would bankrupt a PV system owner.

JOHN SCHAEFFER
and the Real Goods Staff,
Solar Living Source Book

importance was summed up by Mark Twain in his witty aphorism, "Whisky is for drinking, water is for fighting."

Ideal microhydroelectric sites are rare, but good sites are available and, if tapped without damaging streams and the species of plants and animals that depend on them, they represent a valuable supply of energy for homeowners and small businesses. Microhydroelectric is inexpensive, easy to install, and reliable. If you have a site that won't provide all of your electrical demand, consider a hybrid system consisting of PVs and microhydro, or PVs, wind, and microhydro—some economical combination that meets your needs.

EMERGING ENERGY TECHNOLOGIES

Sunlight, wind, and water can all be tapped to produce electricity for lighting lamps, running appliances, and supplying electronic equipment like computers and stereos. But what about cooking? How can you cook food and heat water for a hot shower without conventional power?

In the developed nations of the world, many of us rely on propane or natural gas to cook our meals and to heat water for showers and baths. Natural gas and propane are relatively benign—compared to coal and oil. They burn cleanly and efficiently. However, natural gas and propane are nonrenewable. Their supplies are limited. Their production also exacts a toll on natural systems. Are better options available?

Fortunately, there are hydrogen and biogas. Both gases are combustible and both are derived from renewable sources. Although hydrogen and biogas are minor league players in the energy arena, this situation could easily change.

Hydrogen gas is produced by the breakdown of water, an abundant and renewable resource. When an electrical current passes through water, it produces hydrogen gas in a process called electrolysis. Hydrogen gas ignites easily but unlike fossil fuels burns relatively cleanly. In fact, when hydrogen gas combusts it produces water (regenerating the starting material) and a small amount of nitrogen oxide.

Electricity needed to make hydrogen gas can be produced by wind, solar, and/or microhydroelectric energy. Electricity is then fed into an electrolyzer where the chemical breakdown of water occurs. In recent years, researchers have been experimenting with a new twist on this technology, one in which PV cells are inserted directly into the electrolyzer. In this device, hydrogen gas is generated on the surface of the PV cells, then removed for use.

Gas produced in electrolyzers can be used to power stoves or backup heating systems. It can even be used to run automobiles. However, unlike wind and other renewable energy technologies, this one is in its infancy. If you are interested in learning more, consult Michael Peavey's book, *Fuel from Water*. You can purchase an electrolyzer from Jade Mountain and begin experimenting on your own.

Tens of thousands of small-scale biogas plants exist in India, Korea, Japan, the Philippines, Pakistan, Africa, and Latin America.

Volunteers in
Technical Assistance,
3-Cubic Meter Biogas Plant

Another technology that may prove useful for people living in rural areas with access to livestock manure is biogas. Biogas is methane produced by the anaerobic breakdown of organic matter, especially manure from sheep, pigs, cows, horses, and people. Manure is mixed with water and fed into a fermenting tank daily. Methane and other gases are vented from the top and piped to various end uses.

Far from being experimental, biogas is used in many less-developed nations to provide fuel for internal combustion engines (with some minor modifications), space heaters, lights, and cooking stoves. Small-scale biogas facilities are relatively easy and inexpensive to build and operate. Some simple designs appear in the *3-Cubic Meter Biogas Plant*, published by Volunteers in Technical Assistance of Mt. Rainier, Maryland. Biogas facilities also produce a nutrient-rich effluent that can be used as a fertilizer.

Small-scale biogas facilities require little maintenance and provide a sanitary means of treating fecal wastes. However, small units don't produce much gas. Methane can also ignite and explode, so safety precautions must be taken.

The options for generating electricity at home from renewable resources are many. Consider them carefully. If you are going to install your own system, read more detailed books and talk to the experts. View the videos produced by Scott Andrews and sold by Jade Mountain on residential microhydropower, solar electricity, wind power, and solar water pumping. They're outstanding! (See the resource guide.) And be sure to talk with people who are actually out there tapping safely into the generous supply of wind, sun, and flowing water to help build a sustainable future.

13

ENVIRONMENTALLY SUSTAINABLE WATER SYSTEMS

WATER IS ONE of the most precious resources on the planet. We drink it, cook with it, bathe with it, water our crops and gardens with it, wash our dirty clothes in it, and manufacture many commercial goods with it. In fact, just about everything in modern society from steel to our morning cereal requires water—and lots of it.

As valuable as water is to our lives, indeed to our long-term survival, we hold this invaluable substance in low esteem. Most of us take it for granted. Over the years, our neglect and abuse have turned many of the world's waterways into cesspools. Decades of waste and exploitation coupled with unsustainable population growth have brought water shortages to almost every country on the planet. Today, as population grows and aquifers decline, many cities are beginning to worry about the future of their water supplies.

To be sustainable, a home must have a reliable supply of clean water provided by a system that is gentle on the planet. A sustainable home must also be equipped with environmentally friendly systems to deal with wastewater—systems that do something more earth-enriching with the billions of gallons of wastewater we produce each day than to dump them into surface waters or into those underground pollution time bombs known as septic tanks. One option is to modify home wastewater systems to make them more closely mimic the hydrological and nutrient cycles of nature.

Before describing such a system, let me say that the cornerstone of an environmentally sustainable water system is conservation. Conservation means using only what you need and using it efficiently. Efficient faucets, showerheads, washing machines, and toilets are vital elements of a water system that meets human needs without bankrupting the planet.

CAPTURING RAIN AND SNOW MELT: CATCHWATER

In most regions of the world, tens of thousands of gallons of water fall on the roofs of homes and office buildings, enough to supply our families and businesses with all the water we need. If you live in an area that receives 30 inches of rain a year, you can collect up to 25,000 gallons off a 2,000 square-foot roof. But like solar energy, we rarely think about rainwater and snow melt as resources. Most of us let this abundant and valuable resource go to waste.

Our lack of interest in this source of water is easily understood. Most people living in industrial nations are served by systems that supply pure water with about as much energy on our part as it takes to swat a mosquito. When we turn on the faucet, we get water. In many places, it is relatively clean, too. And it is usually cheap. So, why worry?

Conventional water systems exact a huge environmental price. Behind the scenes, apparently smoothly-operating municipal water systems siphon billions of gallons of water from our rivers, lakes, and streams, sometimes endangering plants and animals, or eliminating aquatic life altogether. To make this possible, cities and towns build huge dams that destroy river valleys. Energy used to pump water and deliver it to our homes contributes to the depletion of global oil supplies and adds to the pollution of the atmosphere. Chemicals like chlorine are required to purify the water we drink. Although studies show little,

It is clear that, if we are to make our planet sustainable, we must examine the way we use water and propose and implement changes in our water use that ensure a wet world, with abundant clean-flowing sources of water, for generations to come.

GARY BECKWITH,
Solar Living Source Book

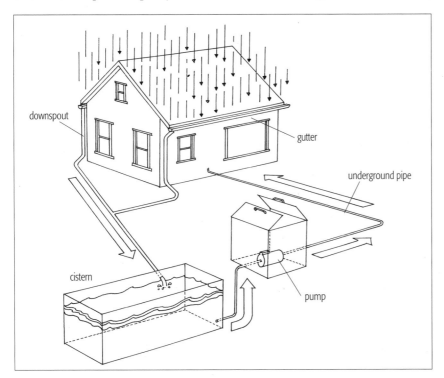

FIGURE 13-1.
Catchwater systems are custom designed for each home, but they all consist of a relatively clean surface to catch water and snow (typically a metal roof), and a means of transporting the water (usually gutters and pipes) to a storage tank located above ground or underground. Water is then pumped into the house where it is filtered and used.

FIGURE 13-2.
This cistern is made from cement rings stacked on top of one another, allowing you to custom make a tank to fit your needs.

Tank Town

if any, effect on human health, many people worry about the possible impact of drinking, cooking, and bathing in chlorinated water. The manufacture and shipping of chlorine disinfectant requires energy, too. Even wells for individual homes require tons of energy and can deplete groundwater supplies. The list goes on.

A much more earth-friendly system is the catchwater or rain catchment system. Figure 13-1 shows a diagram of this simple but elegant idea.

Anatomy of a Catchwater System

A catchwater system consists of several parts. First is a catchment surface, typically a rooftop. To be safe, it must be made of a nontoxic material, such as aluminum or steel, or some of the modern plastic/wood composite shingles. Water striking the roof flows by gravity to the gutters, which channel it to the downspouts. From here it is transported through a series of pipes that lead to one or more water storage tanks or cisterns.

Typically buried underground, although sometimes located in the house itself, the cistern is a home's private reservoir. Used throughout the world, especially on Caribbean islands where fresh water is in short supply, cisterns are located out of the sun to reduce the growth of algae. They are usually covered to eliminate evaporation and possible contamination and to keep kids from climbing in for some fun—and possibly drowning.

Cisterns are made from a variety of materials, including concrete, wood, steel, fiberglass, and plastic. Every expert seems to have a particular favorite. Stu Campbell, author of *The Home Water Supply*, extols the virtues of reinforced concrete tanks. He likes their smooth interior surfaces that make them easy to clean. They are also widely available and can be purchased from septic tank suppliers or manufacturers of precast concrete products. Tanks can be custom made—even built for you on site (see figure 13-2). Companies will pour the concrete into forms to create a seamless tank. According to Campbell, this is the most reliable type of tank you can install.

You can build your own cistern out of packed tires and cement, a technique pioneered by the folks at Solar Survival Architecture (figure 13-3). To make a tank, tires are laid down in successive rows. The first row is placed in a circle near the house. Tires are filled with dirt, then tamped. The next course is set in place, filled, and packed, and so on. When finished, a cement mortar is troweled onto the inner surface of the tire wall to form a watertight surface. Details are given in SSA books and other literature. If you do decide to make your own, remember the advice of Stu Campbell: "Constructing a good cistern is not a project to be taken lightly." Make it well and keep it clean. This structure will hold your drinking water.

Suzy Banks and Richard Heinichen, authors of the humorous and informative *Rainwater Collection for the Mechanically Challenged,* favor fiberglass tanks. Used primarily for above-ground installations, these tanks are coated

FIGURE 13-3.
*The folks at Solar Survival
Architecture build large cisterns
from packed tires. A layer of
cement mortar applied to the
inside surface of the tire
structure makes the tank
waterproof.*

internally with an FDA-approved food-grade resin. To protect them from ultraviolet light, the exterior is coated with a UV-resistant gelcoat that gives the tank an opaque color to inhibit the proliferation of microscopic algae. Fiberglass tanks are available from local suppliers or from Tank Town (listed in the resource guide).

Many people opt for plastic tanks. They are lightweight and easy to transport and install. They are also widely available. Plastic tanks can be used above ground or underground. If you are going to bury your cistern, purchase a tank that is approved for underground use. (They are built to withstand the weight of the dirt that covers them.) Plastic tanks can be purchased from environmen-

BUILDING NOTE

One of the simplest cisterns is a barrel or bucket placed under a downspout. If you want to start experimenting with catchwater, why not give this a try? You can use the water to irrigate plants in your garden.

tank drain

overflow

drain valve

pump vault

sedimentation tank

FIGURE 13-4.
*To hold water, you can acquire
one large tank or string several
smaller tanks together. In this
system, four 1,200-gallon
plastic tanks are joined to
produce a 4,800-gallon cistern.
The system can be expanded
if needed.*

Environmentally Sustainable Water Systems / **319**

tal mail order catalogs such as Jade Mountain or Real Goods, or, usually more cheaply, from a local supplier. Unfortunately, I've only been able to locate buriable tanks that hold 1,200 gallons. For a cistern with 5,000 to 6,000 gallons, you will have to string four to five of these tanks together as illustrated in figure 13-4. This will require a plumber.

If you are planning to install a cistern, underground placement is the best option. This strategy removes the tank from sight and places it out of harm's way. It also protects the tank from sunlight, which is important for plastic tanks. In temperate climates, underground installation helps keep cistern water cool in the summer. If done correctly, it will prevent freezing in the winter, too. Finally, underground placement reduces the growth of algae.

Even though an outdoor tank should be placed underground in most locations, this strategy has some shortcomings. An underground tank is hard to monitor. You will want to keep an eye on water level and periodically assess water purity. It is more difficult to clean than a tank located in your basement or alongside the house. And, burying a tank is more costly and time consuming. It must be done correctly to prevent settling or freezing. If not placed on a solid foundation, a tank will shift. Pipes could bend and crack. Even so, the advantages of a buried tank far outweigh the disadvantages—which leads to another important issue.

Periodic tank cleansing is essential to reduce the buildup of sediment and the proliferation of microorganisms. Stu Campbell says that all cisterns need to be disinfected regularly with a chlorine solution (bleach will do). I add chlorine bleach periodically to my cistern with great reluctance, but to protect myself, my two sons, and visitors, I run the water through a carbon filter to remove the chlorine before we drink the water. One of these days, I'll need to climb down into my tanks and scrub them out. I can hardly wait.

Cisterns should be equipped with drain lines to empty their contents in case you need to clean the interior or remove silt and sediment. They should also be equipped with overflows to accommodate excess water during heavy rains when tanks are brimming with water. Figure 13-5 illustrates the way

FIGURE 13-5.
A cistern should be equipped with an overflow and drain valve.

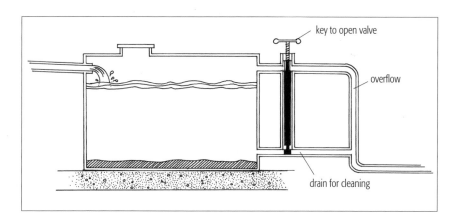

Solar Survival Architecture

FIGURE 13-6.
*Locating a cistern (far left)
outside a home is often the
easiest way to go.*

these features are plumbed in my system. And all cisterns require a childproof
lid to prevent mischievous children from tossing rocks or other things in your
tank or, worse yet, from climbing inside tanks to play.

Locating cisterns on the outside of a house is the most economical
approach (figure 13-6). If you want to place your cistern inside your home, a
basement is a good location. A garage might work well if it doesn't freeze. In
some early Earthships, cisterns were located inside the house in the front of the
buildings. Many of them were left uncovered. These pools added immensely to
the beauty of the homes. Waterfalls designed to aerate the water and keep it
from becoming stagnant added to the aesthetics. Open cisterns supplied
humidity to the interior air, an important benefit in arid climates. However, cisterns positioned inside a house take up valuable floor space, and open cisterns
may overload room air with humidity. Excess moisture may damage insulation, condense on windows in cold weather, and rot wood. Open indoor cisterns present a hazard to pets and small children. They can also become
contaminated with dust and may serve as a growth medium for algae.

Tanks are heavy. A 5,000-gallon tank full of water weighs 40,000 pounds.
Because of this, water tanks always require stable, level pads designed to support their full weight. If not, the tank could settle and cause inlet and outlet
pipes to crack. For more on this topic, see *Rainwater Catchment for the
Mechanically Challenged* by Banks and Heinichen.

Most catchwater systems are designed so that water is delivered to the cistern by gravity. Gravity-feed systems require careful planning, however. The
main goal is to locate the cistern so that it can be fed by gravity flow. This is
not always easy to accomplish. If you can situate the tanks so that they feed
water into the house by gravity as well, that's even better. Earthships do a fine
job of this (chapter 4). In most homes, gravity propels water into the main
storage tank, but a pump is required to transport water into the house.

Unfortunately, gravity feed is not always possible. As shown in figure 13-7,
some homeowners collect water in smaller tanks alongside the house, then

*Ralph Waldo Emerson
must have been talking
about rainwater
collection systems when
he said, "To be simple is
to be great." The fewer
pumps, the fewer twists
and turns in your pipe,
and the more
straightforward your
design, the less you'll
have to hassle with
maintenance.*

SUZY BANKS and
RICHARD HEINICHEN,
*Rainwater Collection for the
Mechanically Challenged*

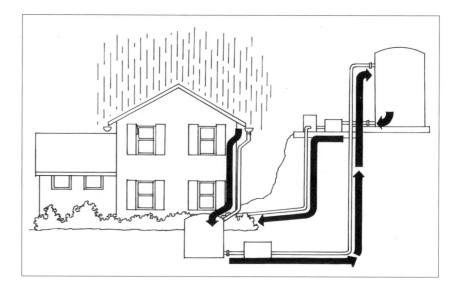

FIGURE 13-7.

Water is sometimes collected by gravity flow in a small tank alongside the house, then is pumped to a large storage tank nearby.

FIGURE 13-8.

For optimal efficiency, the water pump in a catchwater system must be located near the water tanks. It must be protected from the weather, however. Pump vaults protect pumps from rain and snow. Burying the vault in the ground and insulating the lid protects pipes, pump, and filter from freezing in cold climates.

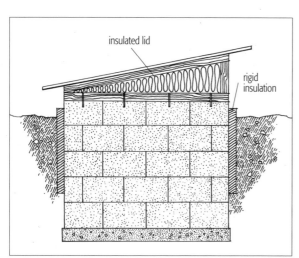

pump it to larger holding tanks located away from the home. Obviously, the more tanks and pumps, the more expensive your system will be—and the more energy will be required.

Water stored in cisterns is pumped to a pressure tank in the house. The pump is activated when pressure in the tank drops below a set level. A pressure tank is a closed steel cylinder ranging in size from a few gallons to 60 gallons or more, depending on a household's needs.

In most systems, pressure inside these tanks is maintained by the pump that transports water to house. Pumps should be located near the water source because they do a better job of pushing water than sucking it. For optimal performance, locate your pump no more than 10 vertical feet above the water source. Pumps should also be protected from the weather (figure 13-8). If you live in a cold climate, insulate the lid to prevent pipes from freezing.

Earthships rely on gravity to transport water from the cistern to the house, further reducing energy demand. Water flows into the house to a small water pump located on a panel known as the Water Organizing Module. It also houses an array of water filters and is available from Solar Survival Architecture in Taos, New Mexico. The module can save you and your plumber a lot of work. As always, research it first to be certain it meets your needs. Consult with your plumber as well to be certain that he or she understands the system and can work with it.

Rainwater is soft and therefore ideal for bathing, laundry, and dishwashing. However, rainwater may contain acids, dust, heavy metals like mercury and lead, and small concentrations of potentially harmful organic compounds, including insecticides, herbicides, and industrial chemicals. Rain also scours dirt from the surface of the roof.

To ensure the purity of the water gathered from rain and melting snow, you should install a hard, nontoxic roof material. Aluminum and steel roofs are ideal. Slate, tile, and terra-cotta roofing are also suitable. However, don't even think about installing treated wood shingles, composite (asphalt) shingles, or rolled roofing if you are going to catch rainwater. Chemical preservatives are used to protect wood shingles. Composite shingles and rolled roofing are made of tar and gravel. They may release potentially toxic substances into your drinking water, unless you install an elaborate and expensive filtration system to remove them. The best choice is a durable, smooth roof finish that won't release lethal chemicals into your water supply.

Ensuring the purity of drinking water requires periodic cleaning or measures to prevent leaves and twigs from gathering in gutters. This reduces contamination of your drinking water; it also prevents clogging of downspouts and loss of water caused when leaves obstruct the flow of water. Gutter protection is especially important for people living in heavily forested areas or in homes surrounded by large trees. Screens placed over the gutters may work, but my experience with them has not been satisfactory. Screens tend to give way under the weight of wet leaves.

The next level of protection is either a sediment trap or a roof washer to remove dirt and debris from rainwater collected on the roof. These are located just upstream from the cistern (figure 13-9). In my system, water from the roof flows down the gutters and through a network of underground pipes to a 200-gallon tank buried in the ground next to the house. The tank allows fine silt and other debris to settle out of the water from the roof. Cleaner water from the upper strata then flows by gravity into a 1,200-gallon tank, the first of four in my system. Although this tank is designed to store water, it also removes silt that manages to escape the 200-gallon sediment tank. The second and third tanks allow additional settling space so that by the time rainwater flows into the last tank it is relatively clean. The pump draws water out of this tank.

Michael Reynolds of Solar Survival Architecture has devised an ingenious sediment trap in his cisterns. As illustrated in figure 13-10, the roof of the cistern contains an inverted cone filled with pumice, a lightweight, highly porous volcanic rock. Rainwater and snow melt are channeled from the roof of the Earthship to the sediment trap where the pumice filters out silt and other debris. Cleansed water then enters the tank through an eight-inch-diameter pipe in the top.

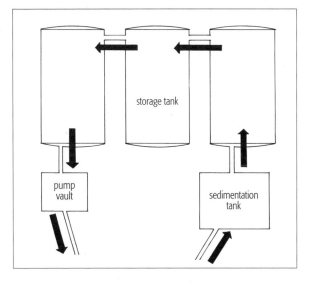

FIGURE 13-9.
Sediment from the roof must be filtered out at several locations. This smaller tank allows sediment to drain out of the water before it flows into the storage tanks.

storage tank

pump vault

sedimentation tank

FIGURE 13-10.
This innovative sediment trap is installed in the newer Earthships. The inverted cone is filled with pumice which filters sediment from the incoming water.

FIGURE 13-11 A, B.
Roof washers remove sediment, bird droppings, and other forms of pollution from the initial water running off a roof. A simple mechanical system (A) consists of a diverter valve in a pipe that must be manually adjusted. The middle drawing shows an automatic system. Rainwater from the roof flows into the rock-filled roofwasher, removing debris and contaminants. One of the simplest designs (B) is used for surface tanks. The initial roof water flows into the vertical pipe. Once it is filled, the remaining water flows into the tank. The valve is opened after a storm to empty the pipe.

Figure 13-11a illustrates two other options for prefiltering catchwater. They are called roof washers. The left side of the drawing shows a manually operated system. It contains a valve and a drain pipe branching off the pipe to the cistern. When rain starts falling, the operator turns the valve so that the initial rainwater collected from the roof—containing bird droppings, leaves, and silt—is shunted away from the cistern, perhaps to a flower bed or garden. After the roof is cleansed, the valve is opened and water flows freely into the cistern.

An automatic roof washer is far more practical. It won't require you to open and close valves during a rainstorm. And you won't have to stand out in the rain, either. The right side of figure 13-11a illustrates this approach. As

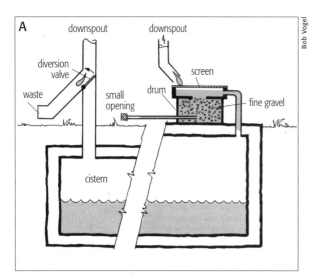

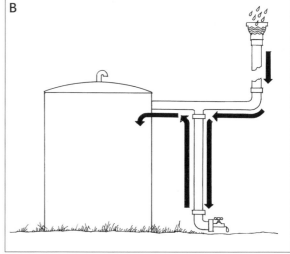

shown, this roof washer consists of a small tank filled about three-fourths of the way to the top with gravel or sand. In this system, initial rainwater containing silt and other contaminants enters the roof washer, filling the rock or gravel bed in lower part of the chamber. Once this section is full and the roof is washed, rainwater begins to flow into the cistern. To recharge the system, the initial roof washwater is slowly drained out of the tank. Figure 13-11b illustrates yet another design.

Roof washers are easy to construct. If you don't want to make your own, you can purchase them from the Tank Town catalogue listed in the resource guide.

Water purity is also ensured by filters and other devices installed in a catchwater system. Filters physically remove inert particles, minerals, bacteria, viruses, acids, and pesticides via screens or highly absorbent carbon particles.

Filters should be placed at strategic locations in the system. It is a good idea to install a sediment filter in the feed line to the water pump in case your roof washer or sedimentation tanks fail or are overwhelmed. This precaution protects the pump from damage and keeps the pressure tank from filling up with silt.

To ensure high-purity water for cooking and drinking, you will need a filter or several filters that remove bacteria, heavy metals, acids, and organic compounds such as pesticides. The cheapest option is to place filters at each major faucet where water is consumed—usually a kitchen sink and maybe a bathroom or two. Another option, albeit more costly, is to install a central water filter, one that filters or purifies all water just before it enters the pressure tank.

The marketplace is flooded with filtration systems. Shop around and be wary. Don't be duped by slick sales pitches and fancy gimmicks. Some systems are not very effective. Many are very expensive. Some water filters use electricity. Others waste a lot of water during filtration. Check out the cost of filter replacements before you buy! Filter cartridges are costly and some don't last very long. Be especially wary of systems sold by door-to-door salesmen or by vendors at home shows. These systems are often grossly overpriced. I have found that homeowners can often purchase effective, inexpensive water filters at all major building materials outlets in my region.

TYPES OF FILTERS/DISINFECTION UNITS

Sediment filters—remove smaller particles, such as suspended dirt, sand, rust, and scale. Improve clarity and appearance of water. Reduce pollution load on other filters, increasing their useful life. Will not remove smallest particles, VOC, pesticides, and biological organisms.

Activated carbon filters—remove compounds that cause disagreeable tastes and odors, including chlorine; eliminate some biological organisms. Generally will not remove dissolved solids, hardness, heavy metals, VOCs, cysts, and coliform bacteria.

Combination filters—contain activated carbon and mesh screens to remove sediment, chlorine, heavy metals, organic compounds, acids, and many biological organisms.

Reverse osmosis—removes dissolved solids, turbidity, asbestos, heavy metals, radium, many dissolved organic compounds, some pesticides, some heavier-weight VOCs, but not small-molecular-weight VOCs. Combine with activated carbon filter for best results.

Distillation—removes microorganisms, sediment, particulates, and heavy metals, but not organic compounds. Often combined with carbon filter.

Ultraviolet light—for disinfection only; destroys bacteria and the potentially harmful microorganisms. However, may not be effective against giardia and other cysts. Should be combined with fine-mesh filtration (0.5 micron) and filters that remove heavy metals, VOCs, pesticides, acids, etc.

Ozone—for disinfection only; destroys bacteria and other potentially harmful microorganisms. Should be used in conjunction with filters that do a more complete job of removing other contaminants.

Chlorination—destroys bacteria and other potentially harmful microorganisms. Adds potentially harmful substance to water however. Should be used in conjunction with filters that do a more complete job of removing other contaminants.

Solar Living Source Book, Tenth Edition. This edition contains a very lengthy and informative section on water contaminants and water filters.

To save on electricity, I suggest filtration systems that do not require electricity. To reduce disposal of filters, look for models that have cleanable filter cartridges. Global Environmental Technologies makes a water filter (TerraFlo) that removes a wide assortment of pollutants, including heavy metals and pesticides. Both models, the countertop and under-the-counter filters, come with a recyclable cartridge. When the filter gives out, you simply slip it into an envelope and return it to the manufacturer for a replacement. The company recycles the parts and offers a 10 percent discount for customers using this service. Filters cost $100 to $120 and replacement cartridges run about $45. According to the company, a filter will produce about 400 gallons, so your cost is about 11 cents per gallon. Jade Mountain sells an in-line household water filter that doesn't require a disposable cartridge. This unit, with a grandiose title of the LifeTime Whole House Water Improvement System, removes all manner of contamination, including heavy metals and microorganisms. Contaminants are filtered out of solution in a bed of sand contained in the unit. Every month, contaminants are purged from the filtration medium by an automatic backwash feature. Contaminants are drained into your septic tank or your community's sewer system. Although the filter medium must be replaced every eight years, it beats having to buy a new cartridge every three to six months. For convenience and guarantee of having pure water, the $900 to $1,400 price tag may be well worth it.

Before you buy this or any filtration system, however, read up on water purification. One excellent source is Real Good's *Solar Living Source Book*. *Rainwater Collection for the Mechanically Challenged* is another wellhead of information on the topic.

Catchwater systems come in many shapes and sizes. They also serve many purposes. Some systems provide all of the water needed to support residents of a house. They may even provide water for irrigation and other outside activities. In other cases, catchwater may be relegated to outside use only. Or it may be used in conjunction with other sources, such as well water or city water. Nearby springs, ponds, lakes, and streams may also be tapped to supplement rainwater. To learn more about obtaining water, see Stu Campbell's book, *The Home Water Supply*.

Should You Install a Catchwater System?

Catchwater is not for everyone. It produces less water than most wells, and certainly much less than a municipal water system. If your system is too small or if a prolonged drought occurs, your cistern may run out of water. Moreover, a catchwater system won't save you much over installing a well, if it saves you anything at all—unless you are located in an area where you have to drill halfway to the center of the Earth to hit water.

Because it produces less water than a conventional system, a catchwater system requires you to take conservation seriously, very seriously. You will need to conserve water every way you can every day of your life. The motto around my house is: Use what you need and use it wisely. You won't see the water running around here while someone is brushing his teeth or shaving! Water-miserly fixtures and water-efficient appliances will be essential in your relentless quest to conserve water. So why do it?

One reason to install a catchwater system is that it puts you in control. You won't have to rely on grid power to run a well pump if your home is solar, and you won't have to rely on the water company, either. Another reason is that a catchwater system requires less energy than a household water system supplied by a well. A small DC pump is all that's required to transport the water from the cistern to the house and maintain adequate pressure in the system. This reduces electrical demand. If you plan to drill a well and pump water to the house using a conventional AC well pump, be prepared to spend a lot of money on PVs to produce the electricity the pump will consume. Another advantage of catchwater systems is that they don't deplete aquifers. With proper filtration, you will produce potable water that local water suppliers would be proud to call their own. And by connecting us closely with our supply, catchwater systems teach us to value water.

To know if you will have enough water from rain and snow melt, you must first determine your daily water consumption. Estimates of water demand vary, but nationwide they are about 100 gallons per day per person. That figure includes all water use, including lawn watering. Indoor use, which is our main concern, is usually less. Water consumption adds up quickly; most people use a lot more water than they realize. If each person in a family of four uses 100 gallons of water per day, a 5,000-gallon cistern won't quite last for two weeks. To operate on a cistern of this size, you must cut water demand, say by taking shorter showers and installing water-efficient faucets, showerheads, appliances, and toilets. Super-efficient appliances not only save water, they save energy.

Water-efficient fixtures and appliances are widely available thanks to new laws that mandate conservation. Several major U.S. manufacturers (Gibson, Frigidaire, Amana, and General Electric) have re-introduced the horizontal axis, or front-loading, washing machine to American markets. These units cut water demand at least in half—compared to standard upright or top-loading washers. Another manufacturer, Whirlpool, sells a 3-cubic-foot vertical-axis clothes washer (Model LSW9245E) that uses 16 to 27 gallons per load, a major improvement over predecessors that used 35 to 50 gallons per load. This model slashes water and energy use by 47 to 56 percent. To learn more about washing machines, read the *Consumer Guide to Home Energy Savings* by Alex Wilson and John Morrill.

WATER-CONSERVING GUYS

My boys and I use about 50 gallons of water a day. Needless to say, we use water very efficiently. We water plants with dishwater carried in a plastic tub from the kitchen sink and graywater collected from the washing machine. We take short showers with a low-flow showerhead. Our toilets are all low-flush models and we subscribe to the western conservation philosophy best summarized in the lyric, "In this land of sun and fun, we don't flush for number one" and the popular admonition, "If it's yellow, let it mellow; if it's brown, flush it down." Our cistern holds 5,000 gallons, which means we can go 100 days on a full cistern without rain, give or take a little.

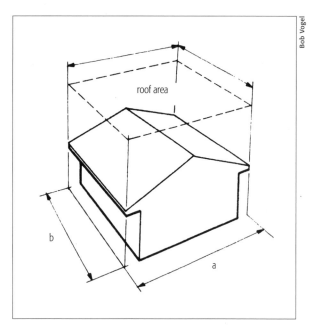

FIGURE 13-12.
The square footage of your roof must be calculated carefully so as not to overestimate your potential water supply.

roof area

b

a

Water-efficient showerheads are widely available, as are low-flush toilets, and are mandated by U.S. law for all new construction. The current models use about 1.6 gallons per flush; the super low-flush toilets use only a pint of water per flush. They are available from Jade Mountain, Real Goods, and other suppliers. Super low-flush toilets work well with septic systems or small in-house composting tanks. To further cut down on water consumption, you might consider using graywater to flush toilets, a strategy discussed in the next section, or a composting toilet, a completely waterless toilet (figure 13-23) discussed under blackwater systems later in this chapter.

After you have determined your water demand—that is, your demand minus the savings you will achieve by adopting efficiency measures—you must determine the amount of water you can collect from your roof. As a rule of thumb, each inch of rainfall provides about 0.55 gallons of water per square foot of roof space. But don't be fooled into believing that the actual square footage of roof represents the roof's potential collecting surface—unless your roof is flat or virtually flat. Roofs are typically sloped and the roof catchment area is the square footage determined by multiplying dimension *a* by dimension *b* in figure 13-12.

If your roof is 2,000 square feet, and you receive 20 inches of rainfall per year, you could theoretically collect 20,000 gallons per year (2,000 square feet x 20 inches x 0.55 inches per square foot = 22,000 gallons). However, it is impossible to collect every drop of rain that falls on your roof. In torrential rains, for

FIGURE 13-13.
The Earthship roof is a rain- and snow-catching machine!

instance, some water may spill over the gutters, although industrial-size gutters and better roof design will reduce this problem. Careful roof design may also increase collection efficiency. The Earthship roof described in chapter 10 is an excellent catchment surface. As illustrated in figure 13-13, the roof is V-shaped. Water (rain and snow melt) is diverted to a central catchment area. From here, it flows laterally toward the cisterns at each end of the building.

Water yield is reduced by roof washers. In addition, some precipitation may come as snow and blow or slide off the roof before it has a chance to melt. Sliding snow is a real problem on metal roofs and on some roof materials made from recycled plastic (discussed in chapter 14). On metal roofs, snow slides can be thwarted by installing metal or plastic fins. These devices grip the snow. Fins were invented to protect people from being crushed by avalanches emanating from roof tops, but they also serve the cistern owner well by retaining snow on the roof until it melts and can be gathered for your water supply.

Expect to collect about two-thirds to three-quarters of the potential roof water. The two-thirds figure may be conservative, but I recommend using this figure anyway. That way, you will be inclined to oversize your cistern. You won't find many people with catchwater systems complaining because they have too much water! In the example I just gave you, you can expect to collect two-thirds to three-fourths of the 22,000 gallons, or about 14,500 to 16,500 gallons over the course of a year—about 40 to 45 gallons per day. If the average daily water supply falls short of your projected need, you may have to consider further water conservation measures, alternative sources, or a larger roof.

The next step is to size the cistern. To do this, determine when precipitation occurs. Are there seasonal patterns? Or, does it rain all year long? If it rains all of the time, you have nothing to worry about. And you probably won't need a very large system. If rainfall is episodic, however, you'll have to gather water when you can and use it sparingly during rainless or snowless intervals. The secret to successful cistern sizing is to determine the average length of dry spells. If dry spells last three months and you use 100 gallons of water a day, you will need a 9,000-gallon storage tank. If they're longer, you will need an even larger cistern to make it through the year comfortably.

PROVIDING WATER FROM A WELL

For those individuals who want well water, yet plan to produce their own electricity via PVs or wind power, don't dismay. Several manufacturers make energy-efficient well pumps that operate with PVs. One popular system is PV-direct. It consists of a submersible DC pump run by a PV array. There are no batteries and no inverter. Whenever the sun shines on the PVs, the pump is activated.

DC pumps draw a small, but continuous supply of water from the ground during daylight hours. Installing a tracking device ensures water is being pumped from sunup to sundown.

A DC water pump in a PV-direct system goes with the flow, pumping more water when the sun is shining brightly, then slowing down when it becomes cloudy. An AC pump overheats if voltage is lower than required.

PV-direct systems avoid the conversion of electrical energy from a PV system to chemical energy in a battery, which is far from 100 percent efficient. Consequently, PV-direct systems pump 20 to 25 percent more water every day than a water-pumping system with batteries.

Anyone starting out from scratch to plan a civilization would hardly have designed such a monster as our collective sewage system. Its existence gives additional point to the sometimes asked question, Is there any evidence of intelligent life on the planet Earth.

G. R. STEWART,
in *The Humanure Handbook*

In these systems, water is fed into a large storage tank located in or near the house. It is then pumped into a pressure tank in the house as needed. If this doesn't provide enough water, you can install a device known as a linear current booster, or LCB. An LCB enables a PV system to operate under low light conditions, for example, on cloudy days. They reportedly boost power output by at least another 20 percent. Twenty percent more power means 20 percent more water.

PV-direct systems offer many advantages over conventional AC-powered well pumps. DC pumps are much more efficient, and because they use electricity derived from a clean, renewable source—sunlight—rather than from coal-fired or nuclear power plants, they are better for the environment. If your storage tanks are big enough, you are assured of a steady supply of water, too.

PV-direct systems also offer many advantages over PV-battery systems. They cost much less because you don't have to purchase batteries, inverters, charge controllers, and so on. They also require less maintenance than a PV-battery system. DC pumps work well with PVs as they can accept variable voltage without problems. In such instances, they simply slow down, whereas an AC pump will overheat if supplied with voltage lower than stipulated by the manufacturer. PV-direct systems work well for single families on a well, but usually require shallower wells. Check that a DC pump will lift water high enough for your well before laying down your hard-earned cash. You may also want to check out wind machines designed specifically for pumping water from wells. For more on the subject, see the *Solar Living Source Book*. You'll be amazed at what's out there.

ENVIRONMENTALLY SUSTAINABLE WASTEWATER SYSTEMS

Fresh water becomes fouled in its brief stay in our homes. But not all water is polluted equally. The water we use to wash our hands and faces is only slightly contaminated with skin oils, dead skin, dirt, and soap. The same goes for shower water. Water from clothing is tainted by dirt and detergent. Kitchen sink water is fouled by grease and food particles. Toilet water, of course, receives the most profane treatment.

To distinguish these types of wastewater, engineers have coined two terms, graywater and blackwater. Graywater is the slightly sullied water from showers, baths, and faucets; it even appears gray. Graywater accounts for 50 to 80 percent of the total wastewater effluent of a household. Blackwater, so named because is contains feces, refers to toilet water, water from bidets, or water from washing diapers. Regulators sometimes include water from kitchen sinks in this category due to its high solids contents.

In Western society, where our understanding of natural cycles seems to have vanished along with covered wagons, gray- and blackwater are viewed as pariahs. Conventional logic dictates a disposal strategy that is fast and efficient with a minimum of human contact. But like so many things we mistakenly label as waste these days, gray- and blackwater are valuable resources, not vile forms of refuse. They contain water and beneficial nutrients needed by plants.

When designing an environmentally sustainable home, the goal is to develop a treatment system that captures and utilizes these valuable nutrients. The following section describes ways to safely collect and use graywater.

REASONS FOR INSTALLING A GRAYWATER SYSTEM

- Reduces demand for fresh water
- Extends life of a septic tank
- Reduces strain on sewage treatment plants
- Provides nutrient-rich water for irrigation
- Provides a source of water for flushing toilets
- Reduces waste flow to septic tanks in areas with slow soil percolation
- Reduces energy demand (needed to pump water to a house)
- Reduces chlorine use if you receive chlorinated municipal water
- Promotes better plant growth
- Recharges groundwater

Designing a Safe Graywater System

The simplest graywater system consists of a hose running from a washing machine to outside vegetation via an open window (figure 13-15). This primitive system works fine in warm climates where windows can be left open to accommodate the hose. Although it is easy to set up, this method does have some drawbacks. Robert Kourik, an expert on graywater systems who was instrumental in the state of California's adoption of a graywater code, writes in the *Solar Living Source Book*, "There's nothing environmentally kosher, groovy, or cool about hurling a bunch of dirty water on top of some poor unsuspecting plants. The graywater may be kept out of the septic tank, but its unregulated dumping will certainly mess up the soil structure, flood, and collapse its pore space, brown root hairs,

BUILDING NOTE

Before you design a graywater system, be sure to check out Oasis Design's Web site. It describes many of the common problems encountered in graywater systems and can save you enormous headaches and costs! They're listed in my resource guide at the end of the book.

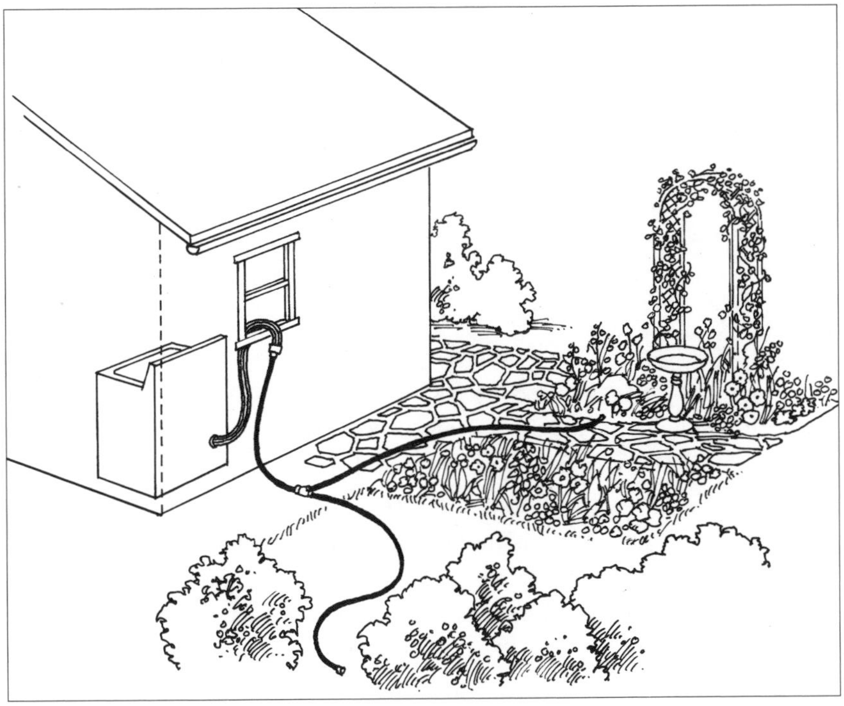

FIGURE 13-14.
The simplest graywater system is a hose leading from a washing machine. This is a crude method that spills a lot of water in one spot. So, be sure to move the hose around so you don't overwater plants, and install some mulch basins around plants to soak up the water fast. You should also use biocompatible detergents or laundry capsules to keep from poisoning your soils with detergent water.

PROPERLY MANAGED, graywater makes plants grow better than either city or well water does. With a slight change in cleansers or detergents used in the house, all the stuff in the graywater, which is often dismissed as filth, dirt, or pathogens, can become wonderful plant food.

ROBERT KOURIK, "A Grayt Way to Water"
in *The Solar Living Source Book*

and make clays more sticky and unworkable." (This is only true for non-biocompatible cleaners.) Art Ludwig doesn't mince words: "This system is really gross, especially over time. It is also highly illegal and unsanitary."

To avoid killing plants and destroying the soil, use biodegradable soaps or, better yet, special biocompatible soaps (available from Jade Mountain)—the only cleaners that are safe for plants and soil. When they biodegrade, they produce plant nutrients. Another fairly effective substitute for laundry detergents is laundry disks. Laundry capsules contain small ceramic beads coated with silver and copper. They ionize the water, much like detergent. This, in turn, pulls the dirt out of clothes and keeps it suspended in solution. Although their effectiveness is controversial, my experience with this product over the past couple years has been fairly good—and I have two young boys who get mighty filthy in the course of their escapades. Laundry capsules are good for 500 to 700 washes and cost around $50. My calculations showed that they cost about half as much as detergent.

Another precaution for those who want to drain graywater from the washing machine directly onto plants is to move the water hose around. Don't dump water continually on one plant. Mulch basins keep the water surges where they belong and prevent unsanitary runoff (figure 13-15).

FIGURE 13-15.
Another simple graywater delivery system.

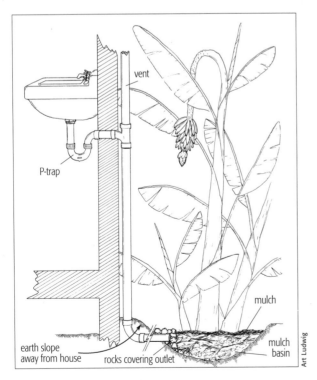

vent

P-trap

mulch

mulch basin

earth slope away from house

rocks covering outlet

Art Ludwig

Figure 13-15 illustrates another means of diverting graywater to plants. As shown, the drain pipe from the sink penetrates the exterior wall and empties directly onto flowerbeds or shrubs that skirt the perimeter of the house. Steps should be taken to avoid excessive water application in one location. Ideally, drain lines should extend away from the house to avoid water buildup next to the foundation. Branching lines could convey graywater to trees and gardens if you have the proper slope (about ¼ inch per foot).

According to Art Ludwig, author of *Create an Oasis with Greywater*, this extremely simple, inexpensive system is used for 90 percent of the graywater systems in the world. To operate optimally, it requires a P-trap and a vent (figure 13-16). The P-trap is an ingenious little invention used in conventional plumbing (take a look at the plumbing under your sink). A P-trap is a U-shaped bow in a drain pipe. When water is drained from a sink, it flows through the P-trap. After the flow stops, a small amount of water remains trapped in the lower portion of the U to prevent gases in septic tanks and sewage lines from seeping into the house. It also blocks gases that might enter the house from decaying material in the drainpipe. In this particular application, the P-trap obstructs the inward migration of insects. Otherwise, this system would be an open highway to your home. The vent pipe carries noxious odors away from the system, usually venting them at roof level.

Sinks and bathtubs can be included in your graywater system. If your house is already plumbed, such additions may be expensive. That's because, in most homes, graywater and blackwater are intermixed close to the source of production. In other words, in most homes pipes draining graywater from sinks and bathtubs join drain pipes carrying blackwater from toilets very close to the fixtures. Graywater and blackwater are then conveyed by a common set of pipes to septic tanks or sewer lines. If you want to isolate the two, do it while the home is under construction. You will need two sets of drain pipes: one to carry blackwater to the septic tank and another to transport graywater outdoors.

Graywater may be applied directly on plants via a movable hose (figure 13-16). Move the hose from plant to plant to achieve adequate irrigation and to prevent overwatering. Or, water could drain into a network of pipes that distributes the water by gravity to different plants. As a cautionary measure, always install a shunt from the graywater to the blackwater drain line just in case you need to divert graywater into the septic tank or sewer. A manually operated diverter valve works fine.

The next level of the sophistication in graywater systems involves the addition of a surge tank. A surge tank accepts water from one or more sources such

Many studies have identified fecal organisms in graywater from homes . . . so it is prudent to conclude that graywater may be routinely contaminated by fecal matter. Treat it with the respect that it warrants.

DAVID DEL PORTO and CAROL STEINFELD, *The Composting Toilet System Book*

FIGURE 13-16.
Sinks and tubs can also be included in a graywater system. Hoses leading to trees and shrubs can distribute the water by gravity.

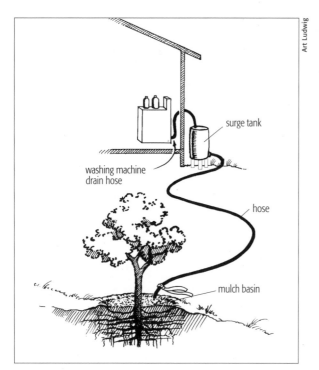

FIGURE 13-17.
The surge tank shown here reduces flooding of plants. It accepts the water from inside sources, then doles it out in amounts that plants can handle. For best results, water should drain into mulch basins around plants, so it percolates into the ground quickly and doesn't pool on the surface.

as washing machines and sinks, then doles it out in reasonable allotments that are better handled by plants (figure 13-17). Rather than dump ten to twenty gallons of water all at once on a plant when a washing machine completes a wash cycle, a graywater surge tank pumps it out at a slower rate, a few gallons per minute. Water soaks into the ground, rather than running off in a flood.

Some homeowners include tanks for graywater storage. The tanks are typically located underground outside the house. They can also be located in a house, for example, in a planter, utility room, or basement. Storage tanks are fed by gravity flow and can be drained by gravity as long as the flowerbeds and gardens they supply are located at a lower elevation. It is more likely, however, that you will have to pump water at least once, either to fill the tank or drain it.

Buried tanks are ideal for cold climates where freezing is a problem. Carefully insulated and properly sized, a graywater tank can store water over the winter for use in the spring, summer, and fall. Buried tanks stay cooler in warmer climates and thus retard decomposition of organic matter in the graywater. But as you shall see, storing graywater can be a big mistake.

Don't let graywater sit more than twenty-four hours. Unless it has been highly purified, it will quickly darken and become more fetid. Bacteria will multiply and convert it into blackwater. For this reason, many graywater advocates strongly discourage the use of storage tanks. In the *Solar Living Source Book*, Robert Kourik offers a series of principles for graywater system design. His first rule is "water in, water out." Translated: "Never store graywater, not even for a day!" He goes on to say, "Storing graywater on warm or hot summer days can quickly produce a bacterial soup that is likely to stink and to breed disease organisms."

DON'T LET GRAYWATER SIT

Graywater that is stored for any length of time quickly goes anaerobic—that is, the bacteria in the solution use up the oxygen as they decompose organic matter in the brew. Anaerobic bacteria, ones capable of living in an oxygen-free environment, begin to multiply and take up where their oxygen-loving buddies left off. Although they successfully decompose organic matter, they don't do so without raising a stink. Pathogenic bacteria and viruses stored in tanks also pose a health risk.

Pumps for graywater systems can be purchased at building supply centers, alternative energy retailers such as Real Goods and Jade Mountain, as well as conventional plumbing outlets. These sources offer a variety of submersible pumps. To learn more, check out the *Solar Living Source Book*. It contains detailed discussion of pumps and information on pricing. Be certain to purchase a pump that is powerful enough for the job. Some pond pumps, which are widely available and favorably priced, have very little lift capacity, which is to say, they can't pump water very high.

Graywater systems recycle a valuable resource. They reduce demand for fresh water and supply nutrients for plants, which may provide food, beauty, protection from the wind, or habitat for wild species. Graywater systems lengthen the life of septic tanks by reducing volume inflow, and they lessen the load on sewage treatment plants.

One drawback of a graywater system is that the water contains a lot of crud that accumulates in tanks, drain lines, or the soil around plants. Organic components of this goo quickly begin to decompose as a result of microbial decay. Dirt, hair, oils and, most important, the bacteria that feed on them, causing decay, produce a rather foul odor. This is especially noticeable in warm climates.

Surge tanks need overflow pipes that allow excess graywater to enter the septic system or sewer line. A plumber can show you how this is done. Another important precaution for systems with pumps or small passages is the installation of a screen or mesh filter that removes solid particles (including hair and lint) contained in the tank inflow. A wire screen prevents the pump inlet from clogging with hair and other debris. Because you will need to clean the filter periodically, be sure that the tank and filter are readily accessible.

If your system is not expressly designed to handle it, divert wastewater from the kitchen sink to the blackwater drain line. This is especially important if your sink has a garbage disposal. Kitchen sink water contains grease, oils, and food particles that quickly decay in graywater systems. Grinding up waste (vegetables and such) in the garbage disposal further adds to the decayable organic matter. Even though you may compost most organic vegetable waste, rinsing plates in a sink produces a lot of crud.

Robert Kourik suggests that you plumb your graywater system so that it can be conveniently diverted to the sewer or septic system, a simple measure that allows you to transfer graywater into the blackwater line when needed. For example, if it has rained excessively, the addition of excess water to your plants might be detrimental. This strategy also enables you to divert graywater when someone in the house has a communicable disease (hepatitis, for example) that could be spread via pathogens (disease-causing microorganisms) in their feces.

Collecting and distributing graywater is fairly straightforward and safe if you follow some basic guidelines (see sidebar on page 336). However, graywater systems are not legal in many jurisdictions. Because of this snag, many homeowners install systems without the knowledge or approval of their local building departments. They do risk being caught and having their systems decommissioned or, in worst-case scenarios, losing their certificates of occupancy.

MANY ADVOCATES of graywater systems assume that they are benign, because excrement is dealt with separately. Unfortunately, that is not the case. People put all sorts of toxic chemicals from cleaning agents to solvents down their drains. Rinsed diapers and fecal matter washed from the body also contaminate shower water. Blood and fluids from raw poultry and meat also enter from kitchen sinks.

DAVID DEL PORTO and CAROL STEINFELD,
The Composting Toilet System Book

FACTORS TO CONSIDER WHEN INSTALLING A GRAYWATER SYSTEM

- Amount of graywater you will be producing
- Characteristics of the soil, including percolation, depth of groundwater, and proximity to wells or surface water
- Distribution system
- Storage requirements
- Plumbing layout of house
- Chemical quality of graywater
- Climate, especially rainfall
- Potential pathogens in the water

With those caveats in mind, the best strategy in new construction (these rogues tell me) is to install the system after final plumbing inspection to avoid arousing suspicion. In a home that is already built, a graywater system can be added with minimal suspicion, although considerable expense may be involved if walls are closed and drain lines are inaccessible.

If you choose to circumvent the building code, be sure to make your system safe. Don't be reckless. Don't cut corners. Don't get carried away by a mood of defiance. Don't flaunt common sense or rules of safety. In areas where graywater systems are not sanctioned, some legitimate plumbers may be willing to assist you. (There are renegades everywhere!) But remember, a plumber can lose his or her license for flagrantly violating code. Approach the topic carefully. Shop around. A few phone calls may turn up a friend of a friend who will put one in.

In areas where graywater is not approved, another option is to blaze the path for those who follow by applying for permission from the local or state health department. They may let you install a system if you propose a design that meets code in another state, for example, California. They may even ask to run tests on the system to monitor water quality and potential health threats for future reference. They will very likely require you to install diverter valves, so that the graywater can be diverted to the blackwater line if the system doesn't perform well or poses a potential health threat.

I encourage you to obtain permits whenever possible. If your system turns out well, building departments are more inclined to grant permits to others.

But the process may be complicated, time consuming, frustrating, and costly. For example, most states require homeowners to have plans prepared or stamped (approved) by a professional engineer or a certified wastewater system designer before submission. That will add to the cost. You may need to hire a professional plumber or an engineer to serve as a liaison between you and the authorities. This hired gun can be your advocate. Knowing the code and being familiar with their concerns, their language, and their BS, your consultant can take the burden off your shoulders. This too will add to the expense. And then there is the time issue. It may take several months, maybe even years, to get approval, so start early. If it looks as if you are not going to get approval or it will cost you a fortune or take till eternity, consider your other options.

In the United States, health and building departments have been reluctant to grant permits for graywater systems. Their major concern is that graywater may contain pathogenic microorganisms such as hepatitis A virus. If graywater from clothing (particularly underwear) or showers ends up on vegetation or the surface of the ground, pathogenic bacteria and viruses could infect anyone who come in contact with it. Are their concerns justifiable?

In my view, concern over the spread of disease is real, but overblown. More important, the potential danger can be lessened dramatically, even avoided, by applying graywater below a layer of mulch. Many people have tried to install graywater drip irrigation systems. Drip irrigation consists of a network of small-diameter plastic tubing that transports water along the surface of the ground to the base of each plant where it leaks out one drip at a time. This minimizes pooling and greatly reduces the potential for human contact. If any pathogens are in the water, they will be destroyed by soil microbes or by exposure to the environment.

Drip irrigation has many benefits. It saves water and doles it out in manageable quantities. As Robert Kourik points, "Gardens with well-designed drip systems display plentiful foliage growth, a tangible increase in bloom, higher vegetable and tree crop yields, and a marked reduction in diseases such as mildew, crown rot, and rust—and all of this with a 30 to 70 percent reduction in the amount of water used." Unfortunately, graywater drip irrigation systems generallly don't work. Of the hundreds of such systems Art Ludwig has seen, every one has been abandoned. Filters and pipes clog with goo. The only exceptions were automated systems manufactured by Agwa, a company that's no longer in business. (Earthstar, a new system, is available, but is as yet unproven.)

Another technique that minimizes exposure to graywater is subsurface or root-zone irrigation. Subsurface irrigation is achieved by installing underground drip tubing (a porous pipe) at the top of the root zone, or about three inches below the surface of the ground. These pipes deliver water evenly along their length right where the water is needed when transplanting clean water but tend to clog when supplied by graywater.

Underground systems require a pump to deliver water under pressure. You will need 5 to 10 psi of pressure to make this system operate successfully. Unfortunately, gravity-fed systems don't work well in underground applications. They just don't have enough pressure. Art Ludwig finds that branched drains to mulch basins work best for subsurface gravity supply.

FIGURE 13-18.
This minileach field consists of a gravel bed and an inverted flower pot to prevent graywater from pooling on the surface.

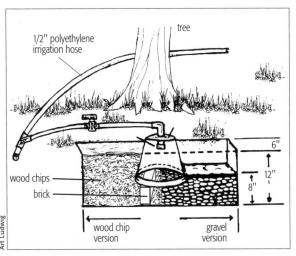

1/2" polyethylene irrigation hose

tree

wood chips
brick

6"
12"
8"

wood chip version

gravel version

Art Ludwig

Figure 13-19 shows an additional option for distributing water safely to trees and shrubs. It is a mini-leachfield, consisting of an upturned flower pot buried in the ground, surrounded by mulch or gravel. The graywater enters the pot and flows into the bed of gravel, then percolates into the ground, supplying the roots of the plant. Another option, the distribution cone manufactured by ReWater Systems operates similarly. Other methods of distribution are described in *Create an Oasis with Greywater* by Art Ludwig and *The Composting Toilet System Book* by David Del Porto and Carol Steinfeld.

Automated Graywater Systems

If you like the idea of graywater but haven't been enamored by the systems described so far, you may want to consider an automated system. One system, produced by Earthstar, consists of a 55-gallon tank and a filter to remove impurities (figure 13-20). Unfiltered graywater drains directly into a temporary storage tank. When the water in the tank reaches a certain level, a float switch is activated. It turns on a small pump that drains the tank,

FIGURE 13-19.
The Earthstar automated graywater system.

forcing the graywater through a sand filter. From here, the water is pumped to a second storage tank or directly to the irrigation system. The effluent is much cleaner than the systems we've been examining. It is so clean, in fact, that code officials may be willing to approve this type of system.

The Earthstar system does require periodic maintenance. Every month or two, you will need to flush the filter. This is accomplished by pouring clean water in the top of the filter, then turning on the back flush switch. As the water flows back through the filter medium, it removes the accumulated debris, flushing it into a septic system or a sewer line. The process takes about five minutes. You won't have to touch the filter or risk exposure to potentially harmful bacteria or viruses.

This automated graywater system is also easy to service. Because it consists of off-the-shelf swimming pool technology, finding someone to fix it shouldn't be a problem. This type of system works well with drip irrigation, as it produces a lint-free graywater that won't clog the openings in soaker hoses or irrigation tubing. And, it is the only acceptable way to irrigate grass. The pump and tank cost about $1,000.

Clivus, the company best known for its composting toilets, also manufactures a graywater system; it costs about $600, and requires minimal maintenance.

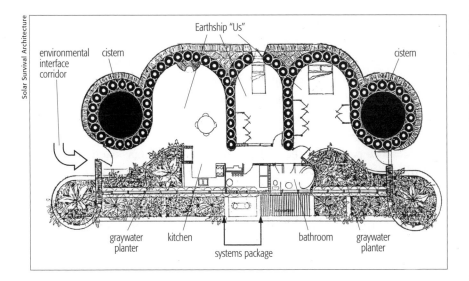

Solar Survival Architecture

environmental
interface
corridor

cistern

Earthship "Us"

cistern

graywater
planter

kitchen

systems package

bathroom

graywater
planter

Indoor Graywater Systems

Michael Reynolds and his colleagues at Solar Survival Architecture have pioneered a seemingly effective indoor graywater system. Although it is designed for indoor plant watering, it also produces water clean enough to flush toilets. Excess water can also be diverted to outside plants.

Figure 13-21 shows a diagram of this innovative design. Unlike systems we have been examining, this one is a self-contained biological treatment center. Its design is based on natural ecosystems. As illustrated above, the graywater treatment units or planters are strategically located in the environmental interface corridor. Each planter consists of four chambers each lined by a thick, rubber liner. The liner prevents chemicals, bacteria, viruses, and liquids from seeping out of the system. Rubber baffles separate the chambers.

BUILDING NOTE

Grease traps can be installed in kitchen sinks to reduce the amount of grease and oils that enter the system. If you have children who are still in washable diapers, you may want to hire a commercial service to clean them. Do not rinse diapers in sinks that drain into your graywater system.

FIGURE 13-21.
The Earthship graywater planter is lined with impermeable rubber. Water passes through a pumice substrate where bacteria and other microorganisms decompose organic nutrients. Inorganic nutrients are absorbed by the roots of the plants growing in the planter.

A household drain is not a waste disposal site. Consider the drain as a conduit to the natural world.

JOSEPH JENKINS,
The Humanure Handbook

Water leaving a sink or shower first empties into a grease and particle filter installed in the upper end of the planter (figure 13-22). This filter contains a screen to remove grease, oil, hair, lint, and other large particles and must be cleaned periodically. The frequency of cleaning depends on the level of use and the amount of debris in the water. The partially filtered graywater then drains downward, percolating through a bed of small rocks designed to prevent water from pooling in the first chamber.

Water then passes into the next chamber through an opening in the rubber baffle that separates the first chamber from the second one. Graywater enters a bed of pumice. Bacteria and other microbes in the pumice quickly go to work on the organic compounds. Water and inorganic nutrients (such as nitrates) are absorbed by the roots of plants that penetrate in the pumice layer. What is left of the graywater, now partially purified, travels into a third rubber-lined chamber. Additional nutrients are removed so that by the time the water flows into the fourth and final chamber it is pretty clean.

Despite what skeptics say, this system works remarkably well. Its success depends on several key features. The baffles in the system increase retention time, that is, they increase the amount of time graywater remains in contact with pumice and plant roots. This, in turn, increases the extent of microbial decay and increases the uptake of nutrients by the plants. Another important component is a peat moss filter. Situated between the third and fourth chambers, it filters impurities from the graywater.

In the final compartment, the now highly purified graywater accumulates in a storage well, a deeper section of the planter specifically designed to collect water. A pipe inserted into the well is used to withdraw water. This water can then be filtered and returned for limited household use, for example, flushing toilets, via a separate cold water feed line. Purified graywater can also be pumped to outside vegetation.

For a detailed description of this indoor graywater system, read *Greywater: Containment, Treatment, and Distribution Systems*, part of the *Earthship Chronicles* series published by Solar Survival Architecture. SSA also provides architectural detail sheets. They will even sell you system components. I strongly recommend that you consider a system like this. But do your research. Plan your system to take advantage of gravity flow as much as possible.

A Word on Cleaning Agents

Graywater systems require the use of biologically friendly soaps and detergents. They also require special cleaning agents to avoid poisoning microbes, plants, and soil. If you are installing a graywater system you must break your chemical dependencies . . . on common household products. Cleaners such as Chlorox bleach, Comet, 409, Lysol, SOS, Boraxo, and any detergent "with bleach" are toxic to plants and soil microbes—highly toxic. For example, many cleaning and personal hygiene products contain antibacterial agents designed to kill pathogens on the body. These may kill beneficial organisms in biological treatment systems and soils. Unfortunately, even the least toxic agents in graywater may accumulate in the soil, eventually building up to levels that are toxic to plants.

As Art Ludwig points out, "When you use graywater for irrigation, you have to think more about what you put down the drain." The first requirement of a household cleaner or detergent is that it is biodegradable—it decomposes. All household cleaners on the market today are biodegradable. But even biodegradable cleaning agents may contain small amounts of harmful substances that can accumulate in the soil. Ludwig notes that "many people's experience is that insensitive plants may be watered for years with graywater containing non-biocompatible cleaning products. However, once damage is finally evident, it can be a really big job to repair the soil."

To address this problem, Ludwig has developed a line of biocompatible cleaning agents. They not only biodegrade, they contain no sodium, chlorine, or boron. And they do not adversely affect the soil pH or soil structure. Information on contacting his company is in the resource guide. You can also order these cleaners through Jade Mountain.

What do you use to bleach clothes if standard bleach is so toxic? One recommendation is liquid hydrogen peroxide. It is less powerful, more expensive, but nontoxic.

Is a Graywater System Right for You?

Despite their central importance in forging more sustainable water use patterns, graywater systems are not for everyone. They may require more effort than you are willing to expend. Or, they may require yard space that you don't have. For retrofits on existing homes, it may be impossible or nearly impossible to access drain pipes or isolate blackwater from graywater. Pipes, for instance, may be entombed in the concrete slab or in stud cavities in finished walls. Another stumbling block is that there may be too much rain in your area. Your outdoor plants just don't need any more water. Or, your climate may be too cold for an outside graywater system. Soil that is extremely impermeable can be an impediment, too; according to Ludwig this problem is more common than too-permeable soil. And of course, there is the legal issue. Your local building department and health department may not approve of graywater. Graywater systems do pose some potential health risks that you may not want to take. Cost may be a factor; certainly, it is easier and cheaper to plumb your house the conventional way.

If you are not dissuaded by these excuses, check out Art Ludwig's *Create an Oasis with Graywater* and *Building Professional's Graywater Guide*. They contain lots of information, including designs for twenty types of systems, the California code, safety precautions, and even information on the chemical contents of various types of detergents. (Ludwig also does design consulting.)

Graywater contains very little nitrogen and is therefore not a very good fertilizer. Add a little nitrogen to your graywater to enhance its nutrient value. One way to do this is to use ammonia-based cleaning agents, such as Neutrogena and Pantene Pro V Shampoo. Small amounts of acidic fertilizer can be added instead.

Additional books that will help round out your knowledge of the subject include *The Composting Toilet System Book* by David Del Porto and Carol Steinfeld, *The Humanure Handbook* by Joseph Jenkins, and Robert Kourik's book, *Drip Irrigation for Every Landscape and All Climates*.

Graywater systems conserve water. They assist the independent-minded homeowner in making the most out of water gleaned from a roof catchment system or other potentially sustainable sources. They also recycle vital nutrients required by plants. The plants, in turn, provide food and fiber to ease the burden on the world's farmland. All in all, graywater systems are a measure of good environmental stewardship. Although your contribution to the global hydrological and nutrient cycles is small, there are a lot of you out there. Together, the "yous" can make a difference. As Ludwig notes, "Taking full responsibility for the little bit of the global water cycle that flows through one's home is a really good feeling."

BLACK WATER SYSTEMS

Blackwater from toilets is one of the most challenging resources. The standard approach, of course, is to pipe blackwater to a septic tank and leach field or dump it into a community sewer line. A far more sustainable option would be to install a composting toilet.

Composting Toilets

Developed in Scandinavia in regions with shallow soils that are unsuitable for septic systems, composting toilets consist of a place to sit and a waste repository, a chamber where fecal matter and urine are deposited. Here human wastes are broken down or composted by a diverse community of aerobic (oxygen-loving) microorganisms. Liquid in the waste is evaporated through a vent pipe.

In a composting toilet, the organic matter in our wastes quickly breaks down to form a dry, fluffy, odorless material—provided the composting chamber is kept at a sufficient temperature. This material is suitable for enriching the soil in flower gardens and around trees. Because decomposition takes place in the presence of oxygen, the compost toilet bears little resemblance to a smelly outhouse. Quite the contrary, composting is a good-smelling process. As the folks at Real Goods note in the *Solar Living Source Book*, "If a composting toilet smells bad, it means something is wrong."

Composting toilets can be grouped into two categories. The first is called a centralized or remote composting toilet system. It consists of a seat in the bathroom and a receptacle located outside the bathroom, usually in the basement (figure 13-22a). The toilet itself may be one of three basic types: waterless, ultralow-flush (microflush), or urine-diverting.

The second option is known as a self-contained unit. This is a toilet seat and waste receptacle that are all part of one assembly (figure 13-22b). Self-contained composting toilets are further broken down into active and passive

Composting toilets are also known as dry, waterless, or biological toilets. They are not to be confused with those smelly outhouses at national forest or state park campgrounds.

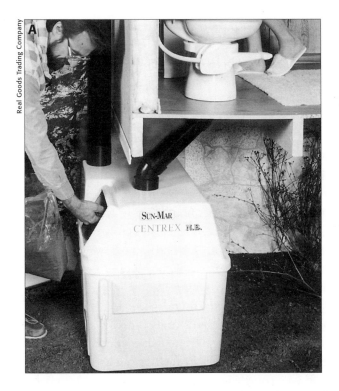

varieties. An active system includes an electric heating element and a fan. The heater accelerates the decomposition of waste. The fan removes any odors and accelerates drying of the fecal matter. Passive systems lack these features.

Composting toilet systems may be purchased from manufacturers or built on site from standard building materials. For a thorough discussion of your options, see *The Composting Toilet System Book* by David Del Porto and Carol Steinfeld. It is a marvelous book well worth reading.

How Does a Composting Toilet Work? The main job of a composting toilet is to contain human waste and create conditions that facilitate rapid decomposition of organic matter, including any pathogenic microorganisms that may be present. This, in turn, depends on removing most of the liquid in the fecal matter. Rapid decomposition also requires adequate levels of oxygen. If conditions are right, composting toilets produce a stable soil-like substance, known as humus (not to be mistaken for hummus). What you do with it depends on where you live. In the United States, it must be buried or removed by a licensed septage hauler, according to state and local regulations.

One of the best composting toilets on the market is produced by Sun-Mar. In this system, waste is deposited into a rotating drum where composting occurs. Turning the drum from time to time promotes oxygen penetration into the organic material, which accelerates decomposition. Oxygen enters through holes in the front of the unit, which establishes an airflow into the system. Odors, if any, are carried away from the unit by a vent pipe.

FIGURE 13-22 A, B.
Composting toilets come in two types, a centralized system such as the Sun-Mar Centrex and Aqua Magic Toilet (A), and a self-contained composting toilet such as Sun-Mar's non-electric composting toilet (B).

The nutrient value of composted yard waste is fairly low, about 10 percent of commercial fertilizers. The nutrient value of composted human waste is much higher, equal to many chemical fertilizers.

The Sun-Mar composting toilet converts human waste, toilet paper, and kitchen scraps into compost in a few weeks, working one hundred times faster than a septic system. The Sun-Mar has been satisfactorily used in thousands of homes in the U.S. Because it has been certified by the National Sanitation Foundation and approved by the Canadian government, it is relatively easy to obtain a permit to install one in the States. So well does it work that the inventor of the system received the Gold Medal for the best invention at the International Environmental Exhibition in Geneva, Switzerland. Sun-Mar toilets are available through Real Goods, Jade Mountain, and other suppliers.

Composting toilets work best at temperatures of 70 degrees Fahrenheit or higher. Below 50 degrees, the biological decomposition slows considerably. Although freezing does not harm the system, it will temporarily halt microbial activity. One question that everyone asks is, "how do you remove the organic matter from the composting toilet? Will I have to scoop it out?"

Manufacturers have been particularly sensitive to this issue. In the Sun-Mar system, organic waste is deposited in the drum, where it accumulates and decomposes. When the drum fills to two-thirds of its capacity, a turn of the crank causes partially composted waste to spill into a drawer in the bottom of the unit. Here the organic matter undergoes its final drying and decomposition. When the process reaches completion, you remove and empty the tray— usually one to four times a year, depending on the level of use.

For optimal performance, the manufacturer recommends the addition of a cup of peat moss per person per day. Peat moss soaks up moisture, but more importantly creates air passages in the waste material, ensuring that the decomposition remains aerobic. Otherwise, operator involvement is minimal.

The Sun-Mar Centrex is a centralized composting system. It consists of an RV-type toilet above a waste repository, as shown in figure 13-22. Together, they cost about $1,150. This unit, like others, comes with a small thermostatically controlled heater and a fan to facilitate drying and accelerate decomposition. Among off-gridders, the Sun-Mar Non-Electric is the most popular composting toilet. Aeration and agitation are provided by the drum. Evaporation occurs passively through a 4-inch-diameter vent pipe. Heat from the compost creates a chimney effect that draws air through the system, minimizing or eliminating odors. To be effective, this unit must be kept at 68 degrees Fahrenheit or higher. A sun-warmed bathroom is sufficient. For more details on passive ways to heat a composting toilet, see *The Composting Toilet System Book*. If you want more information on different models, take a look at the *Solar Living Source Book*.

Sun-Mar is not the only composting toilet on the block. Clivus Multrum, the company that introduced composting toilet technology to North America in the early 1970s, also manufactures a couple of household units that are approved by the National Sanitation Foundation. Their toilets are made from 100 percent recycled plastic. Unfortunately, they cost approximately twice as much as the Sun-Mar models.

Composting toilets have evolved nicely over the past few decades. Even though many early models turned out to be ineffective, designers have made up for past mistakes. "More than a decade of evolution later, the performance of many composting toilets is more commensurate with system designers' and manufacturers' claims. And more and more composting toilets are appearing in mainstream bathrooms," write Del Porto and Steinfeld.

Composting toilets appear to be safe and reliable. While some readers may be concerned about the potential for catching a disease from a composting toilet, there appears to be little to worry about. "According to current scientific evidence, a few months' retention time in just about any composting toilet will result in the death of nearly all human pathogens," says Joseph Jenkins, author of *The Humanure Handbook*. Human pathogens don't last long outside the body. In a composting toilet, they are destroyed by other bacteria, acidity, and higher temperatures. By retaining human waste for sufficient time, conditions in composting toilets become inimical to disease-causing organisms that might be contained in human feces. For a detailed description of the disease issue, see *The Composting Toilet System Book*. It will not only explain what happens to these organisms, but also present protective measures to ensure your safety.

Jenkins suggests that, when properly composted, human feces can be used to fertilize fruit trees, flower beds, and even vegetable gardens. *The Humanure Handbook* is a must-read for anyone interested in capturing the valuable resources contained in human waste. "Severe fecophobics," Jenkins argues, "do not want to use humanure for food growing, composted or not. They believe that it's dangerous and unwise to use such a material in their garden . . . It is well known that humanure contains the *potential* to harbor disease-causing microorganisms (pathogens). The potential is directly related to the state of health in the population which is producing the excrement. If a family is composting its own humanure, for example, and it is a healthy family, the danger in the production and use of compost will be very low."

Jenkins offers an affordable alternative to expensive composting toilets, which he claims rarely reach high enough temperatures to ensure destruction of certain pathogens such as round worms (which very few people in industrial nations have anyway). His alternative is the sawdust toilet combined with a

IT IS IRONIC that composting, so lately embraced in many economies, is one of the oldest forms of recycling known to humankind. As societies become reacquainted with this practice, its value as a natural solution to problems, from overflowing landfills to anemic soils, will become apparent. Then, with proper institutional and economic incentives, composting could become as commonplace as the recycling of cans, newspaper, or paper is today.

GARY GARDINER, *Recycling Organic Waste: From Urban Pollutant to Farm Resource*

FIGURE 13-23.
Sawdust toilet. This may seem crude, but it is an effective, environmentally friendly option to standard toilets that is growing in popularity thanks to Joe Jenkins and his Humanure Handbook.

Composting is easier to do than to describe, and, like lovemaking, magic when you do it well.

SIM VAN DER RYN,
The Toilet Papers

compost system. Currently being used in many urban settings by rebels who are tired of polluting our world and wasting a valuable resource, the sawdust toilet collects waste (urine and feces) in a 5-gallon plastic bucket in a specially built throne (figure 13-23). After each deposit, the user simply adds a handful or two of sawdust or some similar material. This eliminates odors and keeps insects from accessing the waste. When the bucket is filled, it is carted to a 5-foot by 5-foot by 40-inch (high) compost bin. It is carefully deposited in the center of the pile, then covered with a generous layer of straw, or hay or some similar material. When the pile is full, a new one is begun in an adjoining bin. When the second bin is full, usually about a year later, the first pile is ready for application. By then the waste has been broken down and fecal bacteria, viruses, and parasites, if any, have been destroyed. According to Jenkins, it doesn't take much heat to destroy potentially harmful microorganisms. In fact, a compost pile only needs to reach 120 degrees Fahrenheit for a day or about 114 degrees for a week to kill off any potentially harmful microorganisms. Studies on his own compost pile show that temperatures inside his humanure compost pile are high enough long enough to destroy all pathogens.

Feces and urine mixed with sawdust and straw and kitchen scraps decay to form a marvelously rich organic material that, when added to soil, gives it the nutrients it needs to produce generous crops of edible fruits and vegetables and gorgeous flowers. If you want to learn more, get a copy of Jenkins' award-winning book, *The Humanure Handbook*.

Blackwater Treatment Systems

If you are hooked on a flush toilet or must install a more conventional sewage treatment system because your city or county requires it, you may want to follow in the footsteps of maverick innovator Tom Watson, who devised a pumice-based planter system to treat blackwater. It is called the Watson Wick System. So impressed were the folks at Solar Survival Architecture that they adopted the technology, modifying it here and there to meet their needs. Their system consists of an elevated, solar-heated septic tank, that is, a septic tank equipped with a glass front to provide heat to the unit (figure 13-24). This speeds up the decomposition of wastes. Like all septic tanks, solids accumulate and decompose in the bottom of the tank while liquids drain off the top. Rather than entering a standard leach field, however, the liquid drains into a series of rubber-lined cells, actually exterior planters that are typically positioned along the front of the Earthship. The rubber lining keeps leachate from leaking into the ground.

The number and size of the cells are determined by the projected use. The more use, the larger the system. Like their graywater system, the blackwater system filters the wastewater through a pumice base. Plant roots that penetrate into the pumice draw off moisture and nutrients. Solar Survival Architecture hopes someday to purify the water so much that they can reintroduce some portion of it back into the cistern for drinking water. It may sound gross, but it is the way nature deals with wastewater. Every drop of water we drink was once the waste of some other animal.

In Rico, Colorado, builder Keith Lindauer installed one of these innovative filtration systems on an Earthship with excellent results. The local health department insisted on testing the effluent and was shocked by the purity of the water coming out of the system. In fact, the nitrate levels were so much lower than effluent at the sewage treatment plant, says Lindauer, "that they [the health department] didn't believe them." Repeated tests are demonstrating the superiority of these systems to conventional waste management.

The solar septic system with blackwater cells (planters) is a potentially great way to treat wastewater from a single dwelling. It contains, cleans, and uses the effluent to nourish ornamental plants, but not vegetables. This system has been conditionally approved for limited use in New Mexico and parts of Colorado. As in so many other innovations, this one is bound to undergo evolutionary improvements, so stay tuned. Check out the SSA Web site or give them a call for the latest information. Or contact Tom Watson, who is listed in the resource guide.

In some locations, blackwater is being treated by artificial wetlands located near the house. Widely used for municipal waste treatment, artificial or constructed wetlands are fed by blackwater (and graywater, if necessary). The waste enters the upper end of the wetland and then flows by gravity or with

Solar Survival Architecture

FIGURE 13-24.
The septic tank (bottom left) in the Earthship blackwater system protrudes slightly above ground and is equipped with a window for solar heating. Leachate drains into a lined planter (beneath windows) for final treatment.

the aid of pumps through the constructed wetland, a complex biological filtration system. Containing both anaerobic and aerobic decomposition zones, the artificial wetland removes pathogens, organic material, and nutrients thanks to the activity of naturally occurring bacteria, fungi, and aquatic plants such as reeds (bulrushes).

Constructed wetlands exist in two basic types. The first is known as a surface flow wetland. It is a pond or marsh that treats the wastewater. The second is a subsurface flow wetland. In this type of system, water flows entirely through a bed of gravel that is overlain by a healthy bed of plant matter rooted in an organic growing medium containing soil and mulch. Some artificial wetland systems contain features of both types.

Subsurface systems provide many advantages over surface flow systems. Because wastewater flows underground, there is no chance of a human or animal contacting contaminated waste. Odors are contained as well. And, of course, this system lowers the risk of increasing mosquito habitat. Subsurface systems also increase the rate at which wastes break down due to the exposure of wastewater to a large surface area containing many beneficial microorganisms. Plants whose roots extend down into the gravel below the soil also aid in the treatment process by absorbing nutrients. Some plants apparently absorb organic molecules in the wastewater as well. These systems work so quickly that a two-to-six day retention time is all that is needed to completely eliminate impurities.

Surface systems or combinations of surface and subsurface systems work well, too. Rooted and floating plants such as water hyacinth and duckweed help purify the water. Some plants, such as water hyacinths, even take up heavy metals and organic pollutants. Like their subterranean cousins, surface systems do a good job of purifying the wastewater entering them, so well in fact, that hundreds of large systems are now in use worldwide. In some locations, the effluent of these systems is used to recharge groundwater supplies. In others, the water is being discharged into surface waters (both lakes and streams). Surface systems may require periodic harvesting of vegetation, a resource that can be used for compost, mulch, feedstock for methane digesters, or animal feed, provided it is free of toxic substances.

A single-bedroom house will require about 120 square feet of wetland about 1 foot deep. That may sound like a lot, but it is only 10 feet by 12 feet. Experience with constructed wetlands is limited. Conventional wisdom suggests that they perform best in warmer climates year-round. However, much to my surprise, some studies suggest that subsurface systems may work pretty well all year-round, even in cold climates. That is, if the system is prevented from freezing, biologic activity does not decrease substantially during the winter.

To prevent pollutants from escaping, constructed wetlands are often lined with an impervious liner or sealed with bentonite clay. Constructed wetlands require little, if any, maintenance once they are up and running. The Center for

Alternative Technology offers a number of publications that you should read if you are thinking about a system such as this. I also refer you to one of my favorite references, *The Composting Toilet System Book*.

The epitome of constructed wetlands is seen in the indoor systems, usually housed in greenhouses for year-round use and usually reserved for commercial purposes, such as ski areas. Designed and patented by John Todd, indoor biological treatment systems (or "living machines") consist of a series of tanks containing a variety of microorganisms and plant species. As the blackwater passes through the system, organics are decomposed and incorporated into living matter. Nutrients are absorbed as well.

Obtaining permits for an alternative blackwater system can be tricky. As engineer Mike Hannigan of KIS Engineering in Durango, Colorado notes, "Problems with permitting alternative systems are usually related to the personality of the permitter. If the person is open to exploration, you have a better chance of receiving an experimental permit." You will very likely be required to monitor the system to determine its performance over a period of a few years, too.

That concludes our exploration of alternative water and waste systems. As you can probably tell, I'm very excited about the possibilities. But do be cautious, my friends. Read the books I've listed in the resource guide and, if at all possible, talk to the people who are living with these systems to get an honest appraisal. I wish you the best of luck.

14

GREEN BUILDING MATERIALS
Creating The Ultimate Environmentally Friendly Home

The toxicity of many conventional building materials and household products impacts everyone associated with them: the workers in the factories and warehouses, the builders on the construction site, and the inhabitants of the poisonous end products.

MICHAEL G. SMITH
"The Case for Natural Building"
in *The Art of Natural Building*

BUILDING A TRULY SUSTAINABLE home begins with the use of natural building materials such as cob, straw bale, or adobe to fashion the shell of the house. To be truly sustainable, however, a home should attain a high level of energy independence. Complete emancipation from the utility companies is ideal. As outlined in chapter 11, autonomy can be achieved by passive solar heating and cooling techniques. Electricity supplied from renewable sources, alternative water supplies, and more ecologically sound wastewater treatment systems also advance our efforts to create a more sustainable living space. While these measures result in a home that outshines most conventional houses, they are still not enough to create a truly sustainable place to live. Why? Because they fail to address a major contribution to environmental deterioration—the use of unsustainable building materials. In this chapter, we will explore a cornucopia of green building materials and environmentally benign paints and finishes. In order to make this material more useful to you, I've organized the chapter in the approximate sequence in which a home is built, starting with the foundation, and working my way to the roof.

GREEN BUILDING MATERIALS

In recent years, many manufacturers have climbed onto the green building bandwagon. In their zeal to tap into this rapidly expanding market, some manufacturers have introduced products of rather dubious merit. Some of these products work poorly. Others offer little gain over conventional materials. Some involve ridiculous tradeoffs. For example, at a green building conference in Denver, Colorado, in 1999, sponsored by the National Association of Home Builders, one manufacturer (who shall remain nameless) was hawking a device to ensure instant hot water for household use. The device eliminated the need to wait for shower water to heat up. When you turned on the shower the water was hot. It could save thousands of gallons of water, the salesman claimed. However, it achieved this feat by circulating warm water through the pipes of a house 24 hours a day, 365 days a year. When asked if his device would raise the consumption of fuel, he admitted that it would. He didn't know by how much, however. I imagine it would be significant.

The first rule is buyer beware. Think clearly. Examine manufacturers claims. Look for the loopholes in logic, the tradeoffs, and the potential weaknesses of every product. Request data to prove an assertion. A manufacturer's prideful proclamations of environmental benefits may be nonsense, hype, and hysteria.

To assess the greenness or sustainability of a product, you may want to review the principles of sustainable design presented in chapter 1. Many of these can be used to screen green building products. The sidebar on this page provides a quick summary. The products you buy should be produced by companies that care about the environment and take actions to manifest their commitment. Pollution prevention programs are an important sign of environmental responsibility. Production of materials from recycled scrap or sustainably harvested forests is another. Also purchase from companies that treat their employees well; it indicates they truly care about the health and well-being of the people who are making them wealthy—or at least keeping them in business.

Decisions about products are complex. Some materials may meet the criteria, but cost too much—either to purchase or to ship to your site. Others may not be as aesthetically appealing as standard building materials. Still others may not provide the structural strength you require. (In some cases, they could actually be stronger than standard materials.) Sometimes a product may meet only one or two sustainable criteria; nonetheless, it is an improvement over conventional building materials. Some may offer huge detrimental tradeoffs, like the instant-hot-water contraption I just mentioned. Some products may require special skills for installation and repair that are not available in your area. As Sam Clark points out in *The Independent Builder*, it is "not enough for the technology to be possible. Delivery and support systems have to be there too, including . . . local tradespeople who can install and service products."

No one should tell you that choosing environmentally friendly products is going to be simple. But I can say with assurance that selecting a suitable material is getting easier every day. Access is improving. Selection is broadening. Information on alternatives is being disseminated by more and more sources.

My advice is to select the products and materials that meet the greatest number of green criteria possible. Don't expect perfection. You won't find it.

CRITERIA FOR EVALUATING GREEN BUILDING PRODUCTS

- Low embodied energy
- Promote efficient use of resources
- Produce little or no pollution
- Produced with minimal pollution output
- Readily recyclable when their useful life is over
- Made from recycled or natural materials
- Produced by socially and environmentally responsible companies
- Produced from sustainably harvested materials
- Produced or manufactured locally
- Nontoxic
- Durable
- Low life-cycle cost
- Aesthetically appealing
- Meet building requirements
- Competitively priced

Virtually all products have some shortcoming. Choose materials that offer the greatest gain for the environment and the health and welfare of the occupants of the house. One useful strategy is to concentrate on products that are used in great quantity, such as framing lumber, insulation, tile, concrete, and drywall. Purchase as many green building products as you can afford. This way, you will have a larger impact.

You may be surprised to find that many green building products are cost competitive. In my home, the tile, insulation, carpeting, carpet pad, paint, and cabinetry were all cost competitive. If a product does cost more, however, don't dismiss it. The price difference may be offset by savings on another product or by reductions in energy costs. Higher costs may increase comfort levels or reduce health impacts. (How much does it cost to treat cancer?) Keep track of materials on which you save money, then apply savings to items that cost more.

Another bit of advice: Don't let builders who are unfamiliar with the wide array of environmentally friendly materials dissuade you. Builders are a conservative lot. They aren't inclined to experiment. It is too costly if they make a mistake, and they don't want homebuyers calling them to complain about some product that failed to live up to a manufacturer's claims.

Do your own research. Check out the materials. Talk to others who have used them. Although a number of excellent books on green building materials have been published, one of the most useful and most comprehensive resources is John Hermannsson's *Green Building Resource Guide*. This book lists a wide assortment of green building products and briefly describes each one. It also includes a price index that compares the price of each product to the conventional building material it replaces. For example, Pittsford Lumber Company in Pittsford, New York, offers sustainably harvested tropical hardwoods that have been certified by the Rainforest Alliance as "Smart Wood." Its price index is 1.0, meaning it costs the same as unsustainably harvested hardwoods. International Cellulose in Houston, Texas, offers Celbar, a residential insulation made from recycled newsprint with a borate added as a fire retardant. Its price index is 0.5 to 0.8 compared to fiberglass batt. This means that recycled insulation from this company costs one-half to four-fifths as much as fiberglass batt. Hermannsson also maintains a Web site where you can find updates on new products not listed in his book.

Another excellent resource is *The Sustainable Design Resource Guide* produced by the American Institute of Architects (AIA; Denver Environmental Committee) and Architects, Designers, and Planners for Social Responsibility (ADSR; Colorado chapter). This book is a great complement to Hermannsson's book.

Another superb resource is the Austin Green Building Program's Web site, www.ci.austin.tx.us/greenbuilder/. The folks who maintain this site do a terrific job of describing various green building products. They also rate materials on the basis of availability of suppliers, cost, and reliability.

The market in eco-products has become big business and is moving more mainstream.

DAVID PEARSON, *The New Natural House Book*

The National Association of Home Builders' Web site is another valuable resource. Their research center is actively field testing new products to help people like you avoid costly mistakes. Their site describes how products are supposed to work, how they should be installed, and limitations of each product. One of the most useful features is the results of field tests. If you are worried about obtaining a building permit, you will find their section on code issues very useful.

Yet another source is *Environmental Design and Construction*, a magazine dedicated to promoting economically and environmentally successful construction. It publishes great articles on new products and contains ads from companies offering a wide variety of products from sustainably harvested hardwood to arsenic-free pressure-treated lumber to innovative, energy-efficient skylights.

With the products described in these resources and a few others listed in the resource guide, you can build an entire house from state-of-the art, environmentally sensitive materials, including a floor mat for your front steps made from recycled automobile tires. Hardly a product in use today can't be replaced by a more environmentally sound alternative, and the variety of green building materials only promises to get better. If you are building a natural home, be sure to consider alternative materials for foundations, walls, floors, and windows. If you are going to build a more conventional wood-frame house or just an addition, give careful thought to ways that you can use alternative materials.

I want to remind you that mention of a product in this chapter should not be construed as an endorsement. Many of these products are new to me, too. I will clearly indicate those that I've used and tell you about my experiences, but even so, conditions under which I used a product may differ from yours. Bottom line: use extreme caution when approaching any alternative material. Check it out thoroughly.

REDUCING THE USE OF WOOD

Wood has played an important role in home construction throughout much of human history. In the United States, 90 percent of all houses are currently built of wood. In these homes, wood forms the rigid internal skeleton of roofs and walls in stick-frame houses. Wood fiber is also used for the "skin" of many houses, for example, to make exterior sheathing for walls and decking for floors and roofs. Furthermore, wood is used to manufacture shingles, siding, and flooring, doors, stair cases, trim, and cabinets.

Although most natural, sustainable homes use less wood than a conventional stick-frame house, most still require lumber for interior partition walls, roofs, doors, and cabinets. With some research and planning, however, you can cut your wood demand substantially—no matter what type of home you are planning to build. Is this a goal we should be striving to achieve?

Although wood is a renewable resource, the amount of wood required for construction purposes is taxing the regenerative capabilities of this resource, as well as depleting a critical component in ecological balance.

AIA and ADPSR,
The Sustainable Design Resource Guide

Wood is a renewable resource with low embodied energy, much lower than plastics, concrete, and metals. However, many environmentalists believe that the long-term health of the planet depends on finding ways to reduce timber cutting. Opponents argue that plenty of wood is available. Old-growth forests may be in danger, they say, but there are plenty of smaller trees to sustain the building industry. My view is that the danger to the world's forests is very real. Huge tracts of forest are laid bare each year, causing extraordinary damage to the land and to wildlife. Tropical forests are especially hard hit. Temperate old-growth forests such as those in the Pacific Northwest have also been under assault. Although forests do regrow, in the tropics large clear-cuts often fail to recover. Soil is washed away or baked in the sun to a bricklike hardness. Temperate forests regrow, too, but their rate of recovery is slow. Destruction of these forests tears apart the intricate web of life, spelling doom for species that depend on this ecosystem.

Projections of wood demand further validate the environmentalists' call to reduce pressure on the world's forests. Population growth will cause demand to skyrocket. If we don't find ways to reduce our demand for wood, we will lose the many ecological free services provided by the forests such as flood control, groundwater recharge, oxygen production, insect control, and recreation. But how can we cut our demand for wood?

One of the first options is to build walls out of straw bale, cob, adobe, rammed earth, or some other natural material. Building homes with these materials will dramatically reduce the need for lumber. However, not all natural building technologies are as kind to forests as others. Straw-clay, cordwood homes, and straw bale homes are often framed with large posts and beams. Rammed earth homes require wooden forms. Log and cordwood homes require dozens of trees. Many natural building techniques also require wood for interior walls, roofs, flooring, and decks. Therefore, if you are building a natural home, you have to be concerned with wood use.

Whether building a conventional home or a natural home, one way to reduce lumber requirements is by building with engineered lumber. Engineered lumber is made from recycled or reconstituted materials. Some manufacturers are even using straw to make engineered wood. Primeboard in Wahpeton, North Dakota, manufactures a substitute for particle board that is made from straw and a nontoxic, emission-free binding agent (according to the manufacturer). Primeboard is lighter than particle board and more resistant to moisture. It offers superior machining qualities (which means it is easy to cut). The Neil Kelly Company in Portland, Oregon, is now using formaldehyde-free, straw-based particle board to manufacture its cabinets. The company also produces a line of "total resource responsible and healthy cabinets" made from straw-based particle board and certified woods—lumber derived from sustainably managed forests. The company also uses environmentally friendly low-VOC (volatile organic compound) water-based finishes.

Some companies are using industrial hemp in place of wood to produce fiberboard. Industrial hemp belongs to the same genus as marijuana, but contains very little active ingredient (THC). Currently grown by farmers in many countries including Canada, industrial hemp is used to manufacture a wide range of products such as clothing, paper, soap, animal bedding, insulation, fiberboard, cement blocks, stucco, and mortar. (It is even being used to produce foods.) For more on the subject, read John Roulac's, *Hemp Horizons*.

Trees can be chipped, shredded, and peeled to produce the raw materials for a form of engineered lumber known as reconstituted wood. This process uses wood fiber from fast-growing trees and wastes less than conventional sawn lumber production. One example is the I-joist, used to frame floors and roofs. I-joists vary from one manufacturer to another, but most are made from OSB (oriented strand board) sheathing. OSB contains large wood chips bonded together by plastic resins. The horizontal components of the I-joist (the flanges) consist of laminated lumber made from thin layers of wood peeled from logs.

Another reconstituted wood product that has been around longer than OSB is particle board. Made from sawdust, particle board is covered with a veneer of hardwood to manufacture cabinets, doors, and paneling. The process uses waste (sawdust) to make the bulk of the structure with a minimal amount of hardwood, for an appealing end product.

Engineered lumber can be used to build an entire home from studs to beams to roof trusses to floor joists to sheathing (figure 14-1). Be careful. Check out the glues that are used to make various engineered wood products before you buy them. OSB, for example, is made from a glue that contains formaldehyde. Formaldehyde is a substance used to embalm cadavers and preserve biological specimens. It is known to cause serious health problems in people. To avoid exposure, purchase low-formaldehyde brands and apply a sealant to the sheathing before you use it.

A third component of our forest-protection strategy is the use of composite and finger-jointed lumber in place of solid dimensional wood products. Composite lumber is made by gluing dimensional lumber together to produce larger beams and posts. One example is the lam (laminated) beam used for headers (lintels). Lam beams consist of 2 x 4s or 2 x 6s that

INDUSTRIAL HEMP is grown commercially in France, Germany, the United Kingdom, China, and numerous other countries around the world, but not in the United States. Although hemp is not illegal to grow, possess, or sell in the U.S., it is heavily regulated—so overly regulated as to be nearly impossible to grow legally. This, in spite of rigorous efforts on the part of U.S. legislators who drafted the first anti-marijuana legislation to exclude hemp, which had been used in the United States for many years to make rope and other useful products.

JOHN W. ROULAC, *Hemp Horizons*

FIGURE 14-1.
Solid, dimensional lumber in home construction can be replaced by a variety of engineered wood products. They're straighter and are a more efficient use of wood than dimensional lumber—so there's less waste. They're made from smaller trees, too. The only potential problem is the glues used to bind the wood fibers together.

Truss Joist MacMillan

are glued together to form thick beams (figure 14-2). Composite lumber is strong and durable and is less prone to warping than larger solid pieces of sawn lumber. Components come from smaller trees. There is no need to cut old growth to produce a roof beam. Composite lumber is also widely available.

Finger-jointed lumber is wood produced by gluing smaller pieces together longitudinally. Snug finger joints between the individual pieces create a secure bond, resulting in boards that can be used for making window and door frames.

A fourth strategy in the quest to reduce wood consumption is to use salvaged lumber—that is, wood recovered from demolished buildings. Salvaged or reclaimed lumber can be used for roof beams, stairwells, cabinetry, and flooring. According to the Natural Resources Defense Council publication *Efficient Wood Use in Residential Construction*, reclaimed wood is of limited use for stud walls because of performance concerns.

Another valuable source is recovered lumber. Goodwin Heart Pine Company in Gainesville, Florida, for example, recovers heart pine and cypress logs that were lost while being transported to sawmills along rivers during the early 1900s (figure 14-3). After recovery by the company's divers, the wood is dried and cut into boards. The stuff is extraordinarily beautiful. Check out *Efficient Wood Use in Residential Construction* for a list of reclaimed wood dealers in the United States.

A fifth means of decreasing wood use is to recycle waste lumber from building sites. Save useful scraps for later use. Stockpile them by size and type. You never know when you will need a 2-foot section. Grabbing a piece from a scrap pile is far better than cutting one off an 8-foot 2 x 4. If you have leftovers you don't want, post a sign advertising free wood. Someone will gladly haul off your waste. The world is full of scavengers! You may also want to contact manufacturers in your area who use waste wood. One company in Denver grinds scrap wood, then uses it to manufacture fiberboard and garden compost. The company also adds the ground wood fiber to sludge from sewage treatment plants to produce fertilizer.

Number six on the list of forest conservation strategies is plastic lumber. Yes, that's right, plastic lumber. Although it sounds like a contradiction in terms, plastic lumber is a perfectly suitable building material. Made from recycled waste (old milk jugs) and wood fibers, plastic lumber is used to build decks, fences, and playground equipment (figure 14-4). It is ideal for fascia boards. While some readers may have moral objections to using plastic on a natural home, remember that plastic lumber is made from a recycled waste material and is extremely durable. You won't have to stain it or apply wood preservatives. Decks made out of plastic lumber hold up well and won't require the maintenance of a wood deck. You have to look pretty closely to tell they are made out of plastic. Some recycled plastic lumber contains a mix of plastic and sawdust, another waste product.

Another way to reduce wood demand is to design your home to conform to standard-size lumber and sheathing. This simple measure practically eliminates wood waste.

Finally, take care not to overbuild your home. These days many designers and builders use more lumber than is needed. By erring on the conservative side, they drive up the demand for lumber and put unnecessary pressure on the world's forests. Researchers at the U.S. Forest Service report that modern homes could be built with 10 percent less lumber without compromising strength. By spacing studs a little wider and by employing other measures, a builder could reduce lumber demand, save money, and help to safeguard the world's forests.

And lest we forget, smaller houses also save wood. Don't get carried away. Don't let your dream home become an environmental nightmare!

If you use wood, and most of us do, seek out suppliers who provide certified wood from companies

FIGURE 14-4.
Plastic lumber made from recycled plastic milk jugs and wood fibers is durable and easy to work with. You won't have to paint it or repair cracks, and it is safe to walk on and quite nice looking.

Trex Company

that manage and harvest trees in an environmentally sustainable manner. You would be amazed at how many there are. I suspect their numbers will increase, too, especially now that Home Depot, the building products supplier that sells one-tenth of the world's lumber, has announced plans to stock its shelves with wood exclusively from certified producers. For a list of certified wood sources, see *Efficient Wood Use in Residential Construction*, Austin's Green Building Web site, and John Hermannsson's *Green Building Resource Guide*.

Now that we have covered some of the basics, let's see what you can do make the foundation of your building—whether a natural home, a small addition, or even a conventional home—as environmentally sustainable as possible.

STARTING WITH THE FOUNDATION

Many environmentally and structurally acceptable foundations are available for natural homes. These include rubble trench, stone, packed-tire, and railroad tile foundations. Despite this wide variety, many people still build foundations out of concrete. If this is the way you or your builder are leaning, you may want to consider using fly ash concrete. Fly ash is a waste product from coal-fired power plants. It is captured by pollution control devices (electrostatic precipitators) and produced in abundance in many parts of the world. In most locations, it is dumped in landfills. However, fly ash can be substituted for 15 to 30 percent of the Portland cement in a mix, reducing the embodied energy of concrete.

In addition to putting a waste to good use, fly ash concrete is smoother, denser, more workable, and less permeable than conventional concrete. Because the U.S. government recommends the use of fly ash concrete in all federally funded projects, it can be obtained in many parts of the country. Call local concrete suppliers to determine its availability, but before you order any, explore the potential health effects. Some individuals, such as natural, healthy

FIGURE 14-5.

Interlocking insulating concrete forms (made by American Polysteel) are put in place, reinforced with rebar, then filled with concrete. They can be used to build foundations and walls, and provide superior insulation, making substantial reductions in heating and cooling costs.

American Polysteel

FIGURE 14-6.
Believe it or not, this house was built entirely from Polysteel's ICFs.

homebuilder Cedar Rose, view fly ash concrete with suspicion. "Fly ash is nasty stuff," she says. To be on the safe side, the only concrete she uses in her homes is in grade beams on rubble trench or stone foundations, and none of it contains fly ash. Are her and others' fears well founded?

Fly ash contains an assortment of toxic materials, mostly heavy metals. Entombed in concrete, however, these nonvolatile toxicants probably pose little threat to homeowners. I can't guarantee that, so use your own judgment. (I do wonder whether they have any effect on the people who manufacture, pour, and finish concrete.)

Another product that reduces the amount of concrete needed to build a foundation is the insulating concrete form or ICF. Unlike wooden or metal concrete forms, the foam panels are relatively easy to assemble, with pieces snapping in place like toy blocks. Polysteel forms, produced by American Polysteel Forms in Albuquerque, New Mexico, come preassembled (figure 14-5). To build a wall, forms are stacked on top of one another and held in place by interlocking edges. Steel cross bridges resist the outward force exerted on the forms by wet concrete. Polysteel panels come in 1-foot x 8-foot to 4-foot x 8-foot sizes.

ICFs can also be used to build exterior walls (figure 14-6). Unlike wooden forms, ICFs are lightweight and easy to install. Unlike steel or wood concrete forms, they are left in place after the concrete has set up. ICFs reduce the amount of concrete needed to build a wall. Reward forms, manufactured by EPS Building Systems in Broomfield, Colorado, require 25 percent less concrete than an 8-inch poured wall, yet are 50 percent stronger. Because the foam used in ICFs must be relatively thick to resist blowout (caused by wet concrete) and because the foam is left in place after the concrete has cured, foundations made from ICFs have a relatively high R-value, possibly the highest R-value of

BUILDING NOTE

Some manufacturers produce hollow foam blocks with interlocking edges that are assembled on the footing and then filled with concrete.

Dan Chouinard, Rastra

FIGURE 14-7 A, B.
The Rastra block (B) is made from concrete and recycled plastic foam. They are used to make walls and foundations like other ICFs. But unlike most other ICFs, the Rastra block is made from recycled plastic and cement. The results can be stunning (A).

any foundation system in use today. Houses built entirely from Polysteel forms (walls and foundation) have reported savings of 50 to 80 percent on heating and cooling. Foundations and walls built from ICFs reduce the amount of lumber used to build a home, too.

On the downside, however, foam tends to be vulnerable to termites. If you live in a bioregion where termites are a problem, take precautions to prevent termites from boring into the foam. Another problem is that some ICFs are made from extruded polystyrene manufactured with ozone-depleting chlorofluorocarbons (CFCs). Whenever possible, use the expanded polystyrene ICFs; no CFCs are used in their production.

Builders like ICFs because they are quick and easy to install. Because of this, they are cheaper than standard concrete forms. ICFs come equipped with integral fastening strips that facilitate finish work (for example, applying drywall). Homeowners like ICFs because they help keep their homes warm in winter and cool in summer while lowering utility bills. Several companies that manufacture ICFs are listed in *The Sustainable Design Resource Guide*.

Concrete blocks may be used to build foundations. Contact local suppliers to see which ones sell blocks made from fly ash concrete. Furthermore, if you are going to be using bricks for any reason, including firebrick for a fireplace, see if you can locate companies that use fly ash to manufacture them.

A far better option is Faswell blocks. Used in Europe for nearly sixty years, Faswell blocks consist of concrete and wood particles molded into blocks. They are used to construct foundations, exterior and interior walls, and structures such as pump vaults for cisterns. They are lightweight and easy to work with.

Rastra "block," made from cement and recycled plastic foam, can also be used to build foundations and walls (figure 14-7). Like Faswell blocks, Rastra blocks are lightweight and easy to handle. They cut easily with a saw. Both horizontal and vertical rebar may be placed in the forms for structural stability. Stucco adheres well to the porous surface of the Rastra block.

EXTERIOR WALLS

Although your home may have walls of adobe, cob, straw, or logs, it may still require some conventional framing—for example, for south-facing "window walls." Most people frame these walls out of 2 x 4 or 2 x 6s. One alternative to conventional 2 x 4 or 2 x 6 construction is the structural insulated panel or SIP.

Structural insulated panels consist of two layers of sheathing—typically OSB—sandwiching a rigid foam insulation core (figure 14-8). (There are no studs, even in load-bearing walls.) Panels come in 2- to 12-inch thicknesses and vary in size from 4 x 8 feet to 8 x 24 feet. Even larger sheets are available for specialty applications.

SIPs are assembled on the foundation and cut and trimmed to accommodate windows and doors. However, most manufacturers will custom make panels—that is, they'll provide you with precut structural panels tailored to your design. These panels even contain door and window openings. All you do is submit your plans to the company and they will manufacture, trim, and number the SIPs, then ship them to your building site ready for assembly. Although this option costs a bit more than ordering standard panels, it minimizes cutting and trimming on site, reduces waste, and saves time.

FIGURE 14-8.
Structural insulated panels consist of two (outer) layers of oriented strand board and a foam core, ranging from 2 to 12 inches thick. They can be used to build exterior walls, roofs, and floors. These SIPs are made by PanelPros.

Panel Pros, Inc.

SIPs are locked together by special joints. To make your work even easier, some manufacturers produce panels that have sheetrock or drywall mounted on the inside face and/or siding on the exterior, so you can assemble a complete wall with a few adroitly placed nails.

TechNature, an innovative company based in Redmond, Washington, has come up with a rather ingenious twist on the idea of structural wall panels. This company produces small hexagonal units consisting of foam sandwiched between two layers of OSB. The hexagons are fitted together like pieces of a puzzle and can be used to build walls, floors, and roofs. The pieces are then secured to a sheet of ½-inch-thick OSB to form a 4-inch-thick wall, or fitted together two units thick to form a 6-inch-thick wall. Hexagons are fastened by special joint fasteners and can be assembled into a variety of shapes. Drywall and siding are attached to the wall with ease. Lightweight and easy to assemble, this system has the benefit of being easily disassembled for remodeling. Like other SIP manufacturers, TechNature will work from your blueprints to provide precut, numbered pieces for easy assembly.

Manufacturers use three different foams to make SIPs: urethane, extruded polystyrene (XEPS), and molded expanded polystyrene (MEPS). The manufacture of urethane and XEPS uses ozone-depleting chemicals, CFCs. Molded expanded polystyrene production does not. In addition, MEPS is cheaper. Because of these features, MEPS is the manufacturer's foam of choice, which means it is the most commonly sold product.

Although polystyrene is a petroleum-based product, it is not as bad as some people would have you believe. According to one manufacturer, polystyrene is produced by waste gases released at oil refineries. For years, these gases have been carelessly discharged into the atmosphere or burned off (flared). Today, however, many refineries capture the gases, then run them through a catalyst to produce polystyrene.

If the idea of using a petroleum-based product still offends you, but the idea of building walls with minimal labor appeals to you, check out Agriboard, in Fairfield, Iowa. Listed on Austin's Green Building Program Web site, this company manufacturers a structural insulated panel containing a straw core.

One of the main advantages of SIPs is their energy efficiency. R-values of the foams used in SIPs range from 4 to 7 per inch, depending on the type of foam. Therefore, a 4-inch-thick SIP has an R-value of 14 to 25. Compare that to a 2 x 4 stud wall containing fiberglass or rock wool insulation that has R-values of 11 to 15, respectively.

Because they lack studs, SIPs eliminate bridging losses that occur when heat escapes directly through studs. In a standard stud wall, bridging losses can reduce R-values by as much as 25 percent. SIPs also reduces air infiltration. Because of these features, SIPs save homeowners a lot of money over the long haul.

SIP costs vary from 10 percent less to 10 percent more than a conventional stud wall, depending on the manufacturer. Despite higher initial costs, SIPs may be cheaper than standard stud wall construction. In wood-frame construction, for example, builders must make hundreds of cuts and will have hundreds of pieces to assemble. That is not the case with SIPs. The minimal amount of cutting and nailing involved in assembling a wall from SIPs reduces labor costs. Construction waste and disposal costs are substantially lower. Use of SIPs also wastes less material because SIPs are straight, true, and plumb, unlike 2 x 4s and 2 x 6s. You won't be discarding a lot of bowed lumber or spending valuable time sorting through lumber for straight pieces. Because they are more energy efficient than conventionally framed structures, SIPs may save money by reducing the size of heating and cooling systems. More efficient walls reduce energy bills.

SIPs are produced by numerous manufacturers in the United States and are readily available in most areas of the country. They are extremely strong and can be used to construct floors and roofs too, although you will need a crane to lift them onto a roof truss. If you are building a natural home, consider SIPs for your roof and floor. Look at the Austin's Green Building Program's Web site and John Hermannsson's book for a list of manufacturers. Before sending in your order, be sure that the structural panels you buy meet code in your area.

Exterior walls can be framed with engineered 2 x 4 and 2 x 6 studs. Made from strips of lumber that are glued and pressed together in forms, engineered lumber is produced in a variety of dimensions (figure 14-9). It is a remarkably straight and exceptionally strong substitute for standard dimensional lumber. It can be used for headers, top plates, studs, interior columns, and beams. Ron Jones, a custom homebuilder in central New Mexico, now replaces 85 percent of the dimensional lumber in his projects with engineered wood products.

Although it may seem ridiculous, especially in an earth-friendly home built of all natural materials, engineered lumber has numerous benefits worth your consideration. First, engineered studs, beams, columns, trusses, and floor joists are made from the fibers of relatively small trees. In contrast, large dimensional lumber (2 x 10s, for example) comes from much larger, often old-growth trees. (Imagine the size of the tree required to produce a 2 x 10.) Engineered lumber therefore reduces the harvest of old-growth forests and helps protect this valuable ecosystem.

Engineered lumber also makes much better use of trees, compared to traditional methods of making dimensional lumber. Manufacturers of engineered lumber utilize 50 to 80 percent of the wood fiber from a tree. Cutting dimensional lumber from a round tree utilizes only 40 percent of the tree's wood. Because of this, fewer trees are harvested to produce the engineered lumber needed to frame a house.

BUILDING NOTE

When younger trees are used to produce dimensional lumber, such as 2 x 4s the wood that is discarded is older, denser, and stronger than the wood in the interior. The result is that dimensional lumber often consists of young, less dense wood, which is not as strong.

FIGURE 14-9.
Engineered lumber is produced from peeled or chipped logs (here made by Truss Joist Macmillan). The product of this process is placed in forms with glue, then pressed together. The result is a strong, straight piece of lumber that reduces cutting in old-growth forests.

Another important benefit is engineered wood's superior strength. Greater strength results in wider on-center spacing. Builders can use fewer roof trusses or floor joists, saving money and resources. I-joists use half the wood of a solid piece of lumber to achieve the same effect. Bottom line: Less wood does more work! According to Truss Joist MacMillan's (TJM) literature, engineered lumber rivals even the best old-growth wood for strength, size, and consistency. And, it surpasses old-growth lumber in one extremely critical test: availability.

Another advantage of engineered wood is that it is produced with precise specifications. It is straight and uniform in moisture content. When building with standard dimensional lumber, many builders spend a lot of time picking through a lumber bin for straight pieces. Or, if lumber is delivered on site, they may discard a significant amount of lumber because of knots, cracks, twists, and bends. Engineered lumber is free of such defects. In fact, TJM estimates that up to 11 percent of the sawn dimensional lumber delivered to a building site is wasted. For engineered wood, the number is around 1 percent—and most of that is end pieces that are trimmed off to meet the dimensional requirements of the structure.

Engineered lumber is available in extremely long lengths for special applications. Try to purchase a 30-foot 2 x 6. It can't be done. Therefore, when you need to cover long distances and want straight, true walls, engineered lumber may be the product of choice.

The consistent moisture content and the uniform dimensions of engineered lumber also results in less structural distortion. This is important when building walls. Headers (lintels) made from standard dimensional lumber will shrink, bow, and bend. This, in turn, cracks walls or may cause nails in drywall to pop out, necessitating repair.

Although you still may be conflicted by the idea of using engineered lumber in a natural, environmentally sustainable home, these wood products are produced in ways that minimize environmental damage. That, too, adds to their long list of advantages. However, even though manufacturers are striving to produce engineered lumber in an environmentally sound fashion, the wood fiber used to make engineered lumber is typically produced from trees grown on commercial tree farms. Tree farms are intensively managed forests. They are huge, immensely vulnerable monocultures that are frequently doused with chemicals to control insects and thwart the growth of weedy species. Another problem with reconstituted wood may be the glues used to bond wood fibers. Although the verdict on glues is still out, I suspect manufacturers who are sensitive to indoor air quality issues will be making efforts to ensure the use of safer glues.

In the United States, an estimated 300,000 homes are torn down each year. Unfortunately, most of the waste is hauled off to landfills. But reclaimed or salvaged lumber is a valuable building resource. Lumber salvaged from this regretful carnage can be used to construct new homes and offices. Unfortunately, it is often cheapest if purchased in bulk. Check out local suppliers or contact builders who may have purchased large quantities. They may be willing to sell you some of their stock.

Environmental salvage is another option. John Abrams, a custom homebuilder on Martha's Vineyard, uses salvaged lumber for approximately 90 percent of the exposed interior and exterior wood on his homes. He buys recovered lumber from a supplier who gathers cypress logs that were lost in transit along rivers in Florida. Abrams also uses driftwood for elegant railings and salvaged timber from warehouses to make beams and columns.

USING PATENTED technologies developed over two decades, we can take a tree apart and put its fibers back together to take full advantage of its natural strengths . . . Starting with trees too young and small for solid-sawn lumber, we can produce engineered lumber that's bigger—and better—than anything you can cut from a tree today.

Promotional literature, Truss Joist Macmillan

If reclaimed wood is not available or is too expensive, consider purchasing lumber obtained from sustainably managed forests. Check out the Good Wood Publication listed in the resource guide for a list of sustainably harvested wood producers. Austin's Green Building Program's Web site also lists wood-certifying organizations in the dimensional lumber section.

EXTERNAL SHEATHING

After the frame is built, many homebuilders (both professional builders and owner-builders) use oriented strand board (OSB) for the external sheathing. OSB is made from strands of wood that come from relatively small trees such as aspen. Glued and pressed into 4 x 8 sheets, OSB is also used for roof decking, exterior sheathing, and subflooring. So successful is this product that it has virtually replaced plywood in many parts of the United States. In the future, however, new products may render OSB obsolete. One very promising upstart is an exterior wall sheathing called Thermo-Ply (figure 14-10). Manufactured by Simplex Products Division in Adrian, Michigan, Thermo-Ply is a cardboard-thin sheathing made from 100 percent recycled paper (80 percent postconsumer fiber such as recycled newsprint and 20 percent factory waste). The paper fibers are bound together by polyvinyl alcohol, a bonding agent that, according to the manufacturer's literature, is one of the safest on the market. Unlike other sheathing materials, Thermo-ply contains no asphalt emulsion, formaldehyde, or phenols.

FIGURE 14-10.
The Thermo-Ply sheathing in this house is a lightweight material made from recycled paper. It is easy to work with. It can be overlapped to reduce air infiltration. The outer surface is reflective to help keep summer heat from penetrating a house.

Thermo-Ply has a low embodied energy, lower than both OSB and plywood, because it is made from recycled newsprint. It is very light and easy to work with. It is so light that it can be installed by a single worker. Thermo-Ply can even be nailed to stud walls that are assembled on the ground, and then entire walls are hoisted into place. OSB and plywood, on the other hand, must be applied to stud walls after they have been assembled and erected on the foundation. And OSB and plywood must be applied by two workers. Furthermore, Thermo-Ply can be cut with a knife. You won't need a power saw. This feature makes it relatively easy to create window and door openings after the sheathing has been secured to a stud wall.

Thermo-Ply attaches very tightly to studs and can be overlapped to reduce air infiltration. It is nonporous with a reflective foil (aluminum) and nonreflective poly (polyethylene) that act as a barrier to moisture. Thermo-Ply even meets code in areas with high winds or seismic activity. Thermo-Ply may be a very good substitute for OSB and other exterior wall sheathing materials, and it is cost competitive, as well. (Check out the National Association of Home Builder's Web site to see if they have field tested this item.)

THE GREEN ROOF

The roof of a house requires a tremendous amount of resources, especially lumber. Even the roofs of sustainable homes described in this book require a substantial contribution from the world's forests. Fortunately, a host of green building products can be used to build roofs with less impact on the planet.

You have already been introduced to the I-joist. With its vertical member made of OSB (or some similar product) and its flanges made from laminated veneer lumber, the I-joist uses 50 to 60 percent less lumber than solid dimensional lumber of similar width. I-joists allow builders to create a deep roof cavity with less wood. I-joists can also be used as headers above doors and windows.

Besides reducing wood use, I-joists may cost less than standard roof framing lumber. Several manufacturers sell I-joists that are 10 to 20 percent cheaper than 2 x 10s and 2 x 12s. Shop carefully, though. Prices vary from one manufacturer to another. (I found nine companies that were producing I-joists with price indices ranging from 0.92 to 1.6.) I-joists are widely available. They are sold in most lumberyards. If not, consult with Hermannsson's book, the *Green Building Resource Guide,* to contact manufacturers. They should be able to provide you with the names of local distributors.

Glue-laminated timbers are also used in roof construction for vertical columns and beams. Glu-lam beams (commonly called lam beams) consist of 2 x 4s or larger dimensional lumber glued together They may not save you much money, if any, over the cost of 2 x 12s, but they are a better choice for the environment.

Roof Decking

In olden days, builders used plywood for roof decking. In more recent times, plywood decking has been largely replaced by OSB. OSB and similar reconstituted wood products are made from waste wood and wood chips from smaller fast-growing trees, such as aspen. The fibers and chips are bonded by adhesives containing formaldehyde, which have drawn considerable criticism from those concerned about indoor air quality.

To avoid a potential health problem, stay away from products that contain urea formaldehyde or select formaldehyde-free sheathing. If you can't avoid this chemical, buy low-formaldehyde OSB, then seal it to prevent off-gassing.

OSB is widely available and represents a small step forward in sustainable building. However, manufacturers are producing materials that are bound to give OSB manufacturers a run for their money. One product is a sheathing made from 100 percent recycled newsprint. Manufactured by Homasote Company in West Trenton, New Jersey, this product can be used for exterior sheathing and roof and floor decking. Another exciting prospect is a roof and exterior sheathing material manufactured from waste paper and agricultural fibers such as straw, rice hulls, or peanut shells. It is produced by PanTerre America, Inc. in Arlington, Virginia.

According to the folks at Austin's Green Building Program, many of the alternative sheet materials cost more than conventional materials. Shipping to your site can add to the cost as well. However, these products do come in standard dimensions and are applied with standard fasteners such as nails and screws. They are also cut with regular carpentry tools. You may want to call the companies to see if there is a supplier nearby. Maybe you could make a

BUILDING NOTE

Phenol formaldehyde binder is deemed more acceptable for exterior sheathing because it off-gasses less than urea formaldehyde and because exterior applications have less effect on the interior air quality.

promotional trade—they sell you the product at a cheaper price in exchange for bragging rights and exposure.

Roofing Felt

Roofing felt, the waterproof layer that protects the decking from moisture that may penetrate the shingles, is made from paper fibers impregnated with asphalt. Unrolled onto the roof, it is then overlapped and nailed in place. In an attempt to reduce the environmental impact of this product, several manufacturers are currently making roofing felt out of asphalt-impregnated recycled paper or a recycled paper/sawdust combination. Recycled content varies by manufacturer from 20 to 100 percent. One manufacturer now produces roofing felt made from 90 percent recycled PET plastic (the plastic used to make pop bottles and other containers). Recycled roofing felt is competitively priced, and may even be less expensive than conventional roofing felt, according to the *Green Building Resource Guide*.

Shingles

After the roofing felt has been laid, shingles, tile, or metal roofing is applied. Your choices in this final roofing material are many and varied, so shop carefully. Choose a product that suits your needs and yet provides environmental benefits.

If you are considering shake shingles, check out American Cemwood. They produce a shingle containing two-thirds Portland cement and one-third wood fiber, a waste product from sawmills. While costly, Cemwood shingles are fireproof and durable and come with a fifty-year warranty. Steel and aluminum shingles are also available. Metal shingles are ideal for catchwater systems, but check out the recycled content of the metal roofing products. They can vary significantly. Classic Products produces aluminum shingles made from recycled beverage cans (figure 14-11). Their shingles are 98 percent recycled post-consumer waste and come with either a twenty- or fifty-year war-

FIGURE 14-11.
Aluminum shingles are lightweight, weather resistant, and made from recycled aluminum. Unfortunately, they're also expensive. These come from Classic Products Inc.

Classic Products, Inc.

Re-New Wood, Inc.

FIGURE 14-12.
Eco-shake roof shingles are made from 100 percent recycled wood and plastic. They come with a 50-year manufacturer's warranty!

ranty, depending on which option you select. I like this product, but the cost is high. (It is twice the cost of steel roofing and involves a lot more labor.)

A number of companies are producing shingles similar to the asphalt or composite shingle so widely used in this country. Unlike conventional shingles, however, theirs are made from recycled paper. Like conventional shingles, they are impregnated with asphalt and coated with a thin layer of fine stone. The product performs best in warmer climates, and contains asphalt that is horrible from an environmental standpoint and even worse if you are thinking about installing a catchwater system.

Re-New Wood in Wagoner, Oklahoma, produces a shingle made from 100 percent recycled wood and plastic. Known as the Eco-Shake, this product is available in four colors. It simulates weathered wood shake shingles (figure 14-12). Eco-Shake shingles meet or exceed all the testing standards for composite or wood roofing products. They are fire, wind, and hail resistant. In addition, they are resistant to moisture, freezing and thawing, and ultraviolet light. They can be nailed and sawn like wood shingles and come with a fifty-year warranty.

Crowe Industries produces roof slate tiles that imitate black slate tiles. Made from 100 percent recycled polymers and rubber, this product is also recyclable. They are well priced when compared to slate roofing tiles, and come with an impressive fifty-year warranty. I've never seen the product, so I can't judge its acceptability for catchwater systems. EcoStar in Vernon Hills, Illinois, produces a similar product made from 100 percent post-consumer recycled rubber.

As you can see, when it comes to roofing you have many environmentally sustainable options. Before selecting a roofing material, investigate it at the technology inventory on the National Association of Home Builder's Web site. Some new roofing materials have failed miserably, so be careful. A roof must withstand extreme temperature variations, intense sunlight, wind, rain, snow,

When selecting a roofing material, be sure to compare prices. Some recycled roof products are fairly cost competitive with asphalt shingles.

sleet, and hail. You want your roof to endure the elements to protect the house for many years.

CEILING AND WALL INSULATION

When it comes to wall and ceiling insulation, a builder has so many options that it will make his or her head spin. Before we examine the many commercially available products, let's consider the natural materials, such as sheep's wool, sawdust, straw bales, loose straw, and straw-clay.

Sheep's wool is an excellent natural insulator. It has a slightly higher R-value than fiberglass and is produced naturally with little, if any, fossil fuel energy. Wool insulates when it is wet, something you can't say about most forms of insulation! Wool is naturally flame resistant. Although it can be damaged by moths, it contains lanolin, a naturally occurring oil that protects wool insulation from moths. If the lanolin is left in place, wool insulation holds up well. Some homeowners add cedar shavings and mothballs to wool insulation for greater protection. (I'd be careful about adding mothballs; they contain VOCs that may be harmful to human health.)

Although wool insulation is not widely available, in New Zealand where sheep outnumber people by a wide margin, it is a standard building material. So, if you are in an area where sheep are abundant, you may be able to purchase wool insulation batts. For more on wool insulation, see Michael Smith's book, *The Cobber's Companion*.

Straw bales can also be used for insulation. Be sure to frame your ceiling adequately, however, because straw bales are heavy. Loose straw is used in some homes as ceiling insulation. Loose straw or flakes reduce the need to fortify your roof structure and may be a wise strategy in earthquake-prone regions—you don't want all that weight above your head. Unfortunately, loose straw poses a greater fire hazard than intact bales. For more information, see *Build it with Bales* by Matts Myhrman and S. O. MacDonald.

Straw-clay, made by mixing straw and a watery solution of clay, is used to build exterior walls (see chapter 10). It is packed into forms to create thick, insulated walls. Some builders use straw-clay for ceilings, too. The clay protects against fire and mildew.

Many other forms of environmentally sound insulation are available. To achieve the maximum amount of insulation in your walls and ceilings, it may be necessary to combine two or more types of insulation, such as loose-fill insulation in stud spaces with rigid foam on the exterior surfaces of walls.

One of the best choices in environmentally sound, loose-fill insulation is recycled paper, often referred to as recycled cellulose insulation. Newsprint and cardboard are shredded into a fluffy mass, then sprayed with boric acid to reduce mold and decay and to make it flame resistant and unattractive to insects. Some manufacturers add latex binders to reduce settling. (Settling reduces R-value.)

Insulation falls into three categories: loose-fill, batts, and rigid foam. As a rule of thumb, when installing insulation in an attic or a closed ceiling, loose-fill and batts are preferred. But when applying insulation externally, say on decking or external sheathing, foam board is the product of choice.

Cellulose insulation is very common today. Fed by the massive waste of society, recycled cellulose insulation manufacturers are putting mountains of discarded newspapers and cardboard boxes to use. Cellulose insulation is a little cheaper than fiberglass batts, but can be a little more expensive to install.

If you want an earth-friendly insulation batt, mineral-wool insulation blankets are made from 92 percent recycled mineral slag, ceramic tile, and rocks. Mineral wool "blankets" fit between wall studs or ceiling rafters. The manufacturer, Fibrex, Inc. in Aurora, Illinois, claims that mineral wool insulation is noncombustible and will not absorb moisture, rot, settle, or break down. You can read about it in John Hermannsson's book.

Rigid foam insulation can be used for a variety of purposes, such as exterior insulation on the walls, roofs, and foundations. It can also be cut in strips and placed in between studs in exterior walls. However, some manufacturers produce rigid foam insulation with hydrochlorofluorocarbons (HCFCs), mixed with the foam prior to extrusion. When the foam is extruded, HCFCs expand to form millions of tiny bubbles inside the foam. These tiny cavities reduce heat transfer. Although they are much less damaging to the ozone layer than CFCs, HCFCs do break down in the stratosphere where they react with ozone.

One product designed to eliminate this problem is Insulfoam, produced by a company of the same name in Aurora, Colorado. This forward-looking manufacturer uses steam rather than CFCs to generate bubbles in foam board. At least one manufacturer is making rigid foam insulation comprised of 50 percent recycled material.

Gardeners will recognize this next form of insulation, perlite. Perlite is a light, fluffy mineral derived from volcanic soils. It is added to potting soils to promote soil aeration. While rarely used to insulate homes, perlite is surprisingly effective. It is fairly inexpensive, fireproof, and easy to install. At least one manufacturer (Schuller International in Denver) is using perlite and cellulose fibers from recycled newspapers to make rigid insulation board.

Fiberglass batt is one of the most common insulation products on the market today. While inexpensive and effective, fiberglass may pose a health risk to workers who inhale the microscopic slivers of fiberglass dislodged during installation. These fibers may enter the lungs where they become embedded in the walls of the tiny air sacs known as alveoli. Lodged inside the lung, these daggerlike fibers irritate the delicate lining of the alveoli much like asbestos fibers. They may even shear DNA molecules in the nuclei of cells, causing mutations that lead to lung cancer. This problem can be prevented by wearing a respirator. Some manufacturers have dealt with it by sealing fiberglass batts in plastic. (Plastic also acts as a vapor barrier.) Other important advances are the elimination of formaldehyde (by Johns Manville Corporation) and the use of recycled glass, an alternative that costs about as much as—or maybe a little less than—standard fiberglass batts.

Cotton bats can be used for insulation, but cotton is grown with extensive use of pesticides and may affect chemically sensitive individuals.

Cotton is a more natural form of insulation. While not widely available, cotton insulation is manufactured in two forms: batts and loose-fill. Insul-Cot produces a batt made from cotton and polyester (a synthetic). Unfortunately, this product currently costs more than fiberglass batt insulation.

As environmentally appropriate as cotton may seem, cotton production is one of the most chemically intensive and environmentally harmful areas of agriculture. The soil is sprayed with herbicides to control weeds, and crops are sprayed with insecticides to kill insect pests—often many times during each growing cycle.

If you want to learn more about insulation, read *Home Insulation* by Harry Yost. For more specific information on these products, including phone numbers and addresses of manufacturers, check out the *Green Building Resource Guide*.

WINDOWS AND DOORS

Not too long ago, windows were something of a national disgrace. Today, thanks to advances in design and construction, windows are highly engineered products that play an important part in conserving energy. Modern windows are designed to reduce heat transfer (either into or out of a house, depending on the season) and air infiltration.

BUILDING NOTE

Don't skimp on windows. The window is the weak point in a wall in many respects, but especially in terms of energy loss. Even the best-insulated windows pale in comparison to a well-insulated wall. At most, they have one-third to one-fifth of the R-value of a wall.

One option for reducing resource use is to purchase used windows at local salvage yards or from remodelers. Although the use of recycled windows reduces our consumption of natural resources and decreases the cost of construction, they are not always a bargain. Salvaged windows may have been torn out because they were inefficient or were showing signs of deterioration. Don't buy someone else's discard if it is inefficient or beginning to fall apart. The money you save may be quickly lost by higher heating and cooling bills. You may even have to replace the window in a few years, adding further costs.

That said, if you find good, used double-pane windows, buy them. You could save an enormous amount of money, because windows constitute a very large budget item in many homes. Just make sure that the seals and frame are in good condition.

If you are going to purchase new windows, you have a lot of options to choose from in a wide range of prices. Fortunately, most modern windows are engineered to maximize energy performance. For example, most windows are double glazed. Between the two sheets of glass is a small air space that acts as insulation. The air space is often filled with an inert gas, notably argon, which further increases the insulative value of the window. In addition, clear coatings are available to reduce heat loss. Windows of this nature are called low-E windows (low emissivity) and are essential for optimal energy performance of homes in all climates.

Many manufacturers are offering "warm edges," another feature worth considering. Warm edges are created by placing a nonconductive spacer between the panes of glass around the periphery of the window. This improves the thermal performance of the window and reduces condensation around window edges. Moisture and ice buildup on windows may lead to serious problems. Melting ice drips onto window frames and sills, causing the wood to deteriorate. If your house generates a lot of humidity and is tightly sealed, warm edges will extend the lifetime of your windows.

The next question to answer is the kind of window frame material you should buy. Metal frames are an abomination. They conduct much too much heat out of a room. Don't even think about them. Wood windows are preferred because wood is a renewable resource and a poorer conductor of heat than metal. Some manufacturers are building frames from engineered wood products.

Wood window frames do have drawbacks. Rain and bright sun cause wood to deteriorate. To prevent damage, many manufacturers install metal or vinyl cladding on the exterior surface of the frame, thus making the window virtually weatherproof.

Some manufacturers have replaced the wood with vinyl, sometimes from recycled sources. Vinyl windows are cost competitive, extremely durable, and longlasting. In fact, some types of vinyl are resistant to ultraviolet light and therefore maintenance free. Because vinyl is impregnated with pigment, window frames require no painting. Vinyl is popular in Europe and Canada but is made from polyvinyl chloride (PVC), a known carcinogen. PVC poses a danger to workers, although most factories reduce exposure. I've never seen any evidence to suggest that PVC windows are harmful to homeowners, but one has to wonder.

Mikron Industries of Kent, Washington, has taken vinyl one step further by combining virgin and recycled vinyl with recycled wood fiber. This composite is now being used to manufacture frames for windows and patio doors. Unlike traditional vinyl window frames, the wood/vinyl composite is paintable without having to prepare the surface.

Fiberglass-frame windows offer many of the benefits of vinyl. Fiberglass has low embodied energy and is made primarily from silica, one of the world's most abundant resources. Some people think that fiberglass may become the window frame of the future. However, it is made using a fairly toxic resin. It doesn't present a hazard to the homeowner because the resin has evaporated by the time the window is shipped. However, it does pose a threat to workers. Moreover, fiberglass windows are not as widely available as wood or vinyl windows.

The most important considerations when buying new windows are the window's U-value and the amount of air filtration that will occur. U-value is a measure of heat transmission through a material, simply the reciprocal of R-value. Thus, an R-value of 2 would yield a U-value of 0.5. The lower the U-value, the greater its resistance to heat. As a rule, U-values of less than 0.3 (which is an R-value of 3.3 or better) are acceptable.

On cold days, the surface temperature of interior glass of a low-E window is 10 to 15 degrees Fahrenheit warmer than a double-pane window without coatings and 40 degrees warmer than a single-pane window.

U-values can be decreased by installing insulated window shades, curtains, or shutters (see chapter 11). Simply adding a third pane (made of plexiglass) to double-pane windows will increase the R-value. Although the third pane reduces the transmission of light, it increases heat retention even more. It is also very easy to do and inexpensive.

Another consideration is air infiltration. Air infiltration is determined by the quality of construction and the type of window you buy. Generally, the lower the quality, the greater the leakage. Windows fall into two groups: openable and nonopenable. As a rule, any window that opens will let more air in than a nonopenable window. For example, double-hung windows—that is, windows in which the bottom and top sections slide up—permit more air infiltration than casement windows, windows that open to the outside using a crank mechanism.

Choose openable windows that have the best U-values and permit the least amount of air to enter your home. And, whenever possible, install nonopenable windows. If your house has an open floor design and the windows are strategically placed, you can often reduce the number of openable windows without sacrificing ventilation. However, this strategy may not work well in more traditional designs with a multitude of isolated rooms. Hallways, walls, and doors can obstruct air flow. In such instances, you will need to install a couple openable windows in every room. Choose the most airtight models you can afford.

Proper installation is vital to reducing infiltration. Be sure to fill the gaps between the rough framing and the finished window frame with insulation or caulk or foam before installing window trim.

One of the best places to find interior and exterior doors is a construction salvage outlet or neighbors' garages or barns. You would be amazed at how many perfectly usable old doors are out there gathering dust in people's garages. With a little sanding and some paint or stain, you can refurbish an old beat-up door and save yourself lots of money. (I was even able to locate an expensive fire door to install between my garage and front entryway in a friend's shed!) Trying to find matching doors in sufficient number, however, may be a problem.

Solid wood doors are popular for interior use and very attractive. They are widely available, too, although I'm not aware of any that are made from sustainably harvested lumber. A certified wood dealer in your area may be able to recommend a manufacturer that uses certified lumber for building doors. If you are handy with a saw, you may want to make your own doors out of reclaimed or certified lumber.

Solid wood exterior doors are also available. They add a certain intangible quality to a natural home. However, new solid wood exterior doors tend to be pricey. Exterior doors take a beating in the weather, especially those located on the south side of a house.

BUILDING NOTE

When selecting interior doors, you may want to avoid hollow core doors. They are pretty cheap-looking and inexpensive. In a natural home, you will probably want the look of real wood, not that microscopically thin laminate that manufacturers use to produce hollow core doors for apartments and other similar applications.

Several manufacturers make interior and exterior doors out of engineered wood with foam core insulation. Shop around to see what is available. If you don't mind a painted door, no one will know the difference. Another option is the molded fiberglass door. Clearly a step or two lower on the sustainability scale than wooden doors because of the toxicity associated with their manufacture, molded fiberglass doors hold up well against sunlight. They are durable and well insulated.

Steel doors have high embodied energy, and steel production produces large amounts of pollution. However, like fiberglass doors, steel doors are durable and effective in difficult climates. The foam core provides adequate insulation against heat and cold. Steel doors can be ordered with vinyl or fiberglass facing with wood-grain texture.

SIDING

Many homebuilding techniques described in this book rely on plaster finishes for exterior walls. If you are building a conventional home or if you have decided that you want to install siding on your natural home instead of plaster, you'll have a few environmentally friendly options.

One of them is metal siding made from recycled steel or aluminum. Another option is hardboard panel, lap, and shake siding made from sawdust and plastic resin. This product is much denser than traditional wood siding and therefore resists cracking. Some manufacturers have added a small amount of cement to the wood-fiber mix to produce fire-resistant fiber-cement siding warrantied for fifty years. Molded fiber-cement boards can also be used for soffits, fascias, and gables. If you don't want to hassle with maintenance, this may be a perfect choice.

FLOORS (SUBFLOORING)

If your plans call for a wooden floor, you will first need to install subflooring over the floor joists. OSB and plywood are commonly used. (Remember to use low-formaldehyde products and seal them prior to installation.) You can also use one of the more earth-friendly products such as the structural insulated panels (SIPs), discussed earlier in this chapter. They are sturdy, competitively priced, and insulated. Evanite Fiber Corporation in Corvallis, Oregon, manufactures a hardboard underlayment or subflooring made from waste wood.

INTERIOR WALLS AND CEILINGS

Interior walls made from natural materials such as cob, rammed earth, and adobe bricks provide the thermal mass needed in passive solar homes. The sculptability of cob lends itself to graceful, flowing walls. If you want conventional walls, however, you can make them from environmentally sound materials, for example, recycled wood or engineered lumber.

Sure some healthy, environmentally friendly building materials can be expensive, but so's chemotherapy.

CEDAR ROSE

Environmentally friendly alternatives to these common products are currently available, although they are usually more costly:

Wood primer
Drywall primer
Metal primer
Enamel paints
Latex paints
Shellac
Lacquer
Wood sealers, including deck sealer
Driveway sealers
Foundation sealer
Concrete, brick, and masonry
  sealers

In England and Germany, many builders use drywall (also known as sheet rock) manufactured from sludge produced by pollution control devices (smoke stack scrubbers) on coal-fired power plants. (Sludge contains gypsum.) In the United States, GP Gypsum Corporation in Atlanta manufactures drywall containing recycled and byproduct (waste) gypsum. The facing paper is made of recycled newspaper and cardboard. Louisiana-Pacific in Portland, Oregon, sells a fiber-reinforced gypsum wallboard. It is made from recycled cellulose fiber (from telephone books and newspapers), recycled gypsum, and perlite. This product is manufactured so that panels actually lock together, thus eliminating the need to tape and mud seams. Although this saves an enormous amount of time, L-P's environmentally friendly drywall costs two to two and a half times more than standard drywall and must be shipped from the factory in Nova Scotia, which adds to its cost.

While we are on the subject of drywall, you may want to look into a product known as The Nailer. The Nailer is a gypsum board clip produced by The Millennium Group in Waterloo, Wisconsin. Made from recycled plastic (high-density polyethylene, the same as milk jugs), this device substantially reduces the use of lumber in framed walls, cutting the number of studs by 30 to 35 in a 180-foot-long wall (which typically requires about 90 studs).

PAINTS AND FINISHES

Paints and other finishes (like stains and varnishes) are some of the most notorious violators of indoor air quality in our homes. One reason for their bad reputation is that paints, stains, and other finishes often release (outgas) substantial quantities of volatile organic chemicals (VOCs). Because of outgassing, the early months of inhabitation can be taxing, indeed even harmful, to one's health. For people who are chemically sensitive, the effect can be devastating.

To minimize, or better yet, eliminate the outgassing problem, a number of companies are producing low- and no-VOC finishing products. Virtually any commercial product that is stocked on the shelf of local paint stores—even varnishes and stains—can now be replaced by a low-impact, people-friendly alternative (see sidebar). In addition, most alternative paints and finishes are water soluble, which makes for easy cleaning. Although oil-based paints are still available, concern about their impact on health and the environment are causing the paint industry to phase them out. They'll largely be replaced by latex paints, even for outdoor applications. Oil-based products that do exist often contain natural oils, such as hemp seed oil. For a listing of alternative paints, stains, and finishes, see Austin's Green Building Program Web site, Hermannsson's book, and the Building for Health Materials Center catalogue listed in the resource guide.

To protect yourself and your workers, the very least you can do is to purchase a low- or no-VOC latex paint. Glidden's Spred 2000 is a low-VOC paint. It is readily available in the U.S. and other countries. I've seen it in large homebuilding outlets including Home Base and Home Depot, independent paint stores, and many of the large discount stores such as Kmart—and at fairly decent prices. Another paint to look for is Benjamin Moore's no-VOC paint, Pristine.

Low- and no-VOC paints are great, but many of them still contain mercury or lead. To address this drawback, manufacturers such as Wellborn Paint in Denver and Miller Paint Company in Portland, Oregon, produce low or no-VOC paints made without lead or mercury.

Unbeknownst to many, paints contain chemicals to extend their shelf life. Many paints also contain small amounts of chemical fungicides and mildew-cides. Although the concentrations of these chemicals are relatively small and are believed to pose no health risk, you may want to order biocide-free paints, especially if you are chemically sensitive. Check out the SafeCoat paints and stains produced by AFM Enterprises in San Diego. The company sells a complete line of people-friendly paints, stains, and finishes, as well as sealers, cleaners, and adhesives. Their products contain no formaldehyde, fungicide, or mildewcide, and meet the strictest VOC emissions standards currently in place. Although they cost more, they may be worth the extra expense. Be sure to check out BioShield paints, too. They contain no formaldehyde, heavy metals, or biocides. Other manufacturers are listed in Hermannsson's book and on the Austin Green Building Web site.

For those who want to use all-natural products, the Old-fashioned Milk Paint Company in Groton, Massachusetts, is a family-run company that produces paints from milk protein (casein). Their products contain no biocides, are popular among chemically sensitive individuals, and are available in sixteen colors. Schools and hospitals buy a significant portion of their products.

Unlike other paints, milk paints are shipped in powder form. Four kilograms of the powder, or 8.8 pounds, mixed with water make a gallon and a half of paint, costing about $25. According to John Hermannsson, milk paint tends to be flat and streaky. Walls should be coated first with a water-based sealant to prevent water stains.

Another line of products worth strong consideration are the oil-based paints made by Livos. Don't be alarmed. I did say oil-based, but what I meant was oil from orange peels. Livos manufactures a complete line of paints from undercoat or primer to wall paint

BUILDING NOTE

Large paint companies including Benjamin Moore, Glidden, Kelly Moore, and Sherwin Williams have all come out with environmentally safer paints. But chemically sensitive people may still react to low-VOC paints. According to one source, only one paint has been certified to be free of VOCs: ICI's Lifemaster 2000. If you're chemically sensitive, be sure to test all products, even the so-called VOC-free paints. John Harris, chemist and co-owner of Belcaro Paints in Denver recommends that his customers sleep with a product sample by their bed to determine whether they will react.

BUILDING NOTE

VOCs escaping into the atmosphere from paints and other sources react with sunlight and nitrogen dioxide to produce ground-level ozone. This harmful chemical damages lungs and materials such as rubber. According to the U.S. Environmental Protection Agency, VOCs from paints and finishes contribute an estimated 9 percent of the VOCs that cause ground-level ozone production. VOC concentrations are ten times higher in indoor air than outdoor air. Right after a paint or finish has been applied, indoor levels are 1,000 times higher. Levels the first four days are the highest.

The American Lung Association notes that VOCs cause a number of physical ailments. They irritate the eyes, skin, and lungs. They cause headaches, nausea, and muscle weakness, and may damage the liver and kidneys. They are especially hazardous to the elderly, pregnant women, young children, and people with weakened immune systems or environmental sensitivities.

When using a product with VOCs, be sure to provide adequate ventilation and air out the building afterwards, especially during the first four days.

JEANNIE SHORTRIDGE, "Low-Toxic Natural Paints and Pigments," *Natural Homes* 1 (20: 48-54).

BUILDING NOTE

Milk protein-based paints have been used for many centuries to beautify and protect walls.

BUILDING NOTE

While using recycled paint is good way to reduce waste and environmental contamination, recycled paint is typically made from standard latex paint containing VOCs and mercury. If you are going to use reclaimed paint, stay out of the house for a week or two while it dries.

for interior and exterior uses. The company even sells a natural organic paint stripper.

Another option is reclaimed paint. The Green Paint Company in Manchaug, Massachusetts, manufactures paint that contains 90 percent post-consumer waste—that is, it contains 90 percent reclaimed paint. Reclaimed paint isn't stripped from walls with a putty knife, then reconstituted. No, it comes from local hazardous waste pickup sites operated by private companies and government agencies. Rather than be disposed of, "discarded" paint is filtered, then mixed with virgin paint, repackaged, and sold to willing customers.

The Green Paint Company offers a variety of reclaimed paints, including interior and exterior latex and exterior oil-based primers and paints. Although they only distribute in New England, you can order directly from the company. Even though this will increase the cost, the paint sells for 40 percent less than virgin paint. If you live in the Pacific Northwest, check out the recycled latex paints sold by Rasmussen Paint Company in Beaverton, Oregon. You may find a recycled paint manufacturer in your area, too.

Most interior and exterior paints are water-based. In some circumstances, however, builders prefer to use oil-based paints. When this is the case, select a product that is not formulated with formaldehyde, mercury, lead, cadmium, chromium, or oxides of these metals. Select a low-VOC paint as well. (VOC levels should not exceed 380 grams per liter.) Be sure the paint does not contain any halogenated solvents, either.

Another area of improvement in recent years has been in the production of stains, varnishes, and sealers. Although most products in use today contain a variety of toxic substances, including some pretty heavy hitters such as acetone, lead, and pentachlorophenol, those of us who prefer to minimize our exposure to toxic substances have a number of environmentally friendly stains and varnishes to choose from. Although they cost two to two and a half times more than over-the-counter products, they are extremely pleasant to work with.

Environmentally friendly stains and finishes can be purchased from some local retailers, such as Boulder, Colorado's Planetary Solutions, or from distributors such as Eco Design/Natural Choice, which is located in Santa Fe, New Mexico. This company offers a complete line of wood finishes for allergy-prone and chemically sensitive individuals. One of my favorites is a Livos product known as Kaldet. Manufactured in Germany, this citrus-based finish is used in place of highly toxic varnishes. Although it is a lot trickier to work with and requires a long time to dry, Kaldet gives off a very pleasant odor as it dries and is a delight to work with, especially if you have ever applied any traditional finish products. I have used it to finish tables, wood trim, and doors. Another product that is also a bit tricky to use but safer than standard finishes is a water-based urethane.

FLOOR COVERINGS

The sustainable homebuilder has many options for environmentally responsible flooring, some well within the price range of conventional products that have a much greater environmental impact.

Wood and Bamboo Flooring

If you have your sights set on a wood floor, you can purchase reclaimed or recycled wood. Hermannsson's *Green Building Resource Guide* lists nine companies in various parts of the United States that offer this product. Albany Woodworks, in Albany, New York, acquires old wood flooring from early American buildings that are being torn down to make way for that dubious thing we call "progress." The wood is remilled and sanded to look like new. Because this process is fairly labor intensive, reclaimed wood can be pretty costly. Pine, hemlock, chestnut, oak, and heart pine floorboards acquired from nineteenth-century barns are available from Conklin's Authentic Antique Barnwood in Susquehanna, Pennsylvania. It is pricey, too. All in all, you can expect to pay 20 percent to 200 percent more for reclaimed wood.

Another option, of course, is to use wood from sustainably harvested forests. Eco Timber in Berkeley, California, is one vendor you may want to contact. Their wood is priced quite competitively with wood from not-so-sustainably harvested forests.

If you like hardwood flooring, but want an environmentally sustainable product, one option is bamboo flooring (figure 14-13). Discard your notions of what bamboo looks like, however. Bamboo flooring from TimberGrass in Bainbridge Island, Washington, is a laminated floor product that comes in precision-milled 3%6-inch-wide, 6-foot-2-inch-long strips. It is 25 percent harder than red oak and 12 percent harder than rock maple flooring and is more dimensionally stable than its competitors, meaning it won't expand and

Plyboo America, Inc.

FIGURE 14-13.
Bamboo flooring is durable and attractive. It is made from a highly renewable material that can be grown and harvested in a sustainable manner.

contract as much as standard oak or maple floors. Bamboo flooring from this company is laminated using environmentally safe adhesives and comes in tongue-and-groove planks. It accepts stain, paints, and varnishes. Moreover, this highly renewable material is harvested from the same plants every four years, as opposed to oak and maple trees that are cut down once after 50 to 120 years of growth. Bamboo is not only renewable, it has a rapid renewal cycle. (For more on bamboo, see chapter 10.)

Tile and Vinyl Flooring

Tile is a popular material for earth-friendly homes. It is attractive, durable, easy to clean, and serves as thermal mass in passive solar homes. To obtain the solar benefit, however, choose tiles of a darker color, such as terra cotta.

INTERVIEWS WITH INNOVATORS

David Adamson

How long have you been involved in marketing green building products?
Since 1991.

What got you interested in this endeavor?
I worked for ten years in the environmental nonprofit world and for a senator. I helped to found a large statewide volunteer organization (Volunteers for Outdoor Colorado) and spent five years with the Nature Conservancy. Nevertheless, I sensed that in the nonprofit and political world, we were only reacting to the biggest source of human environmental impact: the economy.

I wanted to get at the source of the problem. I wanted to help give consumers choices that minimized destruction and perhaps even helped restore natural systems. I

picked working with building materials first because construction is arguably the sector with the biggest environmental impact: building construction, remodeling, and home operation consume 40 percent of all natural resources utilized by the U.S. economy.

There are so many ways to use green building materials to make homes, offices, and entire cities more alive, more productive, and more healthful places. I knew there would be a demand for these materials.

Would you say it is easier or more difficult now for people to obtain the products they need to build a truly sustainable home?
Green building products are definitely more available and less expensive than they were a decade

ago. The quality of products is increasing as well.

Can individuals influence the green building market?
Consumers wield enormous power when they tell builders and manufacturers that they want to use green products such as third-party-certified sustainably harvested lumber, or less toxic paints and finishes, and maybe even would forego some square footage in return for more investment in the energy efficiency and the quality of the interior finishes.

The building industry is more resistant to innovation than many, and for some good reasons. It's important to see how materials and systems perform over time . . . but the bottom line is, the preference of consumers is the most concrete thing a builder or supplier has. And they will respond to this.

Tiles are being made from a variety of unlikely materials, including recycled auto and airplane glass, feldspar waste from mining, and recycled waste tile generated at manufacturing plants. Prices for these products vary greatly, with the price index of most falling within the 1 to 1.7 range; a few are as high as 4 to 7. So shop carefully!

Sheet Flooring

You may prefer a linoleum-type flooring for some rooms in your house. Once again, you have a fair number of natural and environmentally friendly products to choose from. DLW Gerbert in Lancaster, Pennsylvania, and Forbo North America in Hazelton, Pennsylvania, both manufacture a competitively priced natural linoleum made from linseed oil, pine tree resins, wood flour, and

How many retail outlets are there in the United States?

I know of approximately six green lumberyard/home center outlets in the country and five or six green interior-related outlets. On the product side, there are many, many choices. The *Environmental Building News* Product Specifier lists over 1,800. I am working very hard to create a single national source of the best environmental building products.

What if I don't live near a retail outlet? How can I get green building products?

We and a number of other outlets ship nationwide. I suggest consumers or builders give us a call or call our friends at the Environmental Home Center in Seattle: 800-281-9785.

Will it cost someone more to build an environmentally sustainable home?

If you compare green materials with the cheapest materials, then yes. Green building materials cost more than these because they are generally quality materials. But if you are comparing the same quality of materials, then green building materials are the same price, or perhaps 5 to 10 percent more. But these green materials also have advantages of health, low maintenance, durability, or energy savings. For example, our 100 percent recycled wood composite decking made from recycled wood and milk jugs costs 10 percent more than redwood, but you never have to stain it and it's guaranteed for twenty years. And it's recyclable back into itself. Recycled "pop bottle" carpeting is the same price as comparable-quality nylon carpeting and it's more stain resistant.

Is the green building movement strong elsewhere, for example, in Europe or Australia?

Europe and Canada are certainly leading the U.S. in official initiatives. For example, a Canadian company, Enermodal Engineering, has a great Web site supported by the Canadian government (www.advancedbuildings.com).

What are common mistakes people make when pursuing this option?

- Assuming that green building products will cost more and therefore not pursuing them.

- Believing underinformed architects and builders who say "that kind of thing" is not practical or affordable.

- Failure to call us to help them (seriously. There is a lot of help available).

DAVID ADAMSON is an energetic and passionate spokesman for green building. He has worked for Planetary Solutions and EcoProducts, both green building materials retailers, and now owns and operates a company called EcoBuild in Boulder, Colorado.

cork. I have used it in my bathrooms and have found it works quite well. Because it is thicker and stiffer than conventional linoleum flooring, it is more difficult to install. Retailers recommend hiring an expert who has experience with this particular product.

Numerous manufacturers also produce recycled rubber and PVC sheet flooring primarily for outdoor use or high-volume traffic areas. If you have ever visited Universal Studios, Disney World, or Busch Gardens in Florida, you have probably walked on this product. Church and skating rink entryways are often covered with this material. Most homeowners won't have much use for such products, except in patios, front porches, mud rooms, breezeways, or workshops.

Carpet and Carpet Pad

If carpeting is going to find its way into your home, breathe easy, many manufacturers are producing environmentally friendly carpet pads and carpeting made from either natural or recycled materials. Several manufacturers also produce low-toxicity carpet adhesives.

For some reason, carpet pad manufacturers have really jumped on the environmental bandwagon. They have produced a wide assortment of environmentally friendly carpet pads (also known as carpet underlayment). Dura Undercushions in Montreal, Canada, manufactures a pad made from 92 percent recycled tire rubber. Fairmont Corporation in San Diego sells a carpet pad made from recycled polyurethane. Sutherlin Carpet Mills sells a pad made from recycled nylon, which is ideal for chemically sensitive individuals. Homasote Company in Trenton, New Jersey, offers a carpet pad made from 100 percent recycled newsprint. Reliance Carpet Cushion in Gardena, California, sells carpet pad made from textile waste and recycled carpet fibers. These carpet pads vary in price. Several are cost competitive with conventional materials while several others are actually cheaper than traditional underlayment.

Carpet manufacturers have also been innovative in the name of the environment and personal health. Several are now producing carpet made from wool. Although wool carpeting costs at least twice as much as conventional nylon carpeting, most likely much more, the investment may be excellent if you are chemically sensitive.

Several companies are manufacturing carpeting that is made from 100 percent recycled PET plastic, the type of plastic used to make plastic soda bottles. I have used this carpet with great success. It is durable, easy to clean, and stain resistant. In fact, one manufacturer claims that recycled PET plastic carpeting does not require stain guard chemical treatments like those used to prevent stains in conventional nylon carpets. Further adding to their appeal, recycled PET plastic carpet is cost competitive with standard carpeting. The price index for Image Carpets in Armuchee, Georgia, is a rather impressive 0.3 to 1.0, meaning you can purchase the carpet at the same price or up to 70 percent cheaper than typical nylon carpeting. Also, if you are interested in recycled carpeting, check out Blue Ride Carpet Mills in Ellijay, Georgia, which sells a brand of carpet made from 100 percent recycled nylon. Check out carpet retailers in your area; Hermannsson's book has a good list.

As you can tell from reading this chapter, the sustainable homebuilder has a cornucopia of options. Virtually every product currently used in conventional homebuilding can be replaced by one that is recycled or natural or offers these and other environmental and health benefits. As I've said several times before, however, check out the performance of alternative building products before you buy them. Make sure they really work as well as the manufacturers or retailers claim!

MOVING TOWARD A CARBOHYDRATE ECONOMY

For years, the world economy depended almost entirely on plant matter, notably carbohydrates. Rope was made from hemp rather than plastic or other synthetic materials such as nylon. Most chemicals came from plants as well. Even soaps, resins, textiles, and inks were provided courtesy of the plant kingdom. And of course, building supplies have long been derived from trees and other members of the plant world. Then came oil. Researchers found that petroleum actually contained dozens of organic chemicals that could be isolated and used to make a wide assortment of products such as plastics and pharmaceuticals, not to mention diesel fuel, gasoline, and jet fuel. It wasn't long before the oil-based economy supplanted the botanical economy of years past.

As concerns about the environment rise and oil supplies dwindle, more and more people are looking for a resurgence in the plant economy. Hemp may be one of the main contributors to this revival. It has long been used as a food for animals. Most bird seeds contain hemp seeds. Hemp is also used to make a high-protein seed cake for cattle, chickens, and household pets. Hemp is a source of food for people. With a versatility as remarkable as soybeans, hemp is now being used to produce nondairy cheese, milk, and ice cream. Hemp seed cake is added to bread, pastas, cakes, and cookies. It is being used to make kitty litter and a superabsorbent bedding for horses and smaller pets, such as guinea pigs, rabbits, and hamsters. The seed can be crushed to liberate oil, as well. BMW is using hemp fiber in its airbag system, and current research shows that industrial hemp can be used to make floor mats, seat covers, and gaskets. Hemp oils have been incorporated in a variety of personal care products such as soaps, shampoos, lip balm, massage oil, and perfumes. Some manufacturers use it to make paper. And, of course, hemp is making its way into the building industry. It is used to make fiberboard, paints, and sealants. Much to my surprise, I recently found out that Sherwin Williams paint company once used hemp oil in its paint products. In fact, until the 1930s, linseed and hemp oils made up the majority of all resins, paints, shellacs, and varnishes sold in the United States.

Although our overseas friends have no problem with hemp, the U.S. government seems quite reluctant to forsake its paranoia and fears about industrial hemp. They still view it as a possible source of drugs. However, as *Hemp Horizons* author John Roulac notes, "if one were to roll leaves from an industrial hemp plant into cigarettes and smoke them, no euphoric effect would be experienced, even if a thousand hemp cigarettes were smoked." Dr. William M. Pierce, Jr., of the Department of Pharmacology and Toxicology at the University of Louisville School of Medicine says, "It is absurd . . . to consider industrial hemp useful as a drug. While a person could choose to use hemp in this way, it is unlikely that he or she would repeat the behavior, due to the unpleasant side effects [headaches]." On the subject of the U.S. government's intransigence on the issue, Roulac says, "Should garden poppies be plowed under because another closely related member of the genus is grown for heroin?"

Hemp grows quickly, has many practical uses, grows in poor soil, thrives in most climates, resists disease and insects, and discourages weed growth even after it has been harvested. This last feature makes hemp an excellent rotational crop. Furthermore, hemp adds nutrients to the soil. Research studies show that growing wheat after hemp actually increases wheat production. If you want to learn more, get a copy of Roulac's book, *Hemp Horizons*.

Hemp is not the only plant with useful applications. Expect others such as jute, flax, kenaf, ramie, and urena to make a contribution to a carbohydrate economy. Like hemp, many of these plants grow rapidly—and some even seem to thrive in poor soils, a feature that could help us reclaim agricultural land that has been ravaged in the past by inferior farming practices. Kenaf, a plant native to Africa, reaches harvestable size in 120 to 150 days. Grown commercially in Switzerland, China, India, and the United States, at least on a small scale, kenaf can be used to make paper and several building products.

In this chapter, I have touched on many of the most important green building products on the market today. You will be amazed by the variety and quality of these products. And you will probably be amazed to learn that there are many more products to choose from. For example, you can purchase nails made from recycled scrap steel from Maze Nails in Peru, Illinois. You can buy environmentally responsible fabrics for window treatments (what we used to call curtains in the old days before the advent of the interior designer). If you think there might be a more environmentally benign substitute for a conventional product, you are probably right. It may cost more or may be hard to get, but it is out there. As we move into the twenty-first century, these products will only become easier to locate and purchase—especially if people like yourself use them.

15

SITE CONSIDERATIONS
Choosing a Site, Protecting It during Construction, and Restoring the Land

OW THAT YOU HAVE SEEN what it takes to build a truly sustainable and largely independent home, it is time to think about the site, the piece of ground that will be home to your home. Do not regard this late coverage as indicating a lack of importance. In reality, siting is one of the most important aspects of sustainable design. It is as important to sustainable design as a keystone is to an arch.

I have saved the topic for last so that by the time you come to it you better understand the complex demands a home—with its various natural life support systems—places on us and the site. With an understanding of this subject, you should be able to choose a site that will work optimally with your home, allowing you to meet your needs while preserving the environment.

Unlike the modern home, which is placed just about anywhere thanks to the brazen power of large earth-moving machines, human arrogance, and insensitivity to the landscape, the sustainable home is nestled in the terrain. This harmonious relationship between the house and the land is critical if we are to live as impact-free and as self-sufficiently as possible.

The difference between an energy-efficient, water-conserving, nontoxic landscape and a conventional one is the difference between working with nature and fighting it. Not surprisingly, working with nature has its advantages: It can save thousands of dollars in utility bills, greatly reduce yard maintenance, increase your enjoyment of your property, and lessen your negative impact on the Earth.

ANNE SIMON MOFFAT,
MARC SCHILER, and the Staff of Green Living,
Energy-Efficient and Environmental Landscaping

This chapter also examines site protection: what you can do to protect a site during construction. Finally, we'll investigate one of the very last issues you will address in building your home, landscaping. Let's begin with site selection.

SITE SELECTION: DOS AND DON'TS

To find an appropriate building site involves many practical decisions. The most common ones have to do with proximity to work, schools, stores, and recreational opportunities. People also frequently consider the location of hospitals, police, and fire protection. Of course, affordability plays a key role in the decision as well.

To build a sustainable home, the list of site selection criteria must include an evaluation of solar access, prevailing wind patterns, rainfall, and water supplies, as well as analysis of the soil type, forest cover, and drainage patterns.

In *Adobe: Build it Yourself*, Paul G. McHenry, Jr., writes, "A cold-blooded analysis should be made before purchasing any particular parcel of real estate." If you already own a large piece of land with several potential building sites, analyze each location to determine which would be best. Bear in mind that a good site will provide many ecological "services" that could add up to thousands of dollars of savings in engineering, energy use, excavation, and drainage. Careful site selection, therefore, not only saves you a lot of money, it can make your life much more pleasant and will lessen your impact on the environment.

Select a dry, well-drained site. Although every aspect of siting is important, one of the most important is drainage. Select a well-drained site—one that will move water away from a house on its own without extensive grading. Natural drainage relies on slope and/or porous soils that permit water to escape either above ground or underground. Clayey soils can be a nightmare.

A well-drained site will save you work, money, materials, and repairs, as demonstrated by a development in Woodlands, Texas (near Houston), that relies entirely on natural drainage. Believing in the need to design with nature, developer George Mitchell built houses and roads in this subdivision on high ground, leaving the natural drainage intact. In so doing, he created an attractive open space that serves as a natural conduit for water. He didn't need to install a storm sewer, which saved him $4 million! Shortly after the development was completed, the natural drainage system was put to the test, as heavy rains began to fall. In nearby developments where little regard had been given to natural drainage, water flows increased by 180 percent, causing considerable flood damage. The streams in Mitchell's development swelled by only 55 percent and no serious flooding occurred.

If the soils drain poorly in the area where you are planning to live, place your home on a slope that will allow natural drainage. If that is not an option, elevate your home through careful grading. Provide drainage around the foundation, too, if necessary.

A dry, well-drained site is essential for several reasons. Water around a foundation may leak into basements, causing them to flood. In addition, in temperate and cool climates water can freeze during the winter. Water expands when it freezes, causing the wet soil to expand. This, in turn, may crack the foundation. Moisture in the summer can cause underlying clays to expand as well, also causing cracking.

Avoid marshy areas. When searching for a site, avoid marshy areas and depressions that may accumulate water and serve as breeding grounds for mosquitoes. Mosquitoes make lousy neighbors—and so do noisy red-winged blackbirds. More important, marshes are valuable wildlife habitat and are swiftly declining in number throughout the world. Development can damage wetlands and in some cases destroy them entirely.

Although wetlands are protected by law in the United States, they are often filled by individual homeowners and developers. Homeowners generally do so with impunity. Developers, however, are required to replace the natural wetlands they destroy with artificial wetlands. Although efforts are made to create complex assemblages of plants that reflect the biological diversity of natural wetlands, efforts often fall short. The result is a little like replacing a rich tropical forest with a tree farm.

In other cases, developers offset the losses by protecting existing wetlands threatened by other forces. Yanty Marsh along the shores of Lake Ontario near Hamlin, New York, is a good example. Walking the site in the summer of 1999 with Ed Evans, a former science teacher of mine who had rallied forces to protect the marsh, I could see how close we came to losing the marsh forever. Rising water levels in Lake Ontario caused by natural and human forces (notably maintaining high water levels in the locks to facilitate ship transport) were rapidly eroding the slip of land that separated the marsh from the cool

Edward D. Evans, Jr.

FIGURE 15-1.
This large rock barrier, the Yanty Marsh Revetment, protects Yanty Marsh along the coast of Lake Ontario, from rising water levels caused by human and natural forces. Wetlands are a vital ecological resource that have declined dramatically in the United States and other countries as a result of agricultural development, home construction, and other factors.

waters of Lake Ontario. With Evans' insistence, an artificial barrier was installed. Funding was provided by the N.Y. Department of Transportation, which needed a project to mitigate wetland losses it had caused elsewhere. Their efforts have saved a valuable marsh that would have been destroyed by the invasion of the cold waters of Lake Ontario.

Destroying a wetland and then protecting another, while laudable, still results in a net loss. And despite what critics say, wetlands are vital to the health of our planet. Too many of them have succumbed to human development. Therefore avoid building on or near wetlands.

Avoid natural hazards. Another very important, though frequently ignored rule is to avoid building in the path of natural hazards. One of the most common errors is building in flood plains. Flood plains are strips of land, sometimes very large strips, that border rivers. They are called flood plains for a good reason: they periodically flood. Excess rain or sudden heavy snow melt can transform a meek, mild-mannered stream into a raging torrent. Making matters worse, human activities may exacerbate flooding. Loss of vegetation due to homebuilding, road construction, and farming increase surface flow. The less vegetation, the less water percolates into the soil and the more runs off into streams.

In arid regions, one common mistake people make is building in or near arroyos, dry gullies. Even though an arroyo may be dry most of the year, or may be dry many years in a row, heavy rainstorms can transform a parched, innocent-looking gully into an ugly, raging torrent that will wreck a home in a flash . . . flood that is. Paul McHenry tells of a church school in Bernalillo, New Mexico. The structure was built many years ago in a "dry" arroyo—a gully that hadn't seen running water for decades—so many years, in fact, that there wasn't a single person in town who could remember there ever being water in it. Unfortunately, all that changed one afternoon as a sudden, heavy downpour erupted. The result was devastating. A flash flood swept through the arroyo and the building so innocently built in the path of destruction was demolished.

In his book *The Control of Nature*, John McPhee tells many stories with a hauntingly similar outcome. McPhee chronicles a tale of people doing battle with mud and water in the San Gabriel Mountains east of Los Angeles. In this region, heavy rains often wash mud from hillsides, burying houses or sweeping them off their foundations and crushing them in a wall of rock and debris. One man reportedly went to retrieve the morning newspaper, heard a sound, turned, and died of a sudden heart attack as a mudslide crushed his house with his wife and two children inside. (Figure 15-2 shows a similar mudslide in Stafford, in northern California.)

Mudslides are particularly troublesome after fires or clear-cuts have devastated trees and vegetation in the watershed. Without such a protective skin,

FIGURE 15-2.
Mudslides like this one in Stafford in northern California and the recurrent slides in the San Gabriel Mountains west of Los Angeles annually claim many lives and much property. Stay out of such areas.

hillsides around these expensive homes are especially vulnerable to the forces of nature.

Because environmental destruction—either natural or human-caused—miles away from a home site can have serious impact on the land you plan to build on, check out drainages and watersheds. A careless developer or improper clear-cutting or overgrazing can enhance flooding in nearby areas.

Water and mud from natural events, forest fires, clear-cuts, and careless development are not the only problems you will have to explore. In mountainous areas, be on the lookout for avalanches. Hard as it may be to believe, builders in the United States and Europe still construct homes in or near avalanche shoots, especially around ski resorts where good building sites are rare. When conditions are right, huge chunks of snow may break loose. Barreling down mountains like a freight train, they recruit additional force, crushing some homes and sweeping others off their foundations. Studies show that deforestation in mountainous areas also increases the risk of avalanches.

Choose a sloping site for earth-sheltering. One of the least-used, but most important ways of building an environmentally sustainable home is to earth-shelter it. Earth-sheltering varies by degree from berming—simply pushing dirt against waterproofed walls—to a nearly complete burial—that is, covering the structure with dirt, all except the south-facing wall. This allows you to create a bright interior environment as far from the cave as modern humans are from Neanderthals. (For more on earth-sheltering, see chapters 4 and 11.)

Sloped land enables you to earth-shelter your home more easily than flat terrain. It reduces the amount of excavation and dirt moving. If you are planning to

There simply isn't anything like earth cover to endow a building with a sense of eternal appropriateness.

MALCOLM WELLS,
The Earth-Sheltered House

FIGURE 15-3.

An earth-sheltered home on a south-facing slope is ideal for energy-independent living. South-facing slopes in the Northern hemisphere tend to be warm and cozy in the winter. Watch out for large expanses of unprotected roof glass, as in this house; while beautiful, they can overheat in the summer!

build a passive solar home, the land should slope toward the sun (south in the Northern Hemisphere, and north in the southern part of the world).

Not only is it easier to earth-shelter a home on sloped property, it is also more aesthetically appealing to dig into a hillside than the alternative required on flat terrain: piling dirt up against the walls. The earth-covered roof is planted with an assortment of grasses, wildflowers, and even garden vegetables, resulting in a structure called a living roof. The living roof reduces the amount of land taken out of service by homebuilders. Malcom Wells, an architect who has designed many earth-sheltered structures, notes in that "all construction causes land damage, but underground architecture can heal the wounds, and in many cases can improve the health of the land, by getting the recuperation off to a good start." In addition to being easier to excavate, south-facing slopes stay warmer in the winter, which can translate into sizeable energy savings and greater comfort (figure 15-3).

Choose a site with good solar access. As noted in chapter 2, solar energy is vital component of a sustainable future. Solar can be used to heat homes in the winter and to provide electricity year-round (chapters 11 and 12). This free and abundant resource can provide a tiny bit of the energy you need, or all of it, depending on your house design and location.

If you live in the Northern Hemisphere, your land should open to the south, providing access to the sun for as many hours as possible, but especially between 10 A.M. and 3 P.M. (In the Southern Hemisphere, of course, you need land with access to sun in the north.)

When considering a site, take a compass with you to determine where magnetic south really is. In the United States and Canada, true south varies between 0 and 22 degrees (east or west) of magnetic south (the measurement your compass will give).

After you have determined true south, determine the path the sun takes through the sky during the year. Do this by finding the highest (summer) and lowest (winter) trajectories, a topic discussed in chapter 11. Pay close attention to obstructions such as trees, cliffs, mountains, and other homes that could block the sun during peak hours. You need full access to the sun year-round if you are going to heat your home with sunlight and install PVs. If you are building a passive solar home, be certain there are no obstructions during the winter, spring, and fall. Be mindful of neighboring lots. Future construction could someday block the sun.

If you aren't sure how the sun "moves" during the seasons, start studying it now. Take notice every day where the sun rises and falls. Pay special attention on December 22 (the shortest day of the year and also the day on which the sun is the lowest in the sky) and June 22 (the longest day of the year and also the day on which the sun is the highest in the sky). Use this information to determine the suitability of your site. Detailed advice on tracking the sun is presented in chapter 11.

Study wind currents and air drainage. Wind can be a benefit as well as a detriment in siting your home. If you are planning to install a wind generator or a wind-powered water pump, you will need a site with good, reliable winds. But remember, winds may also strip warmth from a house, making it harder to heat and less comfortable.

If you are smart, you can have your cake and eat it too. That is, you can have a wind generator to provide electricity, yet protect the house from cold winter winds that will rob you of the heat you need to stay warm. One option is earth-sheltering. Another is to locate your house on the downwind side of a grove of trees, preferably coniferous trees. To capture wind energy at the same time, however, you will need to have a free, uninterrupted wind space, as discussed in chapter 12.

FIGURE 15-4.
Cold air collects in depressions known as frost pockets. Don't build in them or try to grow a garden or shelter animals in one. This house has been built to deflect cold air sliding down the hill up and over the house.

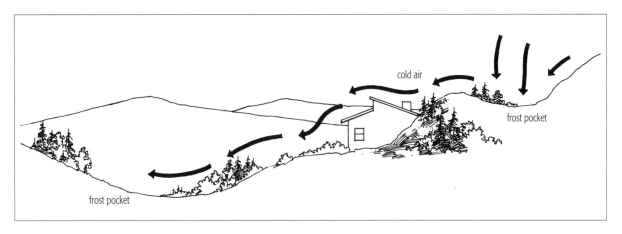

Wind is tricky. Although the prevailing winds in your region may flow from the southwest in summer, they may come from the north in winter. Or they may take alternative routes, channeling through topographical wind tunnels formed by valleys. Take time to study wind patterns in your area. When purchasing land, you may be unable to adequately assess wind potential, but you can at least identify potential obstructions.

Cold air pockets are another important phenomenon to consider when choosing a site. Cold air is more dense than warm air and therefore tends to accumulate in valley floors. On clear nights, cool air flows down slopes like a liquid, collecting in low points. Be on the look out for potential cold air pockets, also known as "frost pockets" because they are the first places to freeze. They make lousy locations for homes, gardens, and animal shelters (figure 15-4).

Select sites with stable subsoils. A house is only as good as its foundation. And a foundation is only as good as the subsoil it rests upon. In most cases, this isn't a problem. In others, subsoils can be extremely troublesome. Bentonite clay, for example, is a highly expansive material found in subsoil. Because it expands and contracts dramatically as it gains and loses moisture, it can cause homes to shift and walls to crack.

To learn about the subsoil in your area, ask locals. Also, dig up a sample and take it to a structural engineer or soil scientist for advice. Or hire someone to come on site to examine your soils. Your building department may also be helpful.

Look for sites with good soil (and sunlight) for growing food and fiber. Another consideration of great importance for sustainability is garden space. Good soil and sunlight are vital to growing vegetables, herbs, flowers, fruit trees, and even trees for fuel and lumber. If you are a gardener, I don't need to tell you the importance of nutrient-rich soil. You can assess the quality of topsoil by its look, feel, and smell. You can also judge its fertility by examining the plants that are growing on the property. If the property is overrun with thistle and other weedy species, it is an indication that the previous owners have abused the land. Topsoil may have been stripped off or the land may have been badly overgrazed or intensely farmed. Remember: Weeds are not a problem, they are a symptom of poor soil typically wrought by poor land management. Weeds are nature's way of filling a gap, invading land that has been overgrazed or laid bare or depleted of its nutrients by poor agricultural practices. The weeds will prevent further erosion and will eventually heal the land, replenishing the topsoil. But that could take decades. You want to plant right away. So select a site with good soil to begin with.

When assessing the condition of your soil, dig down with a shovel to determine the depth of the topsoil. Examine several locations, especially those you think might be turned into gardens. Send some soil samples in for testing.

When bentonite is present in a region, it tends to be found in varying concentrations —some higher than others. For this reason, it is important to obtain engineering and soil analyses of your land.

KIT COHEN, Geologist,
Personal Communication

State extension agencies will run the tests to determine organic content, pH, nutrient levels, and other important parameters—all at a fairly reasonable cost. They will even advise you on ways to improve your soil—for example, what nutrients you should to add. Contact them first to determine the procedure they recommend for soil sampling.

If you are not a gardener, drag along a friend who is and ask him or her to give you an assessment. Or, hire a soil scientist from a local college or university. College teachers are frequently willing to help out a local grower. Although their services may cost a bit, they are very worthwhile.

Select a site that offers building resources. Most homes are built with materials shipped from manufacturers located many miles away. This practice, while popular and cost-effective because of the economy of scale, has its problems. For one, it necessitates the transport of materials. That, in turn, requires fossil fuels, the combustion of which contributes to local, regional, and global air pollution. One way to enhance the sustainability of your structure is to use local materials, especially those obtained from the site itself. Be certain that the site you are considering can supply the resources you will need. If so, can they be obtained with minimal impact? If not, can nearby sources supply your needs? The more you can borrow from your site or nearby sites, the more sustainable your home will be—and the greener our world will be.

Check out access to water and utilities. Water is another key component of development. Without it, few if any households could manage for more than a few days. In selecting a site, assess whether there is sufficient water and the cost of getting it to your home. The most sustainable option for supplying water is a catchwater system (see chapter 13). These systems capture water from rain and snow melt, then store it in a cistern for later use on lawns and gardens or to wash cars. It can also be used to flush toilets. When properly filtered and sterilized, it can be used for drinking, cooking, and bathing. Even in dry climates, you can collect thousands of gallons off a decent-sized roof. With water conservation measures, such as water-efficient showerheads, front-loading washing machines, and low-flush toilets, catchwater can provide 100 percent of household demand.

Another approach is to tap nearby streams, lakes, ponds, or springs. If these sources can be developed without drying up water supplies for wildlife, they will provide water at a fraction of the environmental cost of a full-scale water project. When assessing alternative water sources, you must ask: Is the water potable? Are there any nearby sources of contamination, for example, a farm field that is heavily fertilized or doused with pesticides, or a barn or feedlot where stockpiles of manure could leach nitrates and potentially harmful microorganisms into the groundwater? Are there factories nearby that might be polluting the water? If you are considering water from a spring, you will want to find out if it runs year-round. And, of course, you will want to find out if a potential water source can provide enough water to meet your needs. Four gallons a minute is generally sufficient for most homes, although less will work if you have a way to store it for higher-use periods. Finally, while considering sources of water, determine how easy it will be to transport water to the house. Take time to carefully investigate all of these matters. Send a sample of the water to your local health department for testing. The information they provide will tell you how much filtration the water will require to be potable.

If you are considering a well, determine how accessible the water is or whether you have to drill halfway through the Earth's mantle to find sufficient water. And is the water clean and unpolluted? If you are more conventionally inclined, can you get city water and sewer lines? What is it going to cost? You will need to determine tap fees (the cost of the right to connect) and the actual cost of connecting (excavation, pipe, water meter, and labor).

If you are planning on hooking to the electrical grid, how close are the nearest electric lines? Do you have a right-of-way to run lines to your house? If so, how much will it cost to hook up to the power company's grid? Is phone service available? Will above-ground phone and electric lines be your only option, or can you bury the lines so you don't have ugly telephone poles or electrical poles spoiling the view you are buying the land for?

Be on the lookout for unique microclimates. Although we tend to think of an area as having a fairly uniform climate, that is not always the case. The climate may vary rather dramatically, even in a small area. For example, I live in the foothills of the Rocky Mountains, east of a 14,000-foot mountain that profoundly affects the weather in my region. The mountain creates lots of storms that produce lots of snow and rain that support magnificent ponderosa pines and lush meadows. Ten miles east is shortgrass prairie with 30 percent less rain and trees only along waterways.

Another very intriguing phenomenon is that the valley I live in is often in a pocket of sunshine, especially in the winter when I need the sun the most. Ten miles north, south, and west of me big burly storm clouds bunch together, blocking out the sun, but my property is bright and sunny, permitting me to produce virtually all of my electricity from solar panels (PVs) and heat my home with solar energy.

Throughout the world, there are regions containing unique and highly desirable microclimates, sometimes with dramatically different weather patterns and separated by only a few miles. Even in the rainy Olympia Peninsula of Washington, there is a special place not too far from Port Townsend that is bathed in sunshine, even though a few miles away the rain comes down in a steady drizzle much of the year.

An intimate knowledge of the land will help you avoid adverse microclimates and locate suitable ones. Ask around for those special places. You can sometimes locate them by looking for abrupt changes in vegetation. In Colorado, we call these bizarre little climate zones "banana belts." If you are successful in finding a little slice of heaven on Earth, you will be rewarded many times over, although generally not in bananas.

Be certain you can build a road or driveway to your home without major bulldozing. Years ago, I considered building an off-grid home in a fairly uninhabited location. The site was well drained and ideal for solar energy. My only neighbors would have been a couple of beavers who lived in a self-made swimming pool that stood between the building site and the road. Not wanting to fill any part of the reservoir, even to build a road to an environmentally sustainable home, or span the pond with an ugly bridge, I abandoned my thoughts of the site, leaving it to the beavers who were there first anyway.

For most of us, our homes must be accessible by vehicle. The question we must all ask when considering a site is, "What price am I willing to pay for this access?" You may find a terrific piece of property, but the actual building site is a long way from the road, up a treacherous hillside. Or you may have to carve up a lush meadow or cut down lots of trees to get access. Think about your driveway early on, and do not bulldoze any more land than you have to. It is not only an assault on our planet, it can be extremely costly. If you will need to add road base, costs will go even higher. Furthermore, a long dirt or gravel driveway can be a nightmare to care for in wet or snowy climates. Roadways tend to erode, if not properly cut. This may pollute nearby streams or cut gullies through fields.

Balance view with vital needs. View is a major consideration when selecting a piece of land and then positioning a house on that site. Having purchased a site for its view, many people then logically orient their homes toward the scenery. Tempting as it is, such decisions can commit you to a lifetime of high energy bills or discomfort. One of the most common mistakes in the western United States where I live is to orient houses to the west to capture the view of magnificent mountain ranges. While this may be fine in the winter, during the summer, you are likely to bake in the oven you have created each day as the sun descends toward the horizon. And, in the winter, this orientation will prevent you from taking advantage of the sun's generous supply of free heat. Your magnificent view won't provide much solace as you stand there shivering in your shirt sleeves.

Don't ignore the fundamentals of solar energy when considering a site. In other words, think very carefully about views and solar gain. You may be able to incorporate a view while capturing the solar energy you will need in the winter. As shown in figure 15-5, one way to achieve this dual goal is by restricting the view to a main living area, for example, a living room that faces west, while orienting the rest of the house south to gather sunlight for warmth.

Mountain ranges, river valleys, meadows, forests, ponds, lakes, or any of a myriad of other natural features become the dominant focal point of a piece of land. Unfortunately, solar gain and other important siting considerations may require orientation of a house away from the view.

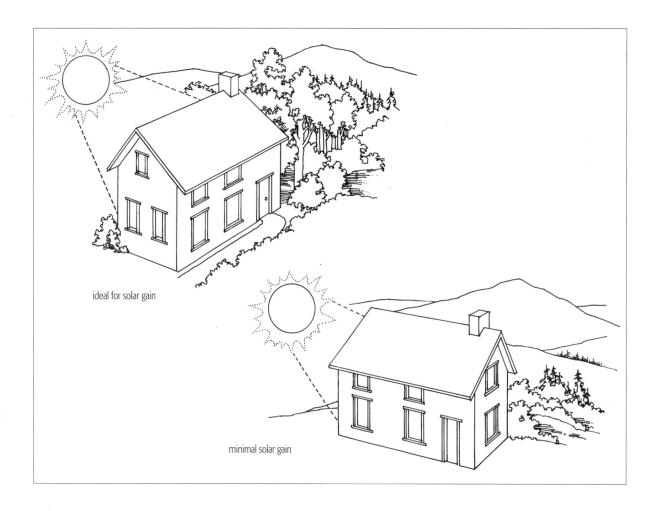

ideal for solar gain

minimal solar gain

FIGURE 15-5.

The view on a home site does not always correspond to the best solar orientation (top). With a little finagling, you can have both.

Don't destroy beauty in your search for it. Beauty is a blessing and a curse. Time after time, it is the attractiveness of an area that turns a place into a hodge-podge of homes, stores, roadways, and billboards. Even Los Angeles was once a beautiful place to live.

Building a sustainable future depends on protecting the beautiful areas. Christopher Alexander and the coauthors of *A Pattern Language* contend that under no circumstance should buildings be sited in the most beautiful places. In fact, they suggest just the opposite. "Leave those areas that are the most precious, beautiful, comfortable, and healthy as they are, and build new structures in those parts of the site that are least pleasant now."

Malcom Wells agrees. He advises us to leave the untouched sites to nature. Buy ugly spots, abandoned lots if necessary, and build earth-sheltered homes that are adorned with a jungle of vegetation. Make beauty.

Check out legal restrictions. Those who have never purchased land before may be surprised to find that a local cable company or a local utility owns certain rights to your property. These rights ensure them free access to your property to string TV cable or electric lines right smack dab in front of your home, dissecting the view for which you paid thousands of dollars. Utility companies aren't the only ones who have legal rights to use your land. Local government, even neighbors, may be entitled by law to build a road through your garden or backyard. Phone companies may also be able to invade your property. They don't even need your permission! How can this be possible?

Land comes with prior access rights called easements for such things as utilities, roadways, and pipelines. An easement is a legal right granted to someone else to use part of your land for some purpose. A former owner may have granted an easement (right of way) on his property to a neighbor who owns an adjacent piece of property. If that land sells, the new owner has the legal right to cut a road through your land to access his or her property. There is not much you can do to stop someone from exercising an easement, either.

Land also comes with other rights that may not be yours. For example, in Colorado you can purchase land but not own a right to the water in a stream that meanders through your property. Why? The water rights are owned by a downstream farmer, rancher, or municipality. You can watch it flow by, fish in it, and swim in it, but don't draw any water out of the stream. It is not yours.

Mineral rights may also be excluded from your property. Whether you like it or not, a mining company that owns the mineral rights on your land can show up any time it pleases with its bulldozers and draglines or drilling rigs to acquire the coal, silver, gold, or oil that lies beneath your vegetable garden—and they are within their legal rights to do so. The moral in all of this: Check out easements and mineral rights very carefully before you buy, so you are certain you know what you are actually purchasing. While you may not own the mineral rights, you can find out whether there are any potentially valuable natural resources on your land—before you buy.

Zoning regulations place restrictions on the use of your property that you should know about before buying a piece of land. They may, for example, prohibit certain kinds of structures, although most zoning regulates only the placement of structures on your property. The most common restriction is a setback, a legal requirement that buildings be "set back" a certain distance from the property line. Although exceptions to zoning regulations are made all the time to developers, especially by pro-growth communities, getting a variance on a zoning regulation may cost you time and money. How do you find out about restrictions?

The seller of the property, real estate broker, or an attorney should supply information on zoning and other restrictions such as easements or rights. Insist on it, in writing, before you purchase a piece of property, no matter how small. Don't wait until the closing to find out the mineral rights on the land are owned by Take-It-and-Run Coal Company. This information is available from county offices or from title companies. In fact, it should be part of the title search that your state requires before the purchase of a home or piece of property.

BUILDING NOTE

Each state has a slightly different set of water laws. Find out what yours are before you buy land or attempt to tap into surface or ground water on land you own.

In some communities, homeowners have banded together to form home-owners' associations. Some homeowners' associations establish architectural review committees that must approve building designs in the subdivision to keep the neighborhood homogeneous. They can be a real pain and may balk at the thought of an alternative home if it strays too far from conventionality. Most of their fears stem from a poor understanding of what an alternative structure will look like when completed. A straw bale home won't look like a pile of straw bales just as a stick-frame home won't look like a pile of 2 x 4s. If you are planning an alternative home, have someone provide a rendering of the home for the committee to see.

Check out future development. Whenever you are thinking of buying property, find out who owns the surrounding land and what they plan to do with it. Are they getting ready to sell the land to a developer who is going to build a shopping mall? Or is the neighboring land slated for building a new prison? Will the land stay in farming? If so, does the farmer spray insecticides on the land or use herbicides carefully? Is your neighbor planning to sell the land to a company that wants to construct a hazardous waste dump on the adjoining 200 acres? Is a neighboring forest going to be cut down?

This kind of research takes a lot of work. The rumor mill may help you, but rumors can also lead you astray. Knowledgeable realtors are your best ally in such instances.

Check out noise, pollution, privacy, and community. Noise is the stench of the ear in modern society. It pervades our society so much that we've learned to tune it out. In addition, many of us are so wrapped up in other considerations that we forget to stop and listen to determine how noisy a lot may be.

Other forms of pollution should be a major concern when siting a home, even in rural areas. Note the location of power plants, factories, and other sources of pollution. Are they of significant concern? If so, will you be locating downwind from them? Another concern is the frequency of temperature inversions that trap pollution at ground level, often for long periods. Temperature inversions are common in low-lying areas such as river valleys. Pollution levels may have serious consequences for people with asthma, pulmonary problems, or cardiovascular disease.

Privacy is worth careful consideration, too. When looking for a home site, many people become caught up in their excitement and fail to look around to see how close their neighbors are and, in some cases, how messy they are. You don't want to build the house of your dreams only to find that you are looking out on a yard full of rusting automobiles.

Where are the closest neighbors? Are they too close or too far for your liking? Remember, not all of us agree on the definition of closeness. Some people don't like to see neighbors, whereas others feel more comfortable with a neighbor in close view. Without them, they feel isolated.

Hard as it may be to think about, consider your own intrusion. Are you going to be disturbing someone who already lives near by? Will your home block their view? Will your noise offend them? Take steps to honor those who are already on the land.

And what about the community? Is the land you are looking at situated among people with whom you are compatible? Or will you be an outcast?

Check out environmental amenities. Recycling facilities, open space, bike paths, and mass transit are environmental amenities that will enhance your life and the life of the planet. Other considerations include access to organic produce, bulk food, vitamins and herbal remedies; and alternative health care practitioners, including herbalists, massage therapists, and homeopathic physicians. If you love music and other forms of artistic expression, a survey of opportunities in this arena is also important.

SITE SELECTION AND THE ZEN OF CHOOSING A MATE

Site selection is like choosing a spouse; you rarely get 100 percent of what you want! The goal in both site and spouse selection is to find the one that you love and one that meets most of your selection criteria. Some modification of both site and spouse is possible, but don't count on too much alteration—unless you are into major earth moving in the case of land acquisition or counseling or surgery in the case of your future spouse.

Choosing the right site means getting to know the land, and carefully analyzing the facts. For those who are building a single home on a piece of property, get to know the land as well as possible before you take the plunge. Learn its hills and valleys and weather patterns. Trace the path of the sun. Learn where the wind comes from. Find out about the soil, especially if you are thinking about an earthen home that will need a clayey subsoil for construction. Try out different locations for your home. Think through the days and the seasons. Imagine how the sun will heat up your spaces at different times of the year. Think about the location of your photovoltaic modules, wind generator, driveway, and garden.

When you have a good feel for the lay of the land, prepare a drawing that shows the various features, including wind direction, solar access, views, location of springs or water supplies, garden spaces, livestock enclosures, driveway locations, and areas to avoid—for example, wetlands, poorly drained soil, or precious scenery. Double check your drawing by taking it to the site and making sure everything is where you placed it on the map. When you are certain you have everything right, prepare a scale drawing of your house and try it on several locations on of the property. See which site gives you the optimal results. Next, check out each potential site to see if it looks as good in reality as it does on paper. For more detailed advice on siting your home, see Sam Clark's book, *The Real Goods Independent Builder: Designing and Building a House Your Own Way.*

City or Country? Some Food for Thought

While many readers will be interested in building in the country, it is important to reflect on your goals and to consider the higher good. If you will still be commuting daily by car to a city or town for work, remember that your automobile will be contributing to local air pollution, making it worse for those who live in the city. The countryside is also under increasing pressure these days from development. Forests and meadows are being sacrificed on the altar of progress. As David Pearson points out in *The New Natural House*, "It is much more ecologically sensitive . . . to improve the environment of the cities we have and live better in them . . . " than to "lose more countryside to yet more development." It's something to think about.

Making Systematic Comparisons

Finding a suitable site is no easy task. To simplify the task, make a list of criteria by which to judge each site. This will allow you to compare different sites. It will enable you to eliminate the worst sites, and probably narrow your choice to a couple of good ones. From that point on, you may have to rely on gut instinct or aesthetics.

Once you have selected the best site, take time for a second and third look. In a fast-moving market, a small down payment (earnest money) may be required to hold the land while you give it a more thorough inspection. Earnest money is nonrefundable if you renege on the deal for personal reasons—for example, you find a better lot. If you find something wrong with the title or the land flunks the perc test needed for a septic tank, you can usually get your money back, but be certain you list reasons for the refund of your earnest money on the Sales and Purchase Agreement so there won't be any misunderstandings.

PROTECTING A SITE DURING CONSTRUCTION

Most home construction involves excavation for foundations, utility service, septic tanks, water lines, landscaping, and driveway construction. If you are going to build a home out of cob, adobe, or some other earthen material, you may have to disturb the land even more to acquire the raw materials to make your walls.

Unless you are planning to totally re-landscape and re-vegetate a piece of land, which is not advised from an environmental standpoint, you must find ways to minimize on-site damage. One of the first measures is to designate an area for parking vehicles or, better yet, ask workers to park along the road. This simple step protects vegetation from damage and reduces compaction of the soil that makes it hard for plants to get the oxygen and water they need to survive. Because workers are forgetful of such things and because many subcontractors and building inspectors may be visiting your site during construction, you may have to post signs designating parking areas and cordon off areas so that earth-moving equipment is not operated in them or parked there when idle.

Loads of sand, gravel, and other building materials may be delivered to a site and should be stored in designated areas as close to the point of use as possible. Designate areas for waste and recyclables. They need to be conveniently placed to enhance worker participation but also to minimize the destruction of vegetation. To encourage workers to presort, mark bins clearly. This will save a lot of time later on.

Protect trees during construction. It makes no sense to cut down a fifty-year-old tree only to replace it with a sapling that will take fifty years to provide the shade and beauty of the one you just bulldozed. Mark trees with brightly colored tape and wrap their lower trunks with cardboard to protect them from heavy equipment. If you are having a house built for you, ask the contractor and all of the subs to be careful when working around your trees, even if it means slowing down a bit. Most builders understand the value of trees—they know how much trees add to the purchase price. However, workers tend to be careless about vegetation. Be vigilant

On most building sites, topsoil is first stripped from the footprint of the house. Stockpile this valuable resource and protect it from wind and water erosion. Topsoil should also be isolated from subsoil; don't mix the two. Each has its own specific applications. Subsoils are used for backfill. Topsoil, on the other hand, is best saved for gardens, lawns, and orchards. Using it for backfill wastes an extremely valuable resource that took hundreds or thousands of years to form. Besides, topsoil is a poor backfill material as it contains a substantial amount of organic material that may decay over time, causing the soil to settle.

While we are on the subject, be certain that topsoil, subsoil, or any fill dirt you have delivered to your site is not piled around trees. This seemingly innocent act can kill a tree, as it cuts off the flow of oxygen to tree roots.

Protecting a job site is to building what preventive medicine is to health care. It makes the task of bringing land back to health easier and less costly. It does require vigilance, however. You will have to sit down with builders and subcontractors to talk over strategies, especially strategies to remind workers. You may want to stipulate site-protection requirements in the contracts, levying fines for damage they do. This will get their attention!

When your home is completed, the subsoil and topsoil you have stockpiled can be spread back over the site and the land re-vegetated, even returned to native species.

BUILDING NOTE

If you are planning to reseed your land or part of your land with native vegetation, you can gather seeds in the fall and plant them after the project is completed.

ENVIRONMENTAL LANDSCAPING

When most of us hear the word *landscaping,* we think about blocking unsightly views, like the view of a neighbor's dented garbage cans; providing shade; and, of course, beautifying the exterior of a home with brightly colored flowers or stately trees. Today, landscaping has taken a whole new direction. It has come to mean preventing erosion and restoring some portion of the Earth we have disturbed while building our homes. In restoring the Earth, we seek not just to create a landscape conducive to human comfort, but one that benefits birds, insects, and other wildlife displaced by homebuilding.

Today, landscaping has assumed even broader environmental goals. With careful design, landscaping can enhance the performance, comfort, and self-sufficiency of a home—and its occupants. As such, landscaping has become a key adjunct to natural building and design, complementing and enhancing architecture to minimize heat loss in the winter, and maximize the preservation of cool interior temperatures in the summer, while reducing our dependence on fossil fuel. Landscaping and environmentally sustainable building design work together, synergizing to enhance human comfort and to reduce environmental impact. A sustainable home design, for example, may provide maximum heat retention through insulation, caulking, weather-stripping, and energy-efficient windows, but fierce winter winds can still steal enormous amounts of heat. To reduce heat loss, a landscape designer may plant trees to direct the wind away from the structure.

Others techniques in the landscaper's bag of tricks serve to deflect hot winds or block the penetrating rays of sun, making a dwelling much more comfortable during the long days of summer. In hot, arid climates, landscapers may plant vines on arbors to shade west-facing walls or plant tall high-crowned trees on the south sides of homes. Both measures reduce the sunlight striking the house, thereby keeping the house much cooler. Simple yet elegant measures such as these can reduce, even eliminate, our consumption of fossil fuels needed to supply elaborate and costly heating and cooling systems.

Like so many ideas we have been exploring in this book, environmental landscaping saves money and reduces pollution and environmental destruction. It creates a win-win-win situation. It is good for people, their pocket books, and the environment.

Environmental landscaping also involves the use of various landforms (natural and human made) and built structures such as fences and retaining walls to beautify a home and protect its occupants from adverse elements. The landscaper, however, must walk a delicate line, for the sun and wind are also great assets. Hot summer sun, for instance, may overheat a structure, but it also provides energy needed to power a solar electric system that runs ceiling fans, televisions, and refrigerators. If the sun is blocked entirely, you will lose the ability to generate electricity.

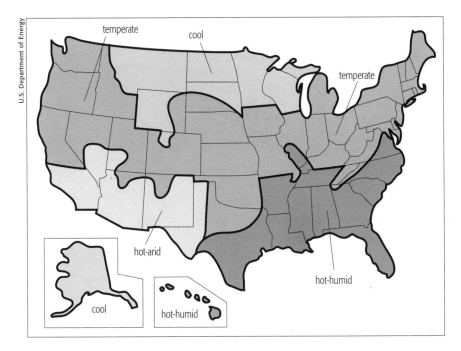

U.S. Department of Energy

temperate

cool

temperate

hot-arid

hot-humid

cool

hot-humid

FIGURE 15-6.
Climate zones of the United States. Each climate imposes special design and landscaping requirements.

Landscaping challenges vary from site to site and from one region of a country to the next. The United States covers a wide range of climatic zones. As shown in Figure 15-6, the U.S. is divided into four climates: hot and arid desert; hot and humid; cool; and temperate. Let's examine each one and explore the design challenges each poses to architects and landscapers. For more specifics on each zone, read Anne Simon Moffat and Marc Schiler's *Energy-Efficient and Environmental Landscaping.*

Landscaping Challenges in Hot, Arid Climates

Hot, arid climates such as the desert Southwest or the Australian outback present enormous challenges to the landscape architect and to owner-builders. In this zone, winters are short but still relatively warm. Temperatures rarely fall below freezing. Most of the year is fairly hot and dry. The main challenge in the desert, then, is to design a home and a landscape that keep the house cool.

Shade trees keep a house cool during the day. If planted on the south side of the home, shade trees will cast shadows on the roofs, walls, and windows. Trees are a friend and ally to those who make their home in the desert (figure 15-7). Unfortunately, the desert is not ideally suited for growing tall shade trees. The lack of moisture makes for slow growth. (This is another reason to protect native shade trees when you are building your home!) Shading on the south face of a home can also be provided by shrubs planted near the home and vines that drape over trellises or grow against exterior walls.

In desert climates, the east and west sides of the house should also be protected by trees and shrubs. They will block the sun early and late in the day.

FIGURE 15-7.

BUILDING NOTE

Mass walls equalize the extremes of desert climate, the hot days and cool evenings, creating a comfortable place to live. If you make the desert your home and want to live as sustainably as humanly possible, consider an earthen home.

Ground vegetation around a home has a cooling effect as well. For example, certain desert plants reflect large amounts of sunlight and therefore cool the space around a home. Plants also lose water by transpiration, the release of water from tiny openings in the leaves of a plant. Transpiration is a lot like perspiration. It draws heat away from the surface of a body. Heat drawn from the environment around a home tends to cool it. (That is why a park in a city is so much cooler than treeless area around office buildings.)

Heat that accumulates from various activities or heat that manages to enter the interior of a house through the roof and walls—or opening and closing of doors—can be dissipated at night by simply opening a few strategically placed windows. As you may recall from chapter 5, in desert climates air temperature often drops dramatically at night because the atmosphere lacks heat-trapping moisture. Opening windows permits cool night air to flush out unwanted heat. Thick mass walls found in stone, adobe, and other earthen homes can aid in the process. They absorb summer heat during the day, but cool off at night, releasing heat into the cool night air. The next day, the walls of the home have cooled enough to permit the cycle to be repeated. Any heat they gain during the next day is released the following evening, and so on throughout the summer months.

In some desert locations, cool breezes blow at night. Careful landscaping will funnel cool air to a house. As shown in figure 15-8, funnels can be formed by trees and shrubs. When planted correctly, they can direct breezes toward a house. They can also concentrate or magnify breezes by the Venturi effect.

Thus, a gentle, almost useless flow of air can be concentrated to produce a much stronger breeze that cools a house at night.

Desert regions can suffer from brutally hot summer winds. Vegetation and landforms such as earthberms may be required to deflect hot winds away from buildings. Check out Moffat and Schiler's book for recommendations on the types of plants that create effective windbreaks in desert climates.

Walkways and driveways in desert climates often become unbearably hot as they absorb sunlight during the day. Heat given off by these surfaces often raises the temperature around a house dramatically. To create a more comfortable home environment, shade should be provided along driveways, patios, sidewalks, and any other paved surfaces by planting trees and shrubs or building walls. Their shadows will reduce heat gain.

In the desert, like all climates, landscaping is an adjunct to house design. Unfortunately, many builders ignore age-old practices designed to cool structures passively. Instead, they rely on technological fixes, notably air conditioners and swamp coolers. Although new energy-efficient models reduce energy consumption, modern homes still consume a lot of energy—a lot more than necessary.

Intelligent design and landscaping are essential to keep houses cool in the desert. If you live in a conventional home in the desert, one designed to let energy-guzzling air conditioners do what nature will gladly do free of charge, simple changes in landscaping can make a huge difference in your comfort level and in the amount of energy your home consumes. In other words, landscaping can compensate for architectural design flaws. By providing shade or funneling cool breezes toward your house, you can make dramatic changes in energy demand; you could save up to 50 percent on your energy bills. Adding houseplants to increase the humidity of the inside air will also make a house more comfortable. As Moffat and Schiler point out, houseplants "are

From an environmental perspective, plants are always preferable to air conditioners for temperature control.

ANNE SIMON MOFFAT, MARC SCHILER, and the Staff of Green Living, *Energy-Efficient and Environmental Landscaping*

Diane Zampino and Diane Divenere

FIGURE 15-8.
Trees and shrubs can be strategically located to funnel wind toward a house. The Venturi effect (funnel effect) can also magnify a slight breeze, cooling a home during summer months.

extremely effective humidifiers and air coolers and, with a minimum of care, are far more reliable than appliances." And, if you currently rely on an air conditioner, shade the unit to cut its workload.

In life, solutions to one problem often spawn another. Solving the daytime heating problem in desert climates by planting trees may block your solar access if the trees grow tall enough to tower over your roof. Proper roof design will ensure a location for PVs that is free of shade year-round. Or place the PVs on a pole or a sun-tracking mount system away from shade.

Deserts are not all heat and dryness. In the winter, some desert regions get downright cold. For a few short months, residents may actually need a little heat. And what better place to get it than from south-facing windows that generously accept sunlight during those months. Earth-sheltering can also help enormously in protecting a desert house during the winter cold. It has the added benefit of cooling the house in the summer, too.

Landscaping Challenges in Hot, Humid Climates

The principal challenge in hot, humid climates is much the same as in the desert: to use landscaping to cool the structure during the dominant hot season. Contending with humidity makes this task extremely difficult, however.

Humidity is especially troublesome because it reduces evaporation from our bodies. When atmospheric humidity is high, evaporation slows down. Heat loss decreases. The net effect is that we feel hotter. To illustrate the importance of humidity, consider the fact that the limit of human tolerance, the maximum temperature one can withstand while working, drops dramatically in the presence of humidity. In a hot, arid climate people can work at temperatures approaching 150 degrees Fahrenheit; in humid climates, the limit of tolerance is 90 degrees.

Another challenge is that nighttime temperatures remain fairly high in hot, humid climates. Unlike the desert, in which moisture-free air allows heat to escape and ambient temperatures to plummet, moisture in the atmosphere tends to retain heat in regions like the southern United States or the tropics. Nighttime temperatures remain uncomfortably high.

Over the years, traditional homebuilders in hot, humid regions of the world have relied on a variety of simple, yet highly effective measures designed to keep houses cool and comfortable on hot summer days with little, if any, energy input. The sustainable house designer and landscaper draw on this wisdom. In hot, humid climates, earth-sheltering a house allows builders to tap into the earth's coolness. Windows can be used to ensure proper ventilation. Openable windows on lower levels or in basements permit cool outside air (coming from a shaded area) to enter a house. The air exits via roof vents or skylights. Open designs allow freer air movement. Ceiling fans powered by PVs or wind energy create air currents that further cool occupants. Covered porches also provide a cool blanket of air around a house.

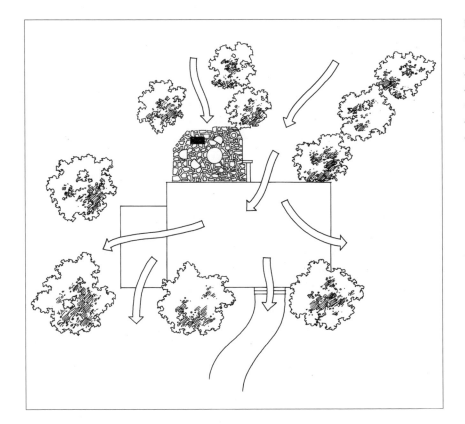

FIGURE 15-9.
Cooling a home in a hot, humid climate is difficult. Among other things, it requires shade, access to breezes, and an open interior design that allows air to circulate freely through a house.

The landscaper plants shade trees that will cast welcome shadows on the roof, windows, walls, sidewalks, patios, porches, and driveways. Trees also create a cool layer of air around the house due to evaporative cooling. The ideal planting consists of tall, high-crowned trees that provide shade but permit breezes to reach the house (figure 15-9). Heavy plantings can block breezes, trapping warm air near a house and making it feel hotter. Although high-crowned trees are desirable, remember, they may block roof-mounted PVs. To block the sun as it descends in the western sky, plant smaller trees and shrubs. Check out some of the references in the resource guide for specific trees or contact a local nursery.

Winter does come to some warm, humid climates. Although it is frequently only a modest cooling trend, winter may bring occasional snow and freezing temperatures; both necessitate some heating. Passive solar once again can provide a little help. Of course, high-crowned deciduous shade trees are required to allow sunlight to enter the structure during the winter.

As in desert climates, cool breezes are a treasured resource in the hot, humid parts of the world. If you are lucky to have a home that lies in the path of cool daytime or nighttime breezes, you can plant trees, erect walls, or modify land forms to funnel them toward your house. Locating a house on a small hill may allow you to tap into breezes not available at other locations. Study air flow on your property. Wind currents may vary from season to season.

Take your time. Plan wisely. A local weather station may will help you determine the direction of breezes at different times of the year.

Another design strategy for the landscaper and builder is the outside living space—for example, patios and porches where you can hang out with friends and family to read, talk, cook, or eat. Moving your cooking tasks outdoors during the summer months goes a long way to maintain interior comfort in hot climates. Shade trees and other vegetation planted around such spaces promote privacy. By designing outdoor living spaces into your home, you expand the square footage of usable space and open yourself up to an indoor/outdoor existence that many find delightful.

In hot, humid climates, the single most important thing you can do to ensure maximum comfort is to protect your home from the midday sun. If you lack funds, get that part of your landscaping underway first. For readers who are stuck in a conventional home that bakes in the summer, the best advice is to plant some swift-growing trees (after consulting with some of the reference books on green landscaping). Careful landscaping can reduce ambient temperature around your home by 7 degrees Fahrenheit or more. Simple steps such as these can result in a substantial decrease in energy bills, lowered impact on the planet, and increased comfort.

Landscaping in the Cool Climates

In the cool northern reaches of the United States and other countries in the Northern Hemisphere, the climate changes dramatically from summer to winter. Nonetheless, the dominant force in such regions is cold and the primary goal of designer, builder, and landscaper is to safeguard internal heat during the cooler times of year. The homebuilder does this by proper siting—for example, locating a house out of the path of the wind. After that, a number of important design features will prove useful. These include reducing surface area to minimize heat loss, earth-sheltering to protect a home, sealing cracks to keep the wind out, and maximizing insulation in walls, ceilings, floors, and foundations. Air locks—two-door entryways or mud rooms—are another essential design element. Passive solar goes without saying.

The landscaper can assist enormously. Wind can be blocked or diverted by creating windbreaks of trees, shrubs, fences, or walls—or some combination thereof. Planted or erected perpendicular to the predominant wind direction, windbreaks are a highly effective means of reducing wind speed, and even of directing it around a house (figure 15-10a). Densely planted windbreaks force wind up and over a home, creating a relatively windless eddy extending approximately five times the height of the windbreak. A dense hedgerow of 30-foot-tall pines, for example, creates a wind shadow that extends about 150 feet downwind. The maximum effect is achieved at a distance two to three times the height of the windbreak, which in this example would be 60 to 90 feet, downwind from the windbreak. The house should be placed in this loca-

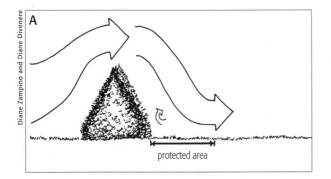

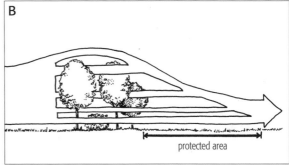

Diane Zampino and Diane Divenere

tion. (But beware of increased snow accumulation around the house due to the snow-fence effect of such windbreaks.)

Less dense windbreaks, known as permeable windbreaks, such as those planted with deciduous trees, also divert the wind over a site, but their primary effect is simply to slow winds down (see figure 15-10b). Although these windbreaks are less effective in reducing wind speed than dense ones, the wind shadow they create extends much further. In fact, the wind shadow will measure ten to fifteen times the height of the windbreak. So, if you need to reduce wind speed over a distance greater than five times the height of a future windbreak, plant a permeable windbreak of conifers or deciduous trees at an appropriate distance upwind from your house site.

If winds come from the south and you want access to low-angled winter sun for passive solar heating, however, be careful. A dense wind break planted on the south side of your home may block the sun and ruin your chances of tapping into the sun. A less-dense barrier, however, can be planted farther from the home, thus permitting the sun to enter unimpeded. Together, the two measures—blocking the wind and permitting the sun to gain access to the house's interior—can greatly reduce heat demand, increase comfort levels, and reduce utility bills.

It should be pointed out that a scattering of trees around a house is generally not sufficient to block wind. In fact, such an arrangement may even increase heat loss by causing air turbulence as the wind blows through the pinball machine–like maze of trees. To be effective, a windbreak must be planted as a hedgerow or shelterbelt with one or more rows of trees (two is usually sufficient) perpendicular to the direction of the wind. As a general rule, dense windbreaks should usually be planted on the north side of the home in cold climates. That is where the winds come from. If you have limited funds, this is *the* most effective landscaping step to take.

To find out how to determine the distance you can safely plant your trees from the house, consult *Energy-Efficient and Environmental Landscaping.* This book provides a lot of useful information, including advice on how to plant a windbreak to obtain results as quickly as possible. Be sure to study wind patterns, too, so you don't make any mistakes.

FIGURE 15-10 A, B.
Dense windbreaks (A) divert winds up and over a house, and thus create a sheltered space for a home. A permeable windbreak (B) slows the wind, but less dramatically than a dense windbreak. However, it provides a larger protected area.

BUILDING NOTE

Note that leafless deciduous trees still cast shadows in the winter and this can reduce incoming sunlight, so place your trees accordingly.

Woody shrubs planted next to a house can protect you from the cold. Shrubs create a dead space around the house that reduces heat loss. In other words, they are a form of external insulation.

Natural and human-made land forms can also be used to divert wind. A slight rise on the windward side of a home can deflect winds up and over your home. Planting trees on a rise will enhance the effect. Garages, barns, sheds, and other outbuildings built in the path of the wind will provide shelter for a home.

Summers in the cool climate vary. While some places are fairly cool and little must be done to shade a house, some portions of the zone, especially the southern and western reaches, can experience searing heat, thus necessitating shading. South and west sides of a house deserve special attention. If you are going to rely on solar energy for winter heating, however, be sure to plant high-crowned, deciduous trees on the south side. They are an excellent source of shade and will permit winter heat gain.

Earth-sheltered design should also be seriously considered in this region. It not only protects a house from wind, it also keeps a structure much warmer by tapping into the Earth's massive reserve of heat. Passive solar design, earth-sheltering, super-insulation, and a windbreak will greatly reduce your demand for fossil fuel.

Landscaping in the Temperate Zone

In between the hot climates of the south and the cold climates of the north lies the temperate zone. Don't be fooled by the name. The temperate zone is anything but mild. In many regions the winters are cold and windy. Summers may be hot and dry or hot and humid, depending on location. Temperatures in the temperate zone often change rapidly. It can be springlike one day, then snow the next.

Because the temperate region is a mixed bag of climatic extremes, it presents enormous challenges to the house designer and landscaper. The two main goals are creating a structure that will stay warm in cold winters and cool in hot summers. In the winter, sustainable design calls for strategies to protect houses from winds while ensuring adequate solar gain for heat. In the summer, shade must be provided to protect a house from the sometimes fierce summer sun. Winds can be channeled to the house to provide cooling.

The key considerations from the architect's or designer's vantage point are proper siting for wind protection and solar gain, and adequate insulation to keep you warm in winter and cool in summer. South-facing windows are essential to solar energy gain. Thermal mass proves valuable for both heating and cooling. Earth-sheltering a home, either with berms or by burying back walls and the roof, also helps create a structure that stays warm in winter and cool in summer with little, if any, boost from fossil fuel.

The landscaper ensures comfort by providing summer shade and respite from heat-robbing winds. Deciduous trees planted along the south side of the house will provide shade in the summer months. Because they shed their leaves in the fall, though, they permit sunlight to warm the house during the colder parts of the year. Don't underestimate the effectiveness of a tree as a means of cooling a house. As Moffat and Schiler point out, a single mature tree gets rid of as much heat as five 10,000-Btu air conditioners, and it does so without adding pollution to our atmosphere. Once again, be careful not to block roof-mounted PVs or flat-plate collectors installed to provide hot water.

Trees and shrubs may be needed to shade a house from early morning and late day sun to the east and west, respectively. This is not a hard-and-fast rule, however. In some portions of the temperate zone, summer heat is often a friend. With cold nights and lots of thermal mass, house interiors can be on the cool side in summer, so a little extra early morning heat from an east-facing window may be very welcome. Consider your climate carefully and think through the heating demands of your house before you block sun from the east or west.

Wind presents a major design challenge in the temperate climate, especially in flat treeless terrain like that found in the Great Plains of North America. Proper siting can shield you from fierce, heat-stripping winter winds, as can a carefully planted shelterbelt of trees. In some instances, it may be advantageous to funnel cool summer winds toward your house. If summer and winter winds come from different directions, this is fairly easy to achieve. If not, you must choose one or the other. Your decision hinges on the provision of greater good. If winter heating load exceeds summer cooling demand, opt for measures to control winter winds. Obviously, you will have to know your site pretty well before you can make this determination.

Intelligent design of a house integrated with thoughtful landscape design is the key to producing a comfortable living environment without expensive mechanical equipment. Together, site selection, proper orientation, earth-friendly design, and landscaping help to create habitat that works well for people and the planet. Graywater from the house or rainwater captured from the roof can even be used to water plants, furthering your contribution to the well-being of our home among the stars.

GOING NATIVE

Another important aspect of ecological landscaping is planting native species. Native grasses, shrubs, and trees have inhabited your bioregion for thousands of years and are acclimated to the environment. They survive extremes of weather, from drought to sub-zero temperatures, usually much better than alien species so often used in landscaping.

In the United States, an increasing number of nurseries have begun to carry native plant species. You can also purchase native plant species through mail order suppliers. A local extension agent may have a list of native species and may know where you can purchase them. In addition, your local water department may provide information on native species if they are at all interested in water conservation. They may even be able to give you a list of local outlets.

Avoid non-native species. Although many tolerate extreme weather conditions as well as natives, they may need periodic boosts in the form of fertilizer and water. This is especially true of non-native grasses. One species of grass that is used throughout much of the U.S. is Kentucky bluegrass. You will see it in lawns in Denver and Boise, Idaho, where rainfall is limited. Native to eastern U.S. where rain is much more plentiful, Kentucky bluegrass may be an excellent choice for western New York where it can get the 35 to 40 inches of rain it needs, but is ill-suited for the arid West where 10 to 15 inches of annual precipitation is common.

Native grasses are a more practical and more environmentally sound choice, but they often are a specialty item in nurseries and other outlets. Thus, they are not only hard to find, they are usually more expensive than less environmentally sound options. I paid $7 to $9 per pound for native meadow grasses, which is three times more expensive than non-native species. (Fortunately, the availability of native grasses is increasing!)

Non-native species may also be prone to insect damage and therefore may require frequent application of pesticides. Although natural pest controls are available, knowledge of them is quite limited. When it comes to saving a plant from insect damage, most people give it all they've got—and that includes some pretty poisonous neurotoxins!

To learn more about plant species indigenous to your bioregion, contact your local water department, your state wildlife department, local nurseries, and wildlife organizations.

Saving Water

Native vegetation usually requires less water than non-native species. Planting native species, therefore, lends itself to water conservation. Although water conservation is not a major concern in all parts of the world, it should be practiced by everyone. Even if a region has abundant water supplies, don't forget that it often takes energy to pump water to treatment plants and end users. Energy is generated by burning fossil fuels in most parts of the world and that creates pollution. And in most places, water is treated with chlorine to reduce the chances of anyone getting sick from viral and bacterial infections. Water is also fluorinated to harden tooth enamel. To meet water demand, dams must often be built. Their impact on the environment can be tremendous. The conservation of water, regardless of its abundance, is therefore a sound strategy to promote environmental protection.

In locations where water is in short supply, water conservation is even more vital, not just to protect natural resources, but to ensure adequate supplies. In Denver, progressive landscapers practice the colorful art of xeriscaping

Like abundant energy, abundant water supplies have undermined our thinking about proper planting.

David Winger

FIGURE 15-11.
Xeriscaping does not mean that your home will be surrounded by cacti and rocks. Many low-water plants are quite colorful and can turn your yard into a beautiful landscape.

(figure 15-11). Coined locally, the term xeriscape (like the word Xerox, for dry copying) means landscaping that minimizes water demand. However, don't make the mistake of thinking of xeriscaping as desert planting. Quite the contrary, many xeriscaped lawns are adorned with beautiful flowers, shrubs, and trees that have one thing in common: They require very little water to survive.

Xeriscaping is based on seven principles: (1) intelligent planning and design, (2) proper soil preparation, (3) use of drought-tolerant plants, (4) reductions in the amount of lawn space dedicated to grass, (5) water harvesting (diverting rainwater to plants), (6) efficient irrigation design, and (7) mulching to reduce water loss from the soil.

According to Patricia H. Waterfall, an extension agent who works in landscape water conservation and whose name couldn't be better suited for her job, the "key to a successful xeriscape is careful planning to provide aesthetic benefits, as well as pleasant spaces for relaxation and play." By carefully grouping plants with similar sun, water, soil, and maintenance needs, a landscaper can increase irrigation efficiency and reduce maintenance time. Grouping plants is also better for them. If a low-water plant is placed next to a high-water plant, neither tends to do well. The low-water plant often gets too much water. Its thirsty neighbor never gets enough! As fellow gardener Kit Cohen reminds me, "Some plants prefer unimproved, lean soils and will do poorly in soils that are too rich." Some plants like clay, others don't. The moral of the story, says Kit, "is to know the needs of the plants you are planting."

Xeriscaping, like any successful planting strategy, requires careful preparation of the topsoil, except in desert climates where topsoil exists only in textbooks. Topsoil should be rich in organic matter, which improves the texture of soil; allows air and water to penetrate to the roots; and like a sponge, holds moisture in the soil, releasing it as plants need it.

If your soil is really unsuitable for the plants you want to grow, rethink your strategy. If your heart says tulips, but your soil and local conditions say yucca, it is easier and wiser to go with what you've got, rather than attempting to rework the soil to satisfy the yearning heart.

A xeriscaper plants native species that require little or no additional watering, that is, no water other than that which Mother Nature supplies. In desert climates, the plants of choice are yucca, ocotillo, and cacti. Bear in mind, however, that even drought-resistant plants require some water to get started. Once they have established a healthy root system, however, their water requirements are minimal and often can be supplied naturally.

Grass is the curse of a xeriscaper, because it loses more water through transpiration than any other plant species. The solution is simple: limit grassy areas. Hard-to-water portions of the lawn, such as the narrow strips of green between the sidewalk and street, can be dug up and planted in ground-dwelling junipers or drought-resistant shrubs. Ground cover such as scented creeping thyme and pussy toes can also be planted in such areas. Once they are established, they will require very little water. In addition to reducing water demand, this strategy reduces the amount of labor required to mow.

Another technique that saves enormous amounts of water is to plant native, drought-resistant grass species such as buffalo grass, blue grama, Bermuda grass, and fairway wheat grass. Reducing fertilizer is a helpful water conservation measure as well. Decreased fertilizer use reduces plant growth and that cuts the need for watering and mowing. In a home I owned in Englewood, Colorado, I dug up my Kentucky bluegrass lawn (actually turned it over with a shovel) and planted a mixture of drought-resistant grasses. Being one who would rather kayak than mow a lawn, I never applied fertilizer or chemical pesticides. (Grass does very well if the soil is fairly healthy and you leave the grass clippings in place so they can decay and return nutrients to the soil.) My lawn filled out nicely and because the species I planted grew slowly and weren't doused with fertilizer, I only had to mow the lawn five or six times a year! While my neighbors pushed mowers and fertilizer spreaders and hauled hoses around every weekend, I was off catching the water that was left in mountain streams!

Planting shade trees and shrubs is another water conservation strategy. Trees and shrubs provide shade and maintain a cooler environment. This slashes moisture evaporation from the soil and transpiration from plants themselves. Early or late-in-the-day watering can also lower water demand by 60 percent or more.

Another strategy that gives a healthier lawn and reduces weeds is to cut your lawn higher—that is, set the mower blade higher. Cutting puts an enormous strain on grass. Setting the blade higher requires more frequent mowing, but the result is a healthier, greener lawn with fewer weeds (because the grass shades sun-loving weeds like crabgrass). The lawn will also require less water. For a terrific discussion of natural and environmentally sound lawn care, see *Energy-Efficient and Environmental Landscaping* and other books on the subject listed in the resource guide.

Xeriscaping uses rainwater, water falling naturally on your plants. The simplest method is simply to channel rainwater flowing in your downspouts directly to your plants. But be careful. Plants with high or moderate water demand may thrive under such treatment, but low-water plants may be drowned by their periodic baths of roof water. Graywater can also be used to water plants. Graywater and catchwater systems can be engineered to deliver any amount of water you want to a plant, and can serve a wide range of plant species. For more on these systems, see chapter 13.

Another key to the success of xeriscaping is mulch. Mulch is an organic material applied in a relatively thick layer on the surface of the ground around plants. Common organic mulches include straw, grass clippings, bark, compost, wood chips, and sawdust. Xeriscapers often use 3/8-inch pea gravel. Mulch is important to gardeners, especially in arid climates, because it reduces evaporation from the soil caused when sunlight strikes the ground and warms the soil. Mulching, in fact, can cut water loss by as much as 90 percent. (I apply a heavy layer of mulch to my flower bed and vegetable garden and rarely have to water after the plants have established themselves.)

By keeping the soil around a plant moist, mulch reduces water demand and makes it much easier to pull weeds out root and all, thus reducing competition for water. A two- to three-inch layer of organic mulch is sufficient. A three- to five-inch layer will eliminate weeds. However, as my friend Kit Cohen pointed out to me, "Excess water held in the soil by organic mulch may cause roots of xeric plants to rot. Therefore, a two-to-three-inch-thick gravel mulch may be preferable if the soil has a high clay content."

Mulch also improves the quality of your soil. This occurs because the bottom layer of mulch decays over time. The decomposition of leaves and grass clippings adds organic matter and nutrients to the soil. The organic matter feeds soil-based microorganisms that recycle nutrients and maintain the health of the soil. Because mulch does break down, however, it must be replenished from time to time.

Proper soil preparation, the use of mulch, diversion of rainwater to plants, and planting low-water vegetation all translate into fairly low water demands for a xeriscaped lawn and garden. You can further reduce water demand by installing root zone and drip irrigation systems, both discussed in chapter 13. Another water-efficiency measure that pays off handsomely is deep watering. This strategy promotes deep root growth. Plants whose roots have grown down into the deeper layers of the soil are more likely to survive dry conditions than plants with superficial root systems promoted by frequent shallow watering.

Shallow watering promotes superficial root growth, creating plants that are addicted to the sprinkler hose.

Deep watering requires less frequent application of irrigation water and a reduction in the amount of water that must be used to promote optimal plant growth. To find out more about xeriscaping, contact your local water department or check out some of the references listed in the resource guide.

PERMACULTURE: TURNING YOUR SUSTAINABLE HOME INTO A COMPLETE LIFE SUPPORT SYSTEM

Some of the best information on environmental landscaping comes from the permaculture movement. The word permaculture was coined in the mid-1970s by two Australian ecologists, Bill Mollison and his student, David Holmgren. The term is a contraction of two phrases, "permanent agriculture" and "permanent culture."

Permaculture relies on ecological principles of design that benefit people and the planet. According to Brad Lancaster, a teacher at the Permaculture Drylands Institute, "By learning to mirror the patterns found in healthy natural environments, you can build profitable, productive, sustainable cultivated ecosystems" that "have the diversity, stability, and resilience of natural ecosystems." Although it began Down Under as a way to develop productive farms and gardens, permaculture has grown substantially since its beginning, and now includes work on all levels, from the backyard to the village to entire bioregions.

Brad Lancaster points out, however, that "unlike market-driven systems that are subservient to short-term cost-benefit analysis, permaculture is guided by a common sense ethical system based on care of the earth" and "care of people of the earth." Permaculturalists recognize that without a sustainable agricultural base, no culture can hope to be permanent. Permaculture strives for a harmonious integration of human dwellings, plants, animals, soil, and

Peter Bane

FIGURE 15-12.

In addition to providing beauty, a permaculture yard provides useful food and fiber, including fish grown in a pond, moving your home one notch higher on the independence and sustainability scales.

water to produce stable, productive communities. It focuses on the relationships created among these elements by the way we place them in the landscape. This synergy is further enhanced by mimicking patterns found in nature.

For the individual, the central idea behind permaculture is to create a landscape you can harvest for food, fiber, and firewood (figure 15-12). Another key idea is the decentralization of food production, that is, to put food production into the hands of individuals. This, too, is a key principle of sustainability: local self-reliance. Permaculture emphasizes multi-use plants, that is, plants that provide vegetables, fruit, fiber, and other resources needed by people and livestock. Livestock can also be raised to provide food (milk or eggs) and other resources (manure for methane production). Animals help in the integration of ecosystem components by grazing on weeds and recycling nutrients.

Over the years, the permaculture movement has grown to encompass human dwellings and waste systems. Energy-efficient buildings, wastewater treatment, and recycling have grown in importance in the permaculture movement. More recently, it has expanded to include economic and social structures, such as cohousing projects and eco-villages, alternative forms of community that suburbanites dream about.

Permaculture can be applied in the city, on farms, and in remote wilderness locations. The brochure from the Bay Area Permaculture Group sums it up nicely: "Its principles empower people to establish highly productive environments providing food, energy, shelter, and other material and nonmaterial needs, including economic. By carefully observing natural patterns characteristic of a particular site, the permaculture designer gradually discerns optimal methods for integrating water catchment, human shelter, and energy systems with tree crops, edible and useful perennial plants, domestic and wild animals, and aquaculture."

Although it may seem radical to urban and suburban builders, the land around a house offers the opportunity to go the next step in sustainability, that is, to provide food, fiber, wood, and other resources, even a pond full of fish—all the staples you need to live. A vegetable garden is a good place to start. Fruit trees are a palatable option. From there, you can expand the self-sufficiency of a home by adding firewood sources. A small fish pond might be suitable. If this idea appeals to you, check out the permaculture resources listed in the resource guide.

Permaculture: The use of ecology as the basis for designing integrated systems of food production, housing, appropriate technology, and community development. Permaculture is built upon an ethic of caring for the Earth and interacting with the environment in mutually beneficial ways.

Permaculture Drylands Institute

LANDSCAPING FOR WILDLIFE

Wildlife are an important element of the ecosystems into which we insert our homes. Even in the suburbs, foxes, squirrels, skunks, and raccoons abound, often wandering around in search of food at night when few of us notice their presence. Although you may not see them, these animals and an assortment of songbirds often live peacefully alongside their human neighbors under porches, in parks, and in cemeteries. In rural areas, wildlife are even more abundant.

Many people like the idea of returning at least part of their land to the creatures they have displaced when building their home. It is relatively easy to do and there are many sources of information on the subject. You will even find a fair amount of technical assistance from local divisions of wildlife, conservation groups, and the National Wildlife Federation.

Don't forget that, during home construction, preventive measures are some of the most important steps you can take to ensure good wildlife habitat afterwards. Protect the land during construction, restrict parking to designated areas on the site, and keep heavy equipment operators on or near the building site. You may want to place colorful stakes to protect certain areas from vehicle traffic. Stake off small trees that could be damaged by heavy equipment and wrap the bark of large trees with cardboard attached by string to prevent such damage. Even an accidental scrape of the backhoe can open up a tree to invasion by pests. And don't let workers pile dirt around the base of tree trunks, which reduces oxygen flow to the roots and can kill trees. Also, clear only what is absolutely necessary, usually no more than ten feet of land beyond the perimeter of the building.

Wildlife require food, water, cover, and places to raise their young. Making the area around your house suitable for wildlife will be successful only if you help to supply these basic needs. Food may be provided by native plants, such as trees and shrubs that produce fruit, nuts, seeds, edible buds, and berries. Water can be supplied by bird baths, small pools, or, if you are really ambitious, ponds. Leave rock piles, brush piles, or dense vegetation, as cover for animals to rest or raise their young (figure 15-13). Because lawn is a sterile open space with few benefits to most forms of wildlife, reduce your lawn size to make your property more suitable for the birds and the bees. Reduce or eliminate pesticide use, too, to protect their food sources.

To design your landscape to benefit wildlife, first determine what critters will be sharing the space with you, then study their habitat requirements. The National Wildlife Federation, your local extension agency, and your state division of wildlife can be of assistance. The National Wildlife Federation, in fact, has a program that recognizes homeowners who have converted some portion of their yard to a wildlife sanctuary. For information on their Backyard Wildlife Habitat program, contact them at the address listed in the resource guide. And remember, designing your landscape to include a place for wildlife doesn't necessarily mean taking steps to attract bears and moun-

FIGURE 15-13.
Making a lawn or yard conducive to wildlife doesn't mean you will be inviting bears to live alongside you. A yard can be designed for butterfly and bird habitat and quiet contemplation by human residents, too.

National Association of Conservation Districts

tain lions who will eat little Gigi, your pet poodle, or Tammy, your lazy but affectionate cat. It can be as simple as planting a bed of fragrant flowers that attracts a colorful array of butterflies to brighten up your day, and theirs.

Postscript

The most incredible thing happened within ten seconds of finishing this chapter. I got up to stretch and glanced out my front window. There on the meadow grass surrounding my home was a magnificent jet black fox, known as a silver fox. As I watched him, he pounced on a vole, not more than fifteen feet from my view. He gobbled down his food, looked at me as I stood in awe, then pranced off in search of more!

16

BUILDING AN ALTERNATIVE HOME
What You Need to Know to Get Your Home Built

UILDING A HOUSE is like putting together a complex three-dimensional puzzle. However, this puzzle is complicated by many factors, including the fact that it can cost tens of thousands dollars, maybe more, depending on the how elaborate your home is. Home construction is further complicated by the fact that the puzzle you are assembling must interact favorably with the environment—wind, rain, snow, shifting ground, hail, and possibly earthquakes. Which is to say, it must endure all the vagaries of climate and geology that your bioregion throws its way. It must do all this yet provide you with an aesthetically pleasing and comfortable living space. And it must, of course, meet local building codes.

Building a home is hard work, involving approximately two to three thousand hours of labor, depending on the complexity of your project and the level of detail. As a rule of thumb, building a house will require about three hours of work per square foot, give or take a little. The more skill you bring to the job, and the faster you work, the less time it will take. Building a smaller home or a simple cottage may reduce labor and save resources. Nonetheless, many parts of a building project are complicated and time consuming. Even simple tasks like cleanup can involve huge amounts of time over the duration of a project. And there is a lot of room for error—mistakes that, if not rectified, you may have to live with for the rest of your life. Some mistakes, such as a faulty foundation, could even threaten the structural integrity of your home and the safety of you and those who live in the house with you.

If your house is to be a sustainable contribution to the housing stock rather than a structure that drains our planet's natural capital, your home must be well sited. It must be efficient in the use of all resources—from water to heat to electricity to building materials. It must rely on renewable energy and other natural forces to stay warm when you want it warm and cool when you don't. To be sustainable, your house should tap into renewable energy such as wind, sun, or falling water for electricity. A sustainable home should be made from green building materials. When the house is finished, the land should be landscaped to prevent erosion and to promote natural, healthy vegetation that serves people as well as birds, butterflies, and a host of other species.

Building a home, no matter how simple, usually involves a cast of helpers to get the job done right, on time, and on budget. You may be dealing with an architect or designer, a structural engineer, a builder (if you hire out the construction), numerous subcontractors, and, of course, building inspectors. You may have neighbors to placate and bankers with whom to negotiate loans. Even if you build your own home, you will still have to deal with designers, structural engineers, and building inspectors. You will have to learn building skills, permitting requirements, and people skills. You may have to spend a lot of time on the Internet or pouring through books to learn about alternative building materials.

No matter which route you take, your project will invariably consume more time and money than you had imagined. Be prepared for both contingencies.

Because sustainability isn't a concept used often in homebuilding, at least in most developed nations, your home will present many challenges that you may have to deal with by yourself. You won't, for example, be able to call on conventional builders in your area to answer key questions like "What's the best mix for a rammed earth home in our area?" or "Where do you buy straw bales?" Except for a small group of sustainable owner-builders and builders, you are pretty much on your own.

This chapter will get you started on the right path and help you complete your project with minimal hardship and expense. Its ultimate goal is to make this important project more enjoyable by helping you avoid a lot of common mistakes. If you want to learn more, I recommend Carl Heldmann's book, *Be Your Own House Contractor*, Sam Clark's *The Real Goods Independent Builder*, and David McGuerty and Kent Lester's book, *The Complete Guide to Contracting Your Home*. They will fill you in on the fine points.

BUILDING YOUR OWN HOME

If you have decided to build your own home, perhaps with the assistance of a spouse, friends, and family, and a few skillful subcontractors to handle specialty labor—concrete pours, electrical, plumbing, sheetrock, tile, or whatever—you are in for the time of your life. As the Eagles say in their song

A very large percentage of my architectural clients over the last twenty-five years have spent more than they could afford, built more than they could use, and then their lives changed radically, so that what they did just build was not what they need now. I have found myself wanting to tell people that maybe they should consider traveling a little lighter.

MICHAEL REYNOLDS,
The Nest Concept

"Hotel California," "This could be heaven, this could be hell." Although the group wasn't talking about homebuilding, the admonition is appropriate for those who venture down the path of owner-builder. The quality of your experience depends on how skilled you are, how well prepared you are, and how much patience you have. Your experience also depends on luck, for example, whether weather cooperates with you or not. Becoming an owner-builder can be very satisfying. It provides the freedom to do things just as you like and provides an opportunity to create and innovate, but it can become stressful. If you deal well with stress, like a good challenge, and have lots of time, the role of owner-builder may be just what you need to wipe away the cobwebs in your head.

Site Selection

The first key to success is having a good site. Be sure you understand the requirements of a good site as outlined in chapter 15. Because the site is so important, a few suggestions are worth repeating. First and foremost, if you are "going solar," your system will require solar access. Be sure there are no obstructions, including other structures, trees, and natural features such as rock outcroppings. Don't compromise on solar access. The more sunlight you can bring into your house on cold winter days, the better. Second, in many locations, a good site provides protection from wind, although landscaping can make up for natural deficits in this area. Another key element of success is a location free from the natural dangers of flooding, rock slides, avalanches, and so on. It goes without saying that you will also want a site that is aesthetically pleasing, a site that speaks to your soul and nourishes your spirit. Be on the look out for human hazards, such as nearby factories and power plants. Be cognizant of highway noise and noisy neighbors.

Good Design, Well Suited to the Site

Ideally, you should have a rough idea of your house design *before* you select your site. Be persistent in your search for a site suited to natural building, passive solar, and earth-sheltering. Many real estate brokers are clueless when it comes to sustainable building. They will show you property that isn't close to being suitable for a sustainable home, even though you painstakingly outline your requirements to them in advance.

When using a real estate agent, prepare a list of major site requirements in order of priority. Let your agent know that this list is of utmost importance to you. But remember, real estate agents are paid on commission and many of them are after a quick sale. The less time they spend with you, the more money they can make. To "sell you" on a piece of property, some agents will focus on the list of amenities given to them by the landowner or the listing agent (the agent who has listed the property for sale). The list may be impressive, but if

Your site is not only the location of your house, but the source of what makes it livable: water, sunlight, air, wood for heat and lumber, soil for a garden, and other necessities of life.

SAM CLARK,
The Real Goods Independent Builder: Designing and Building a House Your Own Way

the land doesn't meet your needs, don't let them sway you with a slick sales pitch. Stick to your principles . . . of sustainable siting, that is.

Most real estate agents know very little about solar energy or other elements of sustainable design. They concern themselves with the usual interests of clients, such as the location of schools, fire protection, and shopping malls. They are into vistas. When faced with a difficult-access site, they may help you envision extensive bulldozing to level a building site and construct a driveway. Anything is possible in the domination-and-control world of land development. Again, don't be swayed; the less heavy engineering and excavating you have to do the better, from an economic and an environmental standpoint.

Remember, too, even though *you* contact a real estate agent to show you land, they're still primarily working for the seller. Yes, that's true. To them you are just a potential buyer; their real boss is the seller. And, it is in the real estate agent's interest to sell you a piece of property for as high a price as possible. The higher the price, the greater the dollar value of their commission.

There is an exception to this rule. Among the sharks in the real estate profession is a subspecies known as exclusive buyer agents. They do not list properties, so there is no conflict of interest. Their job is to work 100 percent for the buyer. They can help with a land purchase or new construction. They are your representative. They still get paid a commission, but their job is to help you get the best deal possible. An exclusive buyer agent, for example, might find out how much the seller paid for the land, which is part of the public record. If the asking price is way out of line with the original purchase price— that is, the seller is trying to make a killing and you are the unlikely prey swimming in the water at the time—a good agent will inform you of the discrepancy and may be able to negotiate a better price. For a buyer agent in your area, contact the National Association of Exclusive Buyer Agents, listed in the resource guide.

A good real estate agent, especially a buyer broker, can help you find a reasonably priced piece of terra firma and can supply you with maps that show boundaries and easements. Check these maps very carefully and walk around the land to get a good feel for it. Visit the property several times and spend a lot of time on it studying sunlight, soils, wind patterns, potential building sites, drainage, and so one, before laying down your hard-earned money.

A good real estate agent can also assist in uncovering other details as well. For example, the neighboring land may be slotted for development or there may be zoning restrictions you should know about on the land you are considering. Ask your agent to research future developments, for example, stores, malls, subdivisions, and hazardous waste dumps. If you are planning to tap into water and sewer lines, ask your agent to determine their accessibility and cost. If you are planning to drill a well, ask your agent to find out how deep you will probably have to go and the expected production. An agent can also research easements and other rights that might be owned by others, a topic discussed in chapter 15.

It is also a good idea to have the soil tested before you purchase a piece of property. Load-bearing tests are required for foundation work. They are not terribly expensive, and your agent can request that they be performed at the seller's expense. Really poor soils can cost a lot of money in foundation work, so check it out first. Even if you have to pay for the tests yourself, the investment is well worth it in the long run. If you put money down to hold a piece of property, be certain your contract includes a refund provision in case the soil tests poorly.

If you are going to be installing a septic tank, you will want to test the soil in advance for percolation. A perc test tells you how well leachate from the septic tank will move through the soil. If your soil fails, you are up the creek unless you want to install a composting toilet and a blackwater treatment system, as described in chapter 13.

A perc test in advance is advisable, but if there are other homes nearby that are served by septic tanks, you probably won't have any troubles. However, be sure to include a provision in your contract that will ensure the refund of your money should your land flunk the perc test after the sale. In some states, this provision is required by law. For more information on septic systems and perc tests, check out Carl Heldmann's book, *Be Your Own House Contractor*.

Because you may not be able to find an ideal site in this highly populated, ever-growing world, keep your house design ideas somewhat flexible. Views and solar access will dramatically affect your design, especially if they are in opposite directions—for example, the view is to the north but the solar access is to the south. Building materials also affect the type of site you should buy. If you want to build a passive solar home out of tires and you want to earth-shelter your home, the best site is one that slopes to the south. Or, if you want to build a cob or straw bale passive home, you will want a relatively flat site with good solar access, because earth-sheltering is out of the question with these building technologies. If your site is relatively flat, be sure that it is well drained so that water won't pool around your house.

If you already own a piece of property, your building options may be restricted unless the property offers several potential building sites. In such instances, look for a location on the land that suits your design requirements. If your plot is small, you may want to redesign your home to better fit the site, or you may want to sell the land and start over.

Some of the biggest "mistakes" in housing, especially in conventional homebuilding, are caused by placing houses that are too large or of the wrong architectural design for the site—or, often, both (figure 16-1). While a house may look good on paper or in a magazine or in a Texas suburb surrounded by other mansions of similar size and style, it may be totally inappropriate in a

desert landscape or on a prairie or on a mountainous region. Dream homes look like hell when they're out of place.

Designing for a Partnership with Nature to Provide Comfort and Convenience

Once you locate a piece of property or select an appropriate location on your property, design work should begin in earnest. This will require special attention to local conditions. Visit the site often to determine from which direction the winds blow, how water drains, and so forth. With a complete understanding of the sun's path, wind flow, natural water flows, wildlife uses, safety hazards, soil types, and topography, you can design a house that fits into the landscape. It should be a gentler form of architecture than we have seen in recent years, a design that doesn't require displacing native plants and animals. Ultimately, your goal is to develop a house that synergizes with the site in a partnership with nature. Most houses today "require" a redesign of nature to accommodate them. What people end up with are unsightly, inefficient, inappropriate, fossil fuel–reliant structures that rob from the Earth. There is no partnership, only a kind of parasitism that is symptomatic of most other human-environment interactions

FIGURE 16-1.
This house sticks out like a sore thumb on the landscape. When siting a home, consider the relationship of the house (its appearance and its size) to the site. Seek balance and harmony.

Building small models of your home is relatively easy to do if you are building with cob, but much harder with other building materials such as tires. Some suppliers, such as Jade Mountain, provide clay and miniature straw bales for those who want to fashion a scaled-down version of their house design before they start on the real thing.

Spend a lot of time on floor plans and traffic flow. Look very carefully at proportions—for example, the size of a window compared to a wall. Look at roof pitches and ceiling heights with a critical eye, not just for function, but for aesthetics. The extra time is worthwhile in the long run.

Estimating the Cost of your Home Before You Begin Design

Before you start sketching your home, you need to estimate your resources, notably how much money and how much time you want to spend on the project. Good advice on estimating costs can be found in *The Real Goods Independent Builder*. This book will also help you develop a floor plan, an art in and of itself, and help you determine how to design a home for a particular site. If you want to build a small home, look at *New Compact House Designs* or *Small Houses for the Next Century*. Remember that a compact house design can actually provide more usable space than a poorly designed larger home.

Finally, use realistic figures for estimating costs of natural homes. Be wary of outlandishly low building costs. Most homes easily run $50 to $100 per square foot. The more elaborate your design and the more elaborate your tastes, the higher the costs. After you have made an estimate of costs, add at least 20 percent. An overrun of 50 percent is not that unusual

Getting Your Project Underway: Blueprints and Building

With a rough sketch in mind, the next step is to prepare blueprints. Blueprints are required by building departments before they can issue a permit. Blueprints also serve as a reference to you and your helpers as you build. If you are going to hire an electrician or a plumber, they will need blueprints to spec your system and prepare an estimate. (As a rule, subcontractors are usually given a simplified drawing of the floor plan to work with.)

Although building department requirements vary from one location to the next, most require blueprints containing the following items: (1) a **site plan** (aka plot plan) that shows where the home and driveway (if any) will be located, where the property lines are, where easements are, and where the street is; (2) a **foundation plan** showing foundation and slab, if any, including cross-sections and placement of reinforcement steel or rebar; (3) a **floor plan** with the location of various rooms, garage, closets, windows, doors, plumbing fixtures, large appliances, and other key features; (4) a **roofing plan** showing the location and size of beams and rafters; (5) **outside elevations** of north, south, east, and west walls (or front, back, and sides of the house and garage); (6) **detail sheets** that include cross-sections of the house to clarify the structure of walls, including insulation, paneling, siding, and even cabinet details; and (7) a **plumbing and electrical plan** showing the location of sinks, tubs, showers, toilets, all electrical lines, fuse boxes, switches, outlets, ceiling fans, and so on.

Your plans should also include a detailed **set of specifications**, a comprehensive list of materials that will be included in your house. This list eliminates the guesswork and potential for conflict over how your house will be built if you are building with a contractor.

In most cases, you will need four to six sets of blueprints, for yourself, lenders, major subcontractors (such as plumbers), and the local building department. For more on plans, see *Be Your Own House Contractor*, which describes each part of the plan in detail, provides samples, and even provides some advice on how to read plans.

Unless you have had special training, hire a professional to draw up a set of blueprints. An architect familiar with your chosen building technology is a good choice—for example, one who knows straw bale construction and has designed a few straw bale homes. Although natural architects are few and far between, the number is increasing. Call around and see who you can find. Some of the references listed in the resource guide may help you locate an architect in your area. Check out the local Yellow Pages and call around to friends. Also check out the organizations involved in different building technologies or find people in your area who have built a home similar to the one you are planning to build.

Architects are a bit pricey. They typically charge 10 to 15 percent of the cost of the project for their services. But an architect can be a valuable ally. He or she transforms your rough drawings and ideas into a more refined sketch for your consideration. Once you and your architect are seeing eye to eye, he or she prepares a detailed set of blueprints.

Another alternative is to hire a drafter/designer to prepare your blueprints. A drafter/designer is usually less expensive than an architect and, if well qualified and experienced, can do the job admirably. A surprising number of builders also do design work and will prepare blueprints for you at a reasonable rate, especially in the off-season (if there is one in your area). Many may be happy to serve as advisors for your project as well.

Besides being less expensive than a full-fledged architect, a builder-designer may bring practical experience to the project—common sense and experience that you may not find in a professional architect or designer. Builders often know what works best on the construction site, which may yield important benefits when it comes to designing a functional, durable building. And remember, hire someone who has lots of experience in alternative building technology, not someone who is hoping to use you as a guinea pig as they venture into the field.

Whomever you hire, be sure it is a person you can work with, someone who will listen to you and not force his or her ideas on you. Check out their fee schedule up front, so there are no misunderstandings. Some experts recommend paying for design services by the hour. You will find that fees vary, but don't fall into the trap of assuming that a lower fee is necessarily better. Just the opposite may be true: A higher fee may result in faster, better work from a more experienced designer. Sam Clark notes that "a good designer might spend a bit more of your money in the design phase, but save larger sums by producing an efficient, easy-to-build design." If you have the necessary skills, you may be able to draw some of your own designs and hire out a professional to do those that you can't draw.

A designer or architect may also be retained as a consultant during the building process. An inspection by the designer may be helpful after the completion of each task by a subcontractor, so you can pay with confidence.

Yet another option for the alternative homebuilder is to purchase a set of plans. Although there aren't as many sources as in the conventional building market, there are some significant sources. Michael Reynolds at Solar Survival Architecture offers extremely detailed blueprints containing approximately thirty pages of drawings for Earthships at a cost of $1,500 to $9,000, the price varying with the complexity of the structure. Nader Khalili offers a few house plans for earthbag homes through CalEarth. And there are books of straw bale designs.

Drawings prepared by a designer will be submitted for your review. Study the blueprints very, very carefully. Pour over the drawings with your spouse or significant other or friends. Imagine the house from every room. Compare room sizes to the sizes of rooms in your current house or apartment. Anticipate problems, such as windows in the wrong location or difficulty getting groceries from the car to the kitchen. Think about every wall. How will sunlight enter each room and at what time of day? Will sunshine bake you while you are cooking a Sunday feast for friends and family? Will it bounce off walls causing glare on TV and computer screens? Will afternoon sunlight cause your home to overheat? Will the plan result in areas with no sunlight at all? Will bathrooms be cold? You should even take the blueprints to the site to see how the house will interact with the sun, wind, topography, and other elements.

Next cut out scale drawings of your furniture and place them in the rooms to see if your furniture will fit and how much room is really available to you. Will furniture impede traffic flow? Will it be bleached by sunlight? Run your plans by someone who is experienced in design work. They may be able to locate potential problems that you won't see until the building is under construction or, even worse, completed. Another set of eyes could help you avoid mistakes or find better solutions. This review process is common practice among architects and designers.

You get the point. Be careful. This is the time to anticipate problems and make changes. Because many architects and builders are using computers to draw their blueprints, changes are fairly easy to make. If you are paying a flat fee, changes occurring at this stage are usually part of the base price. After the building begins, however, changes that require new drawings for approval will probably cost you extra. And, if you are using a builder to construct your home, changes after the bid has been accepted will necessitate a change order, an amendment to the contract that spells out the changes. Change order is synonymous with "Open your wallet, you're gonna shell out more money."

After your blueprints are finished, they will need to be approved by a structural engineer, which is required by most building departments. It helps protect the homeowner and also protects the building department from liability. Structural engineers primarily examine two aspects of the blueprints: foundations and framing (walls and roof). A structural engineer will go over each section in painstaking detail to be certain that the proposed structure is sufficiently strong to withstand live and dead loads and that it also meets local building code. After going over the plans, the engineer may approve them or call for changes before he or she stamps their approval and allows you to move to the next stage, submission to your local building department.

Getting plans approved by your building department varies from one place to the next. Some building departments have had a lot of experience with alternative building thanks to pioneers. Building departments such as these will generally approve your plans without much trouble, so long as they are stamped by a licensed structural engineer. If you are building in one of those areas where alternative building is either in its infancy or openly resisted by building department officials, you may have your work cut out for you.

Don't despair, though. Building departments can be persuaded with videos, copies of approved building codes from other parts of the country, and engineering reports on the particular technology you are pursuing. You can acquire this information from various organizations that promote your technology of choice. These are listed in the resource guide. The Earthship and straw bale folks have a lot of this information. Steve Berlant has gathered up information that may be useful when trying to obtain a permit for a cob house. Alternative builders may also have information that will help you obtain approval.

Another option is to obtain an experimental building permit. In the United States, most building departments have adopted the Uniform Building Code, a nationwide code that allows for the construction of experimental homes. If your building department is balking, a gentle reminder of this provision may help you. If they're still unmoved, a gentle reminder that you have legal recourse in the form of a lawyer might get the ball rolling. If this fails, contact an attorney who knows building codes. He or she will help you get your project approved.

Whatever you do, submit plans that are sensible, sound, and stamped by a structural engineer! And by all means, remain courteous and respectful, even in the face of hostility (although this can be difficult). Show building

1. Staking the lot
2. Clearing the land, removing topsoil, and excavating for foundation
3. Ordering utilities, temporary electric service, and portable toilet
4. Foundation construction (for concrete foundations this means forming and pouring footers and stem walls)
5. Rough in plumbing and electrical if on slab
6. Pour slabs, basement floor, and garage floor
7. Build exterior walls and roof
8. Install windows and doors
9. Apply exterior siding, if any
10. Roofing
11. Build interior walls
12. Rough-ins, install electrical wiring and plumbing in walls, crawl spaces, and attics
14. Insulate walls and ceilings
15. Finish all interior surfaces of walls
16. Prime and paint interior walls
17. Install interior doors and trim around windows
18. Install cabinets and countertops
19. Finish plumbing, electrical, and mechanical (backup heating systems)
20. Install tile, carpeting, hardwood flooring, etc.
21. Driveway construction
22. Landscaping
23. Final inspections
24. Get on with living

department officials that you care about the things they care about, such as the structural integrity of your home and the safety of its occupants. Remember, building departments are here to protect you and a string of owners who over the next century may occupy the house you build. Although they can get carried away, their intentions are good.

The best way to deal with building departments is to show them that you are informed about construction and interested in building a structurally sound home. Strive to maintain a friendly, open relationship. Show them the facts. But don't be afraid to ask questions. If you don't want to deal with them yourself, hire someone to take on the task. If you are building an Earthship, for instance, Solar Survival Architecture will, for a fee, meet with or call local building department officials and provide them with information. Hiring a consultant lets the building department know that you are serious about doing the job right and that you are persistent. The other advantage is that a professional knows the language and technicalities.

Remember that building permits aren't issued for free. There is a fee attached to your general permit, as well as permits for plumbing and electrical. You can build without a permit, but this practice can be risky. In fact, building departments can legally force you to tear down a structure that doesn't have a valid permit. I have seen it happen. But if you live in a remote location where building permits are loosely given and largely ignored, you can proceed without one. If this is a route you take, be certain to follow all of the procedures up to this point. That is, be sure to have good plans drawn up and be sure to have them examined by a structural engineer.

The Next Steps: Developing a Work Plan

After your plans have been approved, you can prepare to build. In temperate and cool climates, your window of opportunity for construction is narrower than in warmer climates. If you live in a cool or temperate climate, begin construction on your home in the spring or at the latest, the early summer. That way you can finish the house or most of it before winter sets in. Building a house can easily take six to twelve months. If you are building with straw, adobe, or cob, starting early is important so you can work around the weather—you want to be building exterior walls in the driest part of the building season. At least try to get the house rainproofed before fall rains come or winter snows begin to fly. Although you can complete inside work after a house is closed in, pouring concrete, applying exterior plaster or stucco, and roofing over the winter is a prescription for unnecessary headaches and much higher costs.

I'd recommend that design work (blueprints and engineering approval) be completed in the late summer to fall of the previous year and submission of plans be scheduled during the fall or winter, a time when the building department is frequently less busy. This will permit you to be ready to break ground in the spring, as soon as you can safely start excavating.

The time between approval of your plans and the start of construction is extremely important for those who are building their own homes. Put it to good use by first making a list of tasks in the order they must be completed. The sidebar on the facing page offers a list of jobs in approximate order. For more details on the steps involved in building a home, check out Heldmann's *Be Your Own House Contractor* and Becky Bee's *The Cob Builder's Handbook*, both of which contain detailed lists that may prove useful to you. *Build it with Bales* by Matts Myhrman and S. O. MacDonald lists the steps required to build a straw bale house and is written in the order of construction. Other major books on alternative building follow a similar approach.

Financing Your Project

After your list is made, it is time to obtain financing, if necessary. You can get loans from relatives or commercial lending institutions. If you need outside financing, you will probably need a construction loan. This is money lent to you, as needed, to build your house.

After approving your construction loan, the bank doles out money as each stage of construction is completed. When they give you the money, you turn around and pay subcontractors and suppliers. Sometimes they pay them directly. However, you are not obligated to pay the bank anything but interest after they start doling out money from the construction loan. Interest payments can even be made out of the construction loan. This arrangement makes it possible to pay rent or mortgage on one house while you are building another without going bankrupt. Once the house is completed, final inspections have been made by the building department and the bank, and you have been granted your certificate of occupancy, your loan is converted to a permanent mortgage. At this point you are responsible for both the principal and interest.

Obtaining a loan for alternative homebuilding can be difficult, although it varies from one location to another. If you are in an area where alternative housing is popular, your chances of getting approval are good. Check around. Fannie Mae, a federally chartered, stockholder-owned company that provides mortgages for homes, now recognizes nontraditional types of housing.

If you have never obtained a construction loan or financed a house, take a look at chapter 3 in *Be Your Own House Contractor*. It walks you through the

IN THE APPRAISAL and underwriting process, special consideration must be given to properties that represent special or unique housing for the subject neighborhood or market area. Mortgages on nontraditional types of housing—such as earth houses, geodesic domes, log houses, etc.—are eligible for delivery to Fannie Mae, provided the appraiser had adequate information to develop a reliable estimate of market value. It is not necessary for one or more of the comparable sales to be of the same design and appeal as the property that is being appraised.

Fannie Mae, "Underwriting Rural Properties"

process and even gives tips on estimating costs, so you know how much money to ask for.

If you are in doubt about financing, contact some banks and get their opinions before you purchase land and go to the expense of having plans drawn up. If you would like to build and remain mortgage free, I highly recommend Rob Roy's wonderful book, *Mortgage-Free! Radical Strategies for Home Ownership*. In this book, described as a "banker's worst nightmare," Roy offers numerous ways to escape from the oppression of a mortgage,

Maggie Remington

Is it possible to get financing to build or buy a natural home?
Yes, it just takes more effort.

How difficult will it be?
It takes a lot of coordinating and cooperation. When seeking financing for an alternative home, it is essential to find lenders and appraisers who are interested in promoting natural building. There are major hurdles to overcome with appraisal requirements. The more down payment involved, though, the easier it will be, and obviously the borrower needs to be creditworthy.

What do I need to obtain financing to build or buy a natural home?
You need to meet the credit criteria for the loan program and, of course, the lender needs to approve the property and the appraisal.

Are there any problems obtaining an appraisal for a home?
Yes. Obtaining an appraisal on a natural home can be a problem.

Most underwriters require one to three sold comparables within the last six months within a radius of one to five miles. Most natural homes are owner-built or custom built for a specific owner; the people love them so much they don't move out. This means there are no current sales to use for the appraisal.

Does it cost more to finance a natural home?
No, typically there are no additional costs. It just takes longer to get everything lined up.

What are some of the roadblocks people should anticipate?
The appraisal for an existing home being approved by the underwriter is the main roadblock, assuming the borrower meets financial criteria. Most investors state they will accept natural homes but know that until there is total market acceptance with current comparables the appraisal will not pass underwriting. So you can't qualify. It is a Catch-22.

How can I make the process go easier?
When buying an already built home, have as much information on your structure as possible available for the appraiser. Also, give the appraiser a list of like structures in the area for him to discuss in the appraisal.

Do you expect things to get better?
Yes. The more natural homes there are and the more that are sold, the easier it will be to obtain financing. Market acceptance is the main concern of lenders. If there is a strong market demand for these homes, the lending institutions will change their policies. Risk is always the main concern. Don't be discouraged. The final product is well worth the extra work required to obtain financing. The fact that so few come on the market for resale is the strongest indicator of how happy the owners are.

MAGGIE REMINGTON is a loan officer at the Bank of Telluride with a strong interest in natural building. Her office is located in Ridgway, Colorado.

spiced with real life tales of individuals who have managed to escape the tyranny of long-term debt.

Lining Up Supplies and Subcontractors

Once you have secured financing, make a list of supplies that will be needed for each job. Note items that may take longer to acquire than others, and place your orders for them early. If you are planning to obtain used windows, doors, cabinets, and other materials from friends, neighbors, or local building salvage outlets, start looking early to ensure you find what you need. The luxury of time will allow you to be pickier, too. There is nothing worse than needing windows or doors in a week and not being able to find them.

This is also the time to make a list of subcontractors you will need to hire. The sidebar on this page lists some of the subcontractors who you may need or who you may be required by your building department to hire. Yes, that's right, some building departments require licensed professional installation of plumbing and wiring. In other words, you may not have an option. Go through the phone book and ask friends and neighbors for recommendations. A good framer with experience in your community is an excellent source of referrals, especially if he or she does quality work. They also know who does superb work and who is reliable. Call potential subcontractors and get a feel for how easy they might be to work with. If they're gruff and discourteous, try someone else. When you have found a couple who seem appropriate, ask for references, then if references pan out, submit plans for bids.

Make sure subcontractors know exactly what you are trying to do. Don't just hand over the blueprints and assume they will grasp the details. Take some time to go over your design and tell them what you are trying to achieve. If they show enthusiasm for alternative building, that's all the better. One technique I have found that helps is to type out your needs, so that subcontractors have a written copy. Keep it simple. Bullet items. Don't write long, detailed sentences. No one is going to read an essay. Lists are also important because people in this business are notorious for forgetting details. And when they give you a bid, make sure they give you details. The more the better. Too many misunderstandings emerge because subcontractors don't spell out details in their contracts. You think you're getting one thing, and they give you another.

POSSIBLE SUBCONTRACTORS

- Surveyor to perform initial survey for plot plan
- Grading and excavation contractor to modify site, remove topsoil, and excavate foundation
- Foundation subcontractor to build forms for footers and stem walls
- Concrete subcontractor to pour slab or any concrete floors in your home and garage
- Carpenter to assist building standard exterior wall framing, interior walls, and roof, as well as installing doors and trim.
- Electrical contractor to install wires and fixtures
- Plumbing contractor to install water and waste lines and fixtures
- Heating contractor to install backup heating system
- Roofing contractor to install shingles or other roofing materials
- Insulation contractor to install insulation in walls, ceilings, and floors
- Drywall contractor to install drywall
- Stucco contractor to apply stucco to interior and exterior walls
- Painting contractor to paint interior and exterior walls
- Flooring, carpet, and countertop contractor to install tile, carpeting, hardwood flooring, and countertops
- Landscape contractor to assist in restoring the land and creating a haven for you and native species
- Cleaning contractor to clean up the job site and haul away what you can't recycle or compost

When hiring subcontractors, don't always go with the cheapest price. That may indicate lower quality workmanship. It may suggest that a person is not that popular in the building community because of poor work or an inability to perform work when they say they will. A subcontractor who shows up a week late can hold up the entire process, causing enormous delays. Their tardiness sends a ripple down the line of subcontractors who are waiting to do their job. If one is late, you may have to hold up everyone else for a week or two, which further destroys your downstream schedule, adds time to the project, and costs more.

I'd recommend that once you have a subcontractor with a good bid, don't be afraid to ask if they can give you a little better deal. I was never once turned down when I asked a sub for a discount on a large bill. There is no harm in asking. I simply asked, "Could you do it for X amount?" And the answer was always, "Sure." A few hundred here and a few hundred there can add up and may offset part of the cost of your solar system.

The rule of thumb in the building business is: Don't trust anyone's memory. Make sure you sign contracts with all of your subcontractors. Contracts should specify precisely what you will be getting. And never, never pay a subcontractor in full until the work is completed. Some may require a partial advance and payment upon completion of various stages of their work. That is totally acceptable, but don't pay anyone fully in advance. And never pay until their work has passed inspection.

If you withhold part of the final payment until you have had time to "test drive" their work, you are much more likely to get someone to fix problems that might arise. Be sure you include this provision in the contract you sign with them. Be sure to ask them to spell out all warranties for materials, fixtures, equipment, and workmanship. You should also be sure that all subcontractors carry liability insurance. This will protect you from any accidents that might occur.

With a list of tasks and bids from subcontractors in hand, set a start date, line up helpers, if any, and then begin site preparation work, assuming the weather is cooperative. You may need to rent a backhoe or hire someone to do some digging for you. When the excavation is finished, it is time to begin the foundation. Next come the walls and roof. The roof is typically covered with felt paper, then some form of roofing. Windows are installed and the structure is considered dried in. The electrical wiring can begin now that the structure is protected from rain.

Working with Building Inspectors

If you are working through your local building department, inspectors will be required to periodically review your construction progress as various stages are completed. The building department will provide you with a list that spells out the process. It will note each stage of the construction that must be inspected *before* you can proceed any further. In most instances, you will need to call a day or two in advance to get an inspector to the site. Pay close attention to the schedule. The sidebar on this page lists some of the inspections you may be required to have. Subcontractors, such as plumbers and electricians, will typically obtain permits, schedule inspections, and be there when the inspector shows up. Make sure it is spelled out in your contract with them.

When a building inspector arrives, be courteous and cooperative. Remember, it is their job to find fault. My experience is that inspectors are often pretty gruff and sometimes downright grumpy. Remember, though, their life is making other people follow the rules. The reason a building inspector may be grouchy is that he just came from a site where he was treated with disrespect and disdain. If you are pleasant and courteous and show respect, you are likely to get better treatment. They may never be overly friendly, but they won't be antagonistic.

Bottom line: Don't turn building inspectors into enemies. They hold your future in their hands. Listen carefully to the concerns of building inspectors. Ask questions to understand why something has to be a certain way. Propose options. Don't be defensive. Don't argue with them. Do what they say with a smile and they will be happy and you will get your certificate of occupancy. In most cases, what they are asking you to correct is minor. In some cases, you may have made a major structural error. Fix these, of course, or your house may fall down on you.

Doing the Work Yourself

Building your own home can be one of the most rewarding experiences of your life, especially if you are skilled in various aspects of homebuilding. Even so, unless you are building a simple, small cottage, you are in for a lot of work.

If you have never built a house before, be prepared to make a lot of mistakes and learn a lot. Be patient. Take your time. Most of the work is logical and can easily be figured out with a little thought. There are tricks that make it easier, so advice is always helpful. You may want to consider hiring a builder to either work alongside you or consult with you as you proceed. A few tips here and there can be very helpful—and can save you a lot of time and money. Proper tools can make your life infinitely easier, as well, and an experienced carpenter will introduce you to the proper tools for each phase of the project.

If you are not terribly experienced, you must be a fast learner and relatively skillful. Even though proponents of various natural building methods sometimes make claims regarding the ease with which one can erect a natural home, don't be swayed into false confidence. Alternative homebuilding requires a lot of skill, especially after the exterior walls have been erected. Roofs, for example, are demanding structures that you can't afford to mess up.

If you are building your first home, read as many books as you can on the subject. Seek advice of qualified builders. Watch videos. Attend a workshop or

INSPECTIONS

- Temporary electrical (saw service)
- Foundation (before pouring concrete)
- Slab (before pouring concrete)
- Electrical and plumbing (rough-in)
- Framing (rough-in)
- Insulation
- Electrical, plumbing, and mechanical (final)

two. And when you are building, be on the lookout for mistakes. If you make one, tear it down and start over. Don't try to hide mistakes, as they may come back to haunt you.

Working with Subcontractors

Electrical and plumbing work are prime candidates for subcontractors. Although it will probably cost you two or three times as much to hire licensed outside help, a good subcontractor can complete a project in a fraction of the time with quality far beyond most beginner's abilities.

Drywall is another candidate for outside help. Although this is a task that can be mastered by many people, a professional can do the job in much less time with much better results. It could take a beginner a few weeks to complete a project that a professional crew can do in three days. Save your time and let the professionals handle the drywall in your house.

Subcontractors must first estimate your job. Be sure you know exactly what you want and where you want it before you ask for a bid. After that, any change you make will cost you more, even if they result in less work. My personal theory of change orders is that builders and subcontractors who bid on homes underbid to get the job, knowing they can make up for their low bid with change orders. It's just a hypothesis, but everyone I've talked to has had such an experience.

The important rule is to make sure you go over the plans in fine detail. Walk through the house in your imagination, try to visualize what each room will look like and anticipate any problems that might arise. Imagine how long it will take hot water to arrive at a faucet. Refine your plans before you submit them for bid. Then, when time comes for a change order, politely challenge the price if it seems ridiculous and ask for an itemization of costs. This might keep the cost down, but don't count on it.

It is always important to sign a detailed contract with all subcontractors, and to check that they have liability insurance, so if one of their workers gets hurt on the job you won't get sued. Make sure that they pay their suppliers and that you aren't liable for costs they incur but fail to pay.

Sponsoring a Workshop

One nifty way to get a house built, save money, and reduce your own labor is to sponsor a workshop. The Cob Cottage Company in Oregon, for example, works with a limited number of clients each year, sponsoring hands-on work-

ADVICE FOR A NOVICE BUILDER

I come from an academic background and have had to learn building skills on my own over the years. During this time, I have found that it usually takes me a while to learn a new skill, such as tile or plastering or shingling. By the time I'm a tenth of the way through the job I've got the hang of it, but the work I've done leaves much to be desired. If you are like me, start your jobs in out-of-the-way corners of the house. Start shingling the back of the roof where no one will see it, or start tile in a utility room, not the front hallway! Better yet, volunteer to help a friend who is an experienced tradesperson. He or she can provide you with some instruction and oversight.

You can also call on friends who possess the required skills and trade them for skills you have. For example, if you're an experienced editor, you could volunteer to write a sales brochure for a carpenter who is trying to expand his or her business. Bartering is a great avenue open to virtually all of us.

shops to teach others how to build with cob. Many straw bale homes are also built this way, as are cordwood and log homes.

Workshop participants pay tuition, but the fee typically goes to the workshop organizers and teachers. What you get out of this is free labor. Workshops typically last three to four days, although some folks sponsor longer ones, lasting up to three weeks. Most workshops focus on building exterior walls. The rest of the project is up to you.

Although the idea of free labor sounds great, workshops are not without their challenges. Most participants bring with them very few building skills. While you may occasionally get a skilled carpenter, most participants must be watched over with great care to be certain your home is built correctly. I sponsored a straw bale workshop one weekend when I was building my home. Even though the tuition went to the instructor, I felt that he didn't watch over workers very carefully. The walls were pretty sloppy and my builders and I spent the next day or two straightening things out. And there is always the possibility of injury. Be sure that you are insured.

Timing Your Construction

Building always takes more time than you think and costs more than you project. To minimize cost overruns, build over the spring, summer, and fall. If you live in a temperate or cool climate and end up building over the winter, expect the project to require a lot more time. Expect costs to be higher, too. Because of some extremely bad luck, my builder began work in September. The winter was mild, for the most part, but it still made for slower going. As luck would have it, we ended up pouring the slab for the straw bale portion of the house in January on the coldest day of the year. This required us to build a tent over the 20-by-24-foot floor and heat the slab for a couple of days to protect the concrete from freezing. My builders, bless their hearts, even slept on the site to be sure the heaters didn't go out. Believe me, this made the whole project cost a bundle more.

BUILDING WITH A CONTRACTOR

If building your own house means letting someone else do the lion's share of the work, you will still have an active role in house design and choice of contractor. But this is where your life diverges from the person building his or her own home. Although a growing number of contractors are capable of building with adobe, straw bales, cob, and other alternative materials, your selection is very, very limited in comparison to standard homebuilders. Shop carefully. Picking the right builder is vital to the success of your project and your sanity. Remember, the person you hire will be in charge of spending your hard-earned money. You need someone you can trust, someone who knows how to stay on schedule and on budget.

Building your own home is an ancient tradition, a wonderful adventure, and with good planning, it's a practical and economic choice. It's probably the best way for families to establish a home—at least in rural areas—without taking on a huge mortgage.

SAM CLARK, *The Real Goods Independent Builder*

If you are lucky, you will be able to find a good contractor in your area. Check out the resource guide for names of builders. The Green Building Professionals Search should be of help. A few phone calls may put you in touch with a builder. Be persistent; alternative building is catching on and new builders are entering the field with regularity.

You can also contact the organizations that promote different building technologies for names of builders in your state. If you can't find one, you may have to hire a builder from another area to do the work. This will cost more because you will have to pay to house the builder and his crew for six months to a year. You may also have to serve as the general contractor in such instances. A builder from another area may not want to incur the expense required to apply for a contractor's license in your area or hassle with the paperwork. You can sign up as the general contractor, however, without knowing a thing about building. Permits will be issued in your name, even though your construction experience is limited to building a bird house that blew apart in a strong wind last year. Your responsibilities can be as few or as many as you want. You can oversee the entire job, hire subcontractors, and pretty much be in charge. Or, you can hire a builder to take over.

If you are going to serve as general contractor, read Carl Heldmann's book, *Be Your Own House Contractor*. It will walk you through the entire process. I recommend this book to anyone who is about to build a house, either on their own or with a contractor; it provides an excellent overview of the process.

Finding a good builder may mean that you will have to delay your project until he or she can finish up work elsewhere. My advice is to start looking early, so you can get on a builder's schedule. Look for a builder who is excited about your project and yet won't make decisions that you should be making. You want a builder who works hard and completes work on time. You want a builder who will work well with your architect or designer, if you plan to have the architect or designer oversee your project.

Builders may have their own crews to perform basic tasks, while subbing out plumbing, electrical, and drywall. Or, they may be general contractors who sub out the entire job. Be sure to ask builders or general contractors how they charge for their services. When it comes to subs, most builders and general contractors charge you what the subs charge them with a percentage (10 to 15

REFERENCE QUESTIONS

To find out more about a prospective builder, ask his or her previous clients the following questions:

1. Did the builder start the project when he or she said they would?
2. Did the builder complete the project on time? If not, why? Was this due to unforeseen problems or inefficient use of time?
3. Were you satisfied with the quality of the work? If not, what were the problems? How could they have been avoided?
4. Did the builder provide a detailed, easy-to-understand contract?
5. Did the builder give you a reasonable price and stick to the contract?
6. What were the strengths of the builder?
7. Did you have any problems working with the builder? What were they?
8. Did the builder stay within budget?
9. Did the builder deliver on his or her promises?
10. Was the builder responsive to your suggestions and needs?
11. Was it easy to talk with the builder? Did you understand what he or she said?
12. Did the builder respect the building site and minimize damage?

percent) added for overhead, such as office expenses. Because of this, a builder with his or her own crew is generally a cheaper way to go.

Don't begrudge builders their profit, unless of course it is excessive. Profits often get eaten up in mistakes and cost overruns that the builder ends up absorbing. If you want to learn more about different payment regimes, see Sam Clark's book.

When you find a builder or several builders that you are interested in, check out homes they have built. Talk to the people who worked with them. Ask them lots of questions, such as how hard or easy they were to work with, how good they were at keeping within budget, how responsive they were, how efficient they were, and how skilled they were. The sidebar on the facing page lists some of the many questions you should ask. Don't rely on a contractor's self-assessment, talk to the people who have hired them in the past.

As with subcontractors, you must have a detailed, ironclad contract when working with a builder. Your contract should stipulate costs, construction details, start-up dates, and completion dates. A good clause to include requires the builder to pay you a fee for late completion. Think what it will cost you to stay in a hotel after you've given up your rental unit or sold your house. I stipulated a $50-per-day late fee.

Specify every detail, even the types of doors. The more details you specify in your contract, the fewer problems you will have later on. And, by all means, have an attorney review the contract.

Avoid huge up-front payments. Pay as you go, so you can control the flow of cash. The last thing in the world you want to happen is to have your builder say he has spent all of the advance but hasn't been able to complete the work the payments were supposed to cover.

You can expect time delays and cost overruns, especially if you start changing things. Remember, contracts to build homes are what keep builders in business; it's the change orders that supply them with spending money. Don't fret, there will be changes. As a wall goes up, you may find that it makes a room look too small. Or you may decide that your kitchen will be more efficient if you shift the sink to another counter. You may find a better building product or a nicer finish product that costs more. No matter how good you are at visualizing a structure from blueprints, your mind will change as a building begins to materialize.

Contractors hire subcontractors, so as a rule, you don't have to worry about lining up your subs, but you can save money by finding the subcontractors yourself. Or you may be able to save money by doing some of the work yourself. Talk with your builder about this option, but be specific. If you are going to help out once in a while, don't expect the contractor to give you much of a price break. If you are going to tile the kitchen and paint the interior of the house, you should receive substantial discounts for these services. Wait till they give you a price before volunteering for the job. This way, you can exact as much savings as possible.

When building an alternative home, you may need supplemental power. In standard home construction, electrical power is delivered to the job site almost immediately by the local utility, usually as a temporary box. But if you are going to be running your home on solar and wind energy, you may need a generator on site to supply power tools. The builder should supply a generator. Don't volunteer to buy one for them to use just because you are going to need one later on. Use of your own generator during construction could significantly reduce its life span.

You and your builder will also have to figure out a way to supply water to the site if your home is going to rely on a catchwater system. Many alternative building technologies such as cob, adobe (if blocks are made on site), rammed earth, and straw bale construction are fairly water intensive, especially when foundations are being made or when the walls are plastered. Water can be supplied in temporary storage tanks or 55-gallon drums.

Refer to the books by Sam Clark and Carl Heldmann for more details on hiring designers, subcontractors, and builders. Clark offers especially helpful advice on creating a team approach and collaborative relationship with designers, builders, and subcontractors.

If you have settled on a natural building technique or two that you like, research them more fully. Read the books I have suggested and view videos, attend workshops, and tour some homes made from the natural building materials that interest you. Talk to the experts and to the "ordinary" folks who are living in them. Then go out and do it.

Good luck in your project and thanks for helping to build a sustainable future.

T

HIS RESOURCE GUIDE contains chapter-by-chapter listings of books, articles, videos, magazines, newsletters, builders and suppliers, and organizations–in that order. Rather than list chapter titles, I've chosen to list the major topics covered in each chapter. Because addresses, phone numbers, and Web sites change, I've tried to provide multiple access points for each resource. If a Web site or phone number is no longer in service, you may want to try writing.

Information in this guide is as current as possible, but these things do change. If you discover any changes, please notify me at my Web site at www.chelseagreen.com/Chiras. I'll post updates for readers at this site. If you think that you or a book, company, or other resource you know of should be listed, please contact me through my Web site.

CHAPTER 1 / SUSTAINABILITY AND SUSTAINABLE DESIGN

Books

Alexander, Christopher, Sara Ishikawa, Murray Silverstein, Max Jacobson, Ingrid Fiksdahl-King, and Shlomo Angel. *A Pattern Language: Towns, Buildings, Construction.* New York: Oxford University Press, 1977. Excellent resource for those seeking to build aesthetics and functionality into a home.

Allen, Edward. *How Buildings Work: The Natural Order of Architecture.* 2nd ed. New York: Oxford University Press, 1996. This book offers a great deal of information that will be useful to anyone who wants to build his or her own home.

Barnett, Dianna Lopes, and William D. Browning. *A Primer on Sustainable Building.* Snowmass, Colo.:

Rocky Mountain Institute, 1995. Excellent overview of sustainable design and construction.

Broome, Jon, and Brian Richardson. *The Self-Build Book: How to Enjoy Designing and Building Your Own Home.* White River Jct., Vt.: Green Books, 1998. Contains a lot of good advice on homebuilding.

Chappell, Steve K., and James J. Marks. *The Alternative Building Sourcebook.* Brownfield, Maine: Fox Maple Press, 1998. This reference lists over 900 products and professional services for natural and sustainable building.

Chiras, Daniel D. *Environmental Science: A Systems Approach to Sustainable Development.* 5th ed. Belmont, Calif.: Wadsworth, 1998. This book covers many subjects including sustainable ethics and principles of sustainable design.

_____. *Lessons from Nature: Learning to Live Sustainably on the*

Earth. Washington, D.C.: Island Press, 1992. A primer on sustainability that outlines key principles of sustainability and how to apply them.

Clark, Sam. *The Real Goods Independent Builder: Designing and Building a House Your Own Way*. White River Jct., Vt.: Chelsea Green, 1996. Although geared toward wood-frame construction, this book offers a wealth of information and some sobering words about building too far out on the fringe. Be sure to read the information on foundations, roof construction, and design for ergonomics and accessibility.

Edminster, Ann, and Sami Yassa. *Efficient Wood Use in Residential Construction: A Practical Guide to Saving Wood, Money, and Forests*. Washington, D.C.: Natural Resources Defense Council, 1998. A must for all builders. To order a copy send $15 plus $4 shipping to NRDC Publications Department, 40 West 20th Street, New York, NY 10011.

Fitch, James M., and William Bobenhausen. *American Building: The Environmental Forces That Shape It*. New York: Oxford University Press, 1999. An elegantly written book that will help you understand how buildings function. Good background reading.

Hermannsson, James. *Green Building Resource Guide*. Newtown, Conn.: Taunton Press, 1997. This book lists numerous green building products. Excellent resource.

Kennedy, Joseph F., ed. *The Art of Natural Building: Design, Construction, Technology*. Kingston, N.M.: Networks Productions, Inc. Compilation of presentations made at the Natural Building Colloquia. Avail-

able from the publisher at www.NetworkEarth.org.

Metz, Don, and Ben Watson, ed. *New Compact House Designs*. Pownal, Vt.: Storey, 1991. Contains plans for twenty-seven award-winning small house designs. Although they are conventional homes, you can get good design ideas for a natural home from this book.

National Association of Home Builders Research Center. *Directory of Accessible Building Products*. Upper Marlboro, Md.: NAHB Research Center, 1999. Another great resource. Available through NAHB at 400 Prince George's Blvd., Upper Marlboro, MD 20774. Tel: (800) 638-8556.

Pearson, David. *The New Natural House Book: Creating a Healthy, Harmonious, and Ecologically Sound Home*. New York: Simon and Schuster, 1998. Beautifully illustrated book that covers a wide range of topics.

Potts, Michael. *The New Independent Home: People and Houses that Harvest the Sun, Wind, and Water*. White River Junction, Vt.: Chelsea Green, 1999. Describes many facets of building energy-independent homes.

Videos

A Sampler of Alternative Homes: Approaching Sustainable Architecture. Two-hour video featuring numerous natural building technologies, including rammed earth, adobe, straw bale, tire homes, and a host of others. Gives an overview of the possibilities. To purchase a copy contact: Hartworks, Inc. 713 W. Spruce #89, Deming, NM 88030. Tel: (800) 869-7342, Web site: www.hartworks.com.

At Home with Mother Earth. This video covers ancient and current issues in earth architecture and contains interviews with important earth architects. You can order a copy at the CalEarth Web site. See the section on CalEarth products listed on the home page: www.calearth.org/.

Magazines and Newsletters

BackHome. P.O. Box 70, Hendersonville, NC 28793. Tel: (704) 696-3838. Covers independent living and related topics.

Building Concerns Newsletter. Offers a wealth of information on building issues, sustainable design, and green building products. Available online at www.greendesign.net.

Building with Nature. P.O. Box 4417, Santa Rosa, CA 95402-4417. Tel: (707) 579-2201. A newsletter for design professionals but easy enough to read for anyone interested in sustainable building.

Cohousing Journal. Published by the cohousing network listed under organizations.

Earth Quarterly (formerly *Dry Country News*). Box 23-J, Radium Springs, NM 88054. Tel: (505) 526-1853.Web site: www.zianet.com/earth. A new magazine devoted to living close to, and in harmony with, nature. Covers all aspects of natural life including homebuilding and renewable energy.

Eco-Building Times. Northwest EcoBuilding Guild, 217 Ninth Avenue North, Seattle, WA 98109. Tel: (206) 622-8350. In-depth articles on green building projects. The organization also publishes a *Green Pages*, an annual listing of ecologically sustainable designers, contractors, suppliers, and professional services.

Ecodesign. The British School, Slad Road, Stroud, Glos GL5 IQW, UK (+44) 1453-765575. This is the official journal of the Ecological Design Association and is available to all members.

Environ Magazine. P.O. Box 2204, Fort Collins, CO 80522. Tel: (303) 224-0083. Covers a wide range of topics related to sustainable building and design.

Environmental Building News. 122 Birge Street, Suite 30, Brattleboro, VT 05301. Tel: (802) 257-7300. Web site: www.ebuild.com. A bimonthly newsletter aimed at builders, architects, and owner-builders. Some people consider this to be the leading newsletter on environmental design and construction.

Environmental Design and Construction. 299 Market Street, Suite 320, Saddle Brook, NJ 07663. Tel: (201) 291-9001. Web site: www.EDCmag.com. Extremely valuable resource for anyone interested in sustainable design and building.

The Junction: Books and News from Chelsea Green, Publisher of Books for Sustainable Living. P.O. Box 428, White River Jct., VT 05001. Tel: (800) 639-4099. Chelsea Green is the leading publisher of books on sustainable building.

Organizations

American Institute of Architects, Committee on the Environment. 1735 New York Avenue NW, Washington, D.C. 20006. Tel: (202) 626-7300. Long interested and active in green building issues.

Austin's Green Builder Program. 206 E. 9th Street, Austin, TX 78701. Web site: www.greenbuilder.com/sourcebook/contents.html. Provides a wealth of knowledge on

water systems, energy systems, green building materials, and solid waste. They publish the *Sustainable Building Sourcebook.* Check out their on-line version.

Center for Maximum Potential Building Systems. 8604 F.M. 969, Austin, TX. Tel: (512) 928-4786. Web site: www.center@cmpbs.org. Offers a lot of information and services on sustainable building design.

The Cohousing Network. P.O. Box 2584, Berkeley, CA 94702. Tel: (510) 486-2656. Web site: www.cohousing.org/resources.html. This organization offers great information and a variety of services for those interested in cohousing.

Eos Institute. Lynne Bayless, 580 Broadway, Suite 200, Laguna Beach, CA 92651. E-mail: eos@igc.org. Non-profit educational and resource center for ecological building. Regional resources and referrals.

Smart Shelter Network. 684 6530 Road, Montrose, CO 81401. Tel: (970) 249-2396. Web site: www.smartshelter.com. Studies natural and sustainable building in southwestern Colorado. Documents building performance and provides consulting, slide shows, courses, workshops, and advocacy work with building officials, bankers, insurance companies, appraisers, politicians, and builders.

CHAPTER 2 / RAMMED EARTH

Books

Berglund, Magnus. *Stone, Log, and Earth Houses: Building with Elemental Materials.* Newtown, Conn.: Taunton Press, 1986. Chapters 9 to 12 of this book provide technical information on rammed earth construction and some beautiful photos.

Easton, David. *The Rammed Earth House.* White River Jct., Vt.: Chelsea Green, 1996. An informative, highly readable book. A must for anyone considering this building technology.

King, Bruce. *Buildings of Earth and Straw.* Sausalito, Calif.: Ecological Design Press, 1996. Another essential read for anyone interested in building a rammed earth home.

Middleton, G. F. *Earth Wall Construction.* Bulletin #5. North Ryde, NSW, Australia: CSIRO-DBCE, 1995. A manual on rammed earth showing a unique forming system. Appendices contain structural and insulation calculations.

Videos

Rammed Earth Construction. A 29-minute video produced by Hans-Ernst Weitzel. To order call Bullfrog Films at: (800) 543-3764.

The Renaissance of Rammed Earth. This 31-minute video features David Easton and serves as an excellent introduction to the subject or a companion to *The Rammed Earth House.* Available from Chelsea Green.

Magazines

Adobe Builder Magazine. Tel: (505) 861-1255. Web site: www.adobebuilder.com. This magazine, while focusing primarily on adobe, offers articles on rammed earth from time to time.

Builders/Suppliers

Earth & Sun Construction, Inc. 5105 Cueva Mine Trail, Las Cruces, NM 88011. Tel: (505) 522-5103. E-mail: Earth-sun@zianet.com. A rammed earth and adobe builder.

Huston Rammed Earth. P.O. Box 99, Edgewood, NM 87015. Tel: (505) 281-934. Builds rammed earth homes and walls (as subcontractor). Also consults nationally and internationally. Works in New Mexico, Colorado, Utah, Texas, and some in Arizona.

Rammed Earth Development, Inc. 265 W. 18th Street #3, Tucson, AZ 85701. Tel: (520) 623-2784. E-mail: redinc@azstarnet.com. Offers general contracting, wall construction, and design services. Builds about twelve houses per year.

Rammed Earth Solar Homes, Inc. 1232 E. Linden Street, Tucson, AZ 85701. Tel: (520) 623-6889. Web site: www.rammed.qpg.com/. Offers site evaluation, design and evaluation of customer plans, full or partial contracting, and monthly seminars.

Rammed Earth Works. 101 S. Coombs, Suite N, Napa, CA 94559. Tel: (707) 224-2532. E-mail: Rew@i-cafe.net. David Easton's company offers a variety of services, including design, construction, consultation, site evaluation, soil evaluation, referrals to engineers and energy evaluators, and workshops.

Soledad Canyon Earth Builders. 949 S. Melendres, Las Cruces, NM 88005. Tel: 505-527-9897. E-mail: info@adobe-home.com. Adobe and rammed earth home builders. Pat and Mario Bellestri.

Organizations

The Earth Building Foundation, Inc. Web site: www.nmia.com/~eaci/index.html. Formerly the Earth Architecture Center, International, Ltd. A nonprofit organization whose mission is to help people learn how to utilize earth building, especially adobe and rammed earth. Offers a newsletter, publications, information on building codes, workshops, and training. Especially helpful is an extensive search list of approximately 1,300 references.

CHAPTER 3 / STRAW BALE CONSTRUCTION

Books

Bolles, Bob. *The Straw-bale Workbook: A Guide to Building a House of Straw Bales.* Poway, Calif.: Bale Press, 1996. Has a wealth of information, especially for California residents. Contains cost comparisons between straw bale and conventional construction. Order a copy by calling (858) 486-6949.

Edminster, Ann V. *Strawbale Construction: A Sustainable System for Ownership.* Self-published, 1994. To order a copy contact the author at: 115 Angelita Ave., Pacifica, CA 94044.

Eisenberg, David. *Straw Bale Building and the Codes: Working with Your Code Officials.* Tucson, Ariz.: DCAT, 1996. Valuable resource for those whose building departments have never approved or heard of straw bale construction. To order a copy: www.zianet.com/blackrange/br_pages/order.html.

Farrant, Tim. *How to Build Straw Bale Landscape and Privacy Walls.* Self-published, 1996. To order write author at: P.O. Box 41991, Tucson, AZ. 85717. You can also order through: www.zianet.com/blackrange/br_pages/order.html.

Kemble, Steve, and Carol Escott. *How to Build Your Elegant Home with Straw* (manual and video set). Bisbee, Ariz.: Sustainable Systems Support, 1995. To order: Sustainable Systems Support, P.O. Box 318, Bisbee, AZ 85603. Web site: www/bisbeenet.com/.

King, Bruce. *Buildings of Earth and Straw.* Sausalito, Calif.: Ecological Design Press, 1996. A great book for the technically minded reader. Great information on tests run on straw bale structures.

Lanning, Bob. *Straw Bale Portfolio.* Tucson: Dawn/Out on Bale by Mail, 1996. This 79-page booklet contains numerous illustrations of sixteen house designs. To order a copy: www.zianet.com/blackrange/br_pages/order.html.

Myhrman, Matts, and S.O. Macdonald. *Build It with Bales (Version 2.0): A Step-by-Step Guide to Straw-bale Construction.* Tucson: Out on Bale, 1998. A superbly illustrated manual on straw bale construction.

Steen, Bill and Athena Steen. *Earthen Floors.* Elgin, Ariz.: The Canelo Project, 1997. A 30-page booklet covering all aspects of earthen floors. You can order this directly from the authors: www.deatech.com/canelo/order.html.

Steen, Athena S., Bill Steen, and David Bainbridge, with David Eisenberg, D. *The Straw Bale House.* White River Jct., Vt.: Chelsea Green, 1994. Excellent information on straw bale construction with numerous drawings and photos.

Straw Bale Testing Documents. Hard copy reports on code testing of straw bale walls. May be essential to help your building department understand what you are doing. A must-read for architects and builders and building department officials. To order a copy of all three reports: www.zianet.com/blackrange/br_pages/order.html.

U.S. Department of Energy. *House of Straw: Straw Bale Construction Comes of Age.* U.S. Department of Energy, 1995. Also available on-line at: www.eren.doe.gov/EE/strawhouse/.

Articles

Bolles, Bob. "Exterior Pinning Put to the Test." *The Last Straw* 25 (1999): 26. Fairly detailed and technical account of external pinning.

Jones, Barbara. "Thatching." *The Last Straw* 26 (1999): 29–30. A must-read for anyone considering a thatch roof.

_____. "Working with Lime in England." *The Last Straw* 26 (1999): 22–25. Excellent resource on lime plasters.

Lacinski, Paul. "The Rainscreen Siding Approach." *The Last Straw* 26 (1999): 21. Describes a technique for siding straw bale structures in cold- and wet-weather climates.

Piepkorn, Mark. "Buying Your Bales." *The Last Straw* 23 (1998): 22. An excellent article on buying straw bales.

Videos

Building with Straw. Vol. 1, *A Straw-Bale Workshop*. Black Range Films, 1994. Covers a weekend workshop in which volunteers helped to build a two-story greenhouse addition onto a lodge. To order: www.strawbalecentral.com.

Building with Straw. Vol. 2, *A Straw-Bale Home Tour*. Black Range Films, 1994. Takes you on a tour of ten straw bale structures in New Mexico and Arizona. To order: www.strawbalecentral.com.

Building with Straw. Vol. 3, *Straw-Bale Code Testing*. Black Range Films, 1994. Tours ten straw bale structures in New Mexico and Arizona; presents the insights of the owners/builders; To order: www.strawbalecentral.com.

How to Build Your Elegant Home with Straw Bales. Covers the specifics of building a load-bearing straw bale home. Comes with a manual. To order: www.strawbalecentral.com.

The Straw Bale Solution, narrated by Bill and Athena Steen and produced by Catherine Wanek. Features interviews with architects, engineers, owner-builders. Covers basics of straw bale construction and much more. You can order directly from the Steens by contacting their Web site: www.deatech.com/canelo/order.html or by contacting the producer at: www.strawbalecentral.com.

Magazines and Newsletters

The Last Straw. HC 66 Box 119, Hillsboro, NM 88042. Tel: (505) 895-5400. Web site: www.strawhomes.com. Quarterly journal containing the latest information on straw bale construction. Annual resource issue contains a gold mine of information. This is an absolute must for all straw bale enthusiasts!

Builders/Suppliers

For a comprehensive list of builders, check out the Human Resource List published annually in *The Last Straw* (see previous listing) or contact Out on Bale, 1037 E. Linden Street, Tucson, AZ 85719. Tel: (602) 624-1673. They have an up-to-date list of builders and architects.

Organizations

ArchiBio. Michel Bergeron, 6282 de St. Valier, Montréal, Québec, Canada. Tel: (514) 271-8684.

Strawbale Building Association for Wales, Ireland, Scotland, and England. SBBA, P.O. Box 17, Todmorden, OLI1 8fD, England. Tel: 011-44-1706-818126. Exchanges on information and experience in straw bale construction.

California Straw Building Association. 115 Angelita Avenue, Pacifica, CA 94044. Tel: (805) 546-4274.

Web site: www.strawbuilding.org. This is group is involved in testing straw bale structures. They also offer workshops and sponsor conferences.

The Canelo Project. HC 1, Box 324, Elgin, AZ 85611. Web site: www.caneloproject.com/.Founded and run by Athena and Bill Steen, coauthors of *The Straw Bale House*. They offer workshops, videos, and books on straw bale construction as well as information on building codes and results of tests on straw bale homes.

Center for Maximum Building Systems. 8604 FM 969, Austin, TX 78724. Tel: (512) 928-4786. Working at the cutting edge of building materials, systems, and methods. Led by Pliny Fisk III.

Development Center for Appropriate Technology. David Eisenberg, P.O. Box 27513, Tucson, AZ 85726-7513. Tel: (520) 624-6628. Web site: www.azstarnet.com/~dcat/. Offers a variety of services including consulting, research, testing, assistance with code issues, project support, instruction, and workshops.

European Straw Bale Contacts. See the annual resource issue of *The Last Straw*.

Greenfire Institute. Ted Butchart, 1509 Queen Anne Ave. N #606, Seattle, WA 98103. E-mail: greenfire@delphi.com. Offers straw bale workshops, design consultation, full design, building consultation, and full building options, all using straw or other sustainable materials.

Natural Building Resource Center. Rt. 1, Box 245B, Mauk, GA. Web site: www.gnat.net/~goshawk. Assists and supports owner-builders interested in straw bale, cob, and earthbags.

Straw Bale Association of Nebraska. Active in promoting straw bale construction. 2110 S. 33rd St., Lincoln, NE 68506-6001. Tel: (805) 483-5135.

Straw Bale Association of Texas. Sponsors monthly meetings, publishes a newsletter, and provides a host of other resources. P.O. Box 4211, Austin, TX 78763. Tel: (512) 302-6766. Web site: www.io.com/~whtefunk/sbat.html/.

Straw Bale Construction Association of New Mexico. Contact Catherine Wanek, Route 2, Box 119, Kingston, NM 88042.

CHAPTER 4 / EARTHSHIPS AND TIRE HOMES

Books

Berlant, Steve. *The Natural Builder.* Vol. 3, *Earth and Mineral Plasters.* Montrose, Colo.: Natural Builder, 1999. A superb resource on plasters.

Reynolds, Michael. *Earthship.* Vol. 1, *Build Your Own.* Taos, N.M.: Solar Survival Press, 1990. A must-read for those wanting to understand the basics of early Earthship design. This book contains some outdated information, however, so be careful. Be sure to read the more current volumes and check out the *Earthship Chronicles* for up-to-date information.

_____. *Earthship.* Vol. 2, *Systems and Components.* Taos, N.M.: Solar Survival Press, 1990. Explains the various systems such as graywater, solar electric, and domestic hot water. Essential reading for all people interested in sustainable housing.

_____. *Earthship.* Vol. 3, *Evolution Beyond Economics.* Taos, N.M.:

Solar Survival Press, 1993. Presents many of the new developments. Latest information, however, is always to be learned in workshops, tours of new houses, and the *Earthship Chronicles.*

Videos

Building for the Future. This is a video about the construction of my house. It explains how it was built and describes many green building products. Contact me at (303) 674-9688 or via E-mail at: Danchiras@aol.com.

Dennis Weaver's Earthship. Shows construction of actor Dennis Weaver's Earthship. Well done and very informative. Helpful in securing building permits. Available from Solar Survival Architecture at their on-line store at http://verde.newmex.com/earthship/literature_services.htm.

The Earthship Documentary. Describes the history of Earthship construction, the underlying philosophy behind this unique structure, and building techniques. Available from Solar Survival Architecture at their on-line store (listed above).

Earthship Next Generation. A look at new Earthship designs and constructions. Available from Solar Survival Architecture at their on-line store (listed above).

From the Ground Up. Takes you through the process of building an Earthship. Available from Solar Survival Architecture at their on-line store (listed above).

Magazines and Newsletters

Earthship Chronicles. Published by Earthship Global Operations, P.O. Box 2009, El Prado, NM 87529. Tel: (505) 751-0462. Pamphlets issued periodically to disseminate new information. You will find pamphlets on graywater, catchwater,

blackwater, mass vs. insulation, and an equipment catalogue.

Solar Survival Newsletter. Available from Solar Survival Architecture, P.O. Box 1041, Taos, NM 87571. Web site: Solarsurvival@earthship.org.

Builders/ Suppliers

Earthship Global Operations. P.O. Box 2009, El Prado, NM 87529. Tel: (505) 751-0462. E-mail: earthshp@taos.newmex.com. Web site: www.tasosnet.com/earthship/. Michael Reynold's headquarters. Contact them for books; supplies, such as preassembled PV and water filtration systems; hot water systems; blueprints; consultation; a list of approved builders; and much more.

Keith Lindauer. P.O. Box 113, Rico, CO. Phone: (970) 967-2882. E-mail: sunearth@theriver.com. One of many builders of straw bale homes and Earthships. Keith also sponsors the Colorado Natural Building Workshop, which features many natural building techniques.

Tim Pettet. P.O. Box 1054, Ouray, CO 81427. E-mail: Pettet@independence.net. Consultant and builder who specializes in Earthships, straw bale, adobe, straw-clay, earthen floors, and can/cement walls for planters, showers, and tubs.

Red Pueblo, LLC. 46195 Coal Creek Drive, Parker, CO 80134. Tel: (303) 841-7507. Builder Gary Dillard specializes in Earthships and Earthship Nests.

Solar Survival Architecture. P.O. Box 1041, Taos, NM. Tel: (505) 751-0462. E-mail: solarsuvival@earthship.org. For a wide range of services from design to construction.

Vanguard Homes. 110 S. Weber, Suite 103, Colorado Springs, CO 80903. Tel: (719) 634-2779. Web

site: www.thecybercenter. com/
vanguardhomes. Specializes in
Earthship derivatives, tire homes
with a more conventional outward
appearance, but also builds adobe
and straw bale homes.

CHAPTER 5 / ADOBE
Books

Berlant, Steve. *The Natural Builder.*
Vol. 1, *Creating Architecture from
Earth.* Montrose, Colo.: The Natural Builder, 1999. Discusses many
earthen building technologies, especially adobe. This book is an
extremely valuable resource. Available from www.naturalbuilder.com/.

Berlant, Steve. *The Natural Builder.*
Vol. 3, *Earth and Mineral Plasters.*
Montrose, Colo.: The Natural
Builder, 1999. Detailed discussion of
plasters and plastering. Great reference for anyone wishing to build an
adobe home. Available from
www.naturalbuilder.com.

Khalili, Nader. *Ceramic Houses and
Earth Architecture: How to Build
Your Own.* Hesperia, Calif.: Cal-
Earth Press, 1990. This book is
mostly about adobe and has good
information on kiln-fired adobe
blocks, foundations, walls, roofing,
and other topics as well.

Lumpkins, William. *La Casa Adobe.*
Santa Fe, N.M.: Ancient City Press,
1986. Contains drawings and floor
plans of adobe homes. The solar
structures in the back of the book
provide design ideas.

Mather, Christine, and Sharon
Woods. *Santa Fe Style.* New York:
Rizzoli, 1986. Contains hundreds of
photos of adobe structures and interior design. If you are looking for
ideas on how to finish rooms artisti-
cally, this book will help.

McHenry, Paul G., Jr. *Adobe and
Rammed Earth Buildings: Design
and Construction.* Tucson, Ariz.:
University of Arizona Press, 1984.
Excellent reference, covering history,
soil selection, adobe brick manufacturing, adobe wall construction, and
many more topics.

_____. *Adobe: Build it Yourself.*
Tucson, Ariz.: University of Arizona
Press, 1985. Highly readable and
surprisingly thorough introduction
to many aspects of adobe construction.

_____. *The Adobe Story: A
Global Treasure.* Washington, D.C.:
American Association for International Aging, 1996. Discusses the
history of adobe and dispels common myths about this ancient building technique.

Romero, Orlando, and David
Larkin. *Adobe: Building and Living
with Earth.* New York: Houghton
Mifflin, 1994. Beautifully illustrated
and wonderfully written treatise on
adobe.

Seth, Sandra, and Laurel Seth.
*Adobe! Homes and Interiors of
Taos, Santa Fe and The Southwest.*
Stamford, Conn.: Architectural
Book Publishing Company, 1988.
Largely black and white photo-
graphic collection of virtually every
aspect of adobe homes, including
walls, ceilings, floors, fireplaces,
patios, fences, and gates. Great for
design ideas.

Southwick, Marcia. *Build with
Adobe.* Denver, Colo.: Sage Books,
1994. Quite informative and full of
practical advice.

Stedman, Myrtle, and Wilfred Stedman. *Adobe Architecture.* Santa Fe,
N.M.: Sunstone Press, 1987. Contains numerous drawings of houses,
floor plans, and well-illustrated

basic information on making adobe
bricks and laying up walls.

Videos

Agni Jata. A video on ceramic
houses made in India. Available
from California Institute of Earth
Art and Architecture, 10376 Shangri
La Avenue, Hesperia, CA 92345.
Tel: (760) 244-0614. Web site:
www.calearth.org.

Earth, Water, Air, and Fire. Video of
the building and firing of a school in
Iran. Available from California Institute of Earth Art and Architecture,
listed above.

Sunrise Dome. Video documenting
the building and firing of a ceramic
dome in the United States. Available
from California Institute of Earth
Art and Architecture, listed above.

Builders/Suppliers

Adobe International, Inc. 1847
Libertyville Road, Libertyville, IA
52567. Web site: www.adobe-block.
com/index_main.htm. Sales office of
a New Mexico-based company that
manufactures a portable (trailer-
mounted) machine that makes pres-
sure-stabilized adobe blocks on sites.

**Earth Uprising Adobe Block and
Machine Company.** P.O. Box 122,
Arivaca, AZ 85601. Tel: (520) 622-
7188. Manufactures adobe blocks
made from compressed earth and
subcontracts wall construction.

The Natural Builder. Steve Berlant,
P.O. Box 855, Montrose, CO 81402
Web site: www.naturalbuilder.com/
who.html.On-site construction assistance, consulting services, and workshops on adobe, cob, straw bale,
light clay, natural plaster, and other
natural building techniques.

Soledad Canyon Earth Builders. See
description in rammed earth section.

Organizations

CalEarth, California Institute of Earth Art and Architecture. CalEarth/Geltaftan Foundation, 10376 Shangri La Avenue, Hesperia, CA 92345. Tel: (760) 244-0614. Web site: www.calearth.org. Nader Khalili's nonprofit organization that offers information on ceramic houses.

The Earth Building Foundation, Inc. See description listed in rammed earth section.

CHAPTER 6 / COB (MONOLITHIC ADOBE)

Books

Bee, Becky. *The Cob Builders Handbook: You Can Hand-Sculpt Your Own Home.* Murphy, Ore.: Groundworks, 1997. Amply illustrated and clearly written introduction to cob building.

Berlant, Steve. *The Natural Builder.* Vol. 2, *Monolithic Adobe Known as English Cob.* Montrose, Colo.: The Natural Builder, 1999. Discusses all aspects of cob construction. The unique level of detail in this book will assist you in designing and building a cob structure. Contains source material and engineering data on cob that is helpful when dealing with building departments. Available only from www.naturalbuilder.com.

_____. *The Natural Builder.* Vol. 3, *Earth and Mineral Plasters.* Montrose, Colo.: The Natural Builder, 1999. Detailed discussion of plasters and plastering. Great reference. Available only from www.naturalbuilder.com.

Cob Cottage Company. *Earth Building and the Cob Revival: A Reader.* Grove, Ore.: Cob Cottage Company. Collection of articles on cob.

To order: www.zianet.com/blackrange/br_pages/order.html.

Evans, Ianto, Linda Smiley, and Michael G. Smith. *The Cob Cottage.* White River Jct., Vt: Chelsea Green, forthcoming. A comprehensive guide to cob building by some of the nation's leaders.

Smith, Michael G. *The Cobber's Companion: How to Build Your Own Earthen Home.* 2nd Edition. Cottage Grove, Ore.: Cob Cottage Company, 1998. Well-written introduction to cob; many excellent and useful illustrations.

Videos

Building with the Earth: Oregon's Cob Cottage Co. Great resource. Obtain from The Cob Cottage Company, P.O. Box 123, Cottage Grove, OR 97424. Web site: www.deatech.com/cobcottage/.

Magazines and Newsletters

The CobWeb. The only cob-focused periodical. Published twice yearly by The Cob Cottage Company, listed above.

Builders/Suppliers

Cob Cottage Company. Contact them for information on cob construction, photos, answers to common questions, tours of cob buildings, workshops, cob products, and their apprenticeship program. Contact information listed above.

Groundworks. Web site: www.cpros.com/~sequoia/workshop.html. Becky Bee's company. Offers general information on cob construction, workshops, and more. P.O. Box 381, Murphy, OR 97533.

The Natural Builder. P.O. Box 855, Montrose, CO 81402. Web site: www.naturalbuilder.com. Offers on-site construction assistance, consulting, and workshops on cob, adobe,

straw bale, light clay, natural plaster, earthen floors, and other approaches to natural building.

Organizations

Center for Alternative Technology. Machynlleth, Powys SY20 9AZ. Tel. and fax: 01654 703409. Web site: www.cat.org.uk/. This educational group in the United Kingdom offers workshops on earth building and natural finishes, among other topics.

CHAPTER 7 / CORDWOOD

Books

Flatau, Richard. *Cordwood Construction: A Log-End View.* Merrill, Wisc.: Self-published, 1997. You can obtain a copy from the author at: W4837 Schulz Spur Drive, Merrill, WI 54452, or from Earthwood Building School (listed below).

Henstridge, Jack. *Building the Cordwood Home.* Fredericton, N.B., Canada: St. Annes Point Press, 1978. Documents author's experience building a cordwood home. Obtained via interlibrary loan.

_____. *About Building Cordwood.* Self-published, 1997. You can obtain a copy from the author at RR 1, Oromocto, New Brunswick, Canada E2V 2G2 or from the Earthwood Building School (listed below).

Lansdown, A. M., G. Watts, and A. B. Sparling. *Stackwall: How to Build it.* 2nd ed. Winnipeg, Manitoba, Canada: Northern Housing Committee, 1998. Difficult-to-find book, but well worth the search. Contains all kinds of information from chain saw maintenance to building walls.

Lansdown, A., and K. Dick. *Stackwall: How to Build It.* 2nd ed. Manitoba, Canada: Technical Services,

1997. Available from the publisher at P.O. Box 22, Anola, Manitoba, Canada R0E 0A0.

McClintock, Michael. *Alternative Housebuilding*. New York: Sterling, 1984. Contains a good section on cordwood homes with excellent drawings and photos.

Roy, Rob. *Complete Book of Cordwood Masonry Housebuilding: The Earthwood Method*. New York: Sterling, 1992. Contains a wealth of information on cordwood construction. Can be purchased from the Earthwood Building School (listed below).

Roy, Rob, ed. *Continental Cordwood Conferences Collected Papers*. West Chazy, N.Y.: Earthwood Building School, 1994, 1999. A transcript of twenty-five presentations by leaders in the field. Can be purchased from the Earthwood Building School (listed below).

Roy, Rob. *The Sauna*. White River Jct., Vt.: Chelsea Green, 1996. Fully illustrated book covering sauna lore, history, and the benefits of saunas. Three chapters describe the construction of cordwood masonry saunas.

Shockey, Cliff. *Stackwall Construction: Double Wall Technique*. 2nd ed. Vanscoy, Saskatchewan, Canada: Self-published, 1999. You can obtain a copy from the author at P.O. Box 193, Vanscoy, Saskatchewan, Canada S0L 3J0, or from Earthwood Building School.

Videos

Basic Cordwood Masonry Techniques with Rob and Jaki Roy. An 88-minute video that shows you how to bark wood, mix mortar, build cordwood walls, lay up window frames, and attach door frames. It discusses wood and cur-

ing, how to estimate quantities of wood, and lots more. Can be purchased from the Earthwood Building School.

Cordwood Homes. A tour of seven cordwood homes featuring interviews with owner-builders and several cordwood experts. This is not primarily a how-to video, although it does feature some new techniques. This video explores the whys, whats, and whos of cordwood building. Can be purchased from the Earthwood Building School.

Builders/Suppliers

Earthwood Building School. Offers books, videos, consultation, and workshops on cordwood construction. Contact them for the latest information on this unique natural building technique at 366 Murtagh Hill Rd., West Chazy, NY 12992. Tel:(518) 493-7744. Web site: www.interlog.com/~ewood.

CHAPTER 8 / LOG HOMES

Books

Berglund, Magnus. *Stone, Log, and Earth Houses: Building with Elemental Materials*. Newtown, Conn.: Taunton Press, 1986. Useful information on log home building.

Branson, Gary D. *The Complete Guide to Log and Cedar Homes*. Cincinnati, Ohio: Betterway, 1993. Covers all aspects of building or buying a log home and includes twenty-eight floor plans.

Burch, Monte. *Complete Guide to Building Log Homes*. New York: Sterling, 1984. Excellent book with numerous photos and drawings.

Chambers, Robert W. *Log Building Construction Manual*. River Falls, Wisc.: Self-published, 1999. If you are serious about building your own

handcrafted log home, read this detailed, well-illustrated book. Even comes in a metric version. Contains an extensive list of resources. You can obtain a copy by writing to author at: N8203 1130th St., 54022. Tel: (715) 425-1739. Web site: www.book@logbuilding.org.

Cooper, Jim. *Log Homes Made Easy: Contracting and Building Your Own Log Home*. Harrisburg, Pa.: Stackpole, 1993. This book is written for people who want to build a log home from a kit.

Hard, Roger. *Build Your Own Low-Cost Log Home*. Pownal, Vt.: Storey, 1985. Complete guide to log home construction.

Johnson, Dave. *The Good Woodcutter's Guide: Chainsaws, Portable Sawmills, and Woodlots*. White River Jct., Vt.: Chelsea Green, 1998. Excellent resource.

Log Building Association. *Log Homes: Land to Lock Up*. Bellingham, Wash.: LBA, 1998. Workbook to help those interested in having a handcrafted log home built for them. To obtain a copy, write LBA at P.O. Box 28608, Bellingham, WA 98228-0608.

MacKie, B. Allen. *Building with Logs*. Buffalo, N.Y.: Firefly Books, 1997. A very popular book well suited to those interested in building their own log home.

McRaven, Charles. *Building and Restoring the Hewn Log House*. Cincinnati, Ohio: Betterway. 1994. Although the history section of this book is hard to follow, the how-to portion is extremely well illustrated and thorough.

Renfroe, J. *The Log Home Owners Manual: A Guide to Protecting and Restoring Exterior Wood*. Bellevue, Wash.: Wood Care Systems, 1995.

Includes a list of suppliers. You can order a copy from the author at: 1075 Bellevue Way NE, Suite 181, Bellevue, WA 98004.

Teipner-Theide, Cindy, Arthur Thiede, and Jonathan Stoke. *The Log Home Book: Design, Past, and Present*. Salt Lake City, Utah: Gibbs Smith, 1995. Features hundreds of innovations in log home construction and numerous color photos. A great idea book.

Thiede, Arthur, and Cindy Teipner. *American Log Homes*. Salt Lake City, Utah: Gibbs Smith, 1992. Chronicles the history of log home construction and includes numerous floor plans.

Williams, Robert L. *How to Build Your Own Log Home for Less than $15,000*. Port Townsend, Wash.: Loompanics Unlimited, 1996. Very detailed and thorough discussion of building a log home.

Videos

Log Cabin Video. A 130-minute, inexpensively priced video of a PBS special on log home building. This video takes you through the entire process of handcrafting a log home. You can order a copy from Sunrise Productions, an on-line source of books and videos on building log homes. Web site: www.sunrise-publications.com/index.html.

Montana School of Log Building. Six-video series (11 hours) on log home construction by Al Anderson. Takes you through the process of building a log home from the bottom up. You can order a copy from Sunrise Productions (listed above).

Wonderful World of Log Homes. Produced by Jim Croft, this video offers a wealth of information on log home construction for those interested in handcrafted and log

home kits. You can order a copy from Sunrise Productions (listed above).

Magazines and Newsletters

Log Building News. Published three times a year by the Canadian Log Builders Association and the American Log Builders' Association. Contains a wealth of practical information. Contact Robert Chambers at N8203 1130th Street, River Falls, WI 54022. Tel: (715) 425-1739. E-mail: robert@logbuilding.org.

Log Home Living. Primarily features manufactured log homes. Published by Home Buyer Publications, 4451 Brookfield Corporate Drive, Suite 101, Chantilly, VA 22022. Web site: www.loghome.com.

Builders/Suppliers

The Internet contains Web sites for log home builders. Rather than list them here, I recommend you search the Web or look in the Yellow Pages. You can also contact the American or Canadian Log Builders' Association (listed below) for a list of builders of handcrafted log homes. Contact the Log Home Council (listed below) for a list of log home kit suppliers.

For information on architects and engineers, see Robert Chambers' *Log Building Construction Manual*. This book contains an extensive list.

Schroeder Log Home Supply. Tel: (800) 359-6614. Sells a wide range of books on log building.

Sun Country Logworks. Tel: (800) 827-1688. Web site: www.sun countrylogworks.com. Log home supplies, books, etc. See their on-line catalog for log building tools and equipment, log home finishes, books, and videos.

Organizations

American Log Builders' Association. P.O. Box 28608, Bellingham, WA 98228-0608. Tel: (800) 532-2900. Dedicated to furthering the craft of log home construction. Works in conjunction with the Canadian Log Builders' Association. Contact them for local builders associations. Publishes *Log Building News* and holds annual conferences with educational seminars.

Canadian Log Builders Association, International. #800, 15355 24th Ave., Box 465, White Rock, British Columbia, V4A 2H9 Canada. Tel: (800) 532-2900. Nonprofit organization dedicated to furthering the craft of handcrafted log construction. Jointly publishes *Log Building News* and offers an annual conference with ALBA.

Great Lakes LogCrafters Association. P.O. Box 633, Grand Rapids, MN 55744. Members are mostly from Minnesota, Wisconsin, and nearby states and provinces. Publishes a newsletter, meets twice a year, and sponsors workshops.

Log House Builder's Association of North America. (360) 794-4469. Web site: www.premier1.net/~log-house/. Trade association with 25,000 members worldwide representing handcrafted log home builders. Disseminates information on log homes. Offers a wide range of products and services, including apprenticeships, classes, free referral service, journal, newsletters, international meetings, tours, and a free question-and-answer hot line.

Log Home Council, an industry organization representing manufacturers of log home kits. Contact them at (800) 386-5242 for a brochure and a list of companies.

New Zealand Log Builders' Association. P.O. Box 8, Masterton, New Zealand. E-mail: tree-hut@clear.net.nz. Publishes a newsletter and meets every other year in New Zealand.

CHAPTER 9 / STONE HOUSES

Books

Berglund, Magnus. *Stone, Log, and Earth Houses: Building with Elemental Materials*. Newtown, Conn.: Taunton Press, 1986. Contains useful information on stone home building, including several case studies.

Cramb, Ian. *The Complete Guide to the Art of the Stonemason*. Cincinnati, Ohio: Betterway, 1992. A detailed reference, ideally suited for the beginner.

Frasch, Robert W., and Delia A. Robinson. *Details of Cobblestone Masonry Construction in North America 1825–1860*. Albion, NY: Cobblestone Society, 1993. Well-illustrated account of cobblestone construction.

Lawrence, Mike. *Step-by-Step Outdoor Stonework*. Pownal, Vt.: Storey, 1995. Contains illustrations on twenty different projects such as arches, walls, patios, paths, and steps.

Long, Charles. *The Stone Builder's Primer: A Step-by-Step Guide for Owner-Builders*. Widowdale, Ontario, Canada: Firefly Books, 1998. Contains the most complete instructions on stone home building of the books I've read or reviewed.

McRaven, Charles. *Building With Stone*. Pownal, Vt.: Storey, 1989. Contains great information on collecting stones and building stone walls.

_____. *Stonework: Techniques and Projects*: Pownal Vt.: Storey, 1997. Covers the use of stone to build walls, paths, ponds, steps, and much more.

Schwenke, Karl, and Sue Schwenke. *Build Your Own Stone House: Using the Easy Slipform Method*. Pownal, Vt.: Storey Books, 1991. Great little primer on the subject of stone house construction, featuring the slipform method.

Vivian, John. *Building Stone Walls*. Pownal, Vt.: Storey, 1986. A small, well-illustrated, and well-written book that focuses primarily on stone wall construction.

Organizations

The Cobblestone Society, P.O. Box 363, Albion, NY 14411. Tel: (716) 589-9013. An organization dedicated to the preservation of the craft of cobblestone building.

CHAPTER 10 / EARTHBAGS, PAPERCRETE, CAST EARTH, LIGHT STRAW-CLAY, BAMBOO, AND HYBRID HOMES

Books

Cusack, Victor. *Bamboo Rediscovered*. Trentham, Australia: Earth Garden Books, 1997. Fairly comprehensive guide on growing and harvesting bamboos. Includes some valuable information on building with bamboo.

Farrelly, David. *The Book of Bamboo*. San Francisco, Calif.: Sierra Club and Random House, 1995. Explains the benefits of bamboo and the many uses of bamboo over the years. It is best described as part catalogue, part history.

Janssen, Jules J. A. *Building with Bamboo*. London: Intermediate Technology Publications, 1995. A brief yet concise guide that outlines the basics of building with bamboo.

Laporte, Robert. *MoosePrints: Holistic Home Building*. Santa Fe, N.M.: Natural House Building Center, 1993. The only published source on straw-clay construction. Has some excellent illustrations, but contains only a fraction of the information you will need to learn this technique. To order a copy: www.zianet.com/blackrange/br_pages/order.html.

Solberg, Gordon. *Building with Papercrete and Paper Adobe*. Radium Springs, N.M.: Earth Quarterly, 1998. Collection of well-written, detailed articles on papercrete.

Also see *Build It with Bales* and *The Straw Bale House* listed in chapter 3 and *The Cobber's Companion* listed in chapter 6.

Articles

Steinfeld, Carol. "A Bamboo Future," *The Last Straw* 23 (1998): 25–26. An excellent article.

Steen, Athena, and Bill Steen. "The Straw Bale Earthen House," 1999. This article is posted at the Steens' Web page, www.deatech.com/canelo/sbearth.html.

Magazines and Newsletters

See *Earth Quarterly* listed in chapter 1 references.

Builders/Suppliers

Cast Earth Affiliates. 4022 E. Larkspur, Phoenix, AZ 85032. Tel: (602) 404-1044. Web site: www.build@castearth.com. Offers a variety of services for those interested in building a cast earth home.

Eric Patterson. 2115 Memory Lane, Silver City, NM 88061. Tel: (505) 538-3625. Eric is one of the pioneers in papercrete construction.

Living Systems Architecture and Construction. Web site: http://future. futureone.com/~mfrerkin.This company owned and operated by architect and builder Mike Frerking designs and builds cast earth homes and provides consultation. Contact him if you are a builder interested in training and licensing.

New Vision Building Unlimited. Mike McCain, Box 1331-EQ, Alamosa, CO 81101. Tel: (505) 531-2542. Information on papercrete, including a video and a schedule of workshops.

OK OK OK Productions. 265 E. 100 South, Moab, UT 84532. Tel: (435) 259-8378. Interested in creating affordable, sweat-equity housing using earthbag technology. Kaki Hunter and Doni Kiffmeyer offer workshops, design services, on-site supervision, consultation, and construction services.

Plyboo America, Inc. 745 Chestnut Ridge Road, Kirkville, NY 13082. Tel: (315) 687-3240. Web site: www.plyboo-america.com/html/ about.html. Sells bamboo flooring.

Safari Thatch and Bamboo, Inc. 2036A N. Dixie Hwy, Ft. Lauderdale, FL 33305. Tel: (954) 564-0059. Web site: www.safarithatch. com/. Sells many bamboo products including decorative and structural timber, fencing, flooring, and thatch.

Organizations

American Bamboo. Web site: www.bamboofurniture.com/. This company supplies a wide assortment of bamboo products including flooring, trusses, poles, panels, fiberboards, fences, and furniture.

The American Bamboo Society. Web site: www.halcyon.com/. Visit their Web site to find a list of books and publications, addresses and phone numbers of U.S. chapters, a listing of national bamboo societies, and a wealth of additional information.

The Australian Bamboo Network. Web site: www.ctl.com.au/abn/ abn.htm. A research organization formed to promote the cultivation and use of bamboo. Their Web site lists sources and suppliers, reading materials, and important links.

Bamboo Society of Australia. Bamboo Society of Australia, 1171 Kenilworth Road, Belli Park, Australia QLD 4562. Web site: www.bamboo.org.au/. An organization seeking to promote all aspects of bamboo in Australia.

CalEarth, California Institute of Earth Art and Architecture. Nader Khalili, CalEarth/Geltaftan Foundation, 10376 Shangri La Avenue, Hesperia, CA 92345. Tel: (760) 244-0614. Web site: www.calearth. org. Offers information on earthbag construction, including an on-line newsletter.

European Bamboo Society. Web site: www.bodley.ox.ac.uk/users/djh/ebs/. An informal federation of national European bamboo societies, each with their own administration and membership.

CHAPTER 11 / PASSIVE SOLAR HEATING AND PASSIVE COOLING

Books

Crosbie, Michael. J., ed. *The Passive Solar Design and Construction Handbook*. New York: John Wiley and Sons, 1997. A pricey and fairly technical manual on passive solar homes. Contains detailed drawings and case studies.

Dresser, Peter van. *Passive Solar House Basics*. Santa Fe, N.M.: Ancient City Press, 1996. This brief book provides basics on passive solar design and construction primarily of adobe homes. Contains sample house plans, ideas for solar water heating, and much more.

Freeman, Mark. *The Solar Home: How to Design and Build a House You Heat with the Sun*. Mechanicsburg, Pa.: Stackpole Books, 1994. Fairly useful introduction, although it contains more information on general building than passive solar design and construction.

Harland, Edward. *Eco-Renovation: The Ecological Home Improvement Guide*.White River Jct., Vt.: Chelsea Green, 1999. Information on remodeling and building homes.

Kachadorian, James. *The Passive Solar House*. White River Jct., Vt.: Chelsea Green, 1997. Presents a lot of good information on passive solar heating and an interesting design that's been successful in cold climates.

Moffat, Anne S., Marc Schiler, and The Staff of Green Living. *Energy-Efficient and Environmental Landscaping*. South Newfane, Vt.: Appropriate Solutions Press, 1994. Contains a wealth of information on energy-efficient landscaping. Also contains information on sun angle to help you determine where the sun will rise and set on your property throughout the year.

Potts, Michael. *The New Independent Home: People and Houses that Harvest the Sun, Wind, and Water*. White River Jct., Vt.: Chelsea Green, 1999. Delightfully readable book with lots of good information.

Roy, Rob. *The Complete Book of Underground Houses: How to Build a Low-Cost Home*. New York: Sterling, 1994. A revision of a 1979 best-seller with lots of new information on earth-sheltered homes. Can be purchased from the Earthwood Building School listed in the cordwood section of this guide.

Solar Survival Architecture. "Thermal Mass vs. Insulation," *Earthship Chronicles*. Taos, N.M. Solar Survival Architecture, 1998. Basic treatise on passive solar heating and cooling.

Taylor, John S. *A Shelter Sketchbook: Timeless Building Solutions*. White River Jct., Vt: Chelsea Green, 1983. Pictorial history of building that will widen your understanding of the relationship between form and function in building and open your eyes to some intriguing design solutions to a home's comfort, efficiency, convenience, and aesthetics.

Wells, Malcolm. *The Earth-Sheltered House: An Architect's Sketchbook*. White River Jct., Vt.: Chelsea Green, 1998. Although you won't find a ton of information on earth-sheltered housing in this book, you will be regaled with lots of inspiring designs that will help you see the potential of this design strategy.

Wilson, Alex, Jennifer Ihorne, and John Morrill. *Consumer Guide to Home Energy Savings*. Washington, D.C.: American Council for Energy-Efficient Economy, 1999. Excellent book.

Yost, Harry. *Home Insulation: Do It Yourself and Save as Much as 40%*. Pownal, Vt.: Storey, 1991. Extremely useful book to read for anyone building his or her own home.

Videos

The Solar Powered Home with Rob Roy. An 84-minute video that examines basic principles, components, set-up, and system planning for an off-grid home featuring tips from America's leading experts in the field of home power. Can be purchased from the Earthwood Building School listed in cordwood chapter.

Magazines and Newsletters

Earth Quarterly. See listing in chapter 1.

EREN Network News. Newsletter of the Department of Energy's Energy-Efficiency and Renewable Energy Network. See listing under organizations.

Home Energy Magazine. 2124 Kittredge Street, No. 95, Berkeley, CA 94704. Great resource for those who want to learn more about ways to save energy in conventional home construction.

Inside and Out. Newsletter of the Passive Solar Industries Council. See their listing under organizations.

National Renewable Energy Lab Now. Check out their newsletter on line at: www.nrel.gov.

Solar Today. 2400 Central Ave., Suite G-1, Boulder, CO 80301. Tel: (303) 443-3130. Web site: www.ases.org/solar. This magazine published by the American Solar Energy Society contains excellent information on passive solar, solar thermal, photovoltaics, hydrogen, and other topics. Be sure to check it out for names of engineers, builders, and installers and lists of workshops and conferences.

Builders/Suppliers

Alternative Energy Systems Company. Web site: www.powerisevery thing.com/. Offers a wide range of products and services for those interested in different forms of alternative energy.

Resnet. Residential Energy Services Network. P.O. Box 4561, Oceanside, CA 92052. Tel: (760) 806-3448. Web site: www.natresnet.org/. Nationwide network of mortgage companies, real estate brokers, builders, appraisers, utilities, and other housing and energy professionals.

Solar Works, Inc. Web site: http://home.solarvt.com/SolarVT/index.html. Offers consulting, sales and installation of solar equipment and training.

Organizations

American Solar Energy Society. 2400 Central Avenue, Suite G-1, Boulder, CO 80301. Web site: www.ases.org/solar/. Publishes *Solar Today* magazine and sponsors an annual national meeting. Also publishes an on-line catalogue of publications and sponsors the National Tour of Solar Homes. Contact this organization to find out about an ASES chapter in your area.

Center for Building Science. Web site: http://eande.lbl.gov/CBS/CBS.html. Lawrence Berkeley National Laboratory's Center for Building Science works to develop and commercialize energy-efficient technologies and document ways of improving energy efficiency of homes and other buildings while protecting air quality.

Center for Renewable Energy and Sustainable Technologies (CREST). 1612 K St., NW, Suite 410, Washington, D.C. 20006. Tel: (202) 293-2898. Web site: http://solstice.crest.org. Nonprofit organization dedicated to renewable energy, energy efficiency, and sustainable living.

El Paso Solar Energy Association. P.O. Box 26384, El Paso, Texas 79926. Web site: www.epsea.org/design. html. Active in solar energy, especially passive solar design and construction.

Energy Efficient Building Association. P.O. Box 22307, Eagen, MN 55122. Tel: (651) 994-1536. Web site: www.eeba.org/. Offers conferences, publications, and an on-line bookstore.

Midwest Renewable Energy Association. P.O. Box 249, Amherst, WI 54406. Tel: (715) 824-5166. Web site: www.the-mrea.org. Actively promotes solar energy and offers valuable workshops.

North Carolina Solar Center. Box 7401, Raleigh, NC 27695. Tel: (919) 515-3480. Offers workshops, tours, and much more.

Passive Solar Industries Council. 1331 H. Street, NW, Suite 1000, Washington, D.C. 20005. Tel: (202) 628-7400. Web site: www.psic.org/. This organization has a terrific Web site with information on workshops, books, and publications, and links to many other international, national, and state solar energy organizations. Publishes a newsletter, *Buildings Inside and Out*.

Renewable Energy Training and Education Center. 1679 Clearlake Road, Cocoa, FL 32922. Tel: (407) 638-1007. Offers hands-on training and certification courses for those interested in becoming certified in solar installation. Courses offered outside of the U.S.

Solar Energy International. P.O. Box 715, Carbondale, CO 81623. Tel: (970) 963-8855. Web site: www.solarenergy.org. Offers a wide range of workshops on solar energy, wind energy, and natural building.

CHAPTER 12 / SOLAR ELECTRICITY, WIND POWER, AND MICROHYDROELECTRIC

Books

Batelle Pacific Northwest Laboratory. *Wind Energy Resource Atlas of the United States*. Washington, D.C.: Windbooks, 1992. Published by the American Wind Energy Association.

Davidson, Joel. *The New Solar Electric Home: The Photovoltaics How-To Handbook*. Ann Arbor, Mich.: aatec Publications, 1987. Comprehensive and highly readable guide to photovoltaics, although it is a bit out of date.

Gipe, Paul. *Wind Energy Basics: A Guide to Small and Micro Wind Systems*. White River Jct., Vt.: Chelsea Green, 1999. The subtitle says it all; a great introduction to the subject.

_____. *Wind Power For Homes and Business: Renewable Energy for the 1990s and Beyond*. White River Jct., Vt.: Chelsea Green, 1993. Comprehensive, technical coverage of home wind power.

Jeffrey, Kevin. *Independent Energy Guide: Electrical Power for Home, Boat, and RV*. Ashland, Mass.: Orwell Cove Press, 1995. Contains a wealth of information on solar electric systems and wind generators and it is fairly easy to read.

Komp, Richard J. *Practical Photovoltaics: Electricity from Solar Cells*. 3rd ed. Ann Arbor, Mich.: aatec Publications, 1999. Fairly popular book on PVs.

NREL. *The Colorado Consumer's Guide to Buying a Solar Electric System*. Golden, Colo.: National Renewable Energy Lab, 1998. Provides basic information about purchasing, financing, and installing photovoltaic systems in Colorado that is applicable to many other states and countries as well. Contact NREL's Document Distribution Service at (303) 275-4363 for a free copy.

_____. *The Borrower's Guide to Financing Solar Energy Systems*. Golden, Colo.: National Renewable Energy Lab, 1998. Provides information about nationwide financing programs for photovoltaics and passive solar heating. Contact NREL's Document Distribution Service at (303) 275-4363 for free a copy.

Peavy, Michael A. *Fuel from Water: Energy Independence with Hydrogen*. 8th ed. Louisville, Ky.: Merit, 1998. Technical analysis for the engineers and chemists.

Potts, Michael. *The New Independent Home: People and Houses that Harvest the Sun, Wind, and Water*. White River Jct., Vt.: Chelsea Green, 1999. See description in Chapter 11.

Roberts, Simon. *Solar Electricity: A Practical Guide to Designing and Installing Small Photovoltaic Systems*. Saddle River, N.J.: Prentice-Hall, 1991. Good reference but a bit dated.

Schaeffer, John, and the Real Goods Staff. *Solar Living Source Book*. 10th ed. Ukiah, Calif.: Real Goods, 1999. Contains an enormous amount of background information on wind, solar, and microhydroelectric.

Solar Energy International. *Photovoltaic Design Manual, Version 2*. Carbondale, Colo.: Solar Energy International. A manual on designing, installing, and maintaining a PV system. Used in SEI's PV design and installation workshops.

Strong, Steven, and William G. Scheller. *The Solar Electric House: Energy for the Environmentally Responsive, Energy-Independent Home*. Still River, Mass.: Sustainability Press, 1993. A comprehensive and more technical guide to solar electricity.

Volunteers in Technical Assistance. *3-Cubic Meter Biogas Plant*. Mt. Ranier, Md.: VITA, 1980. This construction manual presents one of many designs for biogas production. It is available directly from VITA at 3706 Rhode Island Avenue, Mt. Ranier, MD 20822 or can be ordered from Jade Mountain.

Articles

Linkous, Clovis A. "Solar Energy Hydrogen—Partners in a Clean Energy Economy," *Solar Today* 13, no. 4, (1999): 22–25. A good, but detailed and somewhat technical article on hydrogen production.

Videos

An Introduction to Residential Microhydro Power with Don Harris. Produced by Scott S. Andrews. P.O. Box 3027, Sausalito, CA 94965. Tel: (415) 332-5191. Outstanding video packed with lots of useful information.

An Introduction to Residential Solar Electricity with Johnny Weiss. Good basic introduction to solar electricity. Source: listed above.

An Introduction to Residential Wind Power with Mick Sagrillo. A very informative video, especially for those wishing to install a medium-sized system. Source: listed above.

An Introduction to Solar Water Pumping with Windy Dankoff. A very useful introduction to the subject. Source: listed above.

An Introduction to Storage Batteries for Renewable Energy Systems with Richard Perez. This is one of the best videos in the series. It's full of great information. Source: listed above.

The Solar-Powered Home with Rob Roy. See listing in chapter 11.

Newsletters and Magazines

EN Network News. See listing in chapter 11.

Home Power Magazine. The Hands-On Journal of Home-Made Power. P.O. Box 520, Ashland, OR 97520. Tel: (541) 512-0201. Web site: www.homepower.com. Excellent resource. Check out their Web site, too.

Inside and Out. Newsletter of the Passive Solar Industries Council. See their listing under organizations.

Solar Today. See listing in chapter 11.

Wind Energy Weekly. Newsletter published by the American Wind Energy Association, listed under organizations.

Builders/Suppliers/Manufacturers

Alternative Choices. P.O. Box 128, Florence, CO 81226. Tel: (800) 784-3603. Lauri Campbell, the owner, is smart, friendly, and good at what she does. She will design a solar and wind system for your home.

Real Goods. 550 Leslie Street, Ukiah, CA 95482. Tel: (707) 468-9292. Order on-line at: www.realgoods.com. Real goods is the nation's leading distributor of renewable energy products, including solar electricity, solar thermal, microhydroelectric, and wind energy.

Solar Works, Inc. See listing in chapter 11.

Jade Mountain. P.O. Box 4616, Boulder, CO 80306-4616. Tel: (800) 442-1972. Web site: www.jademountain.com. Check out their catalogue for a variety of products for sustainable living, including renewable energy systems, lighting, appliances, and heating and cooling products.

Wind Can Do, Inc. 2320 Conc. #4, RR#1, Goodwood, Ontario L0C1A0 Canada Web site: www.windcando.com/. Provides wind generators from 500 watts to 50,000 watts. Their Web site also includes links to other wind energy companies and organizations throughout the world.

Organizations

American Solar Energy Society. See listing in chapter 11.

American Wind Energy Association. Web site: www.ogc.apc.org/awea/. This organization also sponsors an annual conference on wind energy. Check out their Web site, which contains a list of publications, their on-line newsletter, frequently asked questions, news releases, and links to companies and organizations.

British Wind Energy Association. 26 Spring Street, London W2 1JA. Tel: 0171 402 7102. Web site: www.bwea.com/. Actively promotes wind energy in Great Britain. Check out their Web site for fact sheets, answers to frequently asked questions, links, and a directory of companies.

Center for Alternative Technology. Machynlleth, Powys SY20 9AZ. Tel: 01654 703409. Web site: www.cat.org.uk/.This educational group in the United Kingdom offers workshops on alternative energy, including wind, solar, and microhydroelectric.

Center for Renewable Energy and Sustainable Technologies. See listing in chapter 11.

European Wind Energy Association. Web site: www.ewea.org/. Promotes wind energy in Europe. The organization publishes the *European Wind*

Energy Association Magazine. Their Web site contains information on wind energy in Europe and offers a list of publications and links to other sites.

National Renewable Energy Laboratory. 1617 Cole Blvd., Golden, CO 80401-3393. Tel: (303) 275-3000. Web site: www.nrel.gov. Researches and promotes development and implementation of a wide range of renewable energy strategies.

National Wind Technology Center of The National Renewable Energy Laboratory. Web site is: www.nrel.gov/wind/index.html. Their Web site provides a search mode, so you can check out their site and find a great deal of information on wind energy, including a wind resource database.

Solar Energy International. See listing in Chapter 11.

CHAPTER 13 / ALTERNATIVE WATER AND WASTE SYSTEMS

Books

Banks, Suzy, and Richard Heinichen. *Rainwater Collection for the Mechanically Challenged.* Dripping Springs, Tx.: Tank Town Publishing, 1997. Humorous and informative guide to above-ground rainwater catchment systems.

Campbell, Stu. *The Home Water Supply: How to Find, Filter, Store, and Conserve It.* Pownal, Vt.: Storey, 1983. Good resource on water supply systems, although it is dated. Unfortunately, it has very little about catchwater or graywater systems.

Del Porto, David, and Carol Steinfeld. *The Composting Toilet System Book.* Center for Ecological Pollution Prevention: Concord, Mass., 1999. Contains detailed information on composting toilets and graywater systems.

Harper, Peter. *Fertile Waste: Managing Your Domestic Sewage.* Machynlleth, Powys, U.K.: Centre for Alternative Technology, 1998. This brief book offers some useful information on composting toilets and handling urine.

Jenkins, Joseph. *The Humanure Handbook. A Guide to Composting Human Manure.* 2nd ed. Grove City, Pa.: Jenkins Publishing, 1999. Excellent resource. Well worth your time.

Kourik, Robert. *Drip Irrigation for Every Landscape and All Climates: Helping Your Garden Flourish While Conserving Water.* Santa Rosa, Calif.: Metamorphic Press, 1992.

Ludwig, Art. *Creating an Oasis with Greywater: Your Complete Guide to Managing Greywater in the Landscape.* Santa Barbara, Calif.: Oasis Design, 2000. Fairly detailed discussion of the various types of graywater systems.

_____. *Building Professional's Greywater Guide: Your Complete Guide to Professional Installation of Greywater Systems.* Santa Barbara, Calif.: Oasis Design, 1995. Contains a wealth of information on graywater systems, including important information on safety and chemical contents of detergents.

Solar Survival Architecture. "Catchwater," *Earthship Chronicles.* Taos, N.M.: Solar Survival Press, 1998. Focuses primarily on catchwater systems for Earthships, but has ideas that are relevant to all homes.

_____. "Greywater," *Earthship Chronicles.* Taos, N.M.: Solar Survival Press, 1998. Focuses primarily on greywater systems for Earthships,

but has ideas that are relevant to all homes.

_____. "Black Water," *Earthship Chronicles.* Taos, N.M.: Solar Survival Press, 1998. Provides an introduction to the black water systems under development by SSA.

U.S. Environmental Protection Agency. *US EPA Guidelines for Water Reuse.* Washington, D.C. U.S. EPA, 1992. Publication USEPA/USAID EPA625/R-92/004. You can obtain a copy of this document at US EPA National Center. for Environmental Publications, P.O. Box 42419, Cincinnati, OH 45242. Tel: (800)489-9198. Web site: www.epa.gov/epahome/publications.htm.

VanDer Ryn, Sim. *The Toilet Papers.* White River Jct., Vt.: Chelsea Green, 1999.

Videos

Rainwater Collection Systems. A brief, informative video that comes with a 50-page booklet that provides more details on systems and provides information on equipment and suppliers. Available from Garden-Ville Nursery, 8648 Old BeeCave Road, Austin, TX 78735. Tel: (512) 288-6113.

Builders/Suppliers

Jade Mountain. Offers a variety of useful products, including cisterns, water filters, low flush toilets. See listing in chapter 12.

Oasis Design. 5 San Marco Trout Club, Santa Barbara, CA 93105. Tel: (805) 967-9956. Web site: www.oasisdesign.net. Art Ludwig's company. Publishes books and offers design and consultation services on graywater systems.

Real Goods. Offers a wide variety of products for water systems, including pumps, composting toilets, efficient shower heads, and water filters. See listing in chapter 12.

Tank Town. P.O. Box 1541, Dripping Springs, TX 78620. Tel: (512) 894-0861. Offers a variety of materials and devices you will need to install a catchwater system.

For a complete list of international suppliers and manufacturers of composting toilets and related products, see Joe Jenkins, *The Humanure Handbook*, listed in books section for chapter 13.

Organizations

American Water Works Association. 6666 W. Quincy Avenue, Denver, CO 80235. Tel: (303) 794-7711. Web site: www.awwa.org. Concerned with many aspects of water, including water reuse. They publish proceedings from their water reuse conferences that are very informative.

Center for Alternative Technology. See listing in Chapter 12.

Rocky Mountain Institute. 1739 Snowmass Creek Road, Snowmass, CO 81654. Tel: (970) 927-3851. Web site: www.rmi.org. Check out the catalogue of this outstanding organization for publications on water efficiency and water reuse.

CHAPTER 14 / GREEN BUILDING MATERIALS

Books

American Institute of Architects, and Architects, Designers, and Planners for Social Responsibility. *The Sustainable Design Resource Guide.* Denver, Colo.: AIA and ADPSR, 1994. Now a bit dated and apparently out of print, but once a valuable resource. See listing for AIA in chapter 1.

Baker, Paula, Erica Elliot, and John Banta. *Prescriptions for a Healthy House.* Santa Fe, N.M.: Inword Press, 1998. Comprehensive guide to creating a healthy living space.

Bower, John. *The Healthy House: How to Buy One, How to Build One, How to Cure a Sick One.* Bloomington, Ind.: The Healthy House Institute, 1997. Comprehensive resource with an extensive resource guide of its own.

Certified Forest Products Council. *Good Wood Directory.* Beaverton, Oreg.: CFPC, 1996. Web site: www.certifiedwood.org. This book has been updated with a searchable database providing region-specific information.

Chappell, Steve K., ed. *The Alternative Building Sourcebook.* Brownfield, Maine: Fox Maple Press, 1998. Lists over 900 products and professional services in the area of natural and sustainable building.

Edminster, Ann, and Sami Yassa. *Efficient Wood Use in Residential Construction: A Practical Guide to Saving Wood, Money, and Forests.* New York: Natural Resources Defense Council, 1998. Describes how to reduce lumber use by 30 percent without compromising the structural integrity of a home.

Hermannsson, James. *Green Building Resource Guide.* Newtown, Conn.: Taunton Press, 1997. A goldmine of information on environmentally friendly building materials. Reader beware, however, not all building materials in books such as this pass the sustainability test.

Pearson, David. *The Natural House Catalog: Everything You Need to Create An Environmentally Friendly Home.* New York: Simon and Schuster, 1996. Contains a lot of information on building and fur-

nishing a sustainable home, including a list of products and services.

_____. *The New Natural House Book: Creating a Healthy, Harmonious, and Ecologically Sound Home.* New York: Simon and Schuster, 1998. See description in chapter 1.

Roulac, John W. *Hemp Horizons: The Comeback of the World's Most Promising Plant.* White River Jct., Vt.: Chelsea Green, 1997. Well-written and well-informed book on industrial hemp, an agricultural product used to make numerous building products.

Rousseau, David, and James Wasley. *Healthy by Design: Building and Redmodeling Solutions for Creating Healthy Homes.* 2nd ed. Point Roberts, Wash.: Hartley and Marks, 1999. Extremely informative reference.

Spiegel, Ross, and Dru Meadows. *Green Building Materials: A Guide to Product Selection and Specification.* New York: John Wiley and Sons, 1999. The newest entry into the green building materials category. Looks like a great resource.

Magazines and Newsletters

Environmental Design and Construction, This magazine features many articles on alternative building products and has numerous ads by manufacturers of green building products. Also publishes a buyer's guide that lists green building materials and suppliers. See listing in chapter 1.

Builders/Suppliers/ Manufacturers

Because there are so many manufacturers of green building materials, please refer to a local retailer or check out John Hermannson's Green Building Resource Guide *or*

Spiegel and Meadow's Green Building Materials *or check out the Austin Green Building Program's* Sustainable Building Sourcebook. *The same holds true for hemp products. If you want a complete listing, see the Hemp Resources section of John Roulac's book,* Hemp Horizons, *which is listed above.*

Building for Health Materials Center. P.O. Box 113, Carbondale, CO 81623. Tel: (970) 963-0437. E-mail: crose@sopris.net. Offers a complete line of healthy, environmentally safe building materials and home appliances including straw bale construction products; natural plastering products; flooring; natural paints, oils, stains, and finishes; sealants; and construction materials. Offers special pricing for owner-builders and contractors.

EcoBuild. David Adamson. Tel: (303) 444-5332. E-mail: dadamsonson@indra.com. This new company in Boulder, Colorado works specifically with builders, providing consultation and an assortment of green building materials at competitive prices.

Eco-Products, Inc. 1780 55th Street, Boulder, CO 80301. Tel: (303) 449-1876. Offers a variety of green building products including plastic lumber.

Eco-Wise. 110 W. Elizabeth, Austin, TX 78704. Tel: (512) 326-4474. Retail store that carries Livos and Auro nontoxic natural finishes and adhesives.

Environmental Building Supplies. 1331 NW Kearney Street, Portland, OR 97209. Tel: (503) 222-3881. Green building materials outlet for the Pacific Northwest.

Environmental Construction Outfitters. 44 Crosby Street, New York, NY 10012. Tel: (800) 238-5008.

Sells an assortment of green building materials.

Environmental Home Center. 1724 4th Ave. South, Seattle, WA 98134. Tel: (800) 281-9785. Offers a variety of green building materials.

National Association of Home Builders. 1201 15th Street NW, Washington, DC 20005. Tel: (800) 368-5242 or (202) 822-0200 within the Washington, DC metropolitan area. Web site: www.nahb.com. This organization has begun to pursue green building and now sponsors a national conference.

Planetary Solutions. 2030 17th Street, Boulder, CO 80302. Tel: (303) 442-6228. Long-time player in the green building movement. Offers paints, flooring, tile, and much more.

CHAPTER 15 / ENVIRONMENTAL AND ENERGY EFFICIENT LANDSCAPING AND SITING

Books

Clark, Sam. *The Real Goods Independent Builder: Designing and Building a House Your Own Way.* White River Jct., Vt.: Chelsea Green, 1996. Check out the chapters on choosing a site and site planning.

Moffat, Anne S., Marc Schiler, and the Staff of Green Living. *Energy-Efficient and Environmental Landscaping.* South Newfane, Vt.: Appropriate Solutions Press, 1994. Contains an abundance of information on landscaping strategies and plant varieties suitable for your climate zone.

Mollison, Bill. *Permaculture Two: Practical Design for Town and Country in Permanent Agriculture.* Stanley, Tasmania, Australia: Tagari Books, 1979. A seminal work in the field of permaculture.

Magazines and Newsletters

Earthwood Journal. Eos Institute, 580 Broadway, Suite 200, Laguna Beach, CA 92651. Tel: (714) 497-1896. A glossy permaculture magazine published by the Eos Institute and the Permaculture Institute of Southern California. Geared to the professional designer, architect, and land-use planner.

The Permaculture Activist. P.O. Box 1209W, Black Mountain, NC 28711. Tel: (828) 298-2812. Web site: http://metalab.unc.edu/pc-activist/.Publishes articles on a variety of subjects related to permaculture, including permaculture workshops.

Permaculture Drylands Journal. Permaculture Drylands Institute, P.O. Box 156, Santa Fe, NM 87504. Tel: (505) 983-0663. Web site: http://members.aol.com/pdrylands/PDIhome1.htm. Quarterly journal that focuses on the practice of permaculture in arid lands, especially Arizona and New Mexico.

Permaculture Edge. Permaculture Nambour, Inc., P.O. Box 148, Inglewood 6050, Western Australia. Reports on cutting-edge developments in permaculture.

Permaculture International Journal. P.O. Box PG6039, South Lismore, NSW 2480, Australia. Tel: (066) 220-020. Web site: http://nornet.nor.com.au/environment/perma/.Contains articles, book reviews, and news of permaculture events.

Permaculture Magazine UK: Ecological Solutions for Everyday Life. Permanent Publications, Hyden House Limited, Little Hyden Land, Clandfield, Hampshire PO8 ORU, England. Web site: www.gaia.org/permaculture. Quarterly journal published in cooperation with the Permaculture Association of Great

Britain. Contains articles, book reviews, and solutions from Britain and Europe.

Builders/Suppliers

Permaculture Resources. P.O. Box 65, 56 Farmersville Road, Califon, NJ 07830. Tel: (800) 832-6285. Web site: www.jump.net/users/perma/. An educational publisher and distributor of permaculture resources and publications.

Organizations

Appropriate Technology Transfer for Rural Areas. P.O. Box 3657, Fayetteville, AR 72702. Tel: (800) 346-9140. Actively involved in the permaculture movement.

International Permaculture Institute. An international coordinating organization for permaculture activities such as accreditation. Address: P.O. Box 1, Tyalgum, NSW 2484, Australia. Tel: (066) 793 442.

National Wildlife Federation. 1400 16th Street, NW, Washington, D.C. 20036-2266. Their Backyard Wildlife Habitat Program has helped thousands of homeowners improve landscaping for wildlife.

CHAPTER 16 / BUILDING A HOME

Books

Bridges, James E. *Mortgage Loans: What's Right for You*. Cincinnati, Ohio: Betterway, 1997. This book will help you understand mortgages and pick the one that's best for you.

Dickinson, D. *Small Houses for the Next Century*. New York: McGraw-Hill, 1995. One of several books on the subject. Full of good information.

Freeman, Mark. *The Solar Home: How to Design and Build a House You Heat with the Sun*. Mechanicsburg, Pa.: Stackpole Books, 1994. This book contains a wealth of information on building your own home, including many practical aspects.

Heldmann, Charles. *Be Your Own House Contractor: Save 25% without Lifting a Hammer*. Pownal, Vt.: Storey, 1995. Although it is geared toward conventional home building, the book will walk you through the steps of building any home, giving advice on many issues such as permits and working with subcontractors.

McGuerty, Dave, Kent Lester, R. Adam Blake, and Bob Beckstead, ed. *The Complete Guide to Contracting Your Home*. Cinncinati, Ohio: F&W Publications, 1997. An updated version of a very popular book on contracting your own home.

Roy, Rob. *Mortgage-Free! Radical Strategies for Home Ownership*. White River Jct., VT: Chelsea Green, 1998. Wonderful book that should be on the top of your list if you're looking to build a home but wish to avoid the tyranny of mortgage payments.

Wilson, Alex, Jenifer L. Uncapher, Lisa McManigal, L. Hunter Lovins, Maureen Cureton, and William D. Browning. *Green Development: Integrating Ecology and Real Estate*. New York: John Wiley and Sons, 1998. Contains an enormous amount of information for professional and nonprofessional readers.

Organizations

National Association of Exclusive Buyer Agents. For a referral in your area, contact their headquarters at 7652 Gartner Road, Suite 500, Evergreen, CO 80439. Tel: (800) 986-2322. Or contact them via e-mail at NAEBAHQ@naeba.org. You can also visit their Web site at www.naeba.org.

Resnet. Residential Energy Services Network. P.O. Box 4561, Oceanside, CA 92052. Tel: (760) 806-3448. Web site: www.natresnet.org/. A nationwide network of mortgage companies, real estate brokerages, builders, appraisers, utilities, and other housing and energy professionals. This organization is dedicated to improving the energy efficiency of the nation's housing stock and expanding the national availability of mortgage financing.

INDEX

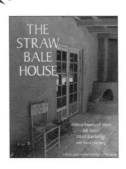
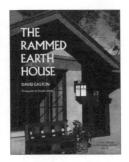
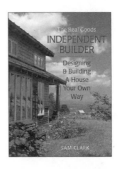

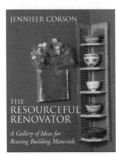

CHELSEA GREEN

Sustainable living has many facets. Chelsea Green's celebration of the sustainable arts has led us to publish trend-setting books about organic gardening, solar electricity and renewable energy, innovative building techniques, regenerative forestry, local and bioregional democracy, and whole foods. The company's published works, while intensely practical, are also entertaining and inspirational, demonstrating that an ecological approach to life is consistent with producing beautiful, eloquent, and useful books, videos, and audio cassettes.

For more information about Chelsea Green, or to request a free catalog, call toll-free (800) 639–4099, or write to us at P.O. Box 428, White River Junction, Vermont 05001. Visit our Web site at www.chelseagreen.com.

Chelsea Green's titles include:

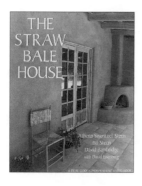  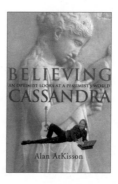

The Straw Bale House
The New Independent Home
Independent Builder:
   Designing & Building a
   House Your Own Way
The Rammed Earth House
The Passive Solar House
The Earth-Sheltered House
Wind Energy Basics
The Solar Living Sourcebook
Mortgage-Free!
The Beauty of
   Straw Bale Homes
The Resourceful Renovator
Serious Straw Bale:
   A Home Construction
   Guide for All Climates

Four-Season Harvest
The Apple Grower
The Flower Farmer
Passport to Gardening
The New Organic Grower
Solar Gardening
Straight-Ahead Organic
Good Spirits
The Contrary Farmer
The Contrary Farmer's
   Invitation to Gardening
Whole Foods Companion
The Bread Builders
The Co-op Cookbook
Keeping Food Fresh
The Neighborhood Forager
Simple Food for the Good Life

Believing Cassandra
Gaviotas: A Village to
   Reinvent the World
Beyond the Limits
The Man Who Planted Trees
Who Owns the Sun?
Global Spin: The Corporate
   Assault on Environmentalism
Seeing Nature
Hemp Horizons
Genetic Engineering, Food,
   and Our Environment
Scott Nearing: The Making
   of a Homesteader
Loving and Leaving the
   Good Life
Wise Words for the Good Life

FIVE
PRACTICES
FOR
EQUITY-FOCUSED
SCHOOL
LEADERSHIP

FIVE PRACTICES FOR EQUITY-FOCUSED SCHOOL LEADERSHIP

SHARON I. **RADD** | GRETCHEN GIVENS **GENERETT** | MARK ANTHONY **GOODEN** | GEORGE **THEOHARIS**

Alexandria, Virginia USA

1703 N. Beauregard St. • Alexandria, VA 22311-1714 USA
Phone: 800-933-2723 or 703-578-9600 • Fax: 703-575-5400
Website: www.ascd.org • E-mail: member@ascd.org
Author guidelines: www.ascd.org/write

Ranjit Sidhu, *CEO & Executive Director;* Penny Reinart, *Chief Impact Officer;* Stefani Roth, *Publisher;* Genny Ostertag, *Director, Content Acquisitions;* Susan Hills, *Senior Acquisitions Editor;* Julie Houtz, Director, *Book Editing & Production;* Joy Scott Ressler & Liz Wegner, *Editors;* Thomas Lytle, *Creative Director;* Donald Ely, *Art Director;* Masie Chong, *Graphic Designer;* Keith Demmons, *Senior Production Designer;* Kelly Marshall, *Manager, Project Management;* Shajuan Martin, *E-Publishing Specialist;* Christopher Logan, *Senior Production Specialist*

All web links in this book are correct as of the publication date below but may have become inactive or otherwise modified since that time. If you notice a deactivated or changed link, please e-mail books@ascd.org with the words "Link Update" in the subject line. In your message, please specify the web link, the book title, and the page number on which the link appears.

PAPERBACK ISBN: 978-1-4166-2975-7 ASCD product #120008 n2/21
PDF E-BOOK ISBN: 978-1-4166-2976-4; see Books in Print for other formats.

Quantity discounts are available: e-mail programteam@ascd.org or call 800-933-2723, ext. 5773, or 703-575-5773. For desk copies, go to www.ascd.org/deskcopy.

Library of Congress Cataloging-in-Publication Data
Names: Radd, Sharon I., author.
Title: Five practices for equity-focused school leadership / by Sharon I.
  Radd, Gretchen Givens Generett, Mark Anthony Gooden, George Theoharis.
Description: Alexandria, Virginia: ASCD, 2021. | Includes bibliographical
  references and index.
Identifiers: LCCN 2020033261 (print) | LCCN 2020033262 (ebook) | ISBN
  9781416629757 (paperback) | ISBN 9781416629764 (adobe pdf)
Subjects: LCSH: Educational leadership—Social aspects—United States. |
  Educational equalization—United States. | Children with social
  disabilities—Education—United States.
Classification: LCC LB2805.R24 2021 (print) | LCC LB2805 (ebook) | DDC
  371.2/011—dc23
LC record available at https://lccn.loc.gov/2020033261
LC ebook record available at https://lccn.loc.gov/2020033262

29 28 27 26 25 24 23 22   4 5 6 7 8 9 10 11 12

We dedicate this book to all those who work in schools, day in and day out, to create a more just and inclusive world; without you, this book would have no purpose. Thank you for living your commitment to engage in the exhausting but urgent work of creating more equitable schools.

SR

To my teachers, those I chose and those who chose me, including and especially my BIPOC teachers (whether you've come as a friend, colleague, student, or formally as an instructor)—I am better because of your patience, care, courage, and commitment in pushing me to see differently and know beyond the confines of my race, gender, class, abilities, and upbringing. I hope you find in this book that your hard work and investment have paid off in some meaningful way.
To Max, Shelia, and Sam—You, too, are my teachers. You have taught me how to truly love. I am inspired every day by your intellect, persistence, courage, sensitivity, and commitment. The world is better with each of you in it.
To Jim, my forever love—You are the wind beneath my wings. I am endlessly grateful for all the ways you love, encourage, support, challenge, nurture, and care for me and our family.

GGG

To William and Gabrielle—Thank you for the hard questions that you ask. You are loving, smart, thoughtful, and tenacious in ways that make me a better educator. I love you to the moon and back.

To Bill—Thank you for all of the love and support you continue to show me day in and day out, year after year.

To my teachers (in formal and informal settings)—I am because of you.

MAG

To Angela—Thanks so much for the incredible support you have provided in your own loving and unique way. Surely I am a better person because of you and your commitment.

To Nia Ayanna—Thank you for bravely continuing to speak up and out. Your wit and humor continue to amaze me, sometimes challenge me, and make me so proud.

To all of my teachers—Though you are too many to name, you have collectively taught me to explore, agitate, dream, discover, and study harder as you encouraged me to continually strive to reach my potential. Asante Sana!

GT

To my teachers from kindergarten to today—Thank you for your patience, insistence, and belief in my abilities, even when those were hard to see.

To those who continue to push me to see the world more fully, to see myself more critically, and to understand the ways we can change the things we cannot accept, thank you.

To Ella and Sam—Thank you for your ebullient spirits and for making lots of lemonade. Remember, we need to choose what is right over what is easy.

To Gretchen—Thank you for sharing your heart and for your fierce commitment to building a more just community... also, thank you for glamping with me.

To Jeanne and Liz—Thank you for your care and the ways you relentlessly make the world better. And bagels, thank you for introducing me to NYC bagels.

FIVE PRACTICES FOR EQUITY-FOCUSED SCHOOL LEADERSHIP

Preface

In early 2013, I called George (a mentor and friend since 2007), then Mark (whom I'd met and whose work I had followed), and then Gretchen (whom I knew of but had never met) and asked them to join me in thinking about the unique practice of school leadership that aims to make meaningful structural and systemic shifts toward educational equity. I will remain forever grateful to each of them for saying yes to the possibility of the four of us working together and, ultimately, for the ways they opened themselves and shared their experiences, energy, wisdom, investment, and hearts to envision how we might offer something of value. That they selected me as lead author of this book reflects only that I made the initial contacts; it does not reflect the true investment of time, knowledge, ideas, and words on the page that we collectively imparted. I am both honored and humbled to be in partnership with Gretchen, Mark, and George and forever changed by the process of, and learning from, our work and relationships together.

We began our four-person team first by talking about ideas and approaches to this type of leadership, and then began to discuss what form our collective work might take, eventually deciding to write a book—this book—together. Our sole purpose was to address an overwhelming need in the field for practical and effective resources and guidance to make systemic change that advances social justice in schools and school districts while building cohesive communities. We wanted to develop a book that would provide leaders with processes and tools to engage in the deeply disruptive nature of equity change with thoughtful attention to fundamental human needs for connection, growth, and homeostasis.

Through our multiyear process of developing content and writing together, we witnessed national events that were changing our world: Ferguson; Philando Castile; the growing power of Black Lives Matter; the unprecedented election of President Donald Trump; the ushering in of the post-truth era; Brett Kavanaugh and the #MeToo movement; escalating climate change; sweeping and inhumane anti-immigrant actions; increasing awareness and acceptance of gender diversity; and the increasing levels of both activism and polarization aided and fueled by the evermore omnipresent role of the internet and social media in our lives. As of this writing, we are months past the initial stay-home orders in most states related to the Coronavirus, and the U.S. count has surpassed 18 million cases of infection, with more than 1.5 million deaths worldwide. These numbers, as well as any report on the economic impact of this crisis, will undoubtedly seem incredibly stunted by the time you read this. We cannot predict the lasting and unimaginable ways this global pandemic—or the political and social polarization that accompanies it in the United States—will change our world on every scale, including in schools and in our daily lives. By the time you read this, although you will have a better idea, the true impact will not be known for years.

Beyond the COVID-19 pandemic, we are witness to social unrest and uprising on the largest scale in decades, ignited by the slow and painful murder of George Floyd, held under the knee of a Minneapolis police officer for over nine minutes, and captured on videotape for the world to see, just a few miles from my home in Minneapolis.

We believe that whatever happens in the months and years ahead, and however our lives change, the processes and tools in this book will be useful for you. We wrote this book such that it could be practical and useful across contexts, and not rely on any particular set of assumptions about the way your school, district, or other educational setting might work, with the exception that yours is a human system, it is focused on learning, and it exists in a centuries-old context of racism, sexism, classism, and other forms of exclusion.

To that end, we know that the management literature and paradigms about how business leaders grow and create more profit through their companies and practices simply cannot be directly applied to schools that seek to serve all students, particularly as they exist in broader systems that structurally guarantee and necessitate inequality and inequity. Educational equity requires the examination and revision of tightly held beliefs about how schools are structured and operationalized, for what purpose, and to what end. The strength of this book originates from the combined array of life, educational, and professional histories we each brought to our work together, as well as our experiences as practitioners, consultants, and professors. We hope it serves as an encouraging guide for you to address the challenge of bringing together individuals with diverse perspectives and life experiences to create an organization that truly fosters meaningful learning in every aspect of its environment.

We hope this book will help you and your team build the types of relationships that we have in our work together. Through our work, we increasingly engaged in open and honest conversations about race, gender, dis/ability, and other aspects of identity and, as a result, our collaboration has required courage, commitment, persistence, humility, compassion, and honesty. We have experienced countless moments as individuals of different races (Mark and Gretchen identify as Black; George and I identify as White) faced with the decision to withhold an opinion or perspective or to step into more honesty, openness, and curiosity. These challenges were complicated and exacerbated by the realities of power and privilege and lived experience that would caution against such honesty and risk. Through our persistent leaning in, we were astounded, not by the ongoing need to address race and racism in our work for social justice, but by our own need to continue learning and talking about race and racism across racialized identities. Gretchen and Mark were skillful and compassionate teachers, and George and I were grateful for new insights and learning but discouraged at times by the magnitude of our own knowledge gaps, despite now decades of targeted learning in attempts to know and understand. Gretchen shared with me on more than one occasion that it was critical that she witnessed George and I struggle and was witness to our subsequent actions as a result of our new learning. As she shared, "The everyday experience of working for justice, to create more equitable educational spaces, is often frustrating and isolating. It was important for me to know that despite knowledge gaps, not only are there White educators committed to and actually doing equity work, but some of us are dedicated to teaching other White educators how it can be done. It was important for me to see that you and George were not just allies in theory, but also in practice." Throughout our teamwork, we developed an unspoken but very-much-alive commitment to hold one another with grace and compassion, while simultaneously (1) facing the stark and inhumane realities of racism, sexism, ableism, classism, and all other forms of exclusion, discrimination, dehumanization, and violence; and (2) struggling through the complexities of systems of oppression and the challenges of intersectionality.

Equity leadership requires that kind of commitment, courage, and self-reflection from all of us. It requires us to imagine equitable spaces and ideas that we have not seen in practice. To do this, leaders require a bold vision, significant knowledge and skills, and collaboration with many people. This book takes leaders and teams through an engaged process that involves deep learning. It creates space for teams to step back from their day-to-day struggles to focus on the big picture, and then it moves them toward action for creating the schools our nation wants and needs. This book invites educational leaders to transform how they think and, ultimately, what they do as a result.

A key feature of this work and thus of this book is a scaffolded use of stories: our stories, the stories of our schools, the stories of communities, the stories we tell, and so on. This book employs our collective understanding that embedded within stories is the unique opportunity to reclaim and reimagine how things might be if we can use stories as impetus for self-reflection and growth, as well as to develop others (Bruner, 1990). As a result, this book provides many opportunities for you to name, reclaim, revise, and expand your equity stories and to place them alongside other stories. Throughout the book we offer a variety of activities designed to encourage you and your team to unearth and improve your equity lenses and leadership by naming, reflecting, and examining the narratives (stories, experiences, and perspectives) that shape your thinking. In the process, we believe that you will learn, as we did, that you are not alone. By engaging in this process together, you will strengthen yourselves as individual leaders and as a team that will be better able to design and lead change.

We are so excited that this book is here now and that you have chosen to pick it up and see what it holds! We believe you will find it useful, instructive, challenging, and affirming. Most importantly, we thank you for your commitment to, and work in, transforming systems to include and truly serve all students, their families, and their communities.

With hope and in leadership,
Sharon I. Radd
With my beloved friends and colleagues
Gretchen Givens Generett, Mark Anthony Gooden, and George Theoharis

PRACTICE I

Prioritizing Equity Leadership: Adopting a Transformative Approach

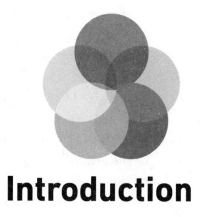

Introduction

Schools in the United States, while providing many great experiences and opportunities, are still grossly inequitable. Consider these facts:

- *School discipline disproportionately targets students of color, students with disabilities, and transgender students.* Black students are suspended or expelled at three times the rate of White students (U.S. Department of Education Office for Civil Rights, 2014, 2016); they represent 16 percent of the student population, but up to 40 percent of students suspended and 27 percent of students referred to law enforcement. Students with disabilities are twice as likely to be suspended than their nondisabled peers. These statistics worsen for students of color with disabilities: Nearly 25 percent of boys of color with disabilities are suspended and approximately 20 percent of girls of color with disabilities receive suspension, compared to 6 percent of students without disabilities. Transgender students are twice as likely to be disciplined than their cisgender peers (Kosciw, Greytak, Palmer, & Boesen, 2014). Those who are suspended are less likely to graduate and more likely to enter the juvenile justice system than those who aren't.
- *White and Asian students are twice as likely as their Black and Latino/a/x peers to attend high schools where the full range of math and science courses—Algebra I, geometry, Algebra II, calculus, biology, chemistry, physics—are available.* Further, Black and Latino/a/x students as well as students with disabilities and English

learners (ELs) are less likely to be in these classes even when they are offered.

- *While students with disabilities represent 12 percent of the student enrollment, they make up 58 percent of students placed in seclusion or involuntary confinement and 75 percent of students who are physically restrained (U.S. DOE OCR, 2014, 2016).* These students comprise 22 percent of students who are retained and only 2 percent of students in an AP class. More than 1.1 million students have minimal or no access to general education and limited inclusion with their peers; in many places, this disproportionately impacts students of color (U.S. DOE OCR, 2016).

- *English learners have graduation rates (Layton, 2014) at approximately 59 percent nationally, with some states graduating only 24 percent of their ELs.* While making up approximately 5 percent of high school students, ELs represent 11 percent of students retained or held back, and 21 percent of ELs are chronically absent (U.S. DOE OCR, 2016).

- *Lesbian, gay, bisexual, transgender, queer or questioning, and intersex (LGBTQIA+) students are targeted at alarming rates.* Seventy-four percent are verbally harassed, 36 percent are physically harassed, and 17 percent are assaulted. While the majority of LGBTQIA+ students miss school and stay away from extracurricular events because they feel unsafe, most (57 percent) do not report it to school officials as they feel nothing will be done. They're not off-base in their predictions: 62 percent of LGBTQIA+ students who did report harassment or assault to school authorities indicated that nothing was done (Kosciw et al., 2014).

- *Since 9/11, Muslim and Arab students and their families have faced heightened concerns about harassment and being targeted.* In 2014, more than 50 percent of Muslim students reported being insulted or abused because of their religion (Blad, 2016; Shah, 2011). Approximately 30 percent of girls who wore hijabs reported that their head coverings were inappropriately touched or pulled. Many have experienced teachers, principals, and fellow classmates profiling them as associated with terrorists.

We believe much can be done to change these circumstances, yet school leaders don't always have the tools, knowledge, and resources to actually do so. We wrote this book because we wanted to offer an actionable framework that individual leaders and school, district, and interorganizational teams can use to address this need.

As an author team, we have experience as practicing preK–12 educators and administrators, preparing preK–12 educators and administrators, and as consultants supporting inservice preK–12 educators and administrators. We have connected with practitioners from across the United States and in other countries to understand the challenges and barriers—and also the opportunities for transformation—that exist in our educational systems. Over the span of our professional careers, we have seen hundreds of attempts at school reform. We have had the fortunate opportunity to participate in, study, and learn from these efforts. We have learned a lot about what works and what doesn't. We want to share that with you!

We also want to share the diversity of perspectives from which this book is written: As an author team, we are a White woman (Sharon), a Black woman (Gretchen), a Black man (Mark), and a White man (George).[1] We are each cisgender, straight, English-speaking, and currently nondisabled and financially secure. We share this information because we believe each aspect of a person's identity is a space where inclusion or marginalization can occur, and that school leaders must seek to be anti-oppressive and commit to full inclusion across this wide spectrum of sociocultural identities. (We'll discuss this in greater detail later in the book.)

That said, you will find that we begin with and center race as we think about educational equity. We do this for a couple of reasons. First, race is similar to other aspects of identity in that it is a social construction; the social construction of race results in profound inequality, and inequality related to this aspect of identity is persistent across history. On the other hand, race is unique in our society because we are deeply segregated based upon this identity. People receiving advantages related to having a White identity (i.e., people who identify as, or are identified as, White) are shielded from seeing or acknowledging their advantages. And, with a few exceptions, race is a fixed aspect of our identity, unlike some other aspects of our identities where we see segregation (e.g., socioeconomic class, religion).

Next, we recognize that this book will most likely be read and used by a majority White audience. While more than 50 percent of school children are of color (National Center for Education Statistics, 2020a), fewer than 20 percent of teachers (National Center for Education Statistics, 2020b), approximately 22 percent of principals (National Center for Education Statistics, 2020c), and 6 percent of superintendents (Kowalski, 2013) are of color. We wrote this book with that reality in mind.

---

[1] Throughout the book, we have elected to capitalize Black and White, based upon the use of these terms to refer to people in a racialized way. In order to reflect our antiracist stance, we do not capitalize white supremacy/ists, white nationalism/ists, white privilege, white fragility, and whiteness.

Last, in our experience, we've found that racism is the most difficult form of inequality for educators to discuss and address. We begin with and center race in our approach because we believe that if you can understand and work on racism in schools, you can understand and work on inequality and inequity in many of its other forms.

Given all of this, we hope to help White educators and leaders become more comfortable thinking and talking about racism and, more importantly, skilled antiracist and anti-oppressive leaders. Further, we think this book is useful for educators and leaders of color as we guide you to examine all aspects of your identity and experiences through strengthening your antiracist and anti-oppressive approaches, and thinking systemically about equity-focused change.

Our framework outlines five meta-practices for building equity-focused systems as illustrated in Figure I.1. We refer to these as meta-practices for three reasons:

- Your professional work is a "practice," and the meta-practices in this book are to be applied in that professional practice.
- We mean that these are literally practices, that you should rehearse and perform them repetitively and cyclically. As opposed to believing that "practice makes perfect," we believe that practice builds new and needed habits and that your new habits will be necessary for effective and authentic equity leadership.
- Each meta-practice contains a robust set of practices for you to use in your professional work.

FIGURE I.1
Five Practices for Building an Equity-Focused System

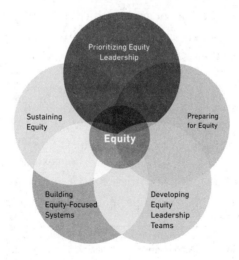

The practices build on one another, so it is important that this book be read from beginning to end, in order, and in its entirety. Further, to make progress in your leadership, you will need to open yourself and your team to emotional and intellectual work. This work will be on yourselves, with each other, and in your schools/organizations. For example, to really learn from the book, you will need to engage with the content and consistently surface your thinking and reflect on your meaning-making, starting here and now, and continue to do so for the rest of your career. We have structured this book so it gives you the foundation, time, and space to do all these things both individually and together with your team, using written reflections, discussion prompts, and online resources. When it's time to reflect in the book, we signal this with the words *Pause and Reflect*. We know it will be challenging, and at times uncomfortable, but taking the time to think deeply, and perhaps differently, is essential if you want to create more equity. When it's time to try a new strategy or action, we signal that with the words *Try This;* to put your learning to use, look for the word *Application*. We do not pretend any of this is easy, nor does this book have all the answers. We offer ideas, strategies, and processes to engage you and your colleagues in the hard labor of working and leading toward equity. We know that although it will be challenging and at times uncomfortable, the promise of greater equity is worth it!

It's important that you engage right from the start. Before you go any further, write your responses to the questions that follow.

Pause and Reflect

- How do you think of yourself in relation to equity leadership? What do you know? Feel? Experience?
- What are you thinking and how are you feeling as you begin this book and learning experience?
- The beginning of the book offers facts for your consideration. However, not everyone will take away the same meaning from these data. Before going further, what is your reaction to the facts presented? What do you think? How do you feel?
- How does your school's current situation compare to the data? What have you learned in past experiences that influences how you see your school?

1

This Is So Hard!
The Challenge and Urgency
of Leading for Equity

We know that schooling and education are powerful gateways to opportunity and quality of life. We know that hundreds of thousands of teachers and administrators give their blood, sweat, and tears to their students every day and have profound impact. We know that the spirit of universal public education is a spirit of equity and opportunity. Nonetheless, in the midst of this noble work and promise, we have gross inequity. This inequity *must* and *can* be changed.

This change requires a systemic and transformative approach. By *systemic*, we mean that the problem lies in the system and the inequities are symptoms and results. In other words, although inequity breeds inequity, it is not the cause but the result of a system that is set up to produce inequities. Therefore, system-based approaches are necessary to create equity. When we say *transformative*, we mean that you have to learn to think and act in some fundamentally different ways to change these historic patterns that are entrenched in the system. This section establishes the need for you, as a school leader, to make a firm and engaged commitment to prioritize equity leadership.

To begin, we illustrate some of the harder truths about how inequity lives, acts, and grows in schools. We visit Ezra in Meadowbrook to explore how systemic inequity manifests. We then present the Levels of Systemic Equity to help you understand inequity as historical, structural, institutionalized, and interpersonal/individual, demonstrating why equity leadership must address systemic causes.

Over the past 10 years, the suburban community of Meadowbrook has transitioned from a farming community made up of mostly White, working class families to a more complex and densely populated exurb. It is now home to an elite gated golf-course community in the southeast corner of the city along the lakefront, as well as an increasing number of Section 8 housing developments in the more commercial areas of the city. New residents are drawn to the city because of the school system's reputation for excellence and high achievement. At all income levels and across the variety of housing options, the city is becoming more racially, ethnically, and linguistically diverse.

Ezra, a White male in his late 30s, has been the principal of Meadowbrook Middle School for eight years. Because the city's 6th through 8th graders attend school there, it is a virtual microcosm of the city. In addition, Ezra's shift from teacher to curriculum specialist to principal began during the implementation of No Child Left Behind, creating increased awareness of racial disparities in opportunity, resources, and achievement. As principal, he was tasked with leading his entire staff to "close the achievement gap" and create a welcoming and inclusive environment for all students.

Now, under the Every Student Succeeds Act (ESSA), Ezra continues to feel pressure to raise test scores and eliminate inequities in his building. Simultaneously, racial and ethnic tensions are increasing throughout the building related to events in the community. Specifically, the local mall management has been under fire related to anti-immigration rhetoric and graffiti that continues there, seemingly without intervention. In addition, the city council is debating policies and services related to the local library and their obligations toward patrons who are immigrants, refugees, or have insufficient financial resources.

Ezra is beginning to realize that all of these things are connected. Due to his long-standing commitment to all students, he has been working at equity for a long time. When trying to address the problems he's facing, he's been encouraged to reach for the nearest, most promising tool. It's been tempting to look to new curricula or teaching strategies, especially those approaches that are "scientifically based" or "scientifically proven." But, as Ezra has experienced, most schools can make some progress with these methods, but then they plateau or worse, regress.

What Is Going On

As you consider how to use your leadership to create equitable outcomes, we encourage you to also consider how the scope of systemic inequity spans historical, structural, institutional, and individual/interpersonal levels (see Figure 1.1).

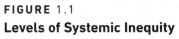

FIGURE 1.1
Levels of Systemic Inequity

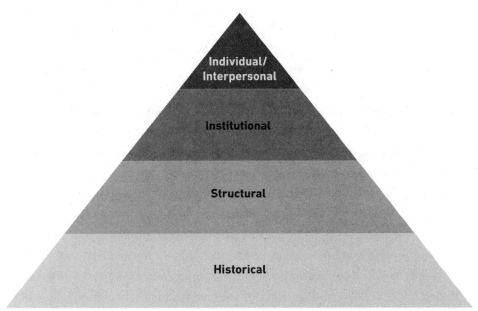

Historical

When we say *historical*, we mean that the problems we face today have their roots in centuries of human experience. In Meadowbrook, many of the families buying homes in this fast-expanding community can trace their economic status back to the GI Bill, when their White grandfathers received federal assistance to attend college, participate in job training, start a new business, and buy a home following their military service in World War II. These benefits were disproportionately available to White male veterans but not necessarily African American or female veterans (Katznelson, 2006; Rothenberg, 2002). Given that higher wages and home ownership are two primary ways to build family wealth over time, 70 years and three generations later, the housing patterns, class differences, and residential segregation in Meadowbrook reflect that history.

Further, people carry their histories. Your histories inform what you think, how you feel, and how you react. This is true of everyone! People who have been taught that they're entitled to a good education assume and expect that, and will be on the lookout to demand it should they suspect the school is falling short. Conversely, people whose lived experience in schools includes a pattern of unfair, disrespectful, and exclusionary treatment toward their community will be on the lookout for signs that the pattern is repeating.

Structural

Inequity is also structural, meaning that the way our system of schooling, and our entire society for that matter, are built and organized predictably lead to the types of disparate outcomes that exist today. It is not a coincidence that neighborhoods and schools are arranged in such a way that children tend to go to school with others who share their race or socioeconomic class. In Meadowbrook, although the secondary schools enroll all grade-level students from the community, the elementary schools are neighborhood-based. As noted, Meadowbrook's housing patterns reflect income and wealth patterns. Nationally, 78 percent of families with middle and upper incomes and wealth tend to live in homes that they are purchasing that help them build wealth and enjoy income tax deductions, while at least 52 percent of families with lower incomes live in rental properties where they face rising rents and the subsequent costs of moving households repeatedly over time. Further, these trends reflect racial divisions as well, with 72 percent of White families living in and purchasing their homes, while 58 percent of African American families and 49 percent of all families of color live in rental properties (U.S. Census Bureau News, 2020). Meadowbrook matches these national demographics and, as a result, the elementary schools' enrollments are economically and racially segregated, as schools are throughout the United States.

This segregation is structural and serves to continue inequity: When students attend schools that are highly resourced, they have access to an array of both tangible and intangible benefits that are rarely available to students in under-resourced areas. From athletic facilities to advanced placement classes to social and professional networks that provide an invitation into schools and employment to "get ahead," these types of structural factors serve to leave current arrangements undisturbed. In Meadowbrook, one can look to the school fundraisers to see one small way this structural inequality creates other forms of inequality. In all the schools, the parent-teacher organizations (PTO) organize fundraisers every year where students sell products to raise money for the schools. In the wealthy quadrant of town, the PTO runs one fundraiser per year, which brings in $65,000 that is used to purchase extra equipment and enrichment activities for students, in addition to providing funds to staff an after-school, high-tech enrichment program. At the elementary school that enrolls most of the students who live in rental housing with federal assistance, the PTO runs two sales each year and five other fundraisers at the school, including a carnival and a bake sale. Because families at this school, and their family and social networks, have far lower disposable incomes than the families at the other school, and despite the fact that people with lower incomes tend to donate a larger percentage of their income than people with higher incomes, all of these efforts result in only $15,000 in extra cash for the school. The school uses these funds to purchase playground supplies and to fund an extra part-time teaching assistant to run the volunteer program with a local business that has adopted the school, sending mentors and reading buddies to work with "students in need."

Institutional

Inequity is institutional as well. In other words, the laws, rules, processes, and organizations we use to engage in schooling and other aspects of our lives all work to continue historical and current patterns of inequity. Decades of tax, finance, and banking policies and practices have been built on top of the GI Bill, such that it is easy to avoid noticing how these institutionalized policies and practices actually serve to reproduce and entrench existing inequities. Housing inequity, for example, contributes to school inequity: Because housing in the United States is racially and economically segregated, and children in the United States tend to go to schools near where they live, children end up going to school with other children of their same race and income level. Because school budgets are funded primarily at the local and state level through property taxes, economic housing segregation leads to inequities in school funding patterns. This pattern is institutionalized through recent court decisions reducing federal involvement in school desegregation efforts; through federal and state housing policies and programs; through banking practices that resulted in families of color losing far greater ground than White families following the 2008 economic and housing crisis; and through local decisions about attendance zones that create neighborhood schools. It's a complex and complicated web, and without certain pieces of historical knowledge, it is easy for one to look at it all and determine that the system is fair and neutral, rather than recognize that inequity has been and is historical, structural, and institutionalized.

Individual/Interpersonal

For many, the default way of defining racism and other forms of discrimination involves overt, intentional acts of individual meanness, exclusion, and unfair treatment. You might even think of organized white supremacists and people who commit hate crimes. It is easy to think that only bad people who carry bad attitudes are the ones who perpetuate racism, sexism, class discrimination, ableism, and other forms of inequity.

Thinking about inequity this way allows you to think that you are not a part of it, and although you might take a role in fixing it, you can continue to think that you are doing nothing to contribute to it. Still, science has concluded inarguably that everyone carries unconscious biases; these kinds of unconscious biases contribute to negative judgment, exclusion, and discrimination.

This happens in many ways. In Meadowbrook, teachers often assume that families who live in low-income housing are "not educated" and do not have the intellectual capacity to help their children learn to read. And conversely, they assume that families in higher-income households are fully engaged in supporting their children's education. They also direct these perceptions to the children. It's not uncommon for the same teacher to say in a professional learning session, "I believe all kids can learn" and then later say to his teaching peer, "I have tried so hard to teach him, but honestly, Jamal is never going to make it past multiplication and division." Other educators say things like, "You can tell in 1st grade which kids aren't going to make it." Or "I'm not being racist, but it sure seems like the kids from the apartments just aren't very motivated, so they don't work very hard." These judgments and opinions can come from a mean-spirited and judgmental place, or they can be benevolent and concerned. Either way, they limit some students' possibilities while giving others the benefit of the doubt.

Why All of This Matters

Recalling that inequity and inequality operate at all four levels of the system, most equity trainings and initiatives address only one or two: They teach about the historical and institutionalized nature of inequity and leave the participant to determine what to do differently at the individual and interpersonal level. Or they provide a technical solution at the individual and interpersonal level, something for the participant *to do,* but do not ensure that the strategy will make an impact in terms of the institutionalized, structural, or historical causes of inequity. This is one primary reason why equity efforts make little progress.

Instead, professional learning and equity leadership need to account for the full span of these levels; we organized this book accordingly. Practices II and III provide a format and process for improving your equity leadership at the individual and interpersonal levels. Practice IV helps you think about transforming systems to change both the institutionalized and structural aspects of inequity. Finally, Practice V describes how you can sustain equity change over time to transform the historical trajectory that influences schools today.

The Levels at Work: Tracking

Ability-based grouping, or "tracking," is a common practice in schools consistently proven to perpetuate inequity, illustrating how inequity in schools spans across the four levels.

First, the practice of sorting students is historical, going back to the very beginning of public education in the United States. At the start, public schools were free as they are now, but they were not intended to educate everyone. Instead, they were intended for the "top 10 percent"—those considered "most educable." As you look at the conversations in schools today related to tracking, this approach continues: The current system of enrichment and advanced placement classes alongside remedial classes perpetuates the idea that students have different levels of ability and their courses should be structured accordingly, despite ongoing research findings that all students can learn at high levels and are best served by enriched and rigorous courses.

In today's schools, programming and policies serve to sort and separate students according to perceived abilities. For example, special education programming, ability tracking, and programming for English learners are supported by an intricate network of systems, processes, tools, and activities as well as underlying "theories of action" and paradigms. In these ways, inequity is both structural and institutional. In the case of tracking in mathematics programs, for instance, schools use standardized and localized tests to assess students' mathematics ability. Structurally, individual performance on these tests is connected to future opportunities for education and, eventually, employment; a school's overall performance dictates the desirability of that school and the home property values around it. Institutionally, tests are built into the school budget and schedule; further, test results impact class placement, staffing, and budgetary decisions. Further ensconcing this inequality is the historical way that teachers learned to teach mathematics in their preparation programs, often learning that the development of math skills is linear, meaning that a student cannot go on to learn the next skill until she has mastered the current skill. These qualities combined create a complex historical, structural, and institutionalized web that keeps both the inputs and the outputs of this system the same. Years of effort under No Child Left Behind served to further entrench this system in many schools, accelerating the sorting of students and schools via schedules, classrooms, practices, programs, and products. Ultimately, this system is based on the paradigm that the development of math skills is a fixed linear process and that the performance of mathematical calculations at a specific point in time reflects a student's fixed intellectual capacity.

Last, the idea that this system is fair, effective, and appropriate is held and carried by individuals (teachers, administrators, parents, students). It is transmitted and acted upon between individuals—that is, interpersonally—perhaps none believing they are creating or perpetuating racial and economic inequality. And yet, this system of tracking consistently divides students by race and family income under the guise of perceived ability, such that White, nondisabled, middle- and upper-income students are over represented in accelerated classes, while students from lower-income families, students with disabilities, and students of color are overrepresented in remedial classes.

Ask yourself: Do I believe that some students are more capable based on their race, economic class, dis/ability, religion, sexual orientation, family background, and so on? Or can I acknowledge that inequity is built into the system at the individual/interpersonal, institutional, structural, and historical levels? If you answer yes to the second question, then the solution is not to "fix" people who have been marginalized and excluded from this system; instead, it is to fix the system!

Consider how this information applies to your context. Here, it is important to continue the reflection you began in the Introduction: Take the time to write your responses.

Pause and Reflect

Consider how inequity is historical, structural, institutionalized, and individual/interpersonal in your context. Too often we are stuck only seeing inequity as individual/interpersonal.

- How do the people who work around you limit the idea of inequity to the individual/interpersonal level?
- In what ways do you see inequity as an individual/interpersonal issue?
- Where do you see historical inequity, structural inequity, institutional inequity, or individual/interpersonal inequity in your school/district/community?
- What barriers do you experience at these levels in trying to advance your equity work?

The Stories We Tell About Why We Don't Do Better

Let's return to Ezra in Meadowbrook, who has been working hard to dig deeper and understand the systemic nature of inequity. He now wants to engage his staff in considering this broader lens and designing a more systemic approach. In a recent faculty meeting, Ezra led a data-focused exercise during which teachers examined student achievement, discipline, attendance, and participation data by race, first language, special education status, and free/reduced lunch status. These data revealed the same types of disparities that exist across the nation. Many teachers felt discouraged and defensive, given that the school had implemented several initiatives to "close the gap." As Ezra facilitated the conversation, he was disturbed by many of the responses he heard:

- "These kids don't want to learn."
- "Their parents don't care."
- "My colleagues aren't capable/cooperative/invested."
- "I don't need this. I 'get it' and I have the outcomes to prove it."
- "I don't need this. It isn't relevant to me."
- "But I have to raise test scores. That's what this is all about."

These are not the responses Ezra expected. He realized the staff had developed these narratives over time to comfort themselves in the face of stalled progress and continuing inequities. His challenge was to respond in a way that would re-engage his staff in thinking more systemically and critically.

As he sought to do so, he remembered that everyone develops and uses stories to explain the events and circumstances in their lives. We know that this is the way the human brain works: People use stories to make sense of what they see and experience. That said, the stories people tell are not the whole story. Instead, stories reflect a person's sense of what matters, what is relevant, what serves the purpose of the story, and what serves the purpose of the telling. For complex and complicated issues, people often tell simplistic stories that make them feel better about the reality of the difficulties they and others experience. As a result, even in their best efforts to be fully honest, the stories people tell involve omissions, inaccuracies, even exaggerations. Sometimes, stories truly misinform and misrepresent what's going on.

Stop and think about that for a minute!

Consider how stories come from within you, are all around you, and reveal a specific message rather than a complete picture or the whole truth.

Pause and Reflect

- What examples can you think of—in your work or outside of school—where stories shape the reality?
- Who benefits from the way the story is told and who is disadvantaged, marginalized, or judged negatively?
- Where do you see competing stories that keep you or others stuck in conflict?
- What keeps you/them from developing a collaborative or mutually acceptable story?

There are no right or wrong answers to these questions, so resist judging yourself or others negatively. The way that stories shape reality and understanding is common throughout the world and across cultures. However, it's important that you begin to notice them and their impact. *Our purpose is to highlight the role of stories and how they can encourage or inhibit your progress toward equity.*

Specific to your work, you use certain stories to understand and explain what is happening in schools related to inequality and inequity. As you proceed through this chapter, consider whether or how the stories you tell discourage and disempower you from taking the actions needed to produce better outcomes for students. In our experience, we hear stories told by well-meaning people that say problems in the neighborhood, the family structure, the child's self-esteem, the child's motivation, the child's resilience ("grit"), the parenting, and so on result in certain students' (children living in poverty, children of color, children with disabilities, ELs, and so on) difficulties in school. This is what we call a *deficit orientation* (Scheurich & Skrla, 2003)—when people blame students, their families, their circumstances, and their communities for the symptoms and results of inequities rather than identify and correct the systemic causes of inequality and inequity. For example, when educators describe parents as uncaring and unwilling to do what is best for their children, or when they focus on the "rough neighborhoods" that children must "endure" when they are not "safe and protected at school," these are deficit-oriented views.

These sorts of deficit views can too easily creep into the stories people tell to make sense of the hard work of schooling. An overwhelming education system that is in constant flux, the enormity of social issues (health, violence, food, housing) that impact students' school lives, and high-stakes pressure all contribute to frustration and discouragement. While these stories serve as coping mechanisms to effectively help teachers keep going, they also perpetuate a deficit orientation that blames students, families, and teachers for not succeeding in a system that is oriented toward their failure. Once these stories become part of the narrative, they circulate in an insidious and powerful way. Recognizing if and how you use a deficit view as a coping mechanism is the first step in reframing it.

To disrupt and reframe these deficit-oriented stories in yourself and others, you will need ongoing reflection and vigilance. Thus, the next step in your equity learning is to make visible the stories you and your colleagues tell, then identify how those stories inevitably create barriers to innovative thinking about how to better serve youth, their families, and their communities. We use this next section to highlight the stories Ezra heard—that educators often tell—about why equity efforts don't work or don't matter. As you read, consider your own connection to and beliefs in these stories; we ask you to log your reflections at the end of the chapter. Then, we explain how Ezra disrupted and eventually shifted these stories.

"These kids don't want to learn."

Over and over, we hear this phrase or variations when well-intentioned and hard-working educators become frustrated with a lack of academic progress. Most often, this story is used to talk about groups of students who are Black, Latino/a/x, or Indigenous, students coming from low-income families, students with disabilities, some groups of immigrant students, and students who are learning English. Seldom (if ever) do we hear this story targeted at middle or upper middle-class White students, even when these students struggle. In some ways, the meaning behind this story is that "these students" *cannot* learn—they are somehow deficient. In part, holding tight to this story is a way of saying "the problem is not my fault; I have done all I can" in the face of these challenges. This story frees educators from responsibility while it simultaneously marginalizes students.

At Green Tree School, just over 50 percent of the students were students of color, 60 percent of the students qualified for free and reduced lunch, and 16 percent of the students were learning English (spanning 12 home languages other than English). With an increasingly diverse student body and a steady decline in student outcomes, Green Tree was sanctioned by federal/state accountability requirements and put on the state list of "failing schools." Over and over in different contexts, staff would say things like, "These kids just don't want to learn."

Over a three-year period, the new principal strategically supported the staff to employ myriad school improvement strategies: balanced literacy; hands-on, standards-based, investigative mathematics; hands-on science; a change to collaborative and inclusive services for students with disabilities and ELs; a schoolwide priority of classroom community building; focused outreach to the community and parents; ongoing discussions and learning about race; and a commitment to democratic and shared leadership between the principal and the teachers.

> Janie, a seasoned, dedicated teacher, expressed her stress and exhaustion about the changes and the needs of her students. She worked incredibly hard with her team and deeply cared for her students—evidenced by decades of Green Tree families of all races, incomes, and abilities who loved her. However, she also regularly said, "Those students just can't learn" about specific students—all of whom were Black, low-income, or with a disability—when progress was slow or imperceptible.
>
> One of Janie's students was an upper-middle class White boy who struggled significantly with literacy and behavior. Janie never said he couldn't learn and always found the patience and ideas to try something new with him. Yet, she was outspoken to the principal that "There is no way we are going to improve on the reading tests (state and local). Too many of my students can't do it. What we're asking them to do is too hard for them."
>
> Janie was a dedicated, talented, and well-respected veteran teacher; AND she also held deep and *problematic* beliefs that some of her students could not and would not

learn. We highlight this tension—not to blame Janie but to illustrate how good and talented educators can do noble work and hold deficit views at the same time.

The school's relentless and multifaceted improvement efforts resulted in significant changes in student outcomes. In looking over the achievement results of her students from the year that Green Tree got off the state "failing schools" list, Janie shared, "I really did not think so many of them would pass the test [pointing to the names of students of color, students with disabilities, and low-income students]." And with a huge sigh of relief and a smile, she continued, "I was wrong when I said they couldn't do it. They did it." Janie's students provided an essential lesson for her about who "can't learn."

We caution you against wanting students to rise above low expectations while you simultaneously cast them in a light of incompetence. We offer you Green Tree's story, and specifically its multifaceted improvement effort, as an important reminder that educators are *wrong* to say, believe, and act as though some students "can't learn" or "won't learn." Recognize this as a deficit view, and acknowledge the complex contradiction in being a good person and holding such views.

"Their parents don't care."

This is another story educators across the nation tell, regardless of context. Many educators believe that successful students, even those who "beat the odds," do so because of their parents' care and support. However, when educators think that way, they simultaneously believe that if students struggle and are not successful, it is because their parents don't care or aren't capable.

Bay Creek is a high-performing urban school where 30 percent of students receive free or reduced-price lunch. The school's sizable upper-middle-class population is enhanced by increasing numbers of students learning English and increasing racial diversity across all socioeconomic strata. Still, the PTO was small, made up entirely of White middle-class parents, and quite influential in school operations. Although Bay Creek was a lovely school in many ways, both staff and active White parents would lament that "the parents don't care" when talking about families living in poverty, families of color, and families learning English. Some staff who were striving to be more empathetic said things like, "Their parents just can't help them—they're working three jobs and don't have the time" or "They don't speak English."

As part of a school improvement effort around inclusive services, Principal Meg and a few staff started ethnically specific Parent Coalitions (PCs) to facilitate communication and enhance relationships with families. These PCs were for parents who, or had students who, identified as Latino/a/x, Hmong, or African American. The Latino/a/x and Hmong group meetings were held in the families' home languages and translated for the principal and other staff.

A number of the active upper-middle-class White families were *very* upset by this effort. They felt excluded, arguing that these distinct groups violated the school's diversity values. Although she was bothered by this resistance, Principal Meg and her staff expressed that the PCs were an important step toward equity; parents and students with marginalized identities also deserve a place where they are centered, and further, different people need different things. While this was controversial to some of the White families, the PCs were an example of differentiated resources as a core equity principle.

In the years that followed, the PCs engaged hundreds of parents who had not previously been actively connected to Bay Creek School. After three to four years, the previously all-White PTO became multiracial, and two parents active in the PCs—an African American parent and a Hmong parent—ran for and were elected to the citywide school board. In addition, everyone in the school began to recognize and value the myriad ways that families show they care about and value a child's education.

We encourage you to recognize how easy it is to rely on deficit-oriented views that suggest that the problems with student learning are problems from outside of school. Remember, it takes leadership, reflection, and action to nurture new spaces for caring and learning. Bay Creek's story shows what is possible when you reject deficit views of parents and instead engage with them in open-minded, respectful, and authentic ways.

"My colleagues aren't capable/cooperative/invested."

Educators who feel they are further along in their thinking about equity, or who possess a particular commitment or passion they believe is not shared by their colleagues, may tell this story. A variety of teacher leaders—some with strong commitments to equity, some with great pedagogical talents, some with deep content or leadership strengths— may tell this story. Sometimes, those with rich life and educational experiences who understand the complexity of equity and diversity issues struggle when others do not appear to think about equity, or have not had important life experiences around equity and diversity. You might hear, "It's not the kids, it's the adults who make life difficult." This deficit orientation is not about students or families but instead about other educators.

Joy is enrolled in a well-regarded educational leadership preparation program and widely respected as an excellent teacher leader. She is deeply reflective, embraces a complex understanding of equity, and possesses a strong commitment to her students. In her graduate classes, Joy naturally reveals her wisdom and skills through conversation and written work, yet simultaneously laments, "[My school] never gets anywhere because my colleagues don't get it. They are not invested in making our school better and are incapable of this work... so I just close my door and get it done with my students." She felt her only path to self-preservation was to leave others to continue inequitable work so she could do her best in her own classroom.

Note the complexity here. This talented equity-focused teacher leader consistently created more equitable and excellent experiences for all of her students; however, this was paired with a disdain for her colleagues who did not have the same skills or perspective. Her approach may seem self-protective, as people like Joy can experience real frustration when working with colleagues who are resistant to equity changes, hold biased and deficit views of children and families, and engage in practices that are not emancipatory. However, creating more equitable schools requires people like Joy to stay engaged in helping others learn and grow.

Through her graduate program, Joy realized her responsibility to lead beyond her classroom. She wrote, "Leadership toward these [equity] issues I care so much about is about my ability to work with other adults. I have to work to see my colleagues as capable, I have to get them invested... that is my job. I need them and so does this community." Joy learned that she needed to reject a deficit orientation about her colleagues if she wanted to create a more equitable school for her students and families.

"I don't need this. I 'get it' and I have the outcomes to prove it."

Some educators feel that their experiences and education have fully prepared them to work effectively with students from all backgrounds, and that they have nothing left to learn about educational equity and how best to serve their students. Sometimes we see this from educators who have a marginalized identity and have lived experience, perhaps coupled with academic learning, about the realities of marginalization and oppression for that identity. We also see this from individuals with many privileged identities—straight, nondisabled, White males for instance—who may have some experience working with students of color, living in poverty, and with disabilities. In either case, these individuals believe that their learning is complete because they already "get it" and know how to apply it.

To push your thinking about this story, we turn to two middle school principals, Betsy and Tomas, who both lead schools with strong performance across state academic measures.

> Betsy is a White woman who has worked her entire career in an urban, racially diverse school district with an increasing number of students living in poverty. Betsy has a social activist background; her commitment to social justice is part of her identity. She is comfortable talking about race in both personal and professional settings, believes that inclusive services are an essential part of a good school, and is committed to an affirming LGBTQIA+ culture—all pillars of equity-oriented leadership. Yet, she has two self-contained special education programs in her school—a separate program for students with autism and a program for students with behavioral challenges. The students in these programs eat lunch where the rest of the students do, but are essentially separated from the other students in every meaningful way. Betsy is frustrated that the district is working with a consultant on improving inclusive special education services, stating, "My school *is* inclusive; we have been doing inclusion for a long time!"
>
> Tomas is a Mexican-American man who has worked in a few districts but has spent the past 10 years in a rural, predominantly White district with an increasing Latino/a/x and Asian population and 45 percent of students receiving free and reduced school lunch. Tomas has worked to create a more inclusive service delivery model for students with special education needs, as well as for the small but growing population of students learning English. Through his deep connections with the Latino/a/x and Asian communities in his district, he has become a strong bridge between the school district, these growing communities, and many White families. People see him as a trusted link who is looking out for all of their best interests. At the same time, he is reluctant to take overt action to eliminate ongoing harassment and bullying of students who are, or perceived to be, LGBTQIA+. He feels the community is "very traditional" and any attempt at an LGBTQIA+ affirming culture would "blow up."

Betsy and Tomas are strong leaders who focus on equity every day. Both are certain their schools are good models of equity and excellence. And both have glaring equity gaps in their school that they allow to continue. Although they both feel that they "get it" and have "results," they still have important learning to do. We say this not to characterize these principals as bad or question their commitment to equity, but to demonstrate that no matter how much you know, equity leadership requires ongoing learning. Remember, it is not possible to ever fully "get it." You have to assume you will always need to stay open to new learning, information, and ways of seeing equity issues.

"I don't need this. It isn't relevant to me."

Too often, school structures and enrollment patterns inhibit authentic inclusion, allowing educators (leaders and teachers) to feel that students with marginalized identities are not their immediate responsibility. We hear this story often in suburban or rural schools with little or no racial diversity among students. We also hear this story from those who do not work directly with students with disabilities or ELs. Educators tell this story as a way to justify why they don't engage in equity-oriented work. This story positions equity as someone else's work and limits it to specific spaces.

Terry Town is a suburban school district that performs well on state measures. Terry Town prides itself on its community support and being a district that is more diverse ethnically and racially than its neighboring suburban districts. Still, during professional learning discussions, teachers and administrators often say things like, "I am all for equity, but it is not relevant in our district. Our community expects excellence."

Through their regional professional learning consortium, the Terry Town administrative team was asked to complete an equity audit (for more on equity audits, see Practice IV) and responded with dismissive eye rolling, stating "data are important but equity is just not the most pressing issue in our district." They proceeded, however, and were dismayed by the results. Although 6 percent of their students are Black, a startling 35 percent of discipline referrals and suspension were for Black students. While the district matched the national average of students receiving special education, students living in poverty were twice as likely as others to be placed in special education classrooms.

There are many dedicated educators in every district, but every school has work to do around issues of equity. Some issues may be more pressing than others at a particular time. But just like individuals must always keep learning, organizations and systems must always keep working to improve their equity efforts and outcomes across *all* groups and identities.

"But I have to raise test scores. That's what this is all about."

The pressure to raise student outcomes, and specifically test scores, has now dominated school improvement efforts for over two decades. In this context, we hear from leader after leader and teacher after teacher, "I have to raise test scores, that's what it's all about."

Principal Natalie came to school administration with a drive and commitment to equity-focused leadership and led her school through significant student achievement gains. She explains that "I have to raise test scores. That's what this is all about" is faulty logic:

> Many of my principal colleagues are so hyper-focused on test scores. Their actions and words say, "Once our scores go up, we'll begin to address our disproportionate discipline referrals, talk about more inclusive special education or inclusive EL structures, work on detracking and access, and create a safer, more welcoming school for our LGBTQIA+ students." They don't see that all those issues are related and addressing the pressing equity issues increases opportunities for student learning. Not an immediate fix, but the right one.

Principal Natalie highlighted the problems with this focus on test scores: First, it reduces the definition of achievement and learning to student test scores. Second, it suggests that equity work is separate from, rather than integral to, achievement and learning. Third, it ignores the cultural nature and systemic context of the tests themselves, as well as the ongoing reality that standardized test scores are far more predictive of a person's zip code and family income than of knowledge or intelligence. Last, it blames students and their families for systemic inequities. Over time, she worked with her staff to explore the connections between student learning and other equity issues: When Black students receive more and harsher school discipline, they miss class time, disconnecting them from teachers and learning opportunities. When students with disabilities and ELs lack access to the general education curriculum and peers, it impacts everyone's social and academic growth. When LGBTQIA+ students feel unsafe and devalued, they often withdraw and disengage. Expanding beyond test scores to consider the full range of equity issues creates real school improvement and improves student learning.

Adopting a Transformative Approach: Individual Activity

Pause and Reflect to Prioritize Equity Leadership

Not all of these stories will resonate with you. However, we reviewed them here so you can see some of the ways you and your colleagues might automatically resist a deeper engagement in equity work. We also offer these stories so you can reflect on them and their impact.

- Which of the preceding stories do you hear regularly from your colleagues?
- In what ways do the people you work with or work for act to embrace these stories, either publicly or privately?

Part of engaging in equity work is understanding how these stories have impacted the way you see school, your roles, and the role of others. When you identify and examine the stories you tell, and the stories that exist around you, you've taken an important step forward in improving your ability to successfully increase equity. In the process, you'll begin to identify your own imperfections and vulnerabilities; this is vitally important to your equity leadership, so don't avoid it! For now, this work will be private, so take the time to deeply engage in your own honest reflection. We'll provide guidance for sharing what you choose later in the book.

- How have *you* told these stories? When have you seen them in yourself? In what ways are you still wrestling with *each of these*? If you are not saying these words, where do you connect or disconnect with these stories?
- What other stories do you tell yourself about why you don't do better?
- What impact do you notice these stories have on you? For your students? In your relationships with your students and their families?

Challenging Old Narratives

Once you identify your own stories, and the stories around you, your next task is to change the story. But what does this look like? In order to provide practical strategies, we return to Ezra, who began a series of reflective conversations with his staff unpacking the stories they were telling themselves about why they couldn't do better. He shared the case studies above, and then asked teachers to work together with him to identify the key takeaways. His teaching staff identified these important considerations for their practice:

Sometimes educators engage in deficit thinking as a coping mechanism. When they have been working so hard and tried so many strategies that work with some students, they find it easier—perhaps natural—to blame the students or their colleagues who are still struggling. Try this:

- Recognize when you engage in deficit thinking toward your students, their families, or your colleagues. Notice times when you say, "They can't, won't, or don't..." as these terms often signal deficit thinking.
- Note that deficit thinking can be malevolent (angry, frustrated, judgmental) or benevolent (empathetic, emotionally generous), but both see students and their families as less than and limit their possibilities.
- Offer empathy to yourself and others about the struggle you all face. Remember that inequity is historical, institutionalized, and structural, and the path to change is to work together. Avoid blame, and instead focus on the strengths and assets that everyone brings to the situation. From there, listen and collaborate with other stakeholders to develop and implement systemic solutions.

Parent involvement has been framed in a culturally specific way that sheds a positive light on the ways many middle- and upper-class White families often show up at school. This framing provides increased access to power and influence for families with privilege while it diminishes all the other ways that parents and guardians can and do support their students' learning. Try this:

- Assume that the school has primary responsibility for student learning, regardless of family background or circumstance. Make sure that students' success in school does not depend on their family's resources or involvement.
- Reframe your ideas of family involvement to be more inclusive. Identify strategies to engage more authentically and respectfully with all families. Be certain to provide all communications to families in their home language, no excuses. Consider these adjustments in your wording to communicate a more inclusive and respectful stance:
  —Rather than parents, use parents, guardians, and caregivers.
  —Rather than mother/father, use parent/guardian/caregiver.
  —Rather than "They don't speak English," use "They speak [insert home language here]." (Remember that everyone has a language, and bilingualism is an incredible asset. When a family speaks a language that is different than the language that is spoken at school, the problem isn't the family's language, but the fact that both the school and the family speak a different language than the other).
  —Whenever you catch yourself saying, "They don't..." or "They won't..." consider how that usually represents a deficit orientation.

- Restructure parent/guardian/caregiver committees to ensure that all families are represented in decision-making groups.

It is easy to get trapped by fear of others' backlash when you take a stand for greater equity. This can cause you to ignore marginalization, oppressive conditions, and persons who are underserved or excluded. Equity leaders need to be concerned for all persons, most especially persons with marginalized identities. Although leaders need to be strategic about when and how to increase inclusion and equity via potentially controversial plans, you can't allow this to deter you from moving forward in a timely way to increase inclusion for all groups. Try this:

- Conduct an Equity Audit (see Practice IV). This process will help you identify those populations and processes that need your attention. The strategies for engaging with stakeholders and building coalitions found in Practices III, IV, and V will help you design and implement plans with a greater likelihood of effectiveness and constituent support.

It's possible for equity leaders to make excellent progress on behalf of one group while ignoring the marginalization, exclusion, or substandard learning conditions for another group. It's tempting to limit your concern to only those students who are in your classroom. Equity leaders, however, need to promote a concern for the entire school community among the entire faculty and staff. A spirit of collectivism has many benefits for advancing equity work. In the case of Ezra's school, it helped the teachers see that they were responsible for all students in the school, not just those in their classroom, and inclusive of students with a variety of marginalized identities. Try this:

- Avoid the "Oppression Olympics" (see Chapter 10 for more on this). Take an anti-oppressive stance, demonstrating concern and commitment for all forms of marginalization and exclusion.
- Promote the spirit of collectivism. Encourage systems and processes that support staff and faculty to take ownership for students throughout the school. Remember, "We all do better when we all do better."

The standardized testing movement has transformed our education system, directing attention to equity gaps. It is also creating an increasingly narrow and bleak education in schools that serve historically marginalized students as they reduce or eliminate art, science, music, and social studies and increasingly spend time on skill-and-drill literacy and math. As a nation, we have a moral and ethical obligation to seek the highest levels of learning for all children and better serve the many students who historically (and currently) have not been offered a robust opportunity to excel. Try this:

- Focus on the conditions of learning and the opportunities that are available across demographic and identity groups. Avoid the temptation to do more of what isn't working, and especially reject any efforts to narrow the curriculum and learning opportunities. Instead, trust that if you create and ensure engaging learning activities, environments, and relationships for all students, they will learn and achieve.

Some educators have done extensive work on themselves to understand key issues, and some educators are newly engaging with these critical topics. It's tempting to think you've learned enough after attending some training sessions. It's even more tempting to feel like you've arrived if you've had a life-changing epiphany about equity. However, moving toward equity is not a linear process, nor is there an endpoint. Equity leadership is not binary (i.e., a person is either an equity leader or not). No matter how much you do or don't know, you will always have more to learn. Try this:

- Step back from certainty and engage with curiosity. Assume you don't know enough. Commit to ongoing learning.
- Remember, there is no neat and tidy destination when it comes to equity leadership; instead, everyone must always stay open to new learning.

*****

Leaders play a vital role in creating excellent schools and ensuring equitable learning opportunities for all students. This includes those with positional authority, such as school principals, superintendents, and department chairs, as well as those with informal authority, such as revered teachers who are viewed as leaders by peers. Effective school leaders set directions for school communities to be academically ambitious and ensure that organizational structures support these ambitions while eliminating systemic inequities. They build the instructional capacity of the faculty and staff, drive organizational learning (Bryk, Sebring, Allensworth, Luppescu, & Easton, 2010), and include stakeholders (e.g., teachers, parents) in decision making (Leithwood & Riehl, 2005). As we move forward, remember that effective equity leadership is not a solitary activity, but instead a collective enterprise in which you consistently engage with others

(Darling-Hammond, LaPointe, Meyerson, Orr, & Cohen, 2007). It is only through this collective approach that you make measurable, sustained, and systemic change to benefit marginalized students, as well as the system as a whole. The remainder of this book will help you develop your skills, mindset, and tools in leading this approach.

Pause and Reflect

- What will be the hardest part of prioritizing equity leadership for you?
- Who is on your team and what is/will their reaction be?
- Who else is willing and able to support your commitment to prioritize equity leadership?

Adopting a Transformative Approach: Team Activity

As you near the end of this first section and before you proceed into the remainder of the book, it's time to gather with your team, review your reflections together, and begin/continue to establish your habit of learning together.

Team Discussion 1

Review the data about inequities described in the Introduction and share your reaction. Discuss with your team the ways in which it mirrors your school/district/organization, what other data matter, and the relevance of the data for your system and your leadership.

Team Discussion 2

Review how inequity is historical, structural, institutionalized, and individual/interpersonal. Too often one gets stuck only seeing inequity as individual/interpersonal. Discuss with your team: What habits exist in your school/district/organization that suggest inequity is an individual/interpersonal issue?

Discuss where you see each of the following in your school/district/community:

- Historical inequity
- Structural inequity
- Institutionalized inequity
- Individual/interpersonal inequity

Team Discussion 3

Share with your team which of the stories from this chapter resonated most with you, why, and what new stories and strategies you will adopt in order to "do better."

Team Discussion 4

What is new learning for you? What new insights do you have that will affect your leadership?

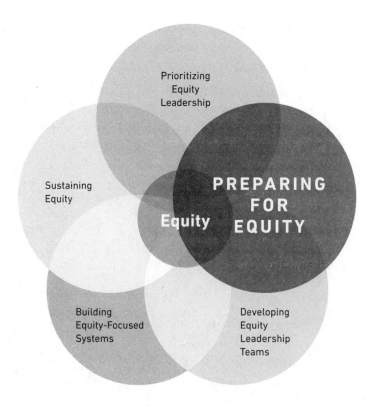

PRACTICE II

Preparing for Equity:
The Ongoing Emotional and Intellectual
Work of Equity Leadership

Preparing to Learn for Equity: Key Concepts and Guiding Principles

Donald

Donald, a White male, was in his third year as superintendent of a midsized, first-ring suburban school district. Others saw him as an up-and-coming superintendent because he had declared himself to be a racial equity champion and had made "closing the achievement gap" the central goal of his superintendency. With his contract renewed for a second term, but with significant concern about a lack of progress on his equity goals, Donald directed his executive cabinet to complete an equity audit. In the process, he became aware of the incredible disproportionality in discipline rates between White students and students of color. Aware of how the suspension rate for African American males contributes to the "school-to-prison pipeline," he decided that the best course of action was to prohibit suspensions in the district's schools. After sharing this information with his leadership team, he convened a press conference to announce the policy change.

Disrupting Punitive Practice and Policies

Susan

Susan, a White female, was a sixth-year principal in the district's most racially diverse middle school and had been leading with a clear focus on equity since her arrival. With several efforts producing positive results, she was concerned about this new policy. She knew that some of her fellow administrators were far too quick to issue

*Out of School
and Off Track:
The Overuse
of Suspension
in American
Middle and
High School*

harsh discipline to African American and other students of color. She was also aware that many other factors might be at play when students act in disruptive ways. She believed that her responsibility was to equip her staff with the tools and skills to effectively engage students in learning, build quality relationships with students and families, and manage behavior to support classroom learning goals.

In the year following Donald's policy change, many schools became more chaotic, and learning outcomes decreased. In addition, conflict increased among school staff, administrators, and the community as many students felt unsafe at school and found their learning opportunities overshadowed by stressful and tumultuous interpersonal and behavioral interactions.

At Susan's school, however, things continued to improve. She had strong relationships with her staff and the school community, founded on gathering input and collective decision making. To respond to Donald's policy change in ways that improved the community and the learning environment, Susan convened a series of conversations with students, staff, and families to discuss their concerns, values, needs, goals, and approaches to social, emotional, and behavioral learning at the school. Stakeholders identified what they needed to learn and do differently to reduce and eliminate suspensions. Topping their list were the following items: help teachers create stronger, more inclusive classroom communities; teach all school staff how to de-escalate tense situations in ways that maintain everyone's sense of dignity and belonging; agree on a set of grounding principles/values to guide all community members in their interactions with one another; create safe spaces and processes for supporting students when their emotions feel particularly intense; create accessible pathways to strengthen the relationships and communications between families and school staff; and adopt a restorative approach for responding to situations where someone has been harmed.

The involvement of so many stakeholders in the process resulted in broad and deep engagement in implementing the plans they created. The process took a significant amount of time and the outcome was not perfect, but the school community continued to improve its approach to supporting students' development and sense of belonging, while also improving their learning.

What happened here? What can we learn from Donald and Susan?

Equity leadership involves working in complex, dynamic, and sometimes hostile educational environments. Further, barriers to equity plague the education profession and schools and can leave you feeling depleted and hopeless in your effort to create change. Donald's approach to leadership in this context is common: It's initially easy and quick to use a hierarchical and isolated approach to decision making. However, doing so often erodes trust and momentum and ultimately leaves organizations and communities in conflict and turmoil. Although less common, Susan's approach is more effective and will not necessarily be quick, but more efficient over time.

For you to be effective in your equity leadership, you must prepare emotionally and cognitively to continuously increase your understanding of equity. This path is not the shortest or easiest. Moreover, you must develop the skills to engage others in this important learning and work. This section walks you through a framework for your individual learning and planning your approach to equity leadership.

Paradigms and Cognitive Dissonance

We use the term *paradigm*[1] to refer to the perspectives through which people view their world and come to understand the activities, items, and relationships within it. They act as complex webs of meaning making that people hold in their subconscious and unconscious mind and that operate and influence them at all times. Paradigms both create and result from social constructions. They evolve from a composite of beliefs, assumptions, values, and theories, and they are influenced over time by a person's experiences, successes, disappointments, and life circumstances. Still, they tend toward simplistic explanations and understandings and thus are incomplete, often containing inaccuracies, paradoxes, and conflicts. Despite these shortcomings, and because most people are usually unaware of their paradigms, the meanings that result from them are often treated as the truth—sturdy, durable, and resiliently so—even in the face of contradictory evidence.

Paradigms involve two intersecting aspects, as shown in Figure 3.1. Horizontally, you see that paradigms are individual and collective. In other words, individuals hold paradigms, which we refer to as "individual self paradigms"; and groups of people, professions, institutions, and so on collectively hold paradigms, which we refer to as "collective systems paradigms." Self paradigms and systems paradigms overlap and influence one another.

In addition, paradigms operate on a continuum from the surface of consciousness, where they may be more visible and flexible, to deep in the unconscious, where they are incredibly difficult to notice and are deemed to be fundamental truths. At that deeper level, people are almost completely unaware of them, do not consider their accuracy or appropriateness, and use them to fundamentally guide how to live and act. Paradigms

---

[1] Various authors and fields have referred to this notion, calling it *mental models* (Senge, 1990), *frames of reference* (Mezirow, 2000), *ideologies* (sociology, political science), or *schemas* (psychology). Although there are likely important distinctions between these terms, in many ways they can be used interchangeably and with our chosen term, *paradigms*. Our use of the term *paradigms* builds on and combines Senge's *mental models*, Mezirow's *frames of reference,* and Brookfield's (2012) *paradigmatic assumptions.*

FIGURE 3.1
Paradigms

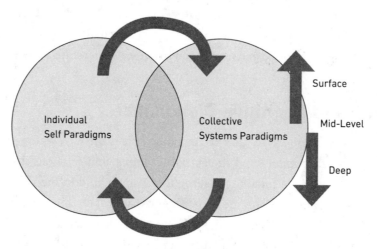

dictate the social norms, codes, and rules that individuals and groups live by. They dictate decisions about how and what should happen, but even in doing so, they are usually unstated and unquestioned. However, within a group of people, they can result in multiple and vastly different ways of seeing a situation and of considering what *should* happen at any given time.

In the case of Donald and Susan, several sets of paradigms are at play, as detailed in Figure 3.2. As you review the table, note where there are similarities and contradictions between and among the various paradigms. Note which paradigms align with your beliefs and which are contrary your beliefs. Take a moment to record your reactions.

In the United States, deep paradigms about education, work, and wealth dictate decisions about who deserves and is entitled to the finest facilities, the most-skilled teachers, and the richest opportunities. These paradigms are fueled by a system of school funding that perpetuates a feeling of scarcity. The No Child Left Behind legislation enacted in 2002 was built on a paradigm that believes competition is the most reliable source of motivation. One of the most powerful paradigms that drives inequality and inequity is a belief that the school system is fair and offers equal opportunity for all who "work hard" to learn, have "success," and have access to further opportunities to achieve their dreams.

We have provided—and will continue to share—information that challenges that paradigm. This information may conflict with your underlying/deep paradigms, creating cognitive dissonance. *Cognitive dissonance* is that disruptive sensation you

FIGURE 3.2

Varying Paradigms Regarding Leadership, Equity, and Suspension from School

	Donald's Individual Self Paradigms	Collective Systems Paradigms	Susan's Individual Self Paradigms
Surface	• If suspension patterns reveal racist tendencies, we must end suspensions. • Suspensions are the problem. • To be an equity champion, I need to take swift and decisive action to end this injustice. • It's my job as superintendent to make this happen. • I'm action-oriented; I can and will get this done. • I will tell the principals to implement this at their schools and they will do what I say.	• Something has to happen to kids who act badly. • Leaders are responsible for fixing problems. • Rules and procedures fix problems.	• If suspension patterns reveal racist tendencies, we must do something to change them. • This is a complex—not simple—problem. • Although suspensions are a symptom, not the problem, they also create other problems. • It's my job as principal to take this on. • Our stakeholders will know what to do. • The way to find and enact a solution is to involve those who are closest to the problem.
Mid-Level	• The way to make this happen is for me to decide. • People need to do what I tell them to do; it's their obligation as my employees. • Taking quick and decisive action will bolster my credibility and reputation as an equity champion.	• Kids who act out should be punished. • Fast, simple solutions are needed and sufficient. • The system is fair; rules and consequences are applied evenly.	• I don't have the answers; *we* have the answers. • You can't assume that simple solutions will solve complex problems.
Deep	• I can, should, and will make important decisions. • Others' input and involvement is rarely needed and might get in the way of doing what I know needs to be done.	• Unequal outcomes are OK as long as White kids don't come out on the losing end.	• When faced with important decisions that will significantly impact others, my job is to gather people together and facilitate a process through which we collectively come up with a solution.

experience—physically and/or emotionally—that causes you to think and feel, in your gut, that "something isn't right." When you feel cognitive dissonance, it is because you perceive that things aren't happening the way you think they should be happening. You have all sorts of tools to dismiss dissonant information and rationalize your reactions and paradigms. Argyris (1985a, 1985b) calls these *defensive routines*, referring to those actions and thought processes that protect your current paradigms, allowing them to remain as is rather than become more accurate in the face of dissonant information. Then, you use the conscious, thinking part of your brain to rationalize your perspective and articulate some sense of logic and consistency.

 For a video on paradigms in education, go to www.youtube.com/watch?v=zDZFcDGpL4U.

Notice when you feel cognitive dissonance and the ways that your mind may internally argue with the information. Begin to build the "discipline of reflection and inquiry skills" (Senge et al., 2012, p. 8), shedding any rigid certainty in order to become curious and contemplative in your response to dissonant information.

The Ladder of Inference

The Ladder of Inference (Senge et al., 2012) is a tool you can use to work through cognitive dissonance and create more accurate and complete paradigms (see Figure 3.3). It provides a schema for understanding how the mind takes in information, makes sense of it, and determines action. For example, your school or district is awash in data, including not only the data that you typically think of, such as test scores or graduation rates, but also data related to areas such as family perspectives, teacher decision making, and student attitudes. Based on this "pool of available data," you make decisions about which data you think matter; these are the data you *select* to pay attention to. Still, as you begin to review the data, some will seem more credible or important than others, leading you to *assign meaning* and *draw conclusions*. From that step, you *take action*, generating more data.

As you may guess, the Ladder of Inference is not just applicable to school-related data. It is at work in each person in every waking moment. In fact, the human brain runs up and down the Ladder of Inference in milliseconds, constantly and repeatedly. Further, this process most often happens below the level of conscious thought, so that when people reach conclusions and take action, it may seem like the only logical path but is actually the result of multiple decisions along the ladder about which things

FIGURE 3.3
Ladder of Inference

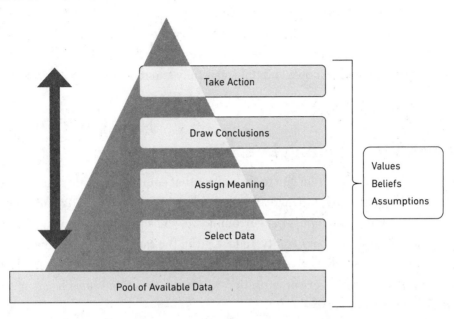

matter and what things mean. Those decisions are highly influenced by people's values, beliefs, and assumptions—their self and systems paradigms—at each step of the ladder. This is one important reason why we advocate that you surface and examine your paradigms to improve them for accuracy and completeness. When you do so, you also improve your actions.

Although Donald's equity audit produced a wealth of data, he chose to focus on suspension rates in his district as well as the injustice of the nationwide school-to-prison pipeline. He determined these were urgent and inexcusable matters, and he decided to eliminate suspensions. Susan, in contrast, agreed that the situation was urgent and inexcusable, but she believed more data should be considered to understand the situation and determine a course of action. Her beliefs also dictated that she include others in her deliberations. By working through the Ladder of Inference collectively, her school community developed a robust, multifaceted plan for addressing the problem.

It bears repeating that most of the time, the meaning-making process happens below your level of consciousness and within a split second. Every step is influenced by underlying paradigms, operating at all times, for each person. There's a lot of room for error! You could miss important data, interpret it inaccurately, draw ineffective or inaccurate conclusions, or take ineffective action. In terms of your equity leadership, this is why you need to slow down and think carefully and repeatedly about each step of the ladder.

The Need for Ongoing Learning

We hope this information has begun to make clear that equity leadership requires ongoing learning—for you, your staff, and all stakeholders involved in the effort. Because of the powerful systems paradigms that surround you, you should assume you will never know all there is to know about equity. In fact, given the rapid rate of information generation nowadays, the historical and entrenched nature of inequality and inequity, and the increasing diversity in our communities, *you must commit to being open to ongoing learning for the rest of your career.*

How do you feel and what do you think about that need for ongoing learning?

As an equity leader, you want to perpetuate this learning attitude, along with the curiosity and humility it requires, among all your stakeholder groups. Accordingly, schools and districts must become learning organizations (Senge, 1990). Your work is to create a climate that is safe for learning but also challenges all members to take risks to move themselves and the organization forward.

In this context, you are asking yourself and others to examine themselves and, often, come to realize that many things they once believed are simply not accurate or true. As this can be quite uncomfortable, they may reveal emotional needs, show a tendency to resist, and find it difficult to be open to dissonant information. These are natural responses to new learning and change. You want to avoid judging people, and instead support and expect them to move through the difficulty and be open to the learning. (Chapter 10 includes suggestions for how to do this as you build effective teams.)

Emotions Ahead: Putting It All Together

Equity learning produces a range of responses within and among individuals. Consider that your emotions, your desire to understand, your desire to feel competent and worthy, and perhaps most of all, your sense of right and wrong are all affected by equity learning. When we share new equity-focused information, we see in the participants— and feel our own—cognitive, emotional, conceptual, physiological, moral, or technical reactions (see Figure 3.4).

FIGURE 3.4
Sphere of Reactions

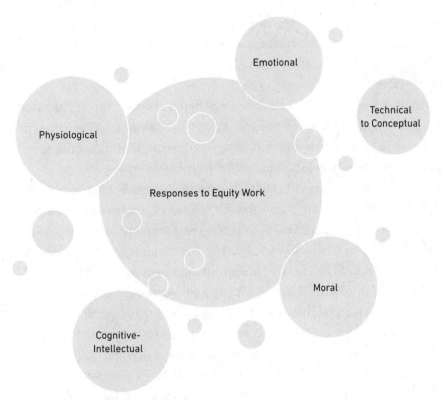

These reactions stem from dissonance related to paradigms. Equity talk touches on self and systems paradigms about what's fair, what's moral and right, what works and what doesn't, and which causes and effects are connected to each other. One common paradigm is that most people see themselves as good and moral people. Further, 85 to 90 percent of Americans hold a collective mental model that racism is a bad thing. At the same time, nearly 75 percent of Americans have an "automatic preference" or unconscious bias that favors Whites over Blacks. Further, this preference or bias is strongly predictive of discriminatory behavior (Banaji & Greenwald, 2013).

Pause a moment to consider this information. Do you want to assume that you are in the 25 percent that doesn't have a preference and that other people make up the 75 percent? If you assume that, what does that mean for you? What does that mean for other people who see themselves similarly to how you see yourself?

What is important is the possible conflict between the idea that you may hold a negative unconscious bias and the idea that you are a good and moral person. When faced with these conflicting thoughts, most people will employ their defensive routines to dismiss the thought that puts them in a negative light. These routines serve as real barriers to equity. They help to explain how we end up in a situation where most Americans believe they are not racist, yet we have overwhelming evidence of racial disparities in the United States.

Your ability to recognize and work through moments of cognitive dissonance while avoiding and minimizing your defensive routines allows you to engage more fully and effectively in equity-focused leadership. Equity leaders begin with the critical first step of reflecting on their experiences and identities, learning about histories and perspectives that differ from their own, exploring how these connect with the broader context of inequity and inequality, and working through their emotional responses to these realities.

How do you do this? Your task is to surface and examine your self paradigms and the systems paradigms to which you subscribe. For example, you are taught from your earliest years whether to believe that life is fair and just. Even when you say, "Life isn't fair," some of you think that it isn't fair in an equal-opportunity sort of way, whereas others think that it isn't fair because of the way society privileges some and disadvantages others. Because fairness is such a complex concept, even if you are taught, or know, that there are inequities in the world related to race, gender, and other factors, you are also taught and believe that certain things about the way things work are logical, fair, and right. These beliefs are a result of, and form, both your collective systems and your individual self paradigms. In other words, they influence everything you think and do in relation to equity leadership. If you're thinking that this means that equity leadership can be fraught with errors and missteps, you're right. If your paradigms are so pervasive and influential but contain inaccuracies, then they are likely to misguide you at least some of the time; and unless you become conscious of this process, and of your specific paradigms, you won't know when this is happening.

As a result, you need to start with your own learning. Remember, each person has individual constellations of paradigms about the way the world works. No matter how you see things, if you fully engage in equity work, at some point it will challenge your fundamental beliefs about what's right and what's wrong. That dissonance, in turn, will create a strong urge to resist the new information or perspective. You might deny it, argue with it, or flat out reject it. Paradoxically, you might presume it to be true and completely reject your prior way of seeing things. However, in such instances, you are not engaging in *learning;* instead, you are just reacting.

To respond constructively to information or perspectives that create dissonance, you must first recognize the dissonance and allow it. Work hard to hold the new information as it confuses, challenges, or upsets you. Once your initial reaction subsides, use the dissonant information as a portal to unearth, surface, and examine the underlying beliefs in your deep paradigms. Ask yourself the *what, why*, and *how* of those beliefs and paradigms, and consider how the new information challenges them. Then consider how you might revise your underlying beliefs and paradigms, and perhaps your understanding of the new information, to gain a more accurate and complete picture.

This process is one of the most fundamental aspects of adopting a transformative approach in equity leadership. And, with some practice, you will get better at it over time.

Pause and Reflect

To begin the process of recognizing dissonance in yourself, take the Harvard Implicit Association Tests on race and skin color. Go to https://implicit.harvard.edu/implicit, select "continue as a guest," read the authorization and select "I wish to proceed," then select "Race." After you complete the test and receive your results, reflect on these questions:

- What do you think? How do you feel?
- Do you want to dismiss this science because it made you uncomfortable or told you something about yourself that you didn't like?
- Take a moment to sit with any discomfort and disappointment you might feel. What happens?
- Now, *look* at your emotions and thoughts as if someone else were feeling and thinking them. What can you learn from this exercise?
- What are your takeaways from Donald and Susan's scenario? Which leader would you prefer to work for? Which leader do you tend to be like? What are the implications for your practice?

Experiences of Inequity

Jvette

Jvette is an African American 11th grader who grew up in a well-to-do family in a wealthy suburb. She is on track to graduate near the top of her class and expects several scholarship offers to top universities. She is often frustrated by being one of only three African American students in her advanced placement classes and feels she has to "act White" to be treated fairly by her teachers. Sadly, she has become used to the fact that when she goes to the local mall, the store clerks often follow her but don't offer her assistance.

Muhammad

Muhammad is a 6th grader who relocated with his family to a large rural town after their request for refugee status was approved. Their sponsor family helped his parents obtain work in the local meat-packing plant. As Muslims, they drive two hours each way to attend Friday prayers at the mosque with other Somali refugee families. At school, Muhammad quickly became fluent in conversational English and is diligent about completing his homework, even though he is still developing his academic vocabulary. He is nervous about going to middle school next year, as he's heard that some of the White students don't like "transplants" and are especially mean to kids with dark skin.

John

John is a White male who grew up living with his mother in a trailer park in a small community. In elementary school, the kids who lived in houses called him "trailer trash" and criticized his clothing. His teachers were quick to assume that his reserved nature meant he was "slow" and placed him in lower-level ability groups for both math and reading. When his mother got married, they moved with her new wife into a house in a bigger town, and their situation became more financially stable. Now, as a high school senior, an accomplished athlete, and an average student, he enjoys acceptance and support from his peers and teachers and looks forward to more of the same in college.

Emily

Emily is a White 2nd grader with cerebral palsy. She uses an electric wheelchair to get around her home, school, and community. Her mother adopted her as an infant and raises her as a single parent, with support from neighbors. Emily uses communicative technology to participate in class, but most of her peers don't interact with her. In addition, her teachers are unfamiliar with how best to involve her in group activities, so she is often left waiting while they get the other students established in their groups before working with her individually.

Donovan

Donovan identifies as Korean American, queer, and agnostic. He was a strong and conscientious student in elementary school, when he felt and acted more constrained about his gender and sexual identities. Now in 10th grade, he is teased by some of his peers about his race and his sexual identity; worse, some of his Christian classmates threaten his safety and say he is going to hell. These interactions make it hard for him to concentrate, and he has recently seen a psychologist about his increasing anxiety. He feels he may need to go to an alternative or charter school to feel safe and accepted again, though he knows this will limit his college choices.

The vignettes describe students who have experienced marginalization and systemic inequity based upon aspects of their sociocultural identity. These identities, also called "areas of difference," include race; disability/ability; socioeconomic class; language; sex, gender identity, and sexual identity; and religion. We define and explore these areas of difference in Chapters 4 through 9. But first, as you think about the stories of Jvette, Muhammad, John, Emily, and Donovan, how would you describe and discuss what is going on? What terms are essential for accurate descriptions and productive discussion?

A Foundational Vocabulary for Talking About Equity

A key capacity of equity leadership is the ability to lead others in discussing challenging topics, which requires a foundation of vocabulary and concepts. Many people are familiar with common definitions of equality and equity: *equality* means that everyone gets the same thing, and *equity* means that everyone gets what they need. We often hear this distinction made as a way to increase the focus on equity and discourage the idea of everyone getting the same thing. We agree these definitions are important, but we want to take them further. We believe it's important to maintain the idea and ideal of equality as we think about equity. As we've shown, schools are subject to and promote the same inequalities that are rampant throughout society. We need equity because inequality is historical and entrenched.

The term *inequality* refers to the absence of equal protections or equal footing, and it relates to *inequity* in two dimensions. First, inequity is the compounding impact of inequality over time; it is unfair and partial, as in the case of the GI Bill, discussed in Chapter 1. The second dimension relates to the ways that systems serve various differences. Stairs provide access to a building for people who move by using their legs but not for people who move using a wheelchair or other mobility aid. Someone may think a building provides equal access because anyone is allowed to enter, but it is not equitable because it does not provide the opportunity or resources for *everyone* to enter.

Thus, equity leadership seeks to provide all individuals with what they need to succeed. Ultimately we envision a society in which every person, inclusive of differences, has equitable access to the full and equal application of the rights of citizenship, including safety and security in all their forms, access to a quality and affirming education, the right to self-determination, and plentiful opportunities to pursue happiness.

Pause and Reflect

Take a moment to use the definitions of *equality*, *equity*, *inequality*, and *inequity* as a lens through which to view the earlier scenarios about student experiences. Ask yourself these questions:

- How would the stories of Jvette, Muhammad, John, Emily, and Donovan be different if they had equitable access to the full and equal application of the rights of citizenship, including safety and security in all their forms, access to a quality and affirming education, self-determination, and the pursuit of happiness?
- What new insights are you developing about the challenges of equity leadership?

Foundational Concepts for Understanding Inequity

Five key concepts can help you understand how "areas of difference" connect to inequality and inequity. People use different terms in their equity leadership work, and sometimes leaders get caught up in arguing about terms in ways that derail progress. Some equity leaders believe that if everyone uses the same terms, that will solve the problem of inequity. We disagree. Instead, we encourage you to consider how you will use the terms we provide and work to understand the underlying concepts. The *understanding* is the important takeaway. We explain these five concepts using the following terms:

- *Social construction*
- *Ism* and *phobia*
- *Privilege*
- *Intersectionality*
- *Race-neutral* or *difference-neutral ideology*

Social Construction

The term *social construction* refers to how we as a society make and give meaning to ourselves and the world around us. It applies to material things such as tools, money, and computers; social agreements such as rules, mores, and customs; and how people think about and identify themselves and one another.

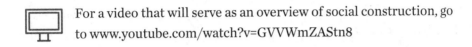 For a video that will serve as an overview of social construction, go to www.youtube.com/watch?v=GVVWmZAStn8

Equity leaders need to understand that a paradigm—either at the self or the system level—is a composite of social constructions. For example, the rules and procedures that govern schools and the expectations and rituals that society applies to students, their families, and the adults in the school are the result of agreements that groups of people developed together. This doesn't mean that social constructions aren't real or that they don't have an impact. In fact, the social constructions of society—and schools—are very real and incredibly powerful. Further, even as the social construction of schooling is governed by a collective paradigm, individuals connected to that system may have differing paradigms as they relate to various aspects of the system.

It's important to understand social constructions for two reasons. First, understanding social constructions means recognizing that the way things are *can* be changed; second, it alerts you to the challenges you face as you attempt to make that change. Equity leaders must understand that the meaning and impact of differences in students' identities are socially constructed, then deeply embedded in systems and self paradigms. In other words, race, disability/ability, socioeconomics, language, gender and sexual orientation, and religion are essential parts of people's identities and produce lived experiences, opportunities, and consequences. Still, the meaning and impact of these areas of difference are not inherent; instead, their meaning comes from collective agreement. Thus, the meanings can change.

Race provides a great example of social construction and how meanings change over time. It is important to note that race is not based on geography or, surprisingly, strictly on skin color. Instead, people in the United States consider a constellation of factors when assessing race, including skin color, eye shape, hair texture, ancestry, vernacular, and so on. However, race is an imprecise and changing way to identify people. For example, at one point in the history of the United States, people of Irish and Italian descent were not considered White, but today, Irish and Italians in the United States are almost always thought of as White people.

Similarly, at one point in recent history, the federal government's definition for a cognitive or intellectual disability was an IQ of 80 or lower. Years later, federal regulations changed this definition to an IQ of 70 or lower. Suddenly, people with IQ scores of 71 to 80 were no longer considered to have an intellectual disability. Even the idea that an IQ test can and should measure a person's intelligence is a social construction.

Lastly, consider gender. For many years, the U.S. construction of a woman's "place" was at home, raising children, cooking, and cleaning. Further, a woman was supposed to be "feminine" in how she performed these roles, and women were not supposed to engage in particular kinds of active or intellectual work. Over time, this construction has changed to a much more expansive understanding of gender. We now know that a gender binary is restrictive and creates inequities (something we explain in Chapter 8).

In each case, the people have not changed, but the social constructions have. Again, remember that these areas of difference are constructed, which is why we use the term *sociocultural identities*. Moreover, society attaches rules and procedures—spoken, unspoken, and unconscious—to these areas of difference, limiting equity. Most important, because these meanings, rules, and procedures were *created*, they can be changed.

Ism and Phobia

We use the terms *ism* and *phobia* to refer to systems of inequality that marginalize and oppress people based on their sociocultural identities. Some of the most common isms include racism, sexism, classism, heterosexism, and ableism. Similarly, the terms *homophobia, Islamophobia*, and *xenophobia* also refer to identity-based systems of privilege and oppression. When we use *ism* or *phobia*, we are referring to a combination of two things: prejudice and power. This is an important point. Anyone can perpetuate or be the target of prejudice, because prejudice is simply a preconceived notion about another person or group, often but not always negative. However, prejudice has a far greater exclusionary impact when it is accompanied by systemic power at any of the four levels—historical, structural, institutional, or individual/interpersonal (as discussed in Chapter 1). Thus, whereas a White male might experience prejudice in a particular setting based on gender or race, that situation does not rise to the level of racism or sexism in the United States because systemic power is currently and durably attached to particular identities: White, cisgender male, heterosexual, English speaking, Christian, U.S.-born, and nondisabled. Conversely, other identities are subject to and marginalized across all four levels of systemic inequity.

Privilege

We define *privilege* as unearned benefits given to individuals with particular socio-cultural identities—white privilege, male privilege, heterosexual privilege, and so on.

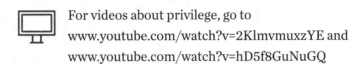 For videos about privilege, go to
www.youtube.com/watch?v=2KlmvmuxzYE and
www.youtube.com/watch?v=hD5f8GuNuGQ

Ferguson (2014) lays out and describes five key aspects of privilege. Here, we reiterate his observations:

Privilege is the other side of oppression. Society disenfranchises and disadvantages people with certain sociocultural identities. This state of affairs is called oppression. On the other hand, the term privilege applies to the people that society doesn't disenfranchise but instead empowers, enables, and supports, usually at others' expense. Privilege and oppression do not exist without each other; they are two sides of the same coin.

We need to understand privilege in the context of power systems. Various power systems, including patriarchy, white supremacy, heterosexism, ableism, and classism, have a significant impact on society. In these systems, the privileged groups and individuals have systemic advantage and power, and further, they are able to use their existing power to secure and expand their power and its benefits. In contrast, female privilege, Black privilege, trans privilege, and poor privilege do not exist, because people with those identities don't have systemic power or advantage. Although individual experiences do not always reflect systemic patterns, equity leaders need to understand privilege as a function of systems of power (we cover this topic in Chapter 10).

Privileges and oppressions affect each other, but they don't negate each other. All aspects of a person's identity—whether oppressed or privileged by society —interact. Many people believe that their experience of oppression means they can't experience privilege. For example, a poor White person might believe that she has no white privilege—that her poverty makes that impossible. But that is not true. (For an example, go to www.huffpost.com/entry/explaining-white-privilege-to-a-broke-white-person_b_5269255.) We address this idea further in the section on intersectionality.

Privilege describes what everyone should experience. Some aspects of privilege operate as special advantages. For example, White people are sometimes assumed to be trustworthy even when they have shown they are not. Treating someone as untrustworthy based on race should never occur, but often people of color—*particularly Black people*—are mistrusted because of unconscious bias or outright prejudice. We see this dichotomy when a White person convicted of a violent crime is given a light sentence because he is deemed worthy of a second chance, whereas a person of color is given a harsh sentence for a nonviolent offense because she is deemed a danger. In addition to special advantages, privilege also represents what everyone should experience but doesn't. These are the sorts of privileges—like the opportunity to rent an apartment when you have a solid credit history or get a job interview when you are qualified for the position—that should be equally and equitably available to everyone.

Privilege doesn't mean you didn't work hard. Many people get upset when others point out some of their privileges. If this is you, we encourage you to pause and examine your reaction. Remember, having privilege "doesn't mean your life is easy or that you didn't work hard. Instead, it simply means that [based upon your sociocultural identities], you don't have to face the magnitude of obstacles others routinely face and have to endure" (Ferguson, 2014), or that you had certain advantages in addition to your hard work.

Learning to think and talk about privilege is an important skill in equity leadership. To prepare, you must work through your own reactions to the term *privilege,* come to understand its full scope of meaning and impact, and develop the capacity to help others think about their own privilege.

Pause and Reflect

You can begin your work on understanding privilege by asking questions such as these:

- What do you think it says about you if you are privileged in some way?
- How does that notion conflict with your identity about yourself?
- How can you help yourself stay open to learning more about what others mean when they say you have privilege?

Intersectionality

Each of us is made up of multiple aspects of identity (such as race, gender, sex, disability/ability, national origin, and religion), not just one. These are integral to our sense of self.

Intersectionality is a term and concept introduced by law professor Kimberlé Crenshaw to describe the complex and compounding impact of a person's marginalized identities. This concept is key to equity-focused leadership because each person's combination contributes to different lived experiences. As described in the scenarios about various students' school experiences, these intersections can result in different and overlapping forms of oppression and privilege.

 For a short, animated video about intersectionality, go to www.youtube.com/watch?v=w6dnj2IyYjE

For example, whereas all women are subject to sexism, the sexism that Black women experience is unique in that it is further affected by racism. Two African American men may experience racism, but if one is a Christian and one is a Muslim, the Muslim may also experience discrimination and oppression based on Islamophobia. Similarly, Lydia Brown, who has autism, and Norman Kunc, who has cerebral palsy, have a shared identity as disability rights activists. But their experiences are different because their disabilities are different, *and* because Lydia identifies as queer, transgender, and Asian American, whereas Norman is heterosexual, male, and White. As noted in the section on privilege, privilege does not mean a person's life is easy. Norman's life has not been easy, and he has had many experiences of marginalization related to his disability; yet he experiences privilege as a White, heterosexual male. Intersectionality addresses the fact that Lydia's experience of oppression is compounded by each of her marginalized identities. Many people use the term *intersectionality* to refer to the variety of identities each person has, but Crenshaw intended for it to be used to refer to the compounding and unique effect of multiple marginalized identities that one person might have.

To hear Kimberle Crenshaw's TED Talk on intersectionality, go to www.ted.com/talks/kimberle_crenshaw_the_urgency_of_inter-sectionality

Race-Neutral and Difference-Neutral Ideology

A *race-neutral* ideology believes that the best way to move through the world and to end discrimination is to see and treat everyone as individuals without regard to race or ethnicity—that is, to not notice race. This stance is at play when you hear someone say, "I don't see color; I just see kids." Similarly, a person who holds a *difference-neutral* ideology believes that the best way to create equality is to treat everyone the same regardless of differences. This difference-neutral ideology is at play when people try to live, think, and act as if equity and social problems are solved best when they do not see, think about, or act on differences. We call this the Myth of Racial and Difference Neutrality1 because we know—from research and experience—that it is simply not true, accurate, or possible to not see and respond to these differences. Moreover, it is undesirable to act as if these differences do not exist.

Specific to race, a race-neutral approach denies the positive aspects of a person's identity and experiences related to race as well as the negative impacts of racism. Further, it is built upon a deficit orientation that suggests there is something wrong with being a person of color, and thus the best way to move, act, think, and behave is to not see or acknowledge the racial differences. Race-neutral ideas and language result in what's called *coded language*. The terms *urban, inner city,* and *underprivileged* are just three of many words used to refer to people of color without specifically naming race. Can you think of others?

For a video (which contains profanity and uses the term *color-blind*) about the myth of racial neutrality, go to www.youtube.com/watch?v=iaqOkutHSpI

---

1 It's important to know that a race-neutral ideology has often been referred to as a color-blind ideology. Although we agree with the concepts described by using the term *color-blind,* we're simultaneously troubled by how it implies there is something wrong with blindness. Disability-studies scholars and disability-rights activists have demonstrated that this type of deficit orientation toward disability is ableist. In our efforts to push ourselves and others to be as inclusive, sensitive, and respectful as possible, we elected to develop and use a new term, the *Myth of Racial and Difference Neutrality,* to represent this idea.

Similarly, the difference-neutral ideology is at play when people try to live, think, and act as if equity and social problems are solved best when they do not see, think about, or act on differences. For example, you may hear this expressed in schools about inclusive classrooms, when the marker of an outstanding inclusive classroom is the statement, "You can't tell who the students with disabilities are." This idea is built upon the myth that the best way to achieve equity and inclusion is to not notice differences, falling back on a deficit orientation that suggests that the differences among children are the problem and are inherently bad.

We know that recognizing and talking about differences can feel uncomfortable. Accordingly, many people shy away from these conversations, believing they lack language and skill. We urge you to push through any discomfort you may have, to avoid the false pretense that race and other differences do not matter. Moving through this discomfort begins with embracing the idea that difference is normal, natural, and positive. Consider that being Black is as natural and good as being White; that using a wheelchair is as natural and positive as walking; that worshiping in a temple is as natural and positive as worshipping in a church or not worshipping at all; that speaking Spanish, Somali, or Hmong is as natural and positive as speaking English. Without naming differences and engaging about difference purposefully, school leaders too often default to a frame of reference—and a whole system—that centers and normalizes White, middle-class, Christian, cisgender, heterosexual students without disabilities who speak English at home as the standard against which other identities should be measured. In working to eliminate race-neutral and difference-neutral mindsets, you must work to see diversity as the norm and engage it as an asset.

Sociocultural identities matter because they are central to peoples' sense of self and lived experience, and because they are the axes upon which inequity and injustice have been constructed. A more equitable and just system relies on people seeing differences as normal while simultaneously dismantling and recreating the structures and mindsets that position difference as "less than."

Pause and Reflect

Take a few minutes to write down your understanding or change in understanding of the foundational terms *social construction*, *isms* and *phobias*, *privilege*, *intersectionality*, and *race-neutral* and *difference-neutral* ideology.

* * * * *

In the discussion of privilege, oppression, marginalization, and intersectionality in this chapter, you probably began thinking about how these terms apply to your experience and identities. Now it's time to dig deeply into your experiences and reflect on questions that likely arose. In the next several chapters, we ask you to work, one by one, through the following areas of difference: race; disability; socioeconomics; language; gender, gender identity, and sexual orientation; and religion. For each, we provide activities, new ideas, and space to reflect on and examine your story as follows: (1) learning about the identity, (2) current realities in schools, (3) the broader context, and (4) disruptive leadership practices to increase equity. We intentionally begin with race because, as we stated earlier, you have to engage with and recognize what race means in the United States, or you can't really provide strong equity leadership.

Exploring Identities: Race

Many people—especially White people—experience race as a confusing and troubling concept. This reaction is not surprising, because race is entirely a *social construction* and has no biological basis. In other words, nothing about human biological variation and genetic makeup determines or results from race.

Even so, race is used (quite powerfully, in fact) to categorize people. The 2020 U.S. Census listed six categories for race: Asian; Black or African American; American Indian or Alaska Native; Native Hawaiian or Other Pacific Islander; White; and Some Other Race. Respondents could select any combination of these categories, and Hispanic/ Latino1 was considered an ethnic category rather than a racial category. We offer this information to illustrate one way that race is confusing. For example, consider how, except for "White," the racial categories are marked by geographic labels. Further, each category includes a broad array of geographic origins, which may or may not also reflect racial differences. Many individuals prefer a more specific way of naming their identity.

---

[1] We use Hispanic/Latino here to accurately restate the terminology used in the 2020 U.S. Census. In the remainder of the book, we use Latino/a/x to reflect the terms most commonly used by U.S.-based people who identify as Latino/a/x, explained further in the following paragraph.

For example, Indigenous, American Indian, and Native are all currently considered respectful terms, but "all citizens of Native nations much prefer that their nations' names in their own language be used" (Dunbar-Ortiz, 2014, p. xiii). Similarly, Latino/a/x is an umbrella term, and a term associated with one's ancestors/own geographic origins (e.g., Cuban, Mexican, Guatemalan), and sometimes combined with race (Afro-Cuban) or citizenship (e.g., Mexican-American) is preferred. Last, *race, ethnicity, culture, heritage,* and *geographic origins* are distinctly different notions, yet these terms are often used interchangeably.

Increasing your understanding of race and how it affects schools and the broader society will help you better understand how to disrupt harmful practices related to race. We begin with a list of definitions and explanations to provide a starting point for expanding your learning. We emphasize that this list can never be complete. For instance, although we have not included terms that would paint a more complete picture of racial categorization, we felt it necessary to include *White* and *Black* as key racial categories that represent two extremes of a complex spectrum.

- *Race:* A pervasive and faulty categorizing system that is variable over time and loosely based on a combination of features that includes skin color and phenotypic features. In the United States, race began with the construction of two categories, "Black" and "White," to quickly distinguish who was to be enslaved. More recently, race has come to be associated—and is sometimes interchanged—with national origin, ethnicity, culture, and so on.

- *Racism:* A system of structural, institutionalized, and interpersonal advantage that privileges one racial category over all others and over time. In the United States, White is the privileged category. This system operates regardless of whether its beneficiaries have good or bad intentions. If educators don't refine their ability to see and disrupt racism in action—that is, if they aren't actively antiracist—then it operates seamlessly.

- *Systemic racism:* An acknowledgment that racism goes beyond individual acts of meanness and prejudice; instead, that discrimination, marginalization, and systems of oppression operate together across historical, structural, institutional, and interpersonal/individual levels.

- *Internalized racism:* An aspect of racism in which people of color collude with their own oppression, such as undermining other leaders of color, maintaining deficit views of students of color, or supporting exclusionary policies and practices under the guise that they are fair for all or merit-based.

- *Cultural racism:* An aspect of systemic racism that regards aspects of advantaged (White) culture, such as beauty standards, art, and language patterns, as superior over those associated with people of color.
- *Prejudice:* A preconceived opinion, often but not always negative, based on limited, monolithic, and therefore inaccurate information about another person or group. When accompanied by systemic power, it is likely to have a discriminatory impact.
- *White:* A racial category loosely based on a combination of features that includes skin color and phenotypic features and is more about placement within the system (privileged) than skin color (the outward marker). Societal systems and structures have been established by and for White people, especially White cisgender men, and the constructs created by and related to Whites and whiteness are often viewed as normative. As a privileged identity, intragroup differences—outspoken, reserved, ambitious, outgoing, shy, and so on—are most often seen as individual qualities unrelated to race, and individual Whites tend not to be seen as representatives for their entire race.
- *Black:* A racial category loosely based on a combination of features that includes skin color and phenotypic features that is more about placement within the hierarchical system (targeted or marginalized group) than skin color (the outward marker). Societal systems and structures in the United States have been established by historically positioning Black people on the bottom of a racial hierarchy. In its marginalized status, Black is a construct often viewed as not normative, but instead, divergent from the norm. Its intergroup diversity—various groups of African Americans, Afro-Latinx, Namibians, Ghanaians, and so on—is made less visible and presented in overly simplified ways in societal systems and structures.
- *People of color:* A way of describing people in the United States who are not accepted, based mainly on skin appearance, to be in the advantaged White category.
- *BIPOC (Black, Indigenous, and People of Color):* This term "highlights the unique relationship to whiteness that Indigenous and Black (African American) people have, which shapes the experiences of and relationship to white supremacy for all people of color within a U.S. context" (thebipocproject.org, n.d.).

Current Realities About Race in Schools

The student population in schools across the United States is becoming more diverse, especially based on race and ethnicity. Students of color are consistently projected to increase to 56 percent of the student population by 2024, whereas White student enrollment in U.S. schools is expected to decrease from 60 percent to 45 percent. In addition, Latino/a/x enrollment will nearly double, from 17 percent to 30 percent (Kena et al., 2014). At the same time, nearly 80 percent of public school teachers and principals identify as White.

In this context, race affects what goes on in schools in many ways. For example, in 2018, White students outscored their Latino/a/x and Black counterparts on standardized math and reading tests by at least 20 points and 26 points, respectively (deBrey et al., 2019). In other achievement data, only 18 percent of White students, and 11 percent of Asian students, did not meet a benchmark on the SAT in 2019, compared to 58 percent of American Indian/Alaska Native and 53 percent of Black/African American students who did not meet a benchmark in the same year (College Board, 2019). Additionally, these racial gaps in "achievement" remain even when controlling for socioeconomic status. These gaps have remained relatively unchanged in the past 20 years.

That said, although educators cannot ignore test scores, relying on standardized tests as the only measure of student success is ill-advised; many other significant race-related disparities affect achievement. For example, 60 percent of schools offer calculus, but only 38 percent of schools serving mostly Black and Latino/a/x students offer calculus (U.S. DOE OCR, 2018b), and in California, only 11 percent of schools serving mostly ELs offer advanced math courses, including calculus (Feldman & Malagon, 2017). Further, students of color are oversubscribed for special education support services and are underrepresented as gifted and talented students (Ford & Moore, 2013; Milner, 2010; Obiakor & Utley, 2004; U.S. DOE OCR, 2012). Keeping in mind that most educators are White, we attribute this racial disproportionality in special education to misassessment, misidentification, and misplacement of children of color (Obiakor, 2001; Obiakor & Utley, 2004).

We argue that these disparities are better described as "opportunity gaps" or part of an "educational debt" (Ladson-Billings, 2006). Compared to "achievement gap" language, reframing to focus on the systemic nature of inequality highlights persistent inequity in opportunities, resource allocation, and school support, as well as belief in and understanding of students of color in comparison to White peers (Carter & Welner, 2013; Diamond, 2006; Milner, 2010, 2012; Singleton, 2015; Taliaferro & DeCuir-Gunby, 2008).

The Broader Context of Race

More than 20 years ago, the American Anthropological Association (AAA, 1998) took a position on race because of frequent misconceptions in popular culture and research. Their statement clarified that race is a social concept and that, in terms of biology, different races of humans do not exist. Accordingly, their position included language stating that race represents "physical variations in the human species [that] have no meaning except the social ones that humans put on them" (para. 4). The concept of race developed alongside European colonialism for the economic purpose of enabling Europeans and White Americans to rationalize their subjugation of non-White people (AAA, 1998).

Race, as a sociopolitical construction, has a history in the United States as the basis for an outmoded hierarchy. In its simplest, crudest, and most often-used form, race refers to skin color with little or no regard to cultural aspects of identity. Describing people by color can be clumsy and inappropriate, and can quickly devolve into offensive language once we move beyond "Black," "White," and "Brown" as descriptors. For example, it's considered offensive to call Indigenous people "red" or "redskin," and Asian Americans "yellow." Unfortunately, America's history with this racial hierarchy has greatly influenced how we identify people and organize ourselves in society, which correlates with who has access to aspects of our education system.

To complicate matters, racial categories are inconsistent now and have been so historically. For example, some ethnic groups (Western and Central European immigrants) moving to America have become a particular race (White) after several generations of living in the United States. Also, paradoxically, Latino/a/x individuals may be of any race but are more often discriminated against if they are darker-complexioned. Race, then, is an imperfect concept based on a problematic history.

So, should we still be discussing race today? Yes, because race cannot (and should not) be ignored, as it is related to several troubling and historically persistent outcomes in education. This simple way of describing and categorizing people affected how students were treated in the past, and it affects how we treat students today. Some educators have expressed beliefs about a correlation between race and intelligence. For example, educators may assume that Asian students are smarter than Black students, using little more than race to reach that conclusion.

The U.S. Department of Education Office for Civil Rights (2018a) collects school race-related data that illustrate national educational disparities and disproportionality. For example, consider the 2015–16 data represented in Figure 4.1.

FIGURE 4.1

Out-of-School Suspensions and Expulsions by Race and Gender

Race	Sex/ Gender	Percent of Students Enrolled	Percent of Out-of-School Suspensions and Expulsions
Black	Male	8	25
Black	Female	8	14
Latino	Male	13	15
Latina	Female	13	6

Source: From "2015 -16 Civil Rights Data Collection: School Climate and Safety," by U.S. Department of Education Office for Civil Rights, 2018a.

Suspension removes students from the educational environment for at least half a school day for disciplinary purposes. Expulsion refers to removing students from their regular school for disciplinary purposes for more than 10 days. Similar disparities exist related to restraining and seclusion practices for Black and Brown students.

These examples point to disproportionality—the idea that these incidences (with the exception of suspension/expulsion figures for Latinas) are occurring more frequently than what is expected relative to these students' percentage by race in the population. Why?

Disruptive Leadership Practices Toward Equity and Race

Well-meaning educators who have earnestly attempted to avoid judging students according to race have found it puzzling that inequities still occur and persist, often along cultural as well as racial lines. This situation is certainly baffling, and we hope you agree that this is a good reason to investigate race further. If you do, then we suggest that you examine your role as an educator in these race-related practices by asking yourself the following questions:

- How am I reproducing versus disrupting the experience for students in terms of their sense of community and social support around race?

- How am I reproducing versus disrupting the culture and learning of the adults at the school (teachers, administrators, other staff) around race to create a more inclusive staff?
- How am I reproducing versus disrupting the structures, policies, and practices around race?
- How am I reproducing versus disrupting the curriculum in terms of race?

Improving the circumstances around race in your school and district can encompass a wide range of actions. As you contemplate your options, consider the following to reduce racial disparities and injustice.

Disrupting Personal Racial Biases and Assumptions

We all have racial biases that may be internally reinforced frequently but discussed infrequently or not at all, especially in mixed-race groups. Good leaders understand this reality and courageously prepare to address it by first becoming more racially aware to surface their assumptions and biases, with the goal of developing a critical consciousness. They challenge their thinking by completing a racial autobiography (see the Personal Timeline activity in the Pause and Reflect section at the end of this chapter). They read relevant content on race that helps them get better at seeing the impact of race in their work (Gooden & O'Doherty, 2015).

Disrupting Negative Student Experiences That Align with Race

Leaders examine their student and school data once they have developed a critical consciousness. They vigorously search for general discrepancies and disproportionality in areas such as curriculum access, special education referrals, discipline data, and hiring data, among others. They then proceed to ask tough questions of themselves and others to disrupt these harmful patterns.

Disrupting Negative Attitudes Associated with Addressing Racial Biases

Leaders anticipate resistance from colleagues but work hard to use such instances as opportunities to grow and disrupt negative energy that slows progress. They work hard to build rich and relevant relationships because they know there will be some tough work ahead with colleagues who may not see the long-term benefits immediately, or at all.

Disrupting Status Quo Thinking That Maintains Racial Inequities

Leaders develop a culture of curiosity around racial inequities and frequently seek to measure progress toward raising awareness and the resulting impact on school culture. They consistently seek ways to invent and reinvent policies and practices that create a strong sense of belonging for all students.

Pause and Reflect

- What has been your experience learning about and discussing race and racial injustice?
- In understanding that a foundation for equity involves taking on racism, where in your experience have you seen, heard, and felt interpersonal, systemic, internalized, or cultural racism that impacted students of color, their families, and their communities?
- Where have you seen racism and racial prejudice in yourself?
- How do the statistics and research on race, racial disparities, and racial opportunity gaps challenge your thinking and practices? How can you open yourself to that information? How does the programming in your school/district support these racial disparities? How does it disrupt the disparities?

Crafting Your Story

Part 1: As you work on your learning around race, you need to be open to new truths about this area of difference. Articulate your takeaways about the following:

- Your thinking/learning about race
- The racial disparities presented in this chapter
- The language and vocabulary of race

Part 2: Working on your paradigms around race is hard and deeply reflective work and requires constant vigilance against your defensive routines. Ask yourself the following questions:

- When and where are you open to talking about race? Can you think of times or spaces where you felt uncomfortable about the topic of race or a discussion about race? What meaning can you make of that?

- What can you do to increase your knowledge, comfort, and language in thinking about and talking about race?

Personal Timeline—Unearthing Paradigms and Developing Understanding About Race

Part 1: Identify multiple points/times/moments across your life and write or discuss your personal story around race. These could be key moments when race was foregrounded, reflections on how race was hidden or manifested itself in a given moment, or a key realization or formative experience that involved race. It is important to look across your development as a child, an adolescent, a postsecondary/college student, an emerging professional, and in recent times.

Part 2: Describe your personal identity regarding race. Next, identify and reflect on experiences you've had of both *privilege* and *marginalization* in this area.

Exploring Identities: Disability

In the United States, special education services in preK–12 schools for the 6.5 million students labeled with disabilities are driven and mediated by the Individuals with Disabilities Education Act of 2004 (IDEA, 2004). The legislation identifies the following categories of disabilities:

- Autism
- Deaf-blindness
- Deafness
- Emotional disturbance
- Hearing impairment
- Intellectual disability
- Multiple disabilities
- Orthopedic impairment
- Other health impairment (OHI)
- Specific learning disability (SLD)
- Speech or language impairment

- Traumatic brain injury
- Visual impairment (including blindness)

IDEA also provides for a category in early childhood (up to age 9) that states may define as "developmental delay."

In an effort to better serve students with disabilities, states track how frequently each disability category is present in a classroom. Along with autism and other health impairment (OHI) labels, emotional disturbance (ED), intellectual disability (ID), specific learning disability (SLD), and speech and language (SL) impairments are "high-incidence disabilities," meaning they are the most frequently occurring across settings. Each state articulates specific guidelines and requirements in order to label a student with a disability and qualify that student for special education services; despite these guidelines, SLD, OHI, ED, and ID are the least concrete and therefore more subjective. One principal summarizes this, "It is cut and dry if a kid has a visual impairment, but it is much more grey about SLD or ED." Across the country, we see the overidentification of students of color and students from low-income backgrounds into these subjective categories, thus compounding inequities both for individual students and the system as a whole.

More info about each category in lay person language

The principles of IDEA and their basic requirements are as follows:

- *Free appropriate public education.* Develop and deliver an individualized education program of special education services that confers meaningful educational benefit.
- *Least restrictive environment.* Educate students with disabilities with non-disabled students to the maximum extent appropriate.
- *Parental participation.* Collaborate with parents in the development and delivery of their child's special education program.
- *Procedural safeguards.* Comply with the procedural requirements of IDEA.
- *Protection in evaluation.* Conduct an assessment to determine if a student has an IDEA-related disability and if a student needs special education services.
- *Zero reject.* Locate, identify, and provide services to all eligible students with disabilities. (Yell, Katsiyannis, Ryan, McDuffie, & Mattocks, 2008)

People with autism answer the question: "What is autism" (excerpt from the film, We Thought You'd Never Ask)

It is important that leaders understand the 13 IDEA categories and recognize the inherently subjective nature of the high-incidence ones. Equally important are the principles that the special education law and its regulations are designed to support and protect.

Current Realities About Disability in Schools

The purpose of special education is clearly stated in IDEA: "to *ensure access* of the child to the *general curriculum*, so that the child can meet the educational standards within the jurisdiction of the public agency that apply to all children" (emphasis added, §300.39(b)(3). Too often this purpose of access is overlooked; the purpose is not about providing a separate education but providing access to what the state and local districts have defined as the "general education," that is, what students without disabilities receive.

In the 1990s and early 2000s, many researchers and practitioners argued that segregated settings benefit students requiring special education services because they provide smaller class sizes and less distraction; higher-quality, individualized instruction at appropriate levels; more-qualified educators who know best how to work with students with disabilities; and more-comfortable settings free from the "assault on self-esteem" that might occur when students with disabilities have opportunities to compare themselves to their nondisabled peers. But over the past three decades, educational research has demonstrated that these justifications for segregated settings simply do not hold up.

An overwhelming body of research shows that segregated settings do not improve short- or long-term outcomes for students with disabilities. Instead, outcomes include poorer IEPs, minimal access to peer models, limited or less rigorous curriculum, significant time spent in nonacademic instruction, significant instruction from paraprofessionals (rather than trained teachers), lower expectations and diminished postsecondary opportunities, environments that are more distracting, limited opportunities for sustained relationships with nondisabled students, social stigma, and racial separation due to the disproportionality of students of color in segregated settings.

In contrast, we know through decades of research in hundreds of schools that inclusion in general education is a critical predictor of both school and postschool outcomes for students with disabilities (Benz, Lindstrom, & Yovanoff, 2000; Carter & Hughes, 2005; Cimera, 2010; Fisher & Meyer, 2002; Hughes et al., 2013; Kurth & Mastergeorge, 2010; Test et al., 2009). Further, we know that students do better academically in inclusive settings. In fact, when compared with their peers in segregated settings, students with disabilities who are included in the general education classroom make greater academic gains overall and increase their academic performance. In a large-scale study,

Cosier (2010) analyzed thousands of students across hundreds of districts and found that more time in general education positively and directly relates to increased academic achievement; further, the relationship is even stronger for students with the most significant disabilities.

We also know that with appropriate supports and services, inclusion in general education settings can help students feel a greater sense of belonging and provide them with meaningful access to friendships and high expectations, leading to greater social development. Additionally, inclusive settings increase positive social-emotional learning for students without disabilities, and research overwhelmingly demonstrates that achievement increases or stays the same for students without disabilities in inclusive classrooms.

Yet despite this abundance of research, segregated classrooms and programs continue to be used throughout the United States. In fact, about 40 percent of students with disabilities who receive special education services spend significant portions of their school day in segregated settings; and nearly 50 percent of students with labels of intellectual disability and multiple disabilities spend at least 60 percent of their school day in segregated settings (U.S. DOE OCR, 2015). Once students from non-dominant racial and linguistic groups are labeled, they are more likely to be in segregated settings than their White classmates with the same disability label (Cartledge, Singh, & Gibson, 2008; U.S. DOE OCR, 2015). This overrepresentation often occurs in the categories of emotional disabilities, intellectual disabilities, and learning disabilities (Parrish, 2002; U.S. DOE OCR, 2015).

The Broader Context of Disability

Society labels various types of abilities to give meaning to experiences that share certain traits. This social construction of disability masks the reality that the range of experiences labeled as disabilities are instead part of the continuum of normal human experience. In seeing disability as part of the normal experience, you need to question the deficit orientation that permeates the United States with the ableist notion that it is better to walk than to roll, scoot, or limp; that it is better to speak than to sign, point, or type to communicate; that it is better to sit quietly and comply than to pace, move, and question. Understanding disability as part of the normal human experience allows you to see that someone in a wheelchair who communicates using modes other than speech is as valuable a member of the community as someone who walks and talks. If you see equal value and validity in these differences, then you can open your mind to an equality of intellect, contribution, friendship, and membership in a community.

Models of disability. This video explains various ways or models of seeing disability

Seeing differences in this way forces you to move beyond a charity or an inspiration mindset about disability, both of which are problematic. A *charity mindset* is feeling sorry for those "poor souls" with disabilities and wanting to be kind because you feel pity. An *inspiration mindset* is feeling that people with disabilities are courageous and brave for engaging in everyday life experiences. Both mindsets position people with disabilities as lesser and outside the "normal" experience. We also encourage you to move beyond the medical model of disability that sees, labels, and defines the experience of people with disabilities as a medical condition arising from a physical and mental limitation that needs to be corrected or fixed—a deficit orientation.

Norman Kunc —The Right to Be Disabled

Keith Jones, a civil rights advocate for people with disabilities, says, "When the baby boomers get old, they're not going to call themselves disabled but [they're] still going to need a wheelchair, an accessible bus, and [an] accessible house" (Habib et al., 2007). His point is that although there are certain realities associated with having a disability (e.g., needing a wheelchair), what matters is what we make of that experience—how we value or devalue it, how we react to and construct it.

Often when people move, communicate, behave, and think in less typical ways, others assume that they are less intelligent and less competent. A key to moving past a deficit orientation about disability is to presume competence. Throughout history, people with disabilities have proven educators, doctors, and others wrong in their ability to take part in and gain from various opportunities. Presuming competence is not making someone prove you wrong; it is assuming from the start that all people (including those with significant needs and challenges) are capable and intelligent and then jointly constructing ways to see and foster that competence.

 For a clip from the movie *Intelligent Lives*, featuring Micah, who has an intellectual disability, go to www.youtube.com/watch?v=jjyENb2qbu0&t=10s

---

The common thinking about someone with an intellectual disability (and promoted by the IDEA definition) is that he has below-average intelligence. That notion is completely situated in a deficit orientation. In the film *Intelligent Lives*, Micah's tenacity and intelligence have forced people repeatedly to see him as a competent adult with real dreams and talents.

Disruptive Leadership Practices Toward Equity and Disability

When considering school equity related to students with disabilities, the body of research on the efficacy of inclusive services must both drive structural organization of services and provide a philosophical center for school leadership and school improvement. (Chapter 14 discusses tools for creating more-inclusive models for special education.) Let's explore this research in relation to educators' paradigms.

A common paradigm about instruction asserts that grouping students by ability will be the most effective and efficient approach; this is particularly common in times that emphasize accountability via standardized testing. These periods of accountability have occurred repeatedly throughout history: the No Child Left Behind (NCLB) act; its successor, the Every Student Succeeds Act (ESSA); and the Race to the Top (RTTT) program have all focused schools on issues of rigor and accountability.

In this context, educators are often led to standardize curriculum and pedagogical approaches, and students are seen as above, at, or below the expected benchmark. With pressure to raise test scores, many educators feel ill-equipped to differentiate and address diverse needs; so-called "ability grouping" becomes commonplace, and students who struggle with math or literacy are then grouped together. Likewise, the ability-grouping paradigm extends to disability—suggesting that separating students with particular needs is the most effective and efficient way to arrange schools and teach students. For students who require the most support and have the most complicated needs, the paradigm constructs a reality in which the only way to serve those students is in a separated setting.

These common responses to addressing diverse student needs are a direct reflection of an inaccurate and incomplete paradigm. This thinking drives the decisions to maintain grouping and segregation based on disability. Without disrupting this paradigm and getting a deeper understanding of disability, schools will continue to design and implement programs that separate and label students. We see this happening with two recent programs—Positive Behavioral Intervention and Support (PBIS) and Response to Intervention (RTI). Both programs were promising and designed to disrupt traditional ways of thinking about and teaching to learning differences. However, as the paradigm to separate and remediate by ability is so pervasive, their implementation has been mostly problematic, relying on sorting, separating, and labeling (Bornstein, 2014; Ferri, 2015).

Perhaps what is most telling about this paradigm is that it is so deeply ingrained that it rejects the vast quantity of research that says inclusive schooling is better—not necessarily easy, but better. Without addressing your paradigm and increasing your personal understanding, and despite the good intention that leads you to like the idea of inclusivity, you will maintain systems that sort and segregate. When you relegate students with disabilities—particularly those with more complicated needs—to more restrictive and segregated spaces, you are sentencing them to a separate life without the intellectual or social opportunities provided to other students. When you understand the implications of such segregation, you will understand the need to answer the call to work on this critical equity and equality issue. Whether or not you have a background in special education, this is your work. It requires disrupting traditional special education paradigms by asking four key questions:

- How am I reproducing versus disrupting the experience for students in terms of their sense of community and social support around disability?
- How am I reproducing versus disrupting the culture and learning of the adults at the school (teachers, administrators, other staff) around disability to foster more inclusion?
- How am I reproducing versus disrupting the structures, policies, and practices around disability?
- How am I reproducing versus disrupting the curriculum in terms of disability?

Improving the circumstances around disability in your school and district can encompass a wide range of actions. As you contemplate your options, consider the following possibilities.

Disrupting Student Isolation

Improving students' experience around community and social support begins with disrupting their isolation. Too often, students with disabilities feel socially isolated because they are not authentically included in a classroom. Many schools and districts have successfully changed that situation by intentionally and systematically training all staff in community-building approaches that are used in every classroom. In inclusive spaces, students provide peer support to each other—an aspect that is particularly important for authentic inclusion of students with disabilities, who otherwise receive most or all of their support from adults. A move to intentional peer support makes a huge positive difference.

Disrupting Paradigms to Support Inclusion

To develop fully inclusive schools, leaders need to help their staff develop more inclusive personal and collective paradigms. One step toward achieving this is through the development of three types of collaborative teams. First, you need a service delivery team that oversees, creates, and maintains inclusive structures and services. Second, creating and maintaining inclusive schools requires building instructional teams of professionals who co-plan and co-deliver instruction. Third, authentic inclusion of students with disabilities requires collaboration with families.

Disrupting Structures, Policies, and Practices to Redesign Service Delivery

Disrupting structures, policies, and practices requires rethinking service delivery to ensure general-education membership for all students. You cannot achieve a more equitable outcome for students with disabilities (as well as ELs, students of color, and students from low-income families who also receive special education and/or EL services) without creating inclusive services (Collier & Thomas, 2012; Frattura & Capper, 2007; Theoharis & O'Toole, 2011). With growing evidence that pullout and separate/segregated programs for special education and EL are stigmatizing and less effective, rethink the traditional ways you serve these students. Inclusive service delivery necessitates creating school structures where pull-out and segregated programs are eliminated and special education, remedial support, and EL services are brought to the students in heterogeneous, general education classrooms. You cannot achieve greater equity for students with disabilities without creating and maintaining inclusive service delivery. Doing so is essential! Chapter 14 has specific strategies for understanding your service delivery and then rethinking how to use the same resources differently to create more inclusive service.

Disrupting the Curriculum by Challenging Perceptions

Inclusive curriculum requires seeing disability as a natural part of the human experience and a form of diversity. It means working with students and adults in your school to disrupt deficit and medical views. Infuse disability into the curriculum. Doing so allows your teams to build on students' strengths.

Pause and Reflect

- What has been your experience with students across the 13 federal categories of disability?
- What do you think are the reasons that students of color and students living in poverty are overidentified for special education services in the categories that are the more subjective and least concrete? In what ways does the overidentification of students of color and students living in poverty for these subjective categories compound inequities and inequalities? What are your experiences about this and what new insights do you have?
- In what ways do you notice people around you speaking and acting as if disability is a deficit or a bad thing? When have you heard either a charity or an inspirational model of thinking about people with disabilities? In what ways do you see the medical model of disability in your school/district?
- In what ways do you see deficit orientations toward disability in yourself?

The Implicit Association Tests

  - What was your reaction to seeing Micah in the film *Intelligent Lives*? How does the concept "presume competence" relate to Micah?
  - Where in your school/district is competence *not* presumed or constructed? In what ways is it? Where is there room for growth and improvement? Where are your barriers to presuming and constructing competence for all students with disabilities?

Crafting Your Story

Part 1: As you work on your learning around disability, be open to discovering new truths about this area of difference. Articulate your takeaways about the following:

- Your learning/thinking about disability
- The research on the efficacy of inclusive services
- The reality that students of color and low-income students with disabilities are more likely to receive segregated special education services

Part 2: Working on your paradigms around disability is difficult and deeply reflective work and requires constant vigilance against your defensive routines. Ask yourself the following questions:

- What sort of dissonance does the idea of inclusive services for students with disabilities create for people you work with?
- What sorts of dissonance do you feel when considering a fully inclusive model for all students with disabilities? How does this dissonance manifest itself?
- How do you convince yourself (and others) to be open to data that support the fact that inclusive services are better?

Personal Timeline—Unearthing Paradigms and Developing Understanding About Disability

Part 1: Identify multiple points/times/moments across your life and write or discuss your personal story around dis/ability. These could be key moments when disability was foregrounded, reflections on how disability was hidden or manifesting itself in a given moment, or a key realization or formative experience that involved disability. It is important to look across your development as a child, an adolescent, a postsecondary/ college student, an emerging professional, and in recent times.

Part 2: Describe your personal identity regarding dis/ability. Next, identify and reflect on experiences you have had of both *privilege* and *marginalization* in this area.

Exploring Identities: Socioeconomics

In many ways, thinking about socioeconomics means purposefully seeking to understand economic and class realities in the United States. Although most Americans consider themselves middle class, their daily realities suggest a different situation.

Gorski (2013) argues that class labels and distinctions are imprecise. Therefore, we need to understand socioeconomics on a continuum that includes *extreme poverty*, *poverty/low-income, working class, middle class, managerial or upper middle class*, and *owning class*, terms that Gorski bases on the ability to have and use economic resources to meet basic needs. (The term *generational poverty* refers to long-term poverty and spans generations; *situational poverty* refers to a more temporary situation.) Here are explanations of the categories related to economic resources:

- *Extreme poverty:* Families of four (two adults and two children) in the United States earning below $12,550, or half the 2018 poverty threshold, are living in deep or extreme poverty.
- *Poverty/low income:* The poverty threshold in the United States is $25,100 for a family of four with two children. Research shows that the same family of four needs double the poverty threshold—an income of $50,200—to meet their basic

needs. Families of four earning less than that amount are considered low income. Approximately 41 percent of children in the United States are part of low-income families, including nearly half (20 percent) who are living below the poverty line and about 9 percent living in extreme poverty (Koball & Jiang, 2018).

- *Working class:* Families in this category can afford basic needs, but if faced with loss of income for a few weeks or months or with an event such as "unforeseen car trouble . . . [or] an unexpected medical condition, the family can find itself suddenly in debt . . . falling behind on rent . . . struggling to afford groceries" (Gorski, 2013, p. 9).

- *Middle class:* Most families in the United States want to believe they are middle class, whether they actually are wealthy or poor. Families that can afford basic needs, have accumulated savings or built some equity in a house, and can survive a few months without income are likely middle class. However, most middle-class families would find themselves in situational poverty if a loss of income lasted beyond a few months.

- *Managerial or upper middle class:* Members of this group likely have advanced degrees, six-figure incomes, and highly valued occupations (e.g., doctors; engineers; lawyers; successful small business owners; directors of nonprofits; administrators in local, state, or federal offices). People in this class run organizations and are largely but not exclusively White. This class is distinct from the next group—the owning class.

- *Owning class:* This group consists of people who receive enough income from investments to support a wealthy lifestyle. Many of them have jobs or once did, but they also may have significant wealth accumulated and passed down over generations (what is often called "old wealth") and own multiple homes. This category includes CEOs of wealthy corporations, heads of financial institutions, and high-ranking officials. This small but powerful population is overwhelmingly White.

Beyond understanding the above terms, it is also important to distinguish between income and wealth and also highlight the importance of networks and communities as a source of resources. *Income* is the amount of accessible money a person earns through work or investments. *Wealth* is a person's accumulated resources (such as property and investments) compared to liabilities (such as debt).

Vast disparities in income and wealth cross all areas of identity. In terms of income, women earn approximately 82 cents for every dollar men earn (Center for American

Progress, 2020), and Blacks earn approximately 74 cents for every dollar Whites earn (Economic Policy Institute, 2020). Income disparities contribute to profound wealth disparities over time. Wealth disparities have increased significantly over the past four decades and are now far greater than income disparities. For example, White families on average have 20 times the wealth of Black families and 18 times the wealth of Latino families (Gorski, 2013).

Income disparities affect the way children and families are able to access and participate in education, but wealth disparities have an even more profound effect. In particular, wealth disparities affect the resources and safety net that families have available should they experience a personal or societal economic crisis. In addition, because the U.S. population is increasingly segregated by socioeconomic status, people tend to have their most binding relationships within, rather than outside of, their economic class. Thus, if individuals living in poverty lose their job or have an expensive health crisis, they are less likely to have a social network that can provide significant financial support to bridge the gap until the crisis is over. Conversely, people in the upper middle class or owning class are unlikely to experience these types of crises because they have more than enough financial resources—including help from their family and network—to move through these events without losing their financial security.

Current Realities About Socioeconomics in Schools

Across the United States, a number of dangerous practices and patterns have emerged in schools related to poverty and income. Although these are often well-meaning and intended to close achievement gaps, in reality they are harmful and contribute to creating unequal and inequitable schooling. Among these practices and patterns are direct instruction, narrowing opportunities, and tracking.

A Focus on Direct Instruction

Hoping to raise test scores and boost achievement among economically disadvantaged students, many educators choose to rely mainly on direct instruction for curriculum and teaching. Such instruction takes many forms, from packaged programs of drill and skill, to computer-based skill programs, to in-class pedagogy focused on adult-led instruction in basic skills. It is not clear that these methods result in higher test scores (which, if they occur, may be short term), but it is clear that an increasing focus on direct instruction pushes aside student-led exploration, a focus on critical thinking and critical literacy, collaborative learning, and a host of other kinds of learning experiences that are essential for a 21st century education. Although all schools can be subject to this dangerous practice, it is most often employed by schools with higher concentrations of students living in poverty or, in more affluent areas, for students from families with low incomes.

Narrowing Opportunities

In this age of accountability measured by high-stakes tests, many schools have focused on helping improve test scores for students from low-income families. Time, staff, and resources are dedicated to preparing students for state literacy and math tests, taking time away from science, social studies, the arts, technology, physical activity and recess, and literacy and math experiences not related to the tests. Because schools are the main venue for students from low-income and working-class families to access opportunities in the arts, music, and other nonacademic pursuits, the result is a significant equity issue.

Tracking

Tracking is not new, and the benefits of eliminating tracking are widely known. Yet, as described in Chapters 1 and 5, educators have found various ways over the past decades to sort and classify students by criteria related to achievement levels, test scores, interventions, and so on. The result for increasing numbers of students from low-income families is that they spend more time in tracked groups and classes. Although tracking is thought of as a middle and high school practice, grouping students for literacy and math by perceived ability level in elementary schools is also tracking. Too often educators use the ideas of intervention and remediation to justify and create tracked systems and tracked opportunities, with a disproportionately negative impact on students from lower-income families.

The Broader Context of Socioeconomics

We know that poverty and the economic challenges that families and communities face have a significant impact on children and their education. The 2016 American community survey (U.S. Census Bureau, 2017) reports the numbers and percentages of each racial group by the categories *low income, poverty,* and *deep poverty* (see Figure 6.1, noting that deep poverty is a subset of both poverty and low income, and poverty is a subset of low income).

The National Center for Children in Poverty information and statistics about poverty

FIGURE 6.1

Number and Percentage of Children in Low-Income and Poor Families by Race/Ethnicity, 2016

Socioeconomic Category	Racial/Ethnic Group										
	White		Black		Latino/a/x		Asian American		Indigenous	
	Number	Percentage	Number	Percentage	Number	Percentage	Number	Percentage	Number	Percentage
Low Income	10,400,000	28	5,700,000	61	10,600,000	59	1,000,000	28	420,000	60
Poverty	4,400,000	12	3,200,000	34	5,000,000	28	500,000	12	250,000	35
Deep Poverty	1,869,000	5	1,600,000	17	1,900,000	11	150,000	4	130,000	18

Source: From "Basic Facts About Low-Income Children: Children Under 18 Years, 2016," by H. Koball and Y. Jiang, 2018, National Center for Children in Poverty, Columbia University Mailman School of Public Health.

Across the United States, poverty and extreme poverty disproportionately affect Black, Latino/a/x, and Indigenous/American Indian families. Poverty also intersects with age (among all age groups, the largest number of people living in poverty are children) and gender (poverty disproportionately affects women) (Koball & Jiang, 2018).

Living in poverty or with a low income brings many challenges and layers of oppression. Research from the ACLU, the Brennan Center for Justice, and the Poor People's Campaign shows the overlap across the United States between concentration of poverty and issues such as voter suppression; low wages; and lack of health care/Medicaid, recreation options, and child care. A wealth of research reports on the challenges to learning, including access and opportunity, that poverty creates for students and families. These realities lead us to embrace what Bryan Stevenson (2014) says: "The opposite of poverty is not wealth; the opposite of poverty is justice" (p. 18).

In the face of the systemic oppression related to poverty, Gorski (2013) argues that our nation embraces several dangerous myths that directly relate to school and families. Here are three examples:

- *Poor people do not value education.* Despite significant research demonstrating the contrary, this stereotype is expressed widely in schools and districts. Many educators hold this stereotype because they conflate valuing education with parents and families attending at-school activities. In fact, low-income families face significant obstacles (e.g., work schedules, transportation, need for child care for younger children) that inhibit attending school events. Yet numerous studies show that poor families do as much to support their child's schooling as wealthier families who are perceived to have high expectations and aspirations for their children.

- *Poor people are ineffective parents.* Examples of this stereotype show up when educators hear that low-income students might be watching a lot of TV and react by saying that poor parents must not be doing a good job. This reaction ignores the wealth of data showing the opportunity gap related to recreation and after-school activities for low-income children. Abundant research demonstrates that poor parents persist through many barriers and are as committed to their children as middle- and upper-middle-class parents.

- *There exists a culture of poverty.* This stereotype was introduced in the 1950s and gained traction in the 1990s and 2000s through the work of Ruby Payne (see, e.g., Payne, 1996). It is built on the assumption that poor people share a unique, consistent, and identifiable set of traits. Research has consistently shown that no such culture exists and that the notion perpetuates deficit ideas about people experiencing poverty and diverts attention from systemic inequities.

Disruptive Leadership Practices Toward Equity and Socioeconomics

Leaders must disrupt dangerous practices and deficit thinking around socioeconomics in order to create more equitable schools and systems. To begin, reflect on these four questions:

- How am I reproducing versus disrupting the experience for students in terms of their sense of community and social support around socioeconomics?
- How am I reproducing versus disrupting the culture and learning of the adults at the school (teachers, administrators, other staff) around socioeconomics to foster more inclusion?
- How am I reproducing versus disrupting structures, policies, and practices around socioeconomics?
- How am I reproducing versus disrupting the curriculum in terms of socioeconomics?

Improving the circumstances around socioeconomics in your school and district can encompass a wide range of actions. As you contemplate your options, consider the following possibilities.

Disrupting Biases and Assumptions

An essential step in disrupting the school staff mindset is to work and keep working to address your biases and assumptions; this section of the book (and this chapter about socioeconomics) offers you a starting place, but more is necessary. It is important to consistently engage yourself, staff, and students in developing critical consciousness around socioeconomics. Maintain vigilance in examining your own and others' thoughts, speech, and action for signs of deficit thinking toward people living in poverty, particularly in contrast to those with greater income and wealth, and offer alternative, strengths-based paradigms.

Relationships with the Community

Developing deep relationships with our communities is essential across areas of difference. Strong relationships with families help leaders learn how schools can best serve and support families, as well as what types of resources are available or needed in the community. One common and often necessary community connection brings health care and community health resources into school to aid in addressing health care gaps.

Universal Early Childhood Education

A vital component of creating more equitable schools around socioeconomics is developing universal early education for young children prior to kindergarten. We know the word gap and other school skills gaps can be reduced through universal early childhood education. This commitment to early education needs to reflect a language- and play-rich program and avoid developmentally inappropriate approaches that force young children to sit and engage in traditional sit-and-get skill practice.

Protect and Expand Opportunities

Creating more equitable schools and systems requires maintaining and growing art, music, recess, science, athletic, technological, and advanced opportunities for children and specifically for low-income students. This requires constant vigilance. Without relentless effort we see recess, music, advanced science, and opportunities quickly erode for those students living in poverty. This is one of the most pressing equity issues around income.

Pause and Reflect

- What has been your experience learning about and discussing socioeconomic status and poverty?
- Where in your experience have you seen, heard, and felt a deficit orientation or systemic class inequities that disproportionately impacted students in families with low income or living in poverty?
- Where have you seen classism and a deficit orientation in yourself?

- How do the research and statistics on wealth, poverty, and income challenge your thinking and practices? How can you open yourself to that information? How does the programming in your school/district support/perpetuate economic disparities? How does it disrupt the disparities?

Crafting Your Story

Part 1: As you work on your learning around socioeconomics, be open to new truths about this area of difference. Articulate your takeaways about the following:

- The data about low-income and poor people by race in the United States
- The definitions of class distinctions
- The data about income and wealth gaps in the United States

Part 2: Working on your paradigms around socioeconomics is hard and deeply reflective work and requires constant vigilance against your defensive routines. Ask yourself the following questions:

- In what ways do you notice people around you speaking and acting as if the stereotypes described in this chapter were true? In what ways do you see these stereotypes about socioeconomics in yourself?
- Where do you see the dangerous practices regarding socioeconomics in your classrooms, school, or district?
- Where can you engage in the key steps toward socioeconomic equity?

Personal Timeline—Unearthing Paradigms and Developing Understanding About Socioeconomics

Part 1: Identify multiple points/times/moments across your life and write or discuss your personal story around socioeconomics. These could be key moments when socioeconomics was foregrounded, reflections on how socioeconomics was hidden or manifesting itself in a given moment, or a key realization or formative experience that involved socioeconomics. It is important to look across your development as a child, an adolescent, a postsecondary/college student, an emerging professional, and in recent times.

Part 2: Describe your personal identity regarding socioeconomics. Next, identify and reflect on experiences you've had of both *privilege* and *marginalization* in this area.

7

Exploring Identities: Language

Bilingualism and multilingualism are an asset! Too often in the United States, students and families who speak languages in addition to or other than English are underserved and undervalued. To begin this chapter, we offer a summary of key research (as described by Collier & Thomas, 2012) most relevant to administrators in order to bolster your equity leadership for English learners (ELs):

- *Language development is natural.* Acquiring a new language follows the same natural process that we use as young children to acquire our first authentic interaction and communication about important and real things. "Children's first language does not 'interfere' with the second language—it is very important and serves as a resource for figuring out the second language" (Collier & Thomas, 2012, p. 160).

- *First language is* critical *to new language.* Our first language is connected intimately to our development and cognition. "Students who continue developing their thinking skills in their first language until age 12 do very well in their second language. Students whose first language development is discontinued before age

12 may experience negative cognitive effects in second language development" (Collier & Thomas, 2012, p. 160).

- *Younger is not necessarily better.* If students have a strong foundation in their first language, they are more efficient learners as they grow older and more likely to reach advanced levels in using additional languages (Collier, 1995; Samway & McKeon, 2007).

- *It's critical to understand the difference between social and academic language.* Cummins (1979, 2000) introduced the difference between social and academic language and found that it takes five to seven years to master academic language, whereas mastering social language takes two to three years. Again, academic mastery of the first language is key to developing academic fluency in (an) additional language(s).

- *Context matters.* Students acquiring a new language need to feel safe, and the culture they bring to the classroom must be respected. This means schools and districts must be free from racial or linguistic discrimination or hostility (including micro-aggressions). Practices such as grouping by ability and segregating students acquiring a new language can cause lasting damage.

Current Realities About Language in School

Students whose first language is not English now account for close to 5 million students or 10 percent of all public school students (National Center for Education Statistics, 2020d). They bring hundreds of different languages to the classroom (Kindler, 2002). Although large numbers of immigrants and refugees have traditionally settled in major urban areas, they are now putting down roots in suburbs and rural areas (National Center for Education Statistics, 2020d). As a result, ELs are now enrolled in schools and classrooms that have not traditionally served a linguistically diverse student body. With the EL population anticipated to double by the year 2050, most, if not all, teachers are likely to teach ELs in the coming years (Meskill, 2005).

Data reveal that the nation's ELs are not faring well academically on state accountability measures. Dropout rates for ELs, especially those born outside the United States, are higher than those reported for other sectors of the school-age population (Cartiera, 2006; Crawford, 2004; U.S. Department of Education, 2018). Still, many schools and districts engage in outdated EL programming, hoping but failing to get equitable and excellent results.

Schools in the United States use a variety of EL programs. Here we describe five main types and some of their variations (Collier & Thomas, 2012).

Pullout and ESL Content Programs

In this model, students at the elementary level are removed from their classroom to get some of their instruction from the English as a Second Language (ESL) teacher each day. At the secondary level, ELs have separate periods in their schedule for classes taught by ESL teachers in segregated classrooms. This practice is sometimes referred to as "sheltered instruction." ESL pullout is the most expensive and least effective model (Collier & Thomas, 2009), yet it is the most common across the United States.

Structured Immersion

Structured immersion is essentially another form of ESL content classes but uses a highly structured and sequenced set of materials. It is problematic because this kind of material does not necessarily match natural language development.

Newcomer Programs

The newcomer model dedicates classrooms to students who have recently arrived in the United States. In best-case scenarios, part of the instruction is conducted in the home language. These programs are often provided in separate schools.

Transitional or Early-Exit Bilingual Education

Typically, transitional or early-exit models provide instruction in students' home language for two to three years, after which students enter mainstream classrooms. Collier and Thomas (2012) describe this as a "remedial, segregated model" (p. 163).

Dual-Language Education

When done well, the dual-language education model focuses on cultural and language support and learning. Dual-language programs, in which students receive instruction in both their first language and their new language, are expanding.

Two-way bilingual programs are one of the best models for language development and learning. Serving both native English speakers and ELs, all students in these programs learn both languages and are experts in their home language and learners in their new one.

One-way bilingual programs are made up almost entirely of students who share the same home language. The most common type in the United States serves students whose home language is Spanish and provides instruction in both English and Spanish. One-way bilingual programs that honor students' home language and culture often reflect the segregated nature of school districts and communities. Notably, we have also seen language-immersion programs that permit only non-native speakers to enroll (e.g., a Spanish immersion program that enrolls only native English speakers). Neither of these models—one-way bilingual nor restricted-enrollment language immersion— provide the full benefits of two-way bilingual programs for English learners nor the language, cultural, and inclusion benefits for students whose home language is English.

Two-way bilingual programs are built on the idea of inclusivity. In other program models working toward inclusive services, schools create heterogeneous groups of students that include ELs, and EL teachers co-plan and co-teach with the classroom and content teachers. Another option uses teachers who are dually certified in content/ EL or general education/EL to teach a heterogeneous mix of ELs and their English-speaking peers.

Many educators' paradigms related to EL services are built on the idea of "remove and remediate." Schools and districts that do not have bilingual programs rely on pull-out/self-contained ESL classes. This is a deficit-oriented practice, but educators who assume this is the best or only way to serve ELs can fall into the trap of rejecting more equitable and inclusive approaches. To illustrate, consider the divergent viewpoints of the directors of EL for two mid-sized urban districts that were similar in terms of size and socioeconomics.

District 1

This district was working under a state improvement plan because of poor outcomes among EL students. Other than one school's bilingual program, the pullout model prevailed. The director saw this model as the only feasible means to serve thousands of ELs and repeatedly rejected the idea and research suggesting that collaborative/ inclusive ESL programs were both possible and more effective. She could not imagine anything other than a separate classroom for students who had just arrived in the United States as immigrants or refugees. Accordingly, in most of the district, ELs were removed from their general education classrooms and peers for significant periods each day. Newcomers were isolated in separate classrooms; even after being in the United States for over a year, they still received only pullout EL programming. The district consistently struggled with achievement for ELs and experienced no gains in student outcomes for ELs.

District 2

The EL director of this district had spent years implementing collaborative ESL services and expected ELs to be integrated into general classrooms served by collaborative teams of EL and classroom teachers. Accordingly, EL teachers and classroom/content teachers throughout the district co-planned and co-taught class- es. The EL teachers were not just taking the ELs to a separate part of the classroom for instruction; instead, both teachers worked with all students throughout the day. Newcomers to the district were immediately integrated into these heterogeneous classrooms. This district saw steady gains in student outcomes for the ELs.

These examples depict the power of more inclusive EL services and the power of the paradigms of school leadership. Dealing with similar circumstances, these two leaders' paradigms related to EL led to the programs and services provided for students—and the outcomes.

The Broader Context of Language

Language orientations, according to Ruíz (1984), are dispositions that mirror one of two conflicting public sentiments: that language is a problem or that language is a right. When principals view language as a problem, they consequently view English learners as having problems that need to be fixed (Crawford, 2004), an orientation that has the potential to negatively affect the quality and type of service they are provided (Reyes, 2006). When educators view language as a right, they work to provide ELs equal access to educational opportunities (Crawford, 2004). Ruíz (1984) proposes an alternative ori- entation—language as a resource—that supports an equity-focused perspective. Leaders who view language as a resource consider ELs' first language skills to be a valuable and relevant asset that contributes not only to their learning but also to the classroom in general.

We know that equitable schooling for ELs "is informed and buoyed by two sources: an asset-based orientation toward language and knowledge of the research on second- language acquisition" (Theoharis & O'Toole, 2011, p. 650). An asset-based language orientation encourages administrators to seek change and helps them develop a sound knowledge base that influences their vision for addressing the educational needs of ELs. The schools in which ELs are most successful have leaders who know and use the research on language (Montecel, Cortez, & Cortez, 2002).

In contrast, many EL programs that separate students perpetuate a deficit orien- tation. Although well intended, such programs raise serious equity issues. First, they isolate students from the rest of the school community. Second, they may convey a stigma of being remedial classes. Third, academic content may be watered down and not delivered by content experts.

Disruptive Leadership Practices Toward Equity and Language

In thinking about leading for equity and language, ask yourself the following questions:

- How am I reproducing versus disrupting the experience for students in terms of their sense of community and social support around language?
- How am I reproducing versus disrupting the culture and learning of the adults at the school (teachers, administrators, other staff) around language to foster more inclusion?
- How am I reproducing versus disrupting the structures, policies, and practices around language?
- How am I reproducing versus disrupting the curriculum in terms of language?

Leadership

One of the most critical attributes of effective schools for ELs is strong school leadership (August & Hakuta, 1998; Reyes, 2006; Shaw, 2003; Walquí, 2000). Leadership may come from a variety of sources within the school community and can most influence the long-term success of programs for ELs (Reyes, 2006). Effective leadership for ELs promotes justice in schools (Shields, 2004), raises issues concerning equity (Cambron-McCabe & McCarthy, 2005), and supports inclusive practices to meet the needs of a diverse student population (Riehl, 2000).

Thus, leadership around EL is essential. It requires substantial knowledge of and advocacy for ELs and their families. Leaders need to take a collaborative approach while embracing a strong vision for improving services for ELs. Leadership that focuses on asset-based, inclusive approaches leverages the opportunity for all students to thrive academically and socially.

Collaborative Process and Professional Learning

The vision for a successful program for ELs cannot reside solely with the principal. Most effective programs for ELs have emerged from comprehensive schoolwide efforts that involve principals as well as teachers and staff. Through informed inquiry and collaborative planning, all of these educators take charge of their educational programs (Shaw, 2003) and customize learning environments for ELs in a way that reflects local contextual factors, as it addresses the learners' diverse needs (August & Hakuta, 1998).

Although exemplary programs for ELs have adopted diverse educational approaches, they converge on a key characteristic: The learner is the priority. This requires educators to redefine their roles and relationships. New relationships among teachers and staff have the potential to ensure ELs' full social and academic participation, erase deficit perspectives of these learners, and create learning opportunities for educators (Freeman, 2004; Haynes, 2007; Mosca, 2006).

Principals in effective programs for ELs respond to the new demands on both teaching and nonteaching staff by offering appropriate and ongoing professional development. They employ a valued- and asset-based approach to develop the dispositions and skills to work in a collaborative and inclusive model. As instructional leaders, principals must also assure that teachers have the time to work together in developing challenging and culturally responsive curriculum and instruction for ELs. Principals find the necessary time for staff to engage in concerted efforts to meet ELs' needs even if it means reconstructing class schedules and instructional time

Family Connections

Equity-oriented EL programs build intentional systems to communicate with EL families in their home language. Leadership and school staff find ways to effectively reach out and listen to EL families whose voices had previously been unheard in the school. This provides families with information about school programs and responds to EL families' needs and desires.

Leaders in equity-oriented schools for ELs place high value on assuring that the school is connected to ELs' families and these families to the school. They facilitate these families' involvement in innovative ways (McLaughlin & McLeod, 1996) and in their home languages (Lenski, 2006; Wenger et al., 2004). Bilingual educators who communicate fully and authentically with ELs' families help them mediate home-school differences and empower the families to be able to participate more fully in their children's education.

Inclusive Services

We cannot achieve schools that are more equitable for ELs without creating and implementing inclusive services (Collier & Thomas, 2012; Frattura & Capper, 2007; Theoharis & O'Toole, 2011). With the growing evidence that pullout and other segregated programs for ELs are stigmatizing and less effective, we must discard traditional ways of serving these students and reimagine new approaches.

Inclusive service delivery for ELs involves valuing students learning English and positioning them, their families, their language, and cultures as central, integral aspects of the school community. It necessitates creating school structures where EL services are brought to the students in heterogeneous general education classrooms, eliminating pullout and separate EL classrooms and services.

In many people's minds, the term inclusion comes from the field of special education. We need to make clear that the needs of ELs have both distinct and intersecting identities with students with disabilities. However, we are *not* constructing language as a disability. We use the idea of inclusive service to exemplify a philosophy that needs to undergird school policy and services. This philosophy is built on the belief that all students should be valued for their unique abilities (e.g., language) and included as an essential part of a school community that is purposefully designed to accept and embrace diversity as a strength rather than a problem. This idea of inclusivity means that educators must work to create community and social/academic support for ELs within general education, demonstrating that membership and instruction with heterogeneous peers matters.

Pause and Reflect

- What has been your experience with ELs? What different languages have your students brought to school?
- In understanding that a foundation for equity involves rejecting a deficit orientation, where in your experience have you seen, heard, and felt deficit and asset orientations about ELs, their families, and their communities?
- Where have you seen deficit and asset orientation about language in yourself? How does the key research on language acquisition challenge your thinking about students and learning languages? How can you open yourself to that information? How does the programming in your school/district support this information? How does it counter this information?
- What service models have you seen in your professional experience? Have you worked in two-way dual-language models? Have you worked in collaborative or co-taught EL models?

Crafting Your Story

Part 1: As you work on your learning around language, be open to new truths about this area of difference. Articulate your takeaways about the following:

- Your thinking/learning about language
- The research on efficacy of program models for ELs
- The idea of more inclusive services for ELs

Part 2: Working on our paradigms around language is hard and deeply reflective work and requires constant vigilance against your defensive routines. Ask yourself the following questions:

- What are the challenges to implementing the more equity-oriented models described in this chapter?
- What sort of dissonance does the idea of inclusive services for ELs create for people you work with?
- What sorts of dissonance do you feel when thinking about an inclusive model for ELs? How does this dissonance manifest itself?
- How do you help yourself (and others) open to the information that inclusive services are better?

Personal Timeline—Unearthing Paradigms and Developing Greater Understanding of Language

Part 1: Identify multiple points/times/moments across your life and write or discuss your personal story around language. These could be key moments when language was foregrounded, reflections on how language was hidden or manifesting itself in a given moment, or a key realization or formative experience that involved language. It is important to look across your development as a child, an adolescent, a postsecondary/college student, an emerging professional, and in recent times.

Part 2: Describe your personal identity regarding language. Next, identify and reflect on experiences you've had of both *privilege* and *marginalization* in this area.

Exploring Identities: Sex, Gender Identity, and Sexual Identity

Much has changed in the United States over the past 30 years in terms of increased public understanding and acceptance of differences related to sex, gender identity, and sexual identity.

To begin our discussion of these identities, we first offer definitions. Biological or birth *sex* is the physical, hormonal, and genetic makeup that a person is born with. (Usually this is male or female, but sometimes people are born with biological aspects of both sexes, referred to as *intersex*.) *Gender identity* refers to how people think of and express themselves. *Sexual identity* refers to an individual's sexual, romantic, and emotional attraction, in relation to each person's sex and gender identity (GLAAD, 2016).

You may wonder why we have chosen to discuss sex, gender, and sexual identity together in one chapter. Although a chapter for each would have been justifiable, we decided to discuss them together because of their deeply entrenched connections to one another and the need to expose how those connections perpetuate inequity. For example, consider how a gay man historically was not seen as masculine, or how a gender-fluid individual was assumed to be gay, or how a woman historically was expected to accept lower pay because she "should" have a man at home to bring in money, or how a person's sexual identity is defined by their own sex and gender identity, as well as the sex and gender identity of the person they are attracted to. We think these connections are important to unpack and explore as we seek to increase equity.

To proceed, we offer these additional definitions:

- *Bisexual* or *bi:* "A person who is attracted to two sexes or two genders, but not necessarily simultaneously or equally. [The term] used to be defined as a person who is attracted to both genders or both sexes, but [because] sex and gender don't operate on binaries . . . , this definition is inaccurate" (http://sexualornot .blogspot.com/2018/03/homsexuallesbian-lgbtqiapd.html).
- *Cisgender:* Individuals whose gender identity aligns with their biological sex.
- *Cisnormativity:* A paradigm that normalizes and preferences people who are cisgender, as well as the gender binary, and marginalizes and excludes people who identify as LGBTQIA+.
- *Feminism:* The idea that women are entitled to equal rights and respect and that disrupting and ending oppression of women is essential, given current and historical unjust realities.
- *Gay*: People who are attracted emotionally, romantically, and/or sexually to other people of their same sex. Women choose either *gay* or *lesbian* to describe themselves, whereas men use only the term *gay*. (Note: Avoid using the term *homosexual*, which is outdated and insensitive.)
- *Gender binary*: A social construction that conflates gender expression and gender identity with biological sex, and poses male and female as opposites. In this paradigm, if a person is biologically female, then that person should be "feminine," and if a person is biologically male, then that person should be "masculine." The gender binary is a restrictive, marginalizing, and oppressive paradigm for everyone, even as it delivers privileges to some.
- *Gender expression:* The way a person expresses his or her gender identity; it can include dress, hair, mannerisms, and so on.
- *Gender fluid or nonbinary*: Used to describe a person and their gender expressions that shift throughout the gender spectrum at any given time or in any given circumstance.
- *Gender nonconforming*: People whose gender expressions do not match traditional gender expressions for their sex.

- *Gender spectrum*: A broad and diverse range of gender expressions and identities; offered as a more accurate, inclusive, and sensitive paradigm in contrast to the gender binary.
- *Heteronormativity*: A restrictive paradigm contending that heterosexuality, predicated on the gender binary, is the norm or default sexual identity.
- *Heterosexism*: Discrimination or prejudice against LGBTQIA+ individuals and groups based on the faulty assumption that heterosexuality is the only normal sexual identity.
- *Homophobia*: Discrimination, prejudice, and other hateful attitudes that exclude and harm LGBTQIA+ persons.
- *Intersex*: The combined anatomy ("sex chromosomes," external genitalia, or internal reproductive systems) of a person that "does not fit the typical definitions" for either male or female (Intersex Society of North America, n.d.). "The existence of intersexuality shows that there are not just two sexes and that our ways of thinking about sex (trying to force everyone to fit into either the male box or the female box) is socially constructed" (http://sexualornot.blogspot.com/2018/03/homsexuallesbian-lgbtqiapd.html).
- *Lesbian*: A woman who is emotionally, romantically, or sexually attracted to other women.
- *LGBTQIA+*: An acronym for lesbian, gay, bisexual, trans, queer, intersex, asexual, and other nondominant identities related to sex, gender, and sexuality. (Note: When referring to sexual identity, avoid using the term *sexual preference*, which suggests that a person's sexual identity is a choice and changeable.)
- *+ (plus)*: Being inclusive of all identities to make all people feel welcome and able to define themselves.
- *Queer*: "An umbrella term to refer to all LGBTQIA+ people . . . a simple label to explain a complex set of sexual [and gender-oriented] behaviors and desires. For example, a person who is attracted to multiple genders may identify as queer" (http://sexualornot.blogspot.com/2018/03/homsexuallesbian-lgbtqiapd.html). Because the word has been used hatefully against LGBTQIA+ persons, some find it offensive, whereas others use it as a political statement that breaks binary thinking, seeing both sexual identity and gender identity as fluid.
- *Sexism*: Prejudice, stereotyping, or discrimination based on sex, used to exclude and harm anyone who is not a cisgender male.

- *Transgender*: Sometimes shortened to *trans* or *TG*, an umbrella term that refers to individuals whose gender identity differs from the sex they were assigned at birth. For example, a trans male was classified female at birth. Transgender also includes people who are transsexual, cross-dressers, transgenderists, gender queers, and who identify as neither female nor male. It is important to note that transgender is *not* a sexual identity; transgender people may have any sexual identity.

Extensive lists of definitions and explanations

Current Realities About Sex, Gender Identity, and Sexual Identity in Schools

Cisnormativity, heteronormativity, and homophobia contribute to profoundly oppressive contexts for students, especially queer youth. Research suggests that, invariably, educators prescribe attributes to students based on biological sex and cisgender paradigms; in doing so, they maintain and replicate sex- and gender-based inequities in schools. For example, heterosexual cisgender males and their perspectives are most represented within curriculum in U.S. schools.

For more on cross-dressing

Across the 20th century—the time during which current teachers and administrators engaged with literature as children—children's books contained, on average, two male characters for every one female character—assumedly all cisgender (McCabe, Fairchild, Grauerholz, Pescosolido, & Tope, 2011). Further, it's well documented that many teachers unconsciously choose cisgender male students as the first students to respond or volunteer to answer questions (Sadker, Sadker, & Zittleman, 2009). In this context, cisgender girls and gender-nonconforming individuals are often silenced and rendered invisible, while cisgender male students receive more feedback and more of a teacher's time and energy.

Although the overall culture and climate toward LGBTQIA+ Americans has made important strides over the past 50 years in the media, at the workplace, in health care, and under the law, cisnormative and heteronormative policies, practices, and thinking patterns remain rampant throughout the United States. LGBTQIA+ youth still face many barriers to school success, including issues of safety and overall mental well-being, that are a direct result of their sexual identity. On the surface, this reality is seen in schools when you hear people saying things such as, "Do you have a girlfriend/boyfriend?," assuming an opposite-sex relationship; "Don't be a sissy," promoting a tough and aggressive masculinity; "You faggot," a derogatory term used to insult a straight male; or "That's not very ladylike" to suggest a restrictive expression of gender and femininity.

At a practical level, one obvious issue is related to school bathrooms: Almost two-thirds of transgender students do not use school bathrooms because they do not feel safe or are uncomfortable (Kosciw et al., 2014). Many students report that they risk harassment, both verbal and physical, regardless of which bathroom they enter. Districts throughout the United States have found themselves in community debates and legal battles over gender-neutral bathrooms that accommodate *all* students.

In this and other ways, schools are often hostile places for students who are, or are perceived to be, LGBTQIA+. Nearly a fifth of all students are physically assaulted because of their sexual identity and more than a tenth because of their gender expression (PFLAG NYC, 2020), and "[78] percent of gay (or believed to be gay) teens are teased or bullied in their schools and communities" (Riese, 2016). These circumstances contribute to poor health care and mental health services, marginalization and exclusion in schools, and disproportionately high rates of mental health issues and suicide. Consider these statistics (PFLAG NYC, 2020):

- LGBTQIA+ students are twice as likely as their non-LGBTQIA+ peers to say that they do not plan to complete high school or attend college.
- Gay teenagers are 8.4 times more likely than their non-gay peers to report having attempted suicide and 5.9 times more likely to report high levels of depression.
- Nearly 70 percent of teachers report handling LGBTQIA+ bullying and harassment and creating a positive atmosphere, but about two-thirds of LGBTQIA+ students indicate they do not feel the adults in their school are addressing the harassment or creating a more affirming culture.
- The average grade point average of students who are frequently physically harassed because of their sexual identity is a half grade lower than that of other students (PFLAG NYC, 2020).

Beyond these statistics, students repeatedly share that LGBTQIA+ topics are not only omitted in schools, but further, are routinely avoided. This avoidance ignores the opportunity to acknowledge the broad range of lived experiences of members of the LGBTQIA+ community and how they, like all students, comprise a diverse group of individuals across race, class, and socioeconomics. When their lives are rendered invisible in this way, it is a daily signal to everyone in that context that LGBTQIA+ students do not belong. These conditions demonstrate that leaders must improve the learning and lived experiences of LGBTQIA+ students in schools.

The Broader Context About Sex, Gender Identity, and Sexual Identity

More broadly, we see how sex, gender, and sexual identity inequities stem from and perpetuate inequities for anyone in the United States who is not a cisgender heterosexual male. Admittedly, we see more openly diverse representations of gender identity in schools than in the larger society, and women are in positions they did not have access to in the recent past. For example, according to the National Center for Education Statistics (n.d.a), in 2018 females attending college in the United States outnumbered males by nearly 2.5 million. These data are consistent with other data showing achievement gains that female students have made in math, literacy, reading, and writing (National Center for Education Statistics, n.d.b). All these data suggest parity among females and males in academic achievement and associated professional outcomes.

Despite this progress, we can use the gender binary to identify inequities. For example, median annual earnings for women working full time are $41,977, compared to men being paid $52,146 (AAUW, 2018). This gap exists across all demographics and is exacerbated by race and age; it exists in every part of the United States and in nearly every line of work—including female-dominated professions such as teaching and nursing. Greater education and higher positions have little effect: Although women in the United States now earn more college and postgraduate degrees than men, women with bachelor's degrees working full time are paid 26 percent less than their male counterparts. And even when women pursue higher-paying occupations, they are paid less across the board, even if they have advanced degrees (AAUW, 2018). A staggering government statistic indicates that the rate of reduction in the pay equity gap has slowed considerably and that "women working full time are still paid, on average, only 80 cents for every dollar paid to a man—a figure that has changed less than a nickel during the 21st century" (AAUW, 2018). Overall, collectively, women lose nearly $500 billion each year because of inequities.

These data have huge implications for the life outcomes of women and their children as they seek to engage in schools and their education. Further, this reality plays out across school systems: Although approximately 77 percent of teachers are women,

54 percent of principals are women (disproportionately at the elementary level), and approximately 24 percent of superintendents are women.

Beyond cisnormative ways of thinking about gender inequality, inequities manifest for LGBTQIA+ individuals. Income inequality is an issue: "LGBT people collectively have a poverty rate of 21.6%, which is much higher than the rate for cisgender straight people of 15.7%. Among LGBT people, transgender people have especially high rates of poverty [at] 29.4%" (Badgett, Choi, & Wilson, 2019). In addition, LGBTQIA+ individuals face increased challenges in accessing respectful, knowledgeable care, and experience poorer physical and emotional/behavioral health outcomes as a result (Cigna, 2017). Finally, LGBTQIA+ individuals are more likely to experience increased violence and victimization (Roberts, Austin, Corliss, Vandermorris, & Koenen, 2010) and the violent murders of trans individuals—particularly Black trans women—has been declared an epidemic by the Human Rights Campaign (2020).

Disruptive Leadership Practices Toward Equity and Sex, Gender Identity, and Sexual Identity

In thinking about leading for equity and sex, gender identity, and sexual identity, ask yourself the following questions:

- How am I reproducing versus disrupting the experience for students in terms of their sense of community and social support around sex, gender identity, and sexual identity?
- How am I reproducing versus disrupting the culture and learning of the adults at school (teachers, administrators, other staff) around sex, gender identity, and sexual identity to foster more inclusion?
- How am I reproducing versus disrupting the structures, policies, and practices around sex, gender identity, and sexual identity?
- How am I reproducing versus disrupting the curriculum in terms of sex, gender identity, and sexual identity?

As you contemplate the wide range of actions you can take to improve circumstances related to sex, gender identity, and sexual identity in your school or district, consider the following possibilities.

Disrupting Language, Concepts, and Messaging to Be Safe and Inclusive

Most school environments are ripe with sexist, heteronormative, and cisnormative language, concepts, and messages, as this type of exclusionary and hurtful communication is rampant on the internet and in other media sources. Earlier we provided examples of how easily oppressive terminology and notions can exist, whether well intended or intentionally harmful, as in bullying. Be on the lookout, relentlessly, for statements and practices that prescribe limited notions of sex, gender, and sexuality; disrupt these practices every time you notice them. Provide training to your staff and students about how to be safe and inclusive in their language and actions. Visibly and vocally advocate for inclusion and equality across the spectrum.

Disrupting Representation Patterns

All students benefit when their schools have adults from across the spectrum in terms of sex, gender identity, and sexual identity. Further, students who identify as gender fluid, gender nonconforming, or LGBTQIA+ need to feel that they are welcome in all spaces of the school. As you hire people, ensure that you have practices that are inclusive. When you bring together groups of students or parents/guardians/caregivers and community members in contexts related to planning and decision making, ensure that you include those who increase the sex, gender, and sexual-identity diversity of the group.

Disrupting Unfamiliarity by Providing Ongoing Learning

Many teachers, school staff, parents/guardians/caregivers, and students may have limited knowledge and few experiences related to the range of diversity and needs regarding sex, gender identity, and sexual identity. Offer support and learning for understanding by providing opportunities to respectfully and persistently share information about this area of difference and related inclusion needs and preferences.

Disrupting Policy and Practice

In the past, many school and district practices and policies were restrictive and exclusionary in specific ways related to sex, gender identity, and sexual identity; some of these practices and policies remain today. From bathroom access, to school dances, to the respectful and inclusive use of pronouns, to safe spaces, establish clear and inclusive policies. Consider starting and supporting a Gay-Straight Alliance group for students or families that provides a safe space and network for LGBTQIA+ students, staff, and families, and that communicates the value and legitimacy of a broad range of identities.

Disrupting Curriculum to Improve Balance

We know that most curricula feature the stories and accomplishments of cisgender heterosexual males. The authors, scientists, mathematicians, artists, and other role models portrayed in fiction and nonfiction works are also likely to be cisgender hetero-sexual males. Be intentional about noticing this imbalance and pointing it out. Continue to add more women, gender-fluid, and LGBTQIA+ individuals to the mix until they are robustly and fully represented.

Pause and Reflect

- What is new learning for you from this chapter? How does it feel?
- As gender diversity and LGBTQIA+ people and issues have become far more visible in the last decade, how do you personally feel about the variety of gender expressions and identities that are now more evident? What about the variety of sexual identities? Are your feelings and thoughts inclusive or marginalizing for some? If so, what resources are available to you to work toward inclusion in your actions and interactions?
- In what ways do you notice that the gender binary and heteronormativity are restrictive and marginalizing in your system?
- Based upon the examples presented in this chapter, which of the inclusion practices would have the most impact at your school?

Crafting Your Story

Part 1: As you work on your learning around sex, gender identity, and sexual identity, be open to discovering new truths about this area of difference. Articulate your takeaways about the following:

- Your learning/ thinking about the definitions related to sex, gender identity, and sexual identity
- The progress made in schools and society around sex, gender identity, and sexual identity
- The glaring disparities girls/women, gender-fluid, and LGBTQIA+ individuals face in schools

Part 2: Working on our paradigms around sex, gender identity, and sexual identity is difficult and deeply reflective work and requires constant vigilance against your defensive routines. Ask yourself the following questions:

- In what ways do you notice people around you perpetuating the disparate realities about sex, gender identity, and sexual identity described in this chapter? Where do you see this in yourself?
- Where do you see sexist, heteronormative, and cisnormative language, concepts, and messages in classrooms, schools, or in your district?
- Where can you engage in making policy, curriculum, and school/district culture supportive and inclusive around sex, gender identity, and sexual identity?

Personal Timeline—Unearthing Paradigms and Developing Understanding Around Sex, Gender Identity, and Sexual Identity

Part 1: Identify multiple points/times/moments across your life and write or discuss your personal story around sex, gender identity, and sexual identity. These could be key moments when sex, gender identity, and sexual identity were foregrounded, reflections on how they were hidden or manifesting in a given moment, or a key realization or formative experience that involved sex, gender identity, and sexual identity. It is important to look across your development as a child, an adolescent, a postsecondary/college student, an emerging professional, and in recent times.

Part 2: Describe your personal identity regarding sex, gender identity, and sexual identity. Next, identify and reflect on experiences you've had of both *privilege* and *marginalization* in this area.

Exploring Identities: Religion

The virtue of religious freedom is often valued as central to the social fabric of the United States. To support this freedom, the U.S. Constitution indicates that the government shall not support, participate in, or promote any particular religion or religious view. Accordingly, public schools are intended to be places that do not promote or support any particular religion. In fact, the First Amendment specifically forbids the establishment of an official religion. Further, public schools may not inhibit the free exercise of religion by students, and to do so would be an abridgement of a student's First Amendment right, recognized in schools since 1969.

To understand the obligations of school leaders to protect this right and ensure freedom from religious discrimination, we begin with a few definitions, including the specific vocabulary of the First Amendment:

- *First Amendment*: The amendment of the U.S. Constitution that states: "Congress shall make no law respecting an establishment of religion, or prohibiting the free exercise thereof; or abridging the freedom of speech, or of the press; or the right of the people peaceably to assemble, and to petition the Government for a redress of grievances."

- *Establishment clause*: The clause in the First Amendment of the U.S. Constitution that prohibits the establishment of religion by Congress. Technically, public school officials should not establish or favor any particular religion, especially as a state governmental agency that receives federal funding.
- *Free exercise clause*: The clause of the First Amendment of the U.S. Constitution that prevents the prohibiting of the free exercise of religion.

Current Realities About Religious Identity in Schools

School leaders must be aware of religious identity and how it shows up in schools, especially when certain identities are marginalized. Leaders must acknowledge that even though the First Amendment prohibits the establishment of a religion, Christianity is currently the majority religion in the United States. As a result, public schools have historically allowed the expression and promotion of Christian views and practices without constraint. Even today, many school practices, rituals, and celebrations tacitly promote this religion as part of a normal way of being within U.S. public schools. For example, most adults need only look to their school memorabilia to see that it wasn't that long ago that their schools held Christmas concerts and parties and included Christmas and Easter vacations in their yearly calendars. In addition, Christian prayer was common at sporting events, graduations, and other school gatherings. In fact, some of these practices continue today.

Although ongoing debates recur as some argue that schools should return to allowing prayer, a set of legal cases offers guidance from the U.S. Supreme Court: Prayers organized by the school, including those delivered by students, are expressly prohibited by the First Amendment, regardless of whether they are delivered at a high school football game, in the classroom, or over the public intercom system (see, e.g., *Engel v. Vitale*, 1962; *Lee v. Weisman*, 1992; *Santa Fe Independent School. Dist. v. Doe*, 2000; *School Dist. of Abington Township v. Schempp*, 1963). Despite the First Amendment and the legal cases that forbid this practice, schools can still feel like places that support one religion over another, and this reality directly affects students whose identities are in the religious minority.

Given this reality, leaders must intentionally ensure an inclusive environment and be prepared to respond when intolerance surfaces at any level. Although we are concerned with every person's right to their own religion, equity-focused school leaders must be particularly concerned with ensuring the rights, safety, and security of those students who are in the religious minority.

How do students who are in the religious minority feel in your school? Are they expected to participate in, and be witness to, majority-religion activities? Consider students who practice Islam, Hinduism, Buddhism, or Judaism: How might they feel about expressing their religious identity? What if a student is agnostic or atheist? If your school requires students to recite the Pledge of Allegiance, Jehovah's Witnesses—whose religion prohibits saluting the flag—will likely feel singled out, disrespected, and/or excluded in your school (*West Virginia v. Barnette,* 1943). Consistently centering and allowing one religion while marginalizing others sends strong messages about the school's position on diversity. School officials can and should be proactive about ensuring a religiously inclusive environment, while maintaining the separation of church and state.

The Broader Context About Religion

Despite the nation's commitment to freedom of religion, many have thought of the United States as a Christian nation in ways that go beyond the earlier examples related specifically to public schools. This perception seems like a logical conclusion: Christmas is a national holiday, and Christian churches exist throughout the country.

Still, the religious landscape of the United States has evolved over time and continues to change. From the early 1990s through the turn of the century, approximately 60 percent of individuals said religion was "very important" to them (Carroll, 2004). In a 2018 report, Gallup noted that 72 percent of Americans consider religion to be "important"; however, a record low percentage (46 percent) say religion can solve all of our problems, down from 75 percent in 1952 and 65 percent in 2002. Overall, 78 percent of Americans say that religion "is losing its influence on American life" (Brenan, 2018). In comparison to 2003, when more than 85 percent of Americans identified as Christian, in 2017 the overall number was 73 percent (De Jong, 2018). A handful of other religions constitute minorities in the United States. Jews make up just under 2 percent of the population, and Muslims, Buddhists, and Hindus are each just slightly under 1 percent (Gallup, 2016; Pew Research Center, 2015). In contrast, the number of Americans who identified as having no religion had increased from 10 percent in 2003 to 21 percent in 2017.

Religious Intolerance

Within this context, and despite the constitutional guarantee of freedom of religion, instances of religious discrimination and violence are on the rise. For example, a Pew Research study (2017) found that a majority of Muslims (75 percent) believe there is significant bias and discrimination in the United States against Muslims, both as individuals and as a group. These perceptions are well grounded: Half of Americans believe that Islam is not a part of U.S. mainstream society, and 41 percent believe that Islam is a more violent religion than others (Green, 2017). In 2017, 48 percent of Muslims had had at least one experience of discrimination. Specifically,

> about a third of Muslims said they [had] been treated with suspicion at least once over the past year. . . . Nearly one in five said they [had] been called offensive names, and a similar share have seen anti-Muslim graffiti in their communities. Six percent said they were physically threatened or attacked. (Pew Center, in Hauslohner, 2017)

Similarly, the rate of anti-Semitism in the United States has risen sharply, with almost 2,000 incidents reported in 2017, the second highest number on record (ADL, 2018).

The degree to which Muslims face discrimination in America

Religious intolerance can take many forms. At the micro level, it can occur when one person assumes another's religious identity and beliefs in ways that demonstrate insensitivity or deny inclusivity to that person. For example, some people insist they have a right to wish another person "Merry Christmas" without knowing whether the person celebrates Christmas; this is akin to wishing another person "Happy Birthday" on *your* birthday. Although this kind of expression is not necessarily forbidden by the First Amendment, leaders must be aware of the effect it can have on the learning and social environment for people in the religious minority. Instances such as this reflect a form of religious intolerance that focuses on attitudes and beliefs and that occurs at the individual/interpersonal level of inequality. These types of insensitivities can create difficulties in relationships and are not inclusive.

Discrimination against Muslims is increasing

Religious Discrimination

Religious discrimination is another level of intolerance. This occurs when someone does not receive the benefits or access to something to which she would otherwise be entitled, based on religion or perception of religion. For example, because Christmas, a

Christian holiday, has been declared a national holiday, many workplaces list it as one of the major holidays for which employees automatically receive a paid day off. Alternately, an employee who works on Christmas Day may receive additional pay for working on a holiday. However, no other religions are accorded a national holiday, and thus those who observe non-Christian holidays must miss work, use vacation time, and/or lose pay to celebrate. These inequities are institutionalized in U.S. laws and organizations.

Hate Crimes

Hate crimes are the most extreme form of religious discrimination and were on the rise in 2017 and 2018. In 2017, for example, reports of religion-based hate crimes increased 23 percent over the previous year and constituted 22 percent of the overall number of reports of hate crimes (U.S. Department of Justice-Federal Bureau of Investigation, 2018). Anti-Semitic incidents account for 58.1 percent of the reports, and 18.7 percent were Islamophobic. These statistics, however, do not tell the whole story, as hate crimes are vastly underreported. The Southern Poverty Law Center tracks hate groups and hate crimes, reporting a rise in anti-immigrant and specifically anti-Muslim hate groups that was especially significant between 2016 and 2020 (Beirich, 2019; Southern Poverty Law Center, 2020).

Hate crime statistics

Religious intolerance is not limited to our institutions and highly visible hate crimes. It affects many facets of society and daily life and exemplifies the cross-sectional nature of equity issues. In 2015, the U.S. Supreme Court ruled that the company Abercrombie & Fitch engaged in religious discrimination when it refused to hire a female Muslim teenager who wore a hijab. In 2018, the Court ruled in favor of the owner of a cake shop who, based on his religious beliefs, refused to bake a cake for a same-sex wedding. As the first Black/mixed-race man to ever rise as a leading candidate for the presidency, Barack Obama confronted questions about his religion, citizenship, and national origin on a regular basis, arguably as a result of racial bias. He was often "accused of being Muslim." Seven years into his presidency in 2015, 29 percent of the U.S. population, including 43 percent of the opposing party, believed that Obama was a Muslim (Bailey, 2015). This was a significant increase from 2008 when he was elected and 10 percent of the population believed he was a Muslim, even after the 2008 controversy surrounding Obama's minister at his Christian church (Pew Research Center, 2008). We raise this example to highlight how it reflects equity issues. Specifically, the very idea of "accusing" someone of being Muslim suggests that being Muslim is a bad thing, and that this "accusation" occurred because of racism.

Tensions Between Religion and Other Forms of Inclusion

One of the key challenges regarding religious inclusion is the moral conviction with which some people hold their religious beliefs. For some, religious beliefs are morally correct views as well as important perspectives. This moral conviction can create tension with others who believe differently.

One area in which this tension arises concerns LGBTQIA+ populations. Some religions believe that LGBTQIA+ relationships and identities are morally wrong. For the equity-focused school leader, we note that students and families are *allowed* to hold this view, and students who do so are *protected* from harassment related to it. However, it is equally important that LGBTQIA+ youth and families are *protected* from all forms of harassment related to their sexuality and gender identity, and *allowed* full access to and inclusion in all programs. School leaders must not *favor or promote* any religious view—including views that exclude or marginalize LGBTQIA+ people and relationships.

Pause and Reflect

- How do you feel about the variety of religious identities present in your school?
- Are your feelings and thoughts inclusive or marginalizing for some? If so, what resources are available to you to work toward inclusion in your actions and interactions?
- How would you describe the climate in your school for students from the religious minority? Those who are atheists or agnostics?

Disruptive Leadership Practices Toward Equity and Religion

In a post-9/11 world, educators have often been confused about how to respond to religious or related First Amendment challenges or questions within schools, especially in light of the media portrayal of Islam and Muslim people. In fact, well-meaning educators are often hesitant to engage in religion-based issues because of fear they may offend, leaving them unsure of how to avoid challenging but necessary discussions. To be clear, the responsibility of the equity-focused school leader is to (1) *allow* religious beliefs, and expressions and (2) *protect* student's rights to safety, religious freedom, and access, both while (3) *not promoting or favoring* any particular religion or religious view.

Our recommendations for ways that school leaders can promote inclusion and equity across religious differences are directed specifically toward public school leaders, where the separation of church and state is a constitutional right. However, they are also useful for school leaders in other settings who wish to be inclusive and equity-focused regarding religious diversity.

To disrupt intolerant attitudes and prevent marginalizing actions related to religion, leaders must accurately understand the religious freedoms and limits guaranteed under the First Amendment. In addition, they must investigate their beliefs and perspectives related to religious equity and inclusion. To work toward religious inclusion and equitable treatment regarding religion, ask yourself the following questions:

- How am I reproducing versus disrupting the experience for students in terms of their sense of community and social support around religion or lack of religion?
- How am I reproducing versus disrupting the culture and learning of the adults at the school (teachers, administrators, other staff) to foster more inclusion?
- How am I reproducing versus disrupting the structures, policies, and practices around religion? How am I ensuring that my staff understands the implications of the First Amendment as applied toward religion in schools?
- How am I reproducing versus disrupting the curriculum (specifically, learning about many faiths) in terms of religion?

Improving the circumstances around religion in your school and district can encompass a wide range of actions. As you contemplate your options, consider the following possibilities.

Disrupting the School Calendar

Most school districts continue to develop a calendar based on the agrarian cycle and, as noted, a vacation around the Christmas holiday. Many districts continue to schedule additional days off surrounding the Easter holiday as well, while not scheduling days off for significant holidays in other religions. In these circumstances, students who celebrate other religious holidays must miss school to participate in their religious practices. At a minimum, federal law requires that school leaders ensure these absences are fully excused and students are afforded support to make up the missed learning and other opportunities. A more inclusive approach is to modify the school calendar in ways that don't favor any one religion but aim to accommodate the school's various religious groups.

Disrupting Policies Related to Prayer

School leaders should ensure that students have the right to pray in school as is practiced in their religion. Still, a fine line must be observed. Although students can voluntarily pray, the act should not be disruptive to others or to the educational activities of the school, nor should it be coercive in any way. At no time should prayer be led by a school staff member, nor can staff members participate in prayer with students at school. It is unconstitutional for a student to lead a prayer at a school event. Prayer should not be offered, allowed, or required in classrooms, on sports teams, at school assemblies, or at sporting events or graduation. To protect student rights to prayer while also protecting all students' right to an education, some schools designate a meditation space where students can go to pray or meditate. Providing such space is allowable but cannot promote any particular religion.

Prayer in public schools

Disrupting Curriculum

School leaders and other educators face two primary issues around curriculum and religion. First, schools can teach about religions in a way that is nonjudgmental, is inclusive of many religions, and does not promote any one religious view. However, public schools cannot promote any sort of religious thought, belief, or ideology. This restriction leads to the second and more controversial aspect of religion and schools, which relates to creation stories. Each world religion has a creation story, but the theory of evolution represents the most current and accepted scientific understanding of the origin of the planet Earth and its inhabitants. Accordingly, public schools can teach the theory of evolution as a scientific theory but cannot promote any one religion's creation story.

Pause and Reflect

- What other areas of identity and diversity stand in tension to religious beliefs in your experience? What does it look like to *allow* a diversity of religious beliefs and expressions, *protect* students' rights to safety, religious freedom, and access, but *not promote or favor* any particular religion or religious view in these instances?
- Reflect on ways that school leaders can include a diversity of religious views without promoting any particular one. For example, you might consider your language, student interactions, celebrations, cafeteria food, and how different religious traditions are impacted by expected dress, locker rooms, and activities during physical education class.

- What other areas can you think of where religious views contribute to conflict and tension in the school building? How does the lens of *allowing* a diversity of religious beliefs and expressions, *protecting* students' rights to safety, religious freedom, and access, but *not promoting or favoring* any particular religion or religious view help you to develop inclusive solutions and environments?

Crafting Your Story

Part 1: As you work on your learning around religion, be open to new truths about this area of difference. Articulate your takeaways about the following:

- Religious intolerance, religious discrimination, and hate in the United States
- Protecting the rights of students to be free from intolerance and discrimination related to their religious beliefs
- The various ways that religious intolerance is enacted

Part 2: Working on our paradigms around religion is hard and deeply reflective work and requires constant vigilance against your defensive routines. Ask yourself the following questions:

- What challenges and dissonance do the ideas of religious inclusivity create for me? For others I work with?
- In what ways do I notice people around me supporting one religion, oftentimes Christianity, over others? Where do I see this in myself?
- Where do I see discrimination or a lack of inclusivity around religion in terms of classrooms, school or district calendars, policy, curriculum, or student activities?
- Where can I engage in making my classrooms, schools, and district affirming and inclusive of students from religious minorities?

Personal Timeline—Unearthing Paradigms and Developing Understanding About Religion

Part 1: Identify multiple points/times/moments across your life and write or discuss your personal story around religion. These could be key moments when religion was foregrounded, reflections on how religion was hidden or manifesting itself in a given moment, or a key realization or formative experience that involved religion. It is important to look across your development as a child, an adolescent, a postsecondary/college student, an emerging professional, and in recent times.

Part 2: Describe your personal identity regarding religion. Next, identify and reflect on experiences you've had of both *privilege* and *marginalization* in this area.

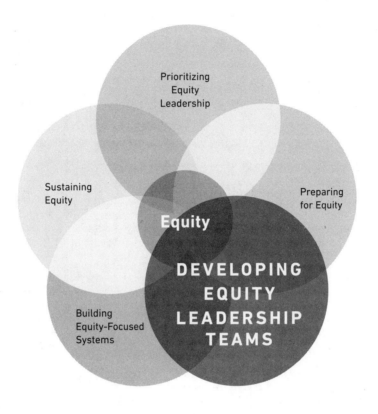

Prioritizing
Equity
Leadership

Sustaining
Equity

Preparing
for Equity

Equity

DEVELOPING
EQUITY
LEADERSHIP
TEAMS

Building
Equity-Focused
Systems

PRACTICE III
Developing Equity Leadership Teams: Essentials for Leading Toward Equity Together

Routines for Building Effective and Cohesive Equity Leadership Teams

Marvin recently became the principal of a public middle school that opened 12 years ago with the specific intent to create a racially and economically integrated learning environment for families wanting that option in their near-suburban community. At the outset, everyone at the school had great energy for innovating in ways specifically intended to increase equity and create a more inclusive school. The founders elected to have a year-round calendar, multi-age classrooms, and an integrated, thematic curriculum all as a means to advance equity among the racially and economically diverse student body. Through targeted efforts, the school was able to hire a teaching staff consisting of 30 percent people of color, despite a state average that was only 4 percent.

As a mixed race, Black-Indigenous male, Marvin was thrilled to begin his principal career at this school. However, as the first few months of his tenure began to unfold, he found that over the years, the spirit of equity-centered innovation and excitement that fueled the opening of the school had turned to conflict. At first, they had professional conflicts about how to "do" equity, with spirited discussions about what to do with curriculum, scheduling, extracurricular activities, teacher assignments, and so on. However, over time these conflicts turned into vehement arguments and an overall breakdown in personal and professional relationships, eroding their ability and will to work together and focus on students. The founders were nursing hurt feelings, believed their relationships were irreparably broken, and maintained their righteousness that the school had fallen short of the "right way" to do equity.

As the months continued, Marvin observed that educators who had joined the school along the way were invited into cliques based upon their philosophy and personality. Importantly, the percentage of teachers of color had dipped below 10 percent. From the outside looking in, there were one or two teachers who were ill-equipped for their jobs and had quit trying, but as a whole, despite the conflict and dysfunction, there were no heroes and no villains among the staff. There had clearly been some triumphs and some failures in what they had tried to do. Overall, though, Marvin observed a high volume of energy directed toward maintaining conflict, energy that was diverted from doing good work and focusing on kids, even as most people thought they were "doing what's best for kids." As he pushed forward each day, he knew that his number one job as principal was to help the staff learn to be a cohesive team focused on equity.

How should Marvin lead?

Now that you, as a leader, have begun the work on your intellectual and emotional growth related to equity leadership as described in Practice II, your next task is to begin modeling for and engaging others in relationship building, learning, and meaningful equity-focused work—that is, to form a leadership team. This approach contrasts with the popular notion of the singular or heroic leader, which is rarely effective over time. Instead, equity-focused leaders need colleagues who are similarly committed. Leadership teams that are focused on transformative systems change must be capable, cohesive, and equipped with the cognitive, relational, and strategic skills to be effective. Practice III offers tools and approaches that can help individuals and teams develop such skills.

Before we begin, we offer a reminder about the powerful influence of habits on behavior and how the habit of traditional leadership models and hierarchies will undermine efforts to advance equity. In our work, we have consistently seen that most people envision leadership in traditional ways, viewing a leader as the person who is in charge, knows the most, and gets to—or has to—make the decisions. This traditional model is not effective for equity-focused work for several reasons. First, leaders using this approach tend to make decisions in isolation, yet feel confident in their decisions due to their underlying moral commitment to equity. Even when leaders engage with a small leadership team to make decisions, the information the team considers and the decisions they make are limited in perspective, and thus inadequate. Second, because of the isolated nature of the decision making, often the people affected are not fully supportive and do not implement the decision with commitment and fidelity. Last, when the system or organization becomes stressed, relationships are not strong enough to weather conflict and endure hardship. When this happens, the leader takes the fall, and the organization must recalibrate and start over. In schools, this ineffective approach hurts students, families, teachers, other staff and administrators, and the community at large.

Our experiences have shown that despite leaders' frustration when others lead in a traditional, hierarchical mode, almost all new leaders revert to this model when *they* are in charge. New leaders often do this because they learn from their predecessors, mentors, and the leadership models that surround them; operating this way is a part of their deep individual and systems paradigms.

For example, in the story that opened this chapter, Marvin believed, based on observations and experiences, that some teachers in the building were "ill-equipped" to do their jobs and were no longer trying. He could respond to these data in various ways. It would be easy to accept the narrative that particular teachers are not invested or committed and start to address this dysfunction by making rules and issuing edicts. He could work to get rid of the "problem teachers" so that they would no longer be "dead weight," and other teachers would learn from their example. These actions would continue traditional paradigms of leadership and reinforce the teachers' habits related to conflict and teamwork. In contrast, Marvin could slow down and think through what the information might mean in terms of the development of individuals and of the building overall. He could recall what made him excited about teaching and what deflated his enthusiasm and use that understanding to try to engage the disillusioned teachers, perhaps finding ways to tap their skill or expertise in an important function. He could create structures and processes that encourage teachers to work more closely and supportively together. Above all, he could use his leadership and facilitation skills to engage everyone in a more positive picture of the future. But to do any of these things, he needs to lead in ways that will help his staff develop the skills to engage with their multiple perspectives positively, inclusively, and constructively.

Two critical concepts here warrant attention. First, when we say "multiple perspectives," we mean convening groups that include numerous and varied paradigms—both surface and deep—to examine, understand, and describe a given situation. Intentionally including multiple perspectives is important because (1) it can improve access to material resources for persons from historically marginalized groups; and (2) it helps organizations become higher performing, more sustainable, and more capable. To be effective as an equity-focused leader, you need to be intentional and persistent about seeking out and engaging with people who bring perspectives and paradigms that differ from yours.

Your purpose in doing so is to learn, build relationships, and share power, thus improving your knowledge, decision making, and implementation of equity approaches.

That said, it takes intentional and concentrated effort to successfully include and manifest multiple perspectives. To do this, you need to develop and use the second critical concept: inclusive processes. Research on organizational decision making and change is clear: The time and effort spent engaging others in understanding the problems you face, and identifying and deciding on solutions, will pay off in higher-quality decisions, more efficient implementation, and broader and deeper organizational support (Roberto, 2013). You should include persons from across the spectrums of sociocultural identities as well as positional and informal power. This must move beyond having a single parent or a single person of color (i.e., token representation), a decision that is pervasive when creating teams. Instead, authentically include stakeholders with multiple and diverse identities, voices, and positions, and meaningfully seek and embrace their participation and contributions.

These dual concepts of multiple perspectives and inclusive processes can be extremely powerful in three regards. First and most important, they add significant information that would otherwise be unknown. Second, when done sincerely and with authenticity, such invitations and conversations build strong relationships between school personnel and other stakeholders. Third, stakeholders who are involved often become liaisons and ambassadors between the district and the community, building support for each with the other.

Specific skills and methods are necessary to succeed in this approach. For example, you will need to put significant effort into building a diverse team whenever hiring

Tools for understanding and running a World Café

opportunities arise. Furthermore, you must prepare to host truly inclusive conversations that surface and value diverse perspectives. We offer suggestions for how to build these types of conversations with your team later in this chapter, but many tools are available to help you plan and host this type of engagement. One model is World Café, which offers flexible formats for having rich, large-scale dialogue. The resources available through Essential Partners (https://whatisessential.org/resources) offer many tools to help facilitators and participants share their thinking and listen carefully to others' ideas. Many districts engage a skilled outside consultant to train their leaders in conducting and facilitating these types of conversations.

You will need be open to—and plan to actually use—the information, perspectives, and recommendations obtained through these processes and dialogues. We can't emphasize this point strongly enough. Individuals and communities who have been marginalized easily and justifiably grow weary of sharing their stories and perspectives, only to have them dismissed or disregarded. Even those whose perspectives have historically been centered are suspicious of experiences where leaders *say* they want to gather stakeholder perspectives but have already reached a decision. Remember, you're not *really listening* unless you're willing to have your mind changed.

This chapter will help you lead in a way that advances your equity goals and aligns with your equity commitments. Individual reflection guided by the following questions can help you identify the ways and times that you use traditional paradigms of leadership in your thinking and decision making, and when you are actually engaging in new ways. Consider these questions thoughtfully and *be honest!* This is not a test or a measure of your character, but an opportunity to identify where you can better leverage your leadership to engage with others and increase equity.

Pause and Reflect

- Consider times when you've experienced traditional hierarchical forms of leadership. What worked and what didn't work with that approach?
- Now reflect on times when you have personally used traditional hierarchical forms of leadership. What were the benefits? What were the costs?
- Consider whether there have been other times when you used more innovative, contextual, and inclusive forms of leadership. What were the benefits? What were the costs?
- Reflect on the current leadership approaches you're involved with. In cases where you're on a leadership team, what is the structure and approach? In cases where you are the ultimate decision maker, what is the structure and approach?
- What conclusions do you reach from these reflections?

A Note About Past, Present, and Future

Before we describe a more equity-focused way to lead, we need to address how the past shows up in the present and the future—another primary way that equity leadership teams can become derailed. In our experience, many people want to say, "That's in the past, we need to let go of that." But in truth, each of us is a product of our past experiences, and so the past is always with us. This point is relevant to the work of equity leadership—for example, when people ask, "Why are we still talking about race?"—and also in terms of past experiences in professional relationships.

Regarding past equity work, many have been involved in projects that left people feeling scarred and angry. These experiences become part of the stories they tell and often complement existing paradigms about what it means to do difficult and complex work. Some may even suggest that their emotions about these experiences taint all subsequent attempts to do equity work because the pain, anger, and disappointment they and others felt overwhelm any focus on the necessary work. It is easy to hold onto these stories for two reasons: prior experiences and the challenge of acknowledging privilege.

In relation to prior experiences, we know that when equity conversations turn to shaming or blaming, or even become demeaning, they taint the learning environment in a way that stops dialogue and discussion, damages relationships, and creates resistance to new information and, ultimately, learning. It becomes easy to associate the *topic* of the learning with the fact that it took place in a negative learning environment. This outcome can happen around any topic, but learning about equity is complex and requires cognitive, emotional, and moral work. In the struggle to make sense of the content and their place in it, some resort to unproductive approaches and responses.

Still, our goal is to fortify your courage and capacity to foster environments in which your team can engage productively and cohesively in this learning and avoid responses that degrade the learning environment. To do this, we distinguish between cognitive conflict and multiple perspectives on the one hand, and affective/interpersonal conflict on the other (Roberto, 2013). Specifically, *cognitive conflict* occurs when ideas don't match with one another. Cognitive conflict is a natural result of bringing multiple perspectives together, and it has the potential to bring out the best organizational outcomes when managed effectively. *Affective* or *interpersonal conflict*, however, occurs when emotion heightens and accompanies cognitive conflict, soon creating interpersonal or relational conflict between colleagues.

Affective/interpersonal conflict is common because many people fear cognitive conflict or lack the skills to engage with it constructively. The work you've done to this point in the book, and the model we share in this chapter, are intended to help you with this situation, as the ability to work with and incorporate multiple perspectives is fundamental to inclusive, equitable leadership.

The second reason people hold onto these stories and their anger is, frankly, because it can be hard to acknowledge your privilege and the impact it has on your life and others. To address this issue, respond to these questions with as much honesty as you can muster:

- If you are White, which aspects of this privilege are you aware of? In what ways, if any, do you feel defensive about your privilege?
- If you are a cisgender male, which aspects of this privilege are you aware of? In what ways, if any, do you feel defensive about your privilege?
- As a school administrator or an educator in another role, what are the various ways in which you have positional decision-making authority that can powerfully affect others' lives? When are you aware of this positional authority? When, if at all, do you forget or ignore it? How often do you consciously think through how you will use your authority to redistribute power and privilege to create more equity?
- What other privileges are hard for you to acknowledge?

It is common, especially if you have many forms of privilege, to focus on your experiences of marginalization while wanting others to focus on their experiences of privilege. If you are someone who has considerable privilege, then you might be tempted to assume others' experiences, opportunities, and options are similar to your own. However, research consistently shows that in the United States, members of some groups (e.g., persons of color) have many more experiences of marginalization, in some cases ongoing and repetitive—even relentless. Similarly, research shows that those with more privilege (e.g., White people, straight cisgender males, those in higher socioeconomic categories) have far fewer experiences of marginalization and can easily find ways to avoid such experiences in the future.

As a learning community, we encourage you to take two precautions. First, avoid engaging in what some call "the Oppression Olympics," in which individuals compete with others to prove whose experiences of oppression and marginalization are worse. Second, simultaneously recognize that, compared with the occasional moments of marginalization that those with more privilege might experience, the isms described in Chapter 3 (e.g., racism, sexism, ableism) are characterized by repetitiveness, severity, duration, and pervasiveness—qualities that make their impact fundamentally different both psychically and materially.

Those with privilege and few experiences of institutionalized or interpersonal marginalization may be tempted to assume that the system and its institutions are life-giving and offer everyone the same optimal experiences they believe they have had. If you hold this view, you might struggle to recognize that others have not had the easy access that you've experienced. You may feel angry when others insist that their experiences are as legitimate as yours and become defensive at any suggestion that the system is not neutral and entirely fair. Or you may be someone who has had an easier time, hears others' experiences of marginalization and exclusion with considerable empathy, but then struggles with guilt or immobilization about how to respond.

If you are someone who has experienced a great deal of marginalization, you may seek acceptance and relationships, enduring micro- and macro-aggressions in the name of getting along. Or you may feel or act angry, critical, and distrustful about ongoing insults and assaults to your personhood. In all cases, you face the difficulty and persistent reality that you are likely to experience future marginalization from any direction.

For everyone, it can be easy to hold onto past harms—and maybe it's even necessary for a while if the harm was particularly egregious. Some conceal their injuries and maintain a cool exterior of politeness, while inside they maintain mistrust of the source of the harm. Add to this the inevitability of conflict in the workplace, and overall most people and organizations are poorly equipped to resolve differences in ways that sustain and restore relationships.

Combined, these factors create a context that can be conflict-ridden by the time you choose to begin or restart your equity-focused leadership. This is certainly what Marvin faced when he realized the full scope and reality of what was going on in his new school. He might conclude that staff members are locked in their positions and unwilling to shift, thus requiring that he take charge by prescribing and enacting solutions to the problems as he sees them. He might decide that conflicts among staff stem from just a couple people and therefore spend his time trying to find a way to transfer those staff to another building. However, that approach would be unlikely to produce a better outcome, especially if it is an early strategy.

We believe that when you *choose* to take up equity-focused leadership, you must step into the responsibility with maturity and willingness to try to heal past harms. Working to heal is different from ignoring past harms, and different from "letting them go." It involves acknowledging truth, reality, and the impact of different perceptions. If you are someone with a great deal of privilege, you need to honestly wrestle with whether any feelings of anger are the result of an unconscious fear that you might lose the position and benefits that you believe you have earned and deserve. If you belong to a historically marginalized population, you may worry that healing yourself may enable the oppressors to continue "as is," such that your community will continue to experience harm.

Further, we encourage you to consider healing as a process and to reject the idea that complete healing is necessary before you can continue or renew your commitment to equity work. If you or someone on your team was part of a bad process—or led a bad process or said something insensitive—you need to commit to bringing your whole self to the effort. You will personally need to engage and work at being a better version of yourself and then expect others to do the same. Simultaneously, be careful not to "freeze" people as the sum of their worst or bad moments. Although it is important to attend to your own and others' discomfort, anger, and disappointment, you shouldn't allow that to delay or keep you from doing equity work.

The Equity-Focused Leadership Team Model

The Equity-Focused Leadership Team Model (Figure 10.1) is intended to guide the development of your leadership team. We explain the model's components in this chapter and the next, along with specific ways to develop and embed its routines and roles as a framework for how your team functions.

FIGURE 10.1
Equity-Focused Leadership Team Model

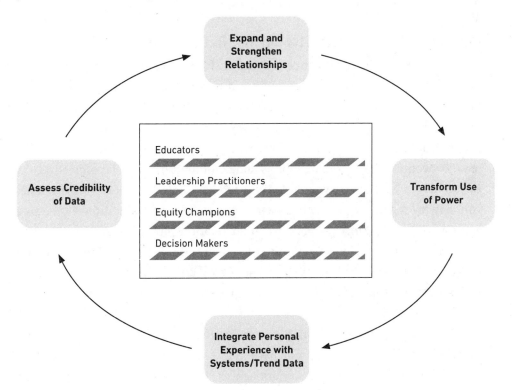

You will need to be diligent in applying this model because doing so likely will require you to revise (and even reject) some well-worn habits that reflect underlying paradigms and to commit to work through past conflicts and resistance. As you proceed, remember that the actions required to implement the model reflect hopes and dreams, and they should be ambitious. You will make mistakes, but your commitment should always be to maintain a clear focus on equity, to align your method with your end goal, and to increase your impact toward the greater good.

Four Routines for Equity-Focused Leadership Teams

The four routines of the Equity-Focused Leadership Team Model are structured procedural habits that your team should engage in frequently and consistently. They are (1) Expand and Strengthen Relationships, (2) Transform Use of Power, (3) Integrate Personal Experience with Systems and Trend Data, and (4) Assess Credibility of Data. The routines provide an important infrastructure to use when you begin practicing the roles (which we explore in detail in Chapter 11).

Routine 1: Expand and Strengthen Relationships

The work you did for Practice II sets the stage for better understanding how your lived experiences shape how you know and interact with the world. Remember that self and systems paradigms are extraordinarily powerful, developing from and in turn creating the stories that tell you how the world works. People are naturally attracted to and build relationships with individuals whose stories complement their own, and they are most comfortable in organizations and communities with complementary stories. Relationships with others who hold differing paradigms may lead to dissonance, and you may try to change each other's paradigms or leave the situation. These reactions are not surprising. Most people have great relationships with people who think like them and do things the way they want them done. Similarly, relationships with people who think differently and do things you do not want them to do present real challenges. Although this situation is logical, it can quickly become problematic for an equity-focused school leader.

The insights you gained from the work you've done for Practice II provide an opportunity to consider how your paradigms and their supporting stories may be sabotaging your ability to build authentic relationships. Recognizing that your paradigms are incomplete and contain inaccuracies is a critical step in doing equity work. Once you understand how they operate and acknowledge their influence, you can begin to work on improving their accuracy and completeness.

Take a moment to reflect: What did you learn from the exercises in Practice II about how you see things? What barriers might that learning reveal about your relationships? How can you use your learning to increase authenticity in your relationships?

When you build strong relationships across differences in race, dis/ability, socio-economic class, language, sex, gender, sexuality, and religion, you expose yourself to different ways of knowing and being in the world. These relationships can improve your paradigms because they expose you to information and perspectives that are otherwise missing, incomplete, or inaccurate for you. Further, having intimate relationships with people from multiple backgrounds makes a person more creative (Schmidt et al., 2019). These relationships are a critical step in learning how to create more equitable and inclusive learning environments. They model transformative ways of being for your colleagues and lay the foundation for doing equity work in difficult times and when conflict arises.

Strong relationships across differences require that you be intentional about diversity (e.g., when hiring staff or engaging consultants). Diversity of identity, experiences, and perspectives is essential to equity leadership. The saying "You don't know what you don't know" is true for everyone. Relationships across areas of difference are critical to each individual's ongoing process of becoming a more equitable leader, because everyone is still building on and expanding their individual paradigms. No one can, for example, become another race or suddenly know what it's like to have been raised in a different socioeconomic class. And although everyone can become more aware, informed, and empathetic, these important qualities do not change personal limitations and imperceptions. Building and maintaining a diverse team is one way to ensure that collectively, as a team, you know more. Put simply, diversity is a way to address limitations and imperceptions.

Expanding and strengthening relationships is demanding work, requiring a high level of maturity and self-awareness; it also requires consistency, a willingness to be vulnerable, and a commitment to stay the course. Here are five things you can do to expand and strengthen relationships:

Make relationships a top priority. Relationships are the foundation of leadership and are integral to working toward equity. Prioritizing relationships begins with your mindset and then needs to translate into behavior and decisions. Make time to build relationships in your meetings. Pause and turn away from your computer when another person needs to talk to you. Do not allow technical plans and processes to overshadow relationships. Remember that the people are the work.

Assume best intentions. This suggestion is easy to state but much harder to do. However, it brings an important attitude, commitment, and engagement to relationships. Catch yourself when you are about to make a negative assertion about why someone did something, and dial back your judgment. Assume that if the person took the action intentionally, she had good reason. Remember that people almost always take the action that they believe is right, justified, or both.

Allow time to share personal and professional stories. Think of the stories that you tell and retell yourself about who you are and how you came to be who you are at this moment. Now think about the stories that you tell and retell yourself about who you are *becoming* and who you *want to be*. These are the backdrop to what you do in professional settings. When you tell your stories to your peers, you learn more about how your stories shape who you are at work. When you hear others' stories, you learn about your colleagues.

Further, teams and organizations cannot overcome their challenges if they fail to understand who they are individually and collectively. When you share stories, your team has a better understanding of who you are collectively. Unfortunately, leaders rarely make time in professional spaces for team members to share stories. It's true that sharing stories can make people feel vulnerable and uncertain, and some think personal stories don't belong in the workplace. Still, it's much easier for people to connect with one another after hearing each other's story. Think of it this way: Human beings want to be accepted and acknowledged for who they are. Remember, stories are always present and everybody has one, whether they are acknowledged or not.

Sharing stories improves your teams in three important ways. First, it helps teams better understand and respond to a story's impact. Second, it enables team members to better support one another in working toward best leadership practice and thinking. Third, it helps team members know one another more deeply and makes the team stronger and more cohesive.

Another benefit of sharing stories is the opportunity to learn how to listen with an open mind. Listening closely reveals that barriers to equity are everywhere. You will learn how to discern which barriers are personal, created by an individual's unique circumstances and context, and which are systemic, created by inequitable policies, procedures, and structures. By listening together across multiple perspectives and with mutual concern and support, a collective wisdom can emerge to show the way to reduce those barriers, personally and systemically.

In the end, taking the time to share personal and professional stories allows each team member to better understand what is involved in working toward the equity goals that you all seek. The team will learn to "negotiate and act on [its] own purposes, values, feelings, and meanings rather than those [that] have [been] uncritically assimilated from others—to gain greater control over [their] lives as socially responsible, clear thinking decision makers" (Mezirow, 2000, p. 8).

Commit to engage in conflict constructively. Team members naturally have differing beliefs about how and why to achieve equity. As a result, conflict is certain, even when team members agree that they want the same goals and outcomes within their system. One group may believe that redrawing attendance boundaries is the next most important initiative, whereas another is certain that detracking is most urgent.

When confronted with this type of conflict, it is more important than ever to constructively exchange and consider multiple perspectives while avoiding interpersonal/affective conflict. Doing so requires that you ensure the conflict is not a personal attack on any person or group. Instead, remember that everyone is limited by their self and systems paradigms. Everyone—including you—is making sense of the world as they know and experience it, and may be unaware or forget that their thinking is incomplete and always evolving. Consistently framing disagreements as cognitive conflict increases everyone's capacity to lean into tensions as they emerge and face them as a normal and natural part of collaborative and productive work.

Many sources are available to help your team improve its capacity to engage in constructive cognitive conflict. For example, the Harvard Negotiation Project has an extensive array of materials: www.pon.harvard.edu/research_projects/harvard-negotiation -project/hnp/.

Here are some tips to make your conflict more constructive, allowing creative new ideas to emerge and avoiding stale, ineffective compromise:

- *Allow space and time for conflict and discomfort.* Be willing to sit with it. This requires structures during meetings to unearth conflict and discomfort and not rush past it. If later you realize that some conflict/discomfort was glossed over, purposefully revisit it. This does not mean allowing disagreement to stop equity-focused engagement, that one conflict should consume a team's time for the year, nor that you sit with conflict without addressing it. But, in our experience working with teams across the country, the number one thing that erodes equity

efforts among otherwise committed individuals is unresolved conflict. Moving
past conflict as if it doesn't exist, agreeing to disagree, or finalizing a decision
without working through the conflict will usually create additional and larger
problems in the future. Make time and space to work through conflict using
an inclusive process to achieve a solution that the team can live with. Use
these strategies:

- *Step back from certainty and step toward curiosity.* Assume you may be missing
  important information.
- *Identify the conflict.* Name the issue, the discrepant views, and why the issue mat-
  ters.
- *Remember to assume best intent.* Doing so now is more important than ever.
- *Listen.* Be willing to practice appreciative listening, committed to fully hearing
  what someone is saying within the context of all that you know about
  that person.
- *Inquire.* Don't assume that you understand why someone holds a perspective.
  Seek to understand it and why it is logical and desirable to that person.
- *Talk.* Share your information, reasoning, and perspective openly and offer it for
  critique. Consider that your teammates can help you improve your perspective by
  adding to it or seeking clarification.
- *Identify actions.* As you make action plans, identify which actions will be most
  inclusive of differences, help redress former inequities, and redistribute power
  more equitably. Further, center the importance of relationships with community
  members—especially those who have been historically marginalized. For example, consider how you might include community members, students, and other
  stakeholders in your deliberations about a potential plan.
- *Follow through with the plan.* It's surprisingly easy to have a big disagreement
  about an issue, discuss it at length, and then develop an action plan that never
  gets implemented. Make sure that the team sees the fruits of conflict by moving to
  resolution and action.

Create hope-filled meetings. Hope and action have a symbiotic relationship
(Hicks, Berger, & Generett, 2005). When people are hopeful, they can take constructive actions that lead toward equity goals. Without hope, people act in ways that
protect them from their fears, or they may even be paralyzed by their surrounding circumstances. Here are some tips for how to remain hopeful and dedicated to
the work:

- *See challenges as opportunities for improvement.*
- *Remember that the entire team is in the process of becoming more equity conscious and better skilled at equity leadership.*
- *Take a moment to assess.* How much hope versus pessimism do you personally have? How much hope versus pessimism exists on your team? How does your attitude influence your actions? Discuss these questions with your team, and set goals about the types of attitudes you wish to foster and maintain individually and as a team.
- *Set a schedule to revisit this topic regularly.*

Routine 2: Transform Use of Power

Building strong, cohesive teams requires transforming how power operates on the team. Power is often thought of negatively because it is frequently used without regard for how it will be received or experienced. Further, power is often used destructively. For example, think of a time when you felt threatened or disempowered professionally. How was power used? Was the existing distribution of power maintained? Bolstered? Or recalibrated to be more inclusive and sensitive?

Although some uses of power are easy to notice, others are not as obvious, even as they maintain or bolster the existing distribution of power. For example, we know that the setup of a room can reinforce dominant perspectives and traditional distributions of power—one person stands in the front of the room and holds all the knowledge while those in the audience are mere receivers of information. Another familiar example is when only certain people speak and there are few opportunities to ask questions, raise concerns, or otherwise contribute to the outcomes of the meeting.

Transforming the use of power means moving from a hierarchical structure and punitive process to an inclusive, engaging, community-building process. It means shared power and ownership in who the team is and the work the team does. It requires that power be used equitably, humanely, for the right purposes, in the right ways, and toward the right ends.

To do this, you must be willing to create new ways of interacting with your team members and others in your system. To be clear, we are not suggesting that as the leader you are no longer in charge. You will be in charge of co-creating processes that model what you say you value, while demonstrating examples of equity in practice. Doing so requires you to learn additional skills, adopt different strategies, and incorporate new tools. Here are five things you can do to transform use of power:

Use inclusive processes to establish group norms. Establish group agreements, practice them, and gather feedback on how they're working. To become a community of educators (i.e., each person as teacher, learner, and listener) your group will need to establish a set of group norms, practice them with intention and focus, and regularly evaluate how they are working. One way to identify group norms that will work for your team is to use the following structured process, which involves plenty of sticky notes and a large whiteboard or piece of butcher paper taped on a wall.

- The facilitator asks the group members to recall the best experience they have ever had as part of a team that conducted high-quality work. The facilitator then asks, "What were the qualities of the group that made it so?" During a period of reflection, each person lists each quality on a separate sticky note.
- The facilitator then asks participants to recall a conversation about work performance in which they had their eyes opened and learned the most about how to improve their work. The facilitator asks, "What were the qualities of the conversation that contributed to your learning and job performance?" Again, each person jots responses on sticky notes, one quality per note.
- The facilitator asks a third question: "If we are fully committed to transformative systems change for equity, what are the qualities we must practice as a leadership team?" Again, each person writes responses on sticky notes, one quality per note.
- When everyone has finished, participants share responses to the three questions in an effort to group them by category or theme. The first person takes his notes to the whiteboard or poster paper, reads the responses to the question, and posts each note in a horizontal row on the paper. Each subsequent person follows suit; however, after the first person, the contributors aim to group their responses with the sticky notes that are already on the paper whenever possible. In other words, the notes become grouped by theme or category. As the process continues, clear themes emerge.
- The team divides into smaller groups. The sticky notes, grouped into themes, are distributed among them, and participants write statements representing each theme. These statements articulate the norms that are emerging.
- Once the process is complete, the larger group reconvenes, reviews the norms, and determines what will need to be added to ensure the group stays focused on equity and maintains its quality as a learning community.
- The norms are collated into one document for the entire team's review and commentary. This step can take place between meetings if needed. Remember

that this can be a working document, so perfection is not necessary. Your goal is to establish a set of group intentions. The group norms should be posted for each subsequent meeting but revisited regularly to ensure they are having a real impact on the team. One way to do so is to regularly evaluate the team's performance based on the norms. For example, set aside 10 minutes at the end of each meeting to ask first for specific examples of when the group demonstrated the norms and then for how the team could improve its use of the norms. Such practices may feel awkward at first. Remember that in any group with long-established personal and collective habits, change requires focused attention and tolerance of some mild discomfort. Regular evaluation will help the team become more comfortable talking about how they work, learn, and communicate together, and will improve their implementation and practice of the norms.

Be thoughtful about meeting preparation. How you prepare for meetings often indicates where power resides. Remember that if you want to be an equity leader, then you must model equity through your practices. Specific to meeting preparation, as a leader you can do the following:

- *Create a consistent template for your meeting agendas.* A template helps make the structure of your meetings predictable, so team members can focus their attention on the content, people, and process of the meeting. Your agenda template could include sections such as Celebrations/Announcements, Relationship Building, Information Sharing, Group Learning About Practice, Discussion/Learning About Problems and Identifying Solutions/Paths of Action, Meeting Evaluation, To-Do List for Next Meeting. Share your template with your team to get their input and suggested revisions.
- *Co-create meetings.* Inviting others to help plan your meetings honors your team's intellect, investment, and humanity. Share goals for the meeting and give others time to contribute not only in the planning, but also in the facilitation and the evaluation processes.
- *Allow sufficient time to cover the agenda.* Your meetings should be a time to collectively consider the tasks at hand, to think and do at both the macro and micro levels, and to support each other in the process. Plan accordingly for success. For example, discussions about organizational mission and statements of beliefs require time for brainstorming, generating ideas, gathering perspectives, prioritizing, and vetting. Making broad, sweeping changes, such as redrawing

boundaries or redirecting the budget, require multiple meetings and time for people to react and process cognitively, relationally, and emotionally during and between meetings. Conversely, receiving an evaluation on an inclusive, successful program takes much less time but should include enough for the report itself, as well as opportunities to show investment in and support for the program, to acknowledge the personnel responsible, and to celebrate the success. In general, your meeting should include what you can only do together and exclude what is most efficiently done individually.

- *Arrange the meeting space for full engagement and equal participation.* The physical space and setup of a meeting room affects our ability to learn as well as our ability to engage with each other. As much as possible, configure meeting space so that it facilitates your work and your relationships. Is the space warm and inviting or cold and distancing? Does the seating foster connection, inclusion, and engagement, or facilitate hierarchy, separation, and independence? Does the room itself feel like it comfortably holds the group and supports its work, or is it too cramped or too large? Plan for how you will make people comfortable and ensure they have what they need to participate fully.

- *Evaluate your meeting plan.* Review your agenda and entire meeting plan using your group norms as a lens. How well does the plan align with the group norms? How will your plan help your group move closer to achieving your aspirations? How well have people completed items on the to-do list?

Make the most of meeting facilitation and participation. Meetings are a perfect opportunity to model and practice what you want to happen in the classroom. Just as you want your staff to create opportunities for growth and development with their students, you should create meeting processes that allow your team members to continuously grow and develop. To that end, we encourage you to think of yourself, in the role of equity leader, as a facilitator and a participant. Specifically, work to include these practices in your meetings:

- *Think of meeting time as an opportunity to practice your aspirations.* Model the behaviors and ways of being that create and support equity in schools. For example, you can take risks, embody the feelings of self and others, share multiple stories (alongside data), and expand participants' toolkit with creative methodologies.

- *Assign and rotate roles and responsibilities.* Every meeting needs a facilitator, a note taker, a time keeper, and an observer. Rotate these roles among team members.
- *Use discussion strategies that foster participation and openness.* Make sure everyone has an opportunity to participate by using strategies such as these:
  - Begin each meeting by having each person respond briefly to a simple check-in question. Large groups—those over 30 or 40 members—can break into smaller groups for this activity. In all cases, align the check-in to a specific purpose, such as helping team members make new or stronger connections, priming their minds for an item on the agenda, or practicing a new skill.
  - Throughout the meeting, consider when you can include a question to which you want each person to respond. This can be done by going around the room, with each person participating according to seating arrangement, or it can be done "popcorn" style, with each person speaking in random order, until everyone has responded.
  - When the group is particularly large or people seem reluctant to speak to the large group, create small groups that then report out to the larger group.
  - Use facilitation strategies that invite critique and divergent perspectives. Ask questions like "What do others think?" or ""Does anyone see it a different way?" or "What other opinions or information exist?" or "What could go wrong?"
- *Allow time to think things through.* It's common to think of success as getting something done or checking it off the list. Do you know people who put things on their to-do list just so they can check them off? We encourage the opposite! Practice slowing down. Commit to setting aside the time necessary to think and work deeply. We have found that when we pay attention to relationships and process, we do our best work, and *it saves us time in the long run.*
- *Reframe challenges and problems.* As the leader, you will have to resist the urge to have the answers and respond quickly. Avoid moving too fast to "solve" the problems you face, as this can lead to ineffective, technically oriented solutions. Be diligent in your commitment to fully understanding the situation. Remember that systems are inherently complex. Real dilemmas may require that you and your team obtain additional support to understand and address them.
- *Recognize resistance as an asset.* Remember that people who criticize are showing their investment and underlying values. Don't assume you are on opposite sides.

Listen carefully to learn, thoughtfully consider how the information they share can help you improve the work you're doing, and think creatively and compassionately about ways to address their concerns without losing sight of your social justice commitment. Remember, if you don't allow people to be part of the solution, they could easily become part of the problem.

Evaluate and reflect on the success, goals, and growth opportunities of your meetings. An old saying from the community-organizing world says, "Any meeting worth having is a meeting worth evaluating." It's important for teams to evaluate the way they work together. Doing so can be a powerful practice to transform the use of power. Here are two evaluation strategies to consider:

- The "one-word" strategy asks each participant to share one word to describe how she is feeling. One person records each word as it is said and then asks for more input on a few of them. It is important to include some of the negative or neutral words in your follow-up, as a means to invite, honor, and show the normalcy and value of critical feedback. To close, the facilitator offers a summary of what was heard and shares a key takeaway for the group's consideration.
- The "exit ticket" strategy uses questions such as "What worked," "What didn't work," and sentence stems such as "Please _____," "I wish you would _____," and "We should _____." Exit tickets have multiple benefits. First, they invite the type of critique that lets team members know that those with more positional power value others' opinion and input. They also provide the leaders, and the entire team, with valuable information for improving practice.

For both strategies, be sure to positively acknowledge any critical feedback to the whole group, and discuss how you or the team might improve in that area.

Employ nonmeeting strategies. Like strategies for meetings, nonmeeting strategies depend heavily on strong relationships. By "nonmeeting," we mean either (1) those times when a team is not formally gathered but important business still takes place through the interactions between team members, or (2) the way decisions are made outside of the team meetings. It's important to practice the team agreements at all times.

Nonmeeting strategies include the following:

- *Take risks and ask questions.* When a teammate does something you don't understand, inquire further. (For example, "Hey, I noticed that you _____. I was curious about why you chose that approach.") Have the conversation!

- *When a conflict (or the potential for a conflict) is emerging, make a phone call or set up a meeting.* Avoid the temptation to have it out via e-mail!
- *Assume your colleague is doing the best s/he can.* In your interactions, come from a place of support rather than criticism. Doing so does not mean that you should agree with or promote everything your colleague does; nor should you turn away when you notice that the person's actions need improvement. Instead, offer your critique with a tone and approach that says, "We are on the same side; we both want our work to be the best it can be" and "We can learn from each other."
- *Make sure to address team and meeting business* during *the meeting, not* outside *the meeting.* If you have a concern that is relevant to the team, put it on the agenda. If you have a concern that is specific to an individual, set a time to talk privately.

Routine 3: Integrate Personal Experience with Systems and Trend Data

Data is a word that is used often in education and is important in our educational systems. Data are used to determine our success and can be used to illustrate that we have lots of room for improvement. However, when we use the term in this book, we are not just referring to data related to evaluation and assessment. Instead, we are referring to data in terms of what information you take into consideration when you make a decision, reach a conclusion, or decide on a course of action

We use the term *data* to refer to all the information around us and in our heads. When we think of data as information, we realize that our lives are inundated with data. Unfortunately, with this overload, we may be accustomed (even desensitized) to some of the real problems in our context because of how they are portrayed in the media. To make matters worse, it is increasingly hard to know the truth in a context flooded with (unrealistic) reality television and news that is increasingly partisan, polarized, and polarizing. When we combine the inherent limits of the mind in this challenging context, we realize that people operate with incomplete sets of facts, stories, and understandings, all of which lead to inaccurate conclusions.

Further, we need to distinguish between *personal experience data* and *systems and trend data.* Despite attempts to separate the personal from the professional, each person's whole self walks into work every day. Past experiences have shaped the lens through which people look at their work. Some of these come from education and training, and others from encounters that have influenced thoughts, feelings, judgments, and actions.

In contrast to personal experiences, *systems and trend data* refers to the information that we typically think of as data—often in the form of spreadsheets and statistics. These data represent the whole of a population or broad trends in a region or society. They refer to things like graduation rates, employment rates, average daytime temperature over time, or the president's approval ratings. The work you will do in Part IV to conduct an equity audit will guide you in gathering and examining systems and trend data specific to your school or district.

Routine 3 guides you in integrating your personal experience with systems and trend data. Doing so is important because sometimes personal experience contradicts trend data. For example, although trend data indicate that African Americans are more likely to live in poverty than Americans of European descent, plenty of White people live in poverty and there are Black people who earn high incomes and have wealth. If a White man who grew up in poverty is sitting next to a Black woman who grew up in a wealthy family, either person might be tempted to say that race or gender privilege do not exist because their personal experiences suggest that race and socioeconomic status do not correlate. Similarly, if you know of someone whose experience doesn't match systems and trend data, you may see that person as evidence that the data are not accurate.

When your personal experience contradicts systems and trend data, it is easy to deny or negate the data in favor of believing in the viability of your story, experience, and observations. However, just as factual trends in the systems data don't negate your experience, neither should you use your personal experience to negate the truth of systems data. Although both can be true, equity leaders must understand how systems data connect to individuals. In other words, we want you to integrate your personal experience with systems data in more nuanced ways.

Throughout your work as an equity leader, you must be vigilant in seeing your personal experience as separate from, but connected to, systems and trend data, and learn to honor the truth of both. Overall, remember that systems are incredibly powerful. You must be vigilant about discerning the distinction and the connection between your personal experiences and what is going on in the system. To make effective and sustainable equity change, focus your lens on changing the system and then consider how each individual actor and teams of individuals can effectively contribute.

Here are four things you can do to integrate personal experience with systems/trend data:

Gather and review data from across the system and across many measures. We offer a full process for this in Practice IV.

Pause when you encounter research, trend, or systems data that contradict or challenge your experiences. Be careful not to reject the data or your story. Instead, spend time considering how both can be true.

Remember that stories are powerful communication tools. As such, consider carefully how you use stories to communicate key messages. Here are two points to consider:

- When speaking in any public setting in your role as a leader, consider how the story you tell can convey the power and impact of the system rather than negate it.
- Practice telling your personal story in ways that honor the truth and power of the broader systems and trend data.

Watch for instances when personal experiences (yours or others') are used to negate or diminish the credibility of systems and trend data. Develop strategies to sincerely honor the truth of those experiences while calling attention to the truth of the systems and trend data.

Routine 4: Assess the Credibility of Data

Building on Routine 3, you already know the importance of data when leading for equity. Data are used to justify decisions that are made for everything that happens in systems, from how class schedules are created to how standardized-test preparation is handled. Data should drive your decision making, pointing you in the right direction. Once you understand that your personal experience and systems and trend data can be true even if they seem contradictory, how do you determine whether the data are credible, valid, and complete?

This is an important question. You may have heard the adage "Statistics can be used to prove anything—even the truth." Your lenses, priorities, and access determine the types of data you get. Consider that for a moment. Are you seeking data that support your beliefs and assumptions, to prove what you already believe? Or are you seeking a robust understanding, such that you continue to assess data for their accuracy and completeness? Are you willing to have your mind changed? This is what we mean when we say, "Assess the credibility of data."

Here are two things you can do (and because they have a dialectical relationship, you should continually loop back through them) to assess credibility of data:

Seek accurate and complete data. Data are available from a variety of sources. You want to gather and understand overall trend data that apply across systems. What do you know about the connection between race and socioeconomic class; LGBTQIA+ status and suicide rates; or disability status and income? How do various identities fare when you consider the social determinants of health? It is important to acknowledge that across the United States, individuals and communities with marginalized and minoritized identities experience adverse outcomes and that these outcomes are the result of systemic inequities, *not* individual failures.

Specific to schools, you can obtain accurate data from many publicly available sources. We recommend the Department of Education's Civil Rights Data Collection (https://ocrdata.ed.gov/) as a credible source for a broad range of information. For example, find out what is already known about the relationship between race and discipline rates, or socioeconomic class and test scores. Some of these trends were covered in the opening chapters of the book. It is important to know the tendencies that are replicated throughout the system nationally.

Next, gather data from across *your* system and across a variety of measures. We encourage you to gather both quantitative and qualitative data, and we provide detailed instructions for doing so in Practice IV. Overall, you want to design processes in which *all* stakeholders can provide feedback. For example, use creative community-engagement methods to garner involvement across stakeholder groups. More specifically, it is important to gather and center perspectives from those groups that are historically and persistently marginalized and underserved across systems.

Consistently assess data to determine accuracy and completeness and to continually improve and refine meaning making. Be transparent in the ways, and with whom, you delve into data. You can use these questions to support your critical examination of data:

- What is the volume of the data? Are the data isolated and limited, or robust and inclusive?
- What is the overall trend of the data? What are the outliers? How do you make sense of the trends in comparison to the outliers? Are there patterns as to who/what is an outlier and, if so, what does that have to do with equity and inclusion?
- What is the source of the data? What are the perspectives, biases, and assumptions that derive from that source?
- To what degree are the data representative of a larger phenomenon?
- How do anecdotal data match or conflict with larger trend data?

- What do your personal paradigms tell you about the credibility of the data? What dominant systems paradigms might encourage you to disbelieve the data? Are these paradigms accurate, or do they need revision? Do the data help you improve these paradigms for greater accuracy and completeness?
- What happens when you disaggregate the data according to various sociocultural identities? What happens when you cross-reference the data with other markers of broader systems, such as access to adequate and preventative health care? Adequate and healthy nutrition? Environmental safety?
- What do you gain if the data are credible? What might you lose if the data are credible?
- Does your meaning making blame people who are underserved or marginalized, or does it seek to identify historical, structural, or institutional factors that contribute to inequality and inequity?
- Who and what informs your meaning making of the data? Who else might provide an alternate or broader perspective?

The Equity-Focused Leadership Team Model is intended to provide you with a way of conceptualizing the development and maintenance of your team in order to perform in a cohesive, informed, and skilled manner toward your equity goals. The routines described in this chapter are structured procedural habits for your team to practice routinely; they serve as an important infrastructure to use when you begin practicing the roles, which we turn to next.

Team Roles

Consider this: Do you behave differently among members of your family of origin compared with those in your current family context? We bet you do, because your role as son/daughter/grandchild/niece/nephew differs from your role as spouse/partner or parent/aunt/uncle/grandparent/caregiver. Our roles shift as our priorities and relationships change. In equity leadership, we want you to be intentional about the role-related behaviors you want to see in yourself and others.

We recommend that equity-focused leadership teams regularly attend to four roles: Educator, Leadership Practitioner, Equity Champion, and Decision Maker. The roles are ways of thinking about how you show up as an equity leader in various spaces and for what purpose.

You might be wondering why Equity Champion is not at the top of this list. To explain, we are intentional about the order we use to present these roles, and we want you to be intentional in how you begin working on them. First, as an Educator, you engage in the teaching and learning process as teacher, listener, and learner. As a Leadership Practitioner, you critically reflect on your professional practice. As an Equity

Champion, you persistently work to increase your and others' awareness of equity. Last, as a Decision Maker, you take an authentically inclusive approach to making decisions.

We recommend that you include these roles on your meeting agendas. The decision to engage in them in a way that advances equity will require commitment and discipline. Some aspects of them may feel uncomfortable and awkward at first, and it will be tempting to put them off or not practice them with fidelity. If you feel that temptation, pause. Remember that the power of the system and the power of old habits are both strong, but if you want to achieve new and better outcomes, you have to be willing to move through the discomfort of changing your action and interactions, and to persist over time.

Educator

Many have long considered a principal's primary responsibility to be instructional leadership, yet we assert that all administrators in a school system need think of themselves as educators first. We use the term *educator* to represent the idea that each person in a learning community is both teacher and learner. Remember, each individual in a classroom—adults and children—brings the capacity and the knowledge to teach others as well as the obligation to learn from them. The same is true on a leadership team. This vision contrasts with what we see as a faulty notion that the classroom has one teacher and everyone else is a student.

For us, this way of *leading together* contrasts with many of our experiences as members of leadership teams. Most educators have engaged in a professional learning experience with their leadership team, and perhaps in those settings they aim to learn from their colleagues. However, in many other team discussions, the sharing of knowledge and opinions is not necessarily an opportunity to learn, as much as it is an exercise in persuasion. People may think, "Do I believe what this person is saying? How does it match my perspective? Where do I disagree? How can I persuade them to see it more as I see it?" In these conversations, listening is not aimed at understanding the problem more fully or even learning from one another, and often the opinion of the person with the most power prevails. The perspectives or opinions of some individuals may be quickly dismissed, whereas others get plenty of consideration. In a matter of seconds, a marginalized perspective is rendered unimportant or irrelevant, while dominant perspectives survive intact, even if they are skewed, inaccurate, or will repeat past and existing marginalizing practices.

Pause and Reflect

- How does this description compare with your leadership team experience?
- Whose perspectives and meaning making tend to dominate the team meetings?
- When and how does the team consider perspectives that are uncomfortable in the way that they challenge the status quo?

Leaders of high-functioning, equity-focused leadership teams create space and safety for each individual's uniqueness to surface, maintain presence, and have influence. Previous chapters described how sociocultural identities contribute to one's experiences and perspectives. Many additional elements also contribute to each person's individuality. When it comes to how you teach, listen, and learn, be sure to consider each individual's broader experiences, the meaning they make of it, their compilation of prior knowledge, their perspective and opinions, their preferred ways of learning, and their interest in and motivation toward the topic.

Equity-focused leaders use the differences among their team members to enhance and strengthen overall knowledge and capacity. In addition to the inclusive dialogue practices we provide elsewhere in the book, we recommend the following ways to create an environment in which each team member listens, learns, and teaches.

Learn Together

Establish learning time—and the expectation of learning together—as a regular part of the group's meeting agenda. To begin, define the purpose of the learning and a process to ensure that the learning materials support the goal of being focused on equity. Then, read books together, watch films (together or as an individual assignment) and then debrief, and invite speakers to address the group.

Practice Dialogue That Supports Learning from Each Other

To further support individual and team learning, employ structured processes that give each person equal time to contribute reactions, ideas, and questions to the discussion. Here's one such process.

After engaging with the learning material (book, film, speaker), each group member spends 5 to 10 minutes recording a reaction, responding to prompts such as these:

- In response to the material, I think ____, and I feel ____.
- I was surprised when ____. I felt affirmed when ____. I felt disrupted when ____.
- One thing I wondered about was ____.

After recording their individual reactions, team members form pairs, triads, or quads and take turns to share and discuss their reactions. This dialogue should be facilitated and structured. For example, one process could follow these steps, adjusting the time suggestions to fit context and purpose:

- The first person shares her reactions, and the others listen without interrupting for three minutes.
- Then, for five minutes, the listeners ask probing and clarifying questions to better understand the speaker's perspective.
- The next person shares his reactions, and the others listen without interrupting for three minutes, followed by five minutes of probing and clarifying questions.
- Throughout this process, listeners do not react to each other's contributions or share their personal opinions or experiences; they only listen and seek clarification.
- After each person in the small group has spoken and responded to the clarifying/probing questions, the group pauses for three minutes. During this pause, the group members record their thoughts and feelings about each other's comments.
- Each group member then shares her thoughts, one person at a time.
- Only after these steps have been completed does the small group engage in cross-talk to discuss the ideas they've shared, as well as their experience of the process.

Although these steps may feel awkward and uncomfortable at first, they will transform how your team members communicate, learn from one another, and work together.

Rotate Responsibilities for Planning and Presenting

Just as you rotate responsibilities for team meetings, you should do so for team learning. Planning involves selecting material; creating an agenda; planning specific, agenda-related activities; deciding on room arrangement; and providing supplies for learning (markers, paper, projector) and comfort (snacks, beverages). Once the learning session begins, team members can rotate responsibilities for presenting material, facilitating discussion, taking notes, reporting learning, and tending to group needs. Rotating responsibilities serves three important purposes: It (1) allows all team members to share their expertise, as well as to learn from one another; (2) helps to balance power as all team members share responsibility for the group in various ways across time; and (3) fosters cohesion by increasing empathy and understanding.

Group into Smaller Learning Communities

Sometimes it is advisable to break up larger teams into smaller learning communities to increase participation and openness. The term *affinity group* refers to a group of people who share a particular quality or identity; this type of grouping provides an opportunity to discuss and debrief common experiences and strategies. Affinity groups can be formed around any of the sociocultural identities you examined in Chapters 4–9. The most common types are those for people of various racial identities and LGBTQIA+/ally groups. Similarly, it can be helpful to group people who are in specific roles or at similar levels in the organizational hierarchy. Facilitators/organizers should think carefully and consult with members of these groups about how best to use this format to build cohesion throughout the whole group.

Leadership Practitioner

To be a Leadership Practitioner means to think of your leadership as a professional practice and to engage in regular, critical reflection to improve it. Notice how we use the word *practice* here. We reject the idea that either leadership or teaching can be performed in a scripted manner that applies to all situations, or that there is one clear formula for how to lead and teach. Although some approaches are reliably more effective than others, ultimately leadership and education are a combination of science, skill, and art. Research—the science—provides vitally important information about how to ensure that practice is more effective. However, the skill is in knowing *how* to apply that research, and the art is in the *when* and with which nuances and modifications. We want you to practice your craft and then reflect on it, with the goal of continual improvement.

The habit of reflecting on a specific case or scenario is common in other professional fields, such as medicine, where doctors at all levels gather regularly to review one or two medical cases and share their learning. One doctor presents the case, providing the relevant details, and then the audience asks questions. The questions are of two types: those that build understanding and those that challenge the presenter's thinking and practice. The overall intention is for practitioners to learn from their practice and improve it.

Although we know that this model has much to offer the field of education, we are also aware that teachers and school administrators are often reluctant to challenge each other's professional thinking and actions. You've likely been on a team in which some or all members assumed their practices were having a positive impact on all students or all staff or all parents. Further, it's easy to assume that any negative impact is the result of something *they* (students, staff, parents) were doing or not doing. Although it is emotionally easier to evaluate the intent of one's work rather than its impact, research on organizational effectiveness is clear that the highest-performing organizations engage in regular and ongoing critical examination of their professional actions and outcomes.

In an equity-focused district, groups of professionals should regularly examine their practice for how it can be improved. Such examination is especially important for equity work. We talked earlier about the power of habits: People tend to keep doing things that don't work or that work only partially, hoping for a different outcome. Without critical examination, you will continue your habits and likely replicate the status quo. Here we use the term *critical* in two ways: First, it represents thoughtful, focused, courageous, and constructive critique; second, it refers to the persistent examination of how power is operating in any given situation, for whose benefit and to whose disadvantage. The focus should be on the *why* and the *how* of what worked well and what can be done better. Team members must agree that the purpose is to help one another improve practice.

Equity leadership requires your critique to centrally consider *equity*. Moreover, it demands that you (1) don't dismiss others' concerns about your impact; and (2) acknowledge and work to address the full range of your impact, especially when it doesn't align with your intent. Remember that impact matters even when your intentions are good. Be especially attentive if someone suggests that you've done something from the perspective of one of your privileged identities (i.e., White, male, straight, nondisabled) that perpetuates one of the isms we've described. In this case, you might be quick to defend yourself by saying that was not your intent. However, rather than defend yourself, we encourage you to pause. Remember that your privilege has been acting to guard you from seeing how the system benefits you at others' expense. Then, listen carefully to learn more and—an important point—to understand your impact. Although you may not have intended to perpetuate an ism or a phobia, the way to align your impact with your intent is to participate in the conversation in a way that shows you want to learn so that you don't repeat your action.

Overall, equity work requires incorporating feedback loops that evaluate impact and demonstrating openness to learning from that evaluation. All teams—teams of teachers, school leaders, district department heads, and district leaders—should undertake this practice. Leaders have the primary responsibility and opportunity to make these types of processes standard with their teams and then cascade them out to broader audiences for increased effectiveness.

Articulate Your Leadership Identity, Beliefs, Values, and Style

Articulating your leadership identity, beliefs, values, and style is both an individual and a collective activity. Ask the following questions, first of yourself and then as a team:

- Who am I (are we) as a leader?
- What key beliefs form the foundation of my (our) leadership practice and approach?
- Which values are most important in shaping my (our) purpose, goals, and actions as a leader?
- How would I (we) describe my (our) leadership style?

Establish a Regular Agenda Item for Critical Examination of Practice, with Guiding Questions

Make time in your agenda—weekly, biweekly, or monthly—for the team to take a deep dive into practice. Here are three options for doing so:

1. One person presents a "case" from her practice to examine with the group. The presenter shares the details of the case, the actions taken, and any related concerns or questions. The group engages in a critical conversation to examine the equity-related implications and to identify potential actions that could improve the impact on equity.
2. The team generates a list of problems of practice and then breaks into small groups, with each group discussing one of the problems. Each group checks the research literature—especially studies with an equity focus—to generate suggested approaches. Often, the research check is done in preparation for the next meeting. Then the groups engage in critical conversation to identify equity implications and more equity-generating actions.
3. The group discusses case studies, such as those found in the *Journal of Cases in Educational Leadership* or even in current events.

Whatever approach you use, establish a set of group guidelines. For example, the group should be clear about the difference between *affective* conflict and *cognitive* conflict (Roberto, 2013) discussed in Chapter 10. By stimulating cognitive conflict, the group can avoid interpersonal/affective conflict. If the group appears to be visibly or ideologically homogenous, then it is necessary to engage specific protocols. You can find online instructions for techniques such as the Six Thinking Hats (www.dau.edu/tools/Documents/Coaching%20Guide/resources/6%20Thinking%20Hats%20-%20

Parallel%20Thinking%20Worksheet.pdf) or the Multiple Perspectives Protocol (www. schoolreforminitiative.org/download/multiple-perspectives-protocol/) that will help your group develop the skills to surface and examine multiple perspectives.

For each case presentation, the group can ask and discuss guiding questions such the following:

- What was done particularly well? Why? Can it be improved further?
- Where were traditional forms of power and privilege at work in this situation? In what ways did the practice replicate the status quo?
- Where did race, socioeconomic class, gender, dis/ability, language, or religion show up? What was the impact? What evidence supports our analysis? What other relevant information might we be missing or dismissing? In what ways could we have been more equity conscious?
- What practices might have improved the outcome?
- What practices could increase the equity impact?
- What, if anything, absolutely should have been different?
- What can we learn from this experience that can inform our future practice?

Beyond these formal reflections, it is valuable for team members to establish relationships with one another that enable this type of critical reflection to occur continually and in the moment. How do you respond when you are in the moment and notice a way that a colleague could improve his practice? What do you do when you need help to improve your own practice in the moment? Here are some questions and conversation starters you can consider that would be useful in these situations:

- Have you considered _____?
- One thing you might try is _____.
- Let's take a moment to pause and think through what we're doing before we proceed.
- What might I be missing?

Overall, remember your core purpose in asking and receiving these questions: You are seeking to improve your practice, and your overall system, to facilitate more equitable outcomes. These questions and conversation starters are not—and should not be used to promote—a personal critique. They are an opportunity to improve work and outcomes. It takes maturity and self-knowledge to remember that engaging in these conversations is not an attack on any one person but provides a powerful opportunity to improve the team's work and, ultimately, its outcomes.

Equity Champion

Equity Champions are individuals who keep equity at the center of their decisions and advance equity through their micro and macro actions and interactions. Although this definition is concise, it is packed with meaning. Stepping into and maintaining the role of Equity Champion requires ongoing, continual, and long-term awareness and commitment. Remember, you don't create equity in a single action or even in a single year. You cannot "do equity" for a time and then move on. Instead, becoming an Equity Champion is a personal, professional, and ongoing commitment.

Equity leadership seeks to create organizations that offer groups and individuals what they need to pursue self-determined goals. As inequity is historical, equity leadership must disrupt systems of inequality and inequity at the other three levels (structural, institutional, and individual/interpersonal) to reverse the compounding impact of inequality over time.

Such change requires wrestling with existing locations, forms, and uses of power. This difficult and complex undertaking begins with a willingness to name, identify, and discuss various forms of power, how they are being used, for whose benefit, and at whose disadvantage. We use the term *critical consciousness* (Radd & Kramer, 2016) to refer to this commitment and ability to continually look for and disrupt power.

Your team's goal is to become a collective of Equity Champions, engaged in critical consciousness and focused equity leadership. This goal is important for two primary reasons. First, as we've already described, equity leadership has no room for the lone heroic leader, who will become the target of those who wish to maintain the status quo. To create real and lasting change, equity leadership teams must stand together and engage others in collective decision making and work to move systems and institutions toward greater equity. Second, as explained earlier, a single person cannot individually know all that is necessary to produce change related to equity; instead, it takes a collective of multiple perspectives.

Engage in Critical Consciousness Focused on Distribution, Use, and Impact of Power

Engagement in critical consciousness focused on the distribution, use, and impact of power should be consistent and persistent. Here are actions to take:

- *Consistently identify your positional and sociocultural power*. You began this process in Chapters 4 through 9, when you examined your experiences and thoughts regarding six of your sociocultural identities. Although power exists in countless other ways, central to equity leadership is the power and privilege that come from sociocultural identities, organizational position, and social and political capital. Inventory your forms of power, then critically examine how you are using them and with what impact. Invite others to join you in this reflection.

- *Beyond your personal and positional power, regularly examine how power operates throughout your system.* Who has power, how is it being used, for whose benefit, and at whose disadvantage? Look for the visible and obvious examples, but also look beneath the surface for consistent patterns of benefit and disadvantage. Avoid the temptation to engage in deficit thinking or to seek heroes and villains. Remember, inequity is historical, structural, and institutional, so identify how history, structures, and institutionalized ways of thinking and doing maintain existing distributions and uses of power and play out in interpersonal interactions.

- *Critically anticipate the equity impact* when making policy, program, and personnel decisions. Ask, "How does this program/decision/action continue or disrupt and redress historical patterns of privilege and disadvantage? Who is most affected by this, and how well have we allowed their perspectives and needs to inform this program/decision/action?"

- *Work with others in a way that shares power and broadens participation in decision making.* Engage with and learn from the people most affected by a problem or change.

Consistently Advance Equity in Ways That Engage Others in Powerful Collective Action

Engaging others in collective action is a powerful tool for advancing equity. Here are actions to take:

- *Model the methods you want others to use.* Reflect on your practice and actions and seek feedback. Treat others with compassion and dignity, even when you disagree.

- *Have the courage and conviction to ask tough questions and point out uncomfortable truths.* Act to interrupt and disrupt inequitable and exclusionary practices, but with compassion and grace.

- *Take personal and professional risks,* knowing you will make mistakes. Be intentional and strategic, but do not let fear of failure constrain you. Critically consider what you have to lose and what might be gained should you take a risk.

- *Persist over time;* don't let hurdles, barriers, setbacks, or resistance keep you from keeping on.

• *Be humble but firm in your commitment.*

Decision Maker

Leaders have profound responsibility in decision making. Inclusive decision-making processes are those that meaningfully and authentically involve the people closest to the problem in defining the problem and discerning the solution. This inclusivity applies to decisions as straightforward as finding the right time and format for professional learning experiences, and as complex and potentially volatile as deciding how to address budget shortfalls or redraw school boundaries.

Often when we propose the idea of collective decision making, leaders' first response is that it sounds like a good idea but is not realistic because it takes too much time. This response prompts us to ask how much time traditional decision-making processes take. Although it seems expedient for a single leader, or a small team of leaders, to make a decision, we all know that once a decision is made, those affected spend valuable time complaining about it. Further, those who were not able to inform the decision are unlikely to implement it with fidelity. These complications consume time, effort, and attention over the long term and often lead to under-performing or failed implementation.

People are most committed to implementing decisions with fidelity when they have had an opportunity to contribute to the decision-making process and hear other opinions and perspectives about the problem to be solved, as well as possible solutions. Even if a decision doesn't go their way, people will support it if they believe their input was heard, had the opportunity to hear their peers' perspectives, and experienced a fair and reasonable decision-making process. Getting feedback early in the process to foster inclusive decision making takes some extra time early on but saves immeasurable time in the end.

Inclusive decision-making models have three important benefits. First, they ensure that the decision is of high quality (Roberto, 2013); in other words, it will be a better decision than one made in isolation. Second, they model the overall goal of inclusion and valuing difference. Last, they lead to higher overall commitment to the decision among all involved, helping to sustain the decision should pockets of resistance arise during implementation.

Before providing you with specific actions to take in this area, we want to offer an important caution about a common pitfall that occurs when leaders attempt to engage in inclusive decision-making processes. Often leaders have a decision in mind before gathering input from those closest to the problem or issue. In that case, their intent is to garner support for their already-decided plan of action. In our experience, stakeholders are highly attuned to when this occurs, and they recognize when their leaders are seeking support rather than input. When leaders use this strategy repeatedly, their followers sense it and experience it as disingenuous. Followers and stakeholders ultimately become cynical of subsequent processes designed by the leaders and skeptical of the leaders overall. In short, leaders must seek input early and authentically, and use it meaningfully and substantially to inform their plans of action.

Even better, and more inclusive, is to work with large groups of people to make decisions collectively, using four steps: (1) identify a problem, (2) gather information to fully understand the problem, (3) generate solutions, and (4) decide on a plan of action. When you engage others in all four steps, your decisions will be of much higher quality *and* will be implemented more quickly and with much more fidelity (Roberto, 2013). Often leaders lament that they don't have time to involve others in all of these steps. However, unless you are truly in an emergency situation, you actually don't have time to deal with the results of doing it any other way.

Decide How to Decide

The decision-making process involves gathering trend, contextual, and local data about the issue at hand and discerning the most effective means forward to disrupt the status quo and create more equity. Resist the temptation to make quick decisions that seem appropriate but are not fully informed. You can follow a process like this:

- *Set a timeline for your decision-making process,* allowing substantial time for gathering input.
- *Concurrently, identify your stakeholders.* Who cares about the issue or problem? Who does the issue or problem affect? Who will be involved in implementing any solutions? Whose support is needed to make it happen? And who has the power (formal or informal) to block a decision?
- *Articulate a plan to exchange information with these stakeholders.* A key step here is the *exchange* of information. You want to gather and share information in ways that support all parties, including the leaders, in learning from one another. At a minimum, you can do this through surveys, focus groups, one-to-one conversations, or large group meetings. Gather and share input that will help you develop a full understanding of the problem or issue, as well as ideas about possible solutions.

- *Communicate early and often.* When people don't have information, they draw conclusions that eventually become unfavorable toward their leaders. Reduce this tendency by sharing as much information as possible, as often as possible.
- *Similarly, ensure that stakeholders know that you are interested in their input.* Respond to e-mails and phone calls, thank people for their input and questions, and sincerely and conscientiously listen with a willingness to let this input change your mind.
- *If appropriate, consider establishing a stakeholder committee with representatives from each stakeholder group.* Stakeholders will have the responsibility to gather information from their peers. Remember, these groups need to have robust representation from across sociocultural identity groups.
- *Develop a plan to make the decision.* Ideally, this development process will occur within a group that can exchange and discuss the information that has been gathered and collectively analyze the possible decisions.

Communicate as Fully, Apparently, Humanely, and Genuinely as Possible

Effective communication requires that you share as much information as possible, making the information *apparent* (visible) rather than *transparent* (assumed), and that you demonstrate your concerns for others and your trustworthiness by being both *humane* and *genuine* in your tone and message. Communicating in this way should occur at all times.

Sticking with Routines and Roles

To ensure that you stick with the routines (as described in Chapter 10) and roles associated with building an effective equity leadership team, take the time to perform a quarterly checkup with your team. You can do this in various ways. For example, you could host a session in which you split your team into small groups and discuss a set of questions, then have them report out to the large group. You could discuss as a whole group or conduct an electronic survey and bring the results back to the team for discussion. Here are some questions that will help you improve the way you work together over time:

- What are our strengths in practicing the routines and roles?

- What routine, role, or practice, if given our full commitment, would make the most difference on our team?
- What must we do differently to achieve success? If our success at achieving equity were guaranteed, what would we do differently?
- Pick one role and one routine to focus on each quarter (or other interval); at the beginning of the time period, ask, "What are the most important things we can do to deepen our practice?"

Remember that the purpose is to focus on and improve how well your team is functioning as an equity-focused leadership team. To achieve your equity goals, you need a strong and cohesive team equipped with the cognitive, relational, and strategic skills to lead substantial learning, growth, and organizational change.

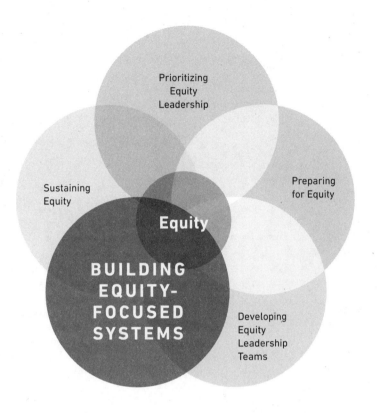

Prioritizing Equity Leadership

Preparing for Equity

Sustaining Equity

Equity

Developing Equity Leadership Teams

BUILDING EQUITY-FOCUSED SYSTEMS

PRACTICE IV

Building Equity-Focused Systems: Identifying Needs and Planning Systemic Change

Conducting a Needs Assessment

Carrie and Layla

Carrie, an African American superintendent, and Layla, a mixed-race assistant superintendent, were committed to equity in their racially diverse, suburban school district. Upon beginning their new roles as equity leaders, they found many interventions and initiatives going on in the name of equity, but overall the district did not have a comprehensive, coordinated approach. In addition, it didn't have adequate information to assess which strategies and changes were making a positive difference, and where more concentrated effort was needed.

Further, Carrie and Layla observed that throughout the district, special services such as reading intervention and special education were negatively affecting students of color and students with disabilities at egregiously disproportionate rates. Similarly, during school visits, they observed few Black and Brown students in advanced courses and fine arts programming.

Unsettled by their observations and experiences, they worked with key stakeholders to determine how best to address these issues. They decided to engage their leadership team in a districtwide equity audit. The team involved many stakeholders in designing and conducting a systematic process for collecting and analyzing the data to identify needs and priorities.

The audit identified a variety of glaring gaps and equity issues. Based on the results, an even broader group worked together to develop a multiyear systemic plan to fundamentally address equity needs and priorities, beginning with changing their service delivery model. Over time, their plan led to authentic change in their district and significant improvement in the lives of students who were previously underserved. The district became a state-recognized exemplar of inclusive and integrated schooling.

We all know that despite attention and effort, many educational systems have consistently failed to improve their equity outcomes. The reality is, most organizational change efforts fail.

We assume that, like Carrie and Layla, you want your change efforts to meaningfully address your district's equity issues and to create sustainable differences. To do so, you need to fully understand the specific equity-related needs and priorities in your setting. Drawing on what we know about organizations, equity, and school improvement, we offer specific strategies for developing effective systemic equity-focused change. We lead you through the process for conducting an equity-focused, in-depth needs assessment and environmental scan—an equity audit—from beginning to end, and then explain how you can use the results to design sustainable systemic change. We highlight the ways and importance of involving stakeholders at every step in the process, and we share more about Carrie and Layla's experience to illustrate these strategies in action.

What Is an Equity Audit?

Equity audits (Frattura & Capper, 2007; Skrla, McKenzie, & Scheurich, 2009) are a key tool used to inform equity-focused change. By systematically collecting and examining data, you can more accurately understand and assess how various policies, practices, decisions, and contexts affect different groups of students. Although an equity audit is a vital needs assessment in overall planning to advance equity, it is not an end goal. Instead, its purpose is to provide a detailed and comprehensive picture of your system as a whole, emphasizing the experience of historically underserved or underrepresented populations. It will help you see where there are gaps and where there is greater equity, so that you can plan accurately and effectively to (1) improve services and conditions for underserved students and (2) increase the overall equity consciousness and equity commitment across your system.

We recommend you think of an equity audit in two main parts. The first part is to assess proportional representation, access, and outcomes (Capper & Young, 2015). This aspect of the audit relies heavily on quantitative data and breaks the system down into granular parts to examine patterns of disproportionality across race, dis/ability, socioeconomics, language, and sex. (See an example in Appendix A.)

Video that further explains the theory behind equity audits

The purpose of disaggregating the data is to analyze natural proportion and proportional representation. *Natural proportion* refers to the percent of a particular group in the general local population where you work.[1] In other words, if Black students make up 30 percent of the school's student population, the natural proportion is 30 percent Black. *Proportional representation* exists when the proportion of students in any given setting matches the natural proportion. For example, the number and percentage of Black students in an AP English class, or with an IEP, should match the local natural proportion. If 30 percent of the student population is Black, you would expect to see that 30 percent of the students in AP English are Black and 30 percent of the students with an IEP are Black. If 10 percent of the AP English class is Black, an equity gap exists because the proportion of Black students in AP English does not match the natural proportion of Black students in the school. Similarly, if 50 percent of the students with an IEP are Black, an equity gap exists because Black students are overrepresented among the special education population.

Beyond examining representation in your local setting, you may also find it useful—or even necessary—to compare the proportion of a particular program to the natural proportion beyond the local school or district. This comparison is valuable because it reveals how your local situation compares to other schools and districts across your region, the state, or the nation.

The second part of the audit is an environmental scan (see Appendix B for an example). In this part of the audit, you explore myriad factors that make up and contribute to your context. You begin by examining your service delivery model—asking teams to create visual representations of how specific services (such as special education or EL) are provided. You review other kinds of contextual data, including a systematic collection of information from stakeholders, that provide key information about the environment for equity. This aspect is particularly important with regard to the experiences of historically and currently underserved/marginalized groups—students and families of

---

[1] Note that the "natural proportion" of students may be skewed by patterns in housing and school district boundaries. For the purposes of your audit, you will initially use the natural proportion in your school or district, and you may later find it useful or necessary to look at patterns of segregation in your broader region.

color, low-income families, students with disabilities, students and families whose home language is other than English, LGBTQIA+ students and families, and students and families who practice faith traditions other than Christianity or have no faith tradition. Last, you identify organizational assets, readiness, and challenges in advancing equity. All these data combine to create an accurate portrait of your full context, thus helping you to construct a doable systemic equity action plan.

Audits are generally accepted as good barometers of an organization using a specific set of indicators. No organization (public or private) would reasonably decide how to spend its financial resources without some understanding of its financial picture and some respect for a set of generally accepted rules and practices related to managing money; this is the reason for financial audits. Similarly, equity audits provide a clearer, more detailed, and highly accurate picture of how schools address inequities. Audits also help leaders understand where best to focus their limited resources and where faculty may need support. Audits demonstrate in a somewhat objective manner where you are as you prepare for change, and as a result can help stakeholders see how you can make the best use of existing resources, assets, and procedures to support equity change. Overall, equity audits offer a foundation for leaders to create strategic priorities and actions plans for making equity-focused changes in their school or district.

Performing an Equity Audit

Across the many available equity-audit models, several steps are common: establishing an audit team, designing the audit, collecting data, analyzing the data, identifying strategic priorities, and developing action plans. Throughout each step, you also need an effective communication plan to inform those stakeholders who are not directly involved about what is going on, and to obtain permission and support as necessary.

In this chapter, we cover communication and the first three steps of the audit process. In Chapter 13 we address analyzing the data, and in Chapter 14 we discuss identifying priorities and developing action plans.

Strategic Communication to Support Your Process

Although sometimes it's true that asking forgiveness is easier than seeking permission, there are many cases in which a strong and effective program is undermined or vetoed by people who learned about it later than they wanted. When it comes to equity change, it's important to be strategic in your communication. You need to assess who needs to know what, when. At every step, consider who should be informed and consulted about your plans and decisions. Whenever possible, obtain the permission, support, and even involvement of key decision makers, as well as those who have strong informal power.

Identifying and Engaging Your Audit Team

One of the most important decisions in pursuing an equity audit relates to who will be involved throughout the process. We recommend a cascading-out process—engaging more people as you proceed. You want to be fully intentional about representation, identifying and engaging stakeholders who can offer multiple perspectives, expanding the group at each step until you have identified a multiyear plan and action priorities. Then, as you proceed with detailed action planning, you will reset and determine new goals and processes for inclusive stakeholder involvement to ensure multiple perspectives.

We cannot overstate the importance of involving a diverse group of stakeholders. In short, it is almost always the right time and situation to involve your stakeholders, including administrators and teacher leaders in the district, as well as the broader set of teachers, other staff, students, family members, and community members. You should include people from across the areas of difference, exceeding proportional representation for those who bring historically marginalized perspectives. Remember, every time you meaningfully and authentically engage stakeholders, you have the opportunity to build relationships, learn more, make better decisions, and build support for the district's action. In addition, stakeholder involvement allows new leaders to emerge, more people to better understand the challenges that the district faces and the resources available to address them, and better engagement in implementing action plans.

As you invite people to be part of your team, remember the process, precautions, and tips we offered in Chapter 10 about how to ensure inclusivity and multiple perspectives. In addition to using mass communications to extend invitations to the entire community, you must phone and meet with people in person to invite and encourage them to participate—particularly those whose voices are not often heard or included.

To make participation viable, provide the necessary supports and accessibility measures. Offer child care, transportation assistance, and translating services whenever necessary. In addition, consider providing food for meetings and events that occur at or near mealtimes. Last, be clear about what you are asking people to do and how much time will be involved both for meetings and for any work outside of meetings.

Once you have gathered your team, use the processes described in Chapters 10 and 11 about routines and roles—even if somewhat truncated—to help your team members become acquainted with one another, and establish agreements for what you will do together and how you will interact.

Designing the Equity Audit

The first charge for your audit design team is to thoughtfully consider how to conduct the audit. Appendix A is a tool you can use to guide data collection, the first part of the equity audit. It is structured around areas of difference, providing a format for gathering and recording disaggregated data in multiple ways. This structure is intentional: Understanding equity gaps requires looking at data through multiple lenses of difference. For some people, purposefully collecting data about race or sexuality can feel uncomfortable, but part of the power of doing an equity audit with a team is to gather concrete data across the range of equity issues, specifically with regard to race, dis/ability, socioeconomic status, home language, and sex. The first part of the audit is structured around those areas.

The second part of the audit includes the components of the environmental scan (see Appendix B). The scan provides questions to use to collect perceptions, experiences, and opinions from students, family members, and district employees through interviews, focus groups, or surveys. Your analysis of these data will help you identify specific areas where the organization can advance equity more quickly through existing processes and procedures, or through individuals and teams who want to take leadership or are eager to experiment with new approaches. It will also help you identify areas of resistance, where you can use relationship building and strategic thinking to build momentum for the necessary changes.

The scan also includes an exercise in which your team creates visual maps of your service delivery models, providing a bird's-eye view of how your school or district provides various services. This exercise is an essential step in rethinking the status quo and moving toward a more equitable service delivery model.

Time is an important element to consider when designing an equity audit. A reasonable timetable will allow for a full semester to collect data, followed by two to four months for analyzing the data and determining strategic directions. The audit sets the stage for action planning for the coming one to five school years.

That said, your team may decide at times to do a partial or focused audit. For example, we have worked with districts that were focused on discipline and did a comprehensive equity audit around that topic, collecting a range of data related to suspensions, discipline referrals, and surveys of students, staff, and parents. Another district conducted an equity audit focused on special education, looking at the range of service delivery models and data on demographics, placement, and achievement.

We recognize that although you will collect a lot of information in the audit, the data are not necessarily all-encompassing. Some teams consider incorporating other existing sources of data, such as previously conducted climate surveys of students, families, and staff.

For some teams, surveying families and students is not a priority during an initial audit. Their reasons might include limited time or resources, or low response rates that will skew the data. However, we offer this caution: It is easy but problematic to find reasons not to collect information and perspectives from stakeholders, especially those who have been marginalized or underserved.

Gathering information from, and building relationships with, these stakeholders is a critical step in learning how best to redress systemic inequities. To do so effectively, some teams hold focus groups with families to ensure that these underrepresented voices are more fully included in the audit. If you choose to gather audit data, be sure to make a targeted effort to reach out to persons in groups including parents/guardians/caretakers of color, families of ELs, families of students with disabilities, and LGBTQIA+ families. In addition, ensure that any data-collection activities are accessible to those you intend to engage. Efforts should include translation into the home languages of your students, and accessibility for those who are deaf, hard of hearing, or sight-impaired or who need other forms of physical assistance to participate. This is just one example of how having a highly diverse planning team will help you take such actions inclusively and with sensitivity.

Once you and your team have a general sense of the process, consider whether you can conduct your audit and subsequent planning as part of an existing practice in the district, such as continuous improvement or strategic-planning cycles. We recommend avoiding this option if it would delay your audit for too long, or weaken or take the focus off equity. But, if you can embed the equity audit into an existing process, it will become a normal part of how the school does business. In addition, you will have several aspects of needed infrastructure in place to support the audit.

Examine the equity audit tools found in the appendices, including the disaggregated quantitative data tables and the environmental scan tools. Articulate a detailed plan for conducting your audit, including who will gather what parts of the data, the timeline, how the information will be securely stored, and under what conditions and with whom it will be shared. Here are some specific questions to guide your team's planning:

- In what ways, if any, can your equity audit process fold into your existing timeline and format for continuous improvement and strategic planning?
- What is the timeline for collecting data?
- What data should you collect?
  - Will you conduct the entire audit? If not, why not? What parts will you focus upon? Why not the other parts? Will you do some parts now and some parts next semester or year?
  - What caused you to choose the data that you did? What reasons or rationale did you use? Why do you think these data are more important and others less so?
- What other data are available and should be considered?
  - What do you gain by including specific data? What, if anything, do you lose by including or not including other data?
  - Beyond the data suggested here, what other data do you need to collect?
  - As you include other data, to what degree do they include or focus on historically marginalized and underrepresented perspectives and experiences?
- Who is responsible to collect which pieces of the data?
  - What support and resources will they need to successfully collect the desired data within the designated timeframe?
  - Do they have the skills they need to collect the data, or will you need to provide additional training?
- How will you ensure that participants provide informed consent, or understand that they do not have to participate? (Note: Many sample "consent to participate" forms are available on the internet.)
- Where and how will the data be stored and shared?
  - Who will and should have access to the data and under what circumstances? Who can add data? Edit it? View it?
  - How will you maintain the security, accuracy, and authenticity of the data?
  - How will you organize the data so that the information is logical and accessible?
- Who needs to be informed and to approve the process?
  - Which formal decision makers need to be informed? When and how will this happen?

—Who else should be informed? When and how will this happen?

—What authentic opportunities for providing input can you plan and provide?

Once you and your team have made these decisions, draft a written overview of your data collection plan to help everyone involved follow the plan that you have established.

Collecting Data for the Equity Audit

Now that you have your data collection plan and process, it's time for implementation. Much of the data in the first section of the audit are nonidentifiable numerical data; be sure to obtain permission from within the district to gather and use these data for a needs assessment and equity audit. If you conduct surveys or focus groups, make sure to obtain informed consent from all participants.

As a team leader, once data collection is underway, you need to keep the process and the team on track. Be available for troubleshooting challenges in access, responses, procedures, bureaucracy, technology, and so on. Keep track of the timeline, and help all those involved in data collection to continue their progress. Ensure that data immediately go into secure storage. Continue to provide communication within and outside of the team so that team members and others have the information they need.

Your data collection process may take many months. Keeping it organized and on track is essential to completing this step and moving on to the next. Even if your plan, or parts of it, becomes delayed, don't be discouraged. Provide the necessary troubleshooting and keep moving forward.

Analyzing Data and Identifying Findings

Carrie and Layla—Analyzing the Audit Data

Carrie and Layla's audit produced a wealth of data. One of their key learnings was that across their reading interventions at the elementary level, more than 95 percent of the students who were identified for intervention in 1st grade remained in intervention across 1st, 2nd, 3rd, and 4th grade—all four years of the district's intervention program. Only three students out of more than 100 were exited from intervention, and only two students' achievement even got close to grade-level proficiency standards. Also, in a district with 50 percent Black and Latino/a/x enrollment, students with those identities made up more than 80 percent of the students in intervention.

Carrie and Layla also found that, as a district, they were excluding 40 percent of their students in special education from general classroom experience by providing services in segregated, self-contained rooms. In comparison, 23 percent of students with disabilities across their state were receiving self-contained special education. More than 60 percent of the students in their district being taught in self-contained special education were Black, even though the overall percentage of Black students was only 35 percent.

Carrie and Layla created service delivery maps of their reading intervention and special education programs that showed structural barriers to inclusion and greater access. They brought in an outside consultant to help gather this information and report it to staff, the board of education, and the community. Carrie and Layla took those opportunities to voice their concerns about the inequitable patterns in the data for students of color and students with disabilities, and they took copious notes to record perspectives from each group to be able to respond over time to staff, the board, and families. They used these forums to do two things: (1) repeatedly highlight the inequities around race and disability present in their district and (2) actively listen to and engage with various stakeholder audiences so they could understand the root causes of these inequities, gather additional information about the factors and impacts of these inequities, and understand the values and priorities of the community in developing goals and action plans.

As the experience of Carrie and Layla illustrates, your data collection activities will produce an abundance of information. Once you have completed your data collection, it is time to organize and analyze that information. We cannot overstate how important it is that you learn from the data and then use your learning to generate new, improved equity-focused actions. Gathering information and then proceeding as though the effort had never taken place is damaging to school-community relationships and a leader's credibility as well as an extraordinary waste of precious resources.

In addition, as you begin the data analysis process, know that it takes a lot of time. You should spend at least as much time and effort, if not more, in the data analysis stage of your audit as you did on data collection. Accordingly, you need to plan processes that allow time for people to sit with the data, explore what the information means, and consider the *why* behind it. This time interval is hard for many people to accept. It's common to feel a strong desire or pressure to jump to solutions in the form of new ideas, initiatives, and programs. You will have to resist any tendency—in yourself or others—to explain away or justify the data, as well as any impulse to develop quick, short-sighted plans. Too often these quick solutions are like trying to put bandages on issues that require much deeper thinking and analysis, and ultimately, take valuable time and resources away from more systemic changes. Instead, your process needs to allow for stakeholders to spend time examining, analyzing, and discussing the data. Remember that data analysis and any subsequent planning processes are iterative by their very nature. Before hosting your first data analysis session, be certain that all the pieces of the session will fit and flow together effectively, and that you have prepared and organized all the necessary materials to support participants in their work.

Preparing Your Data for Analysis

As a leader, you need to ensure that the data are adequately prepared for sharing with your stakeholder groups. Make sure that each person has access to the data you plan to discuss, keeping in mind two vital precautions: (1) You should not share private or individualized student data, and (2) you must consider your rights and needs, including ensuring that the data will be handled properly. When you distribute data for review, think carefully about where it could end up and the consequences of that happening. For example, if you prepare data summaries for participants to read before the session, can you be certain they will not share the information with other parties or even the press? Consider ways to share the data that allow those involved in the analysis to review it carefully but not use it beyond its intended purpose. It is reasonable, and may be necessary, for you to divide the data and assign different portions to different groups for discussion. That said, be sure that you plan a process that allows the groups to gather and discuss the data as a whole (e.g., a jigsaw activity could work well for this process).

In addition, plan to present your agenda using a variety of mediums for maximum engagement and learning. For example, make sure participants can see information as well as hear it. Carefully sequenced presentations (slides, videos, demonstrations) of information allow participants to write down what is shared by trainers or facilitators. Provide participants an opportunity to exchange information through speaking and listening. Lectures and other forms of one-way presentations are useful at times, but discussions and question-and-answer sessions must be integrated in substantive ways. Last, design and use activities that will help participants take in and work with information. Physical movement and tactically working with materials help participants stay engaged, participate richly, and discover new perspectives and insights. Role-plays, engaging learning games, exercises, artistic activities, and simulations are useful formats.

Last, as you prepare your data for sharing, be sure to include space at the end of each section for individuals to record their initial reactions and analysis. For example, you might ask, "What is your reaction to these data? How do you feel about the information?" This query is an important first step in making meaning of the data, as it drives each participant to articulate her perspectives and paradigms for later examination. Or you might ask for even deeper reflection using the Ladder of Inference (described in Chapter 3) as a guide, as in these instructions:

- Notice what is going on in your mind as you review each piece of data: What meaning are you assigning to the data?

- Which data do you think are important and which do you think are unimportant? Why? What is your reasoning?
- What happens if you challenge yourself to believe that all the data are important? Or that the pieces you think are important are not and vice versa? What do you notice about your thinking?"

These questions should help you begin to see how participants' values, beliefs, and assumptions (i.e., their paradigms) influence what happens at each step of the ladder. Be sure your participants understand the importance of reflecting this way. Also be sure to take the time to do this yourself.

Planning for Learning in Your Data Analysis Sessions

As you plan for your data analysis sessions, we encourage you to consider that you are planning for your participants to learn from the data and each other, just as you will plan professional learning sessions later on when you are developing your plan for systemic equity change. Indeed, a vital part of any equity change process is planning for how and what adults will learn together, and how and what teachers will teach in the classroom. For these reasons, we take time here to share information about *leading learning*, because these are processes you will want to consider as you engage others in analyzing the data you collected in your equity audit. Through the data analysis process, you can deepen your understanding of adult-learning principles, practice facilitation strategies and approaches, and support others as they engage in meaning making and become a cohesive equity-minded learning community. This information will continue to serve your equity planning and systemic change strategies in the years to come.

Planning for Dissonance

Here's a recap of the relevant concepts outlined in Chapter 3 that are central to effective systemic equity change:

- Every person, and every group of people, holds paradigms that explain the way things work and should work. Developing more accurate understandings of equality, equity, and justice will conflict with some aspects of your and others' underlying paradigms. These conflicts cause *dissonance*, an uncomfortable emotional and sometimes physiological feeling that says, "something isn't right here."

- As you plan for data analysis and equity learning, it's important that you and your co-learners be prepared to accept dissonance and discomfort, and then move constructively through it (rather than avoid it) to get to the real work ahead. To develop the perspectives, relationships, and knowledge needed to effectively lead equity-focused change, do the following:
    - Whenever you can, proactively share the information about paradigms, dissonance, and their relationship to learning with as many stakeholders as possible. Merely understanding this process—and staying conscious of it—can help increase everyone's ability to engage in the necessary equity learning.
    - Work to cultivate curiosity, humility, and cognitive flexibility in yourself and in others, noting that the best way to cultivate these attributes in others is to consistently demonstrate them yourself.
- An equity leader's learning is never complete. If you never feel dissonance in your equity learning, you may be protecting your personal paradigms.
- Even when you know about paradigms and dissonance, you will still feel or see resistance to the new information. Try not to interpret this reaction, or the people who have it, as adversarial. Instead, remember that everyone must work through their reaction to the material, and your best move as a leader is to support people in doing so.

Planning for Various Learning Preferences

To help you plan how to effectively support your participants as they work with and learn from the equity audit data and move through their dissonance, we encourage you to consider various approaches to learning. Our intent is to provide an accessible framework to use when planning for adults to discover new information (learn) and solve the problem of not knowing. Building off Silberman's work (2006), we describe four Learner Types using the terms *Experimenters, Risk-Takers, Inferers,* and *Investigators* (see Figure 13.1).

Teaching, presenting, and learning about equity requires a variety of activities and approaches geared toward various learner types. Learners need ample opportunities to connect in fun, engaging, and meaningful ways. At the same time, it is important to ensure that the learner types are not used to judge or divide. This is a tricky combination to maintain as a facilitator, but being humble helps. We offer these seemingly opposing

FIGURE 13.1
Learner Types

Learner Type	Attributes
Experimenters	*Experimenters* prefer to solve problems by careful experimentation. For example, when planning for equity change, they would appreciate a well-controlled review of data and research literature to minimize emotions. Although not always possible, they want the opportunity to see what will happen before the stakes are raised and it truly counts.
Risk-Takers	*Risk-Takers* solve problems by engaging in exciting, concrete, and sometimes unpredictable experiences that require them to take action and then examine what happens. When it comes to equity work, they are not afraid to jump into a complex, previously unknown situation. They find it exciting to learn as they go and embrace mistakes as part of their learning.
Inferers	*Inferers* prefer solving problems by inferring a generalized conclusion from smaller, particular instances they encounter and pull together. In the complex context of equity work, they are not comfortable with random exploration and appreciate a structured process to discover the "right" answer. Inferers like having instructions for a complex equity scenario so that they can review them before recommending action priorities and plans.
Investigators	*Investigators* use their own reasoning to arrive at answers. They most desire access to the big picture and benefit from watching others puzzle through equity challenges first. Investigators appreciate a process of thinking through how (or whether) equity can be applied in their own minds before becoming personally engaged.

approaches so that you can consider, and model, all the ways that diversity can manifest. By planning professional learning using a variety of approaches, you are teaching by doing and will learn much more in this process. Your overall purpose is to structure a learning session in which each participant can grow and support the upcoming changes.

As you design your data analysis session or any professional learning session, how will you incorporate structures that allow all learners to feel comfortable with their preferred approach? How will you present information and format activities that allow participants to engage in a variety of ways, including auditory, visual, and kinesthetic? As you lead your session accommodating for all these differences, at times different participants will feel comfortable and others will feel stretched. Remember that the goal is to balance each person's opportunity and responsibility to feel both comfort and stretch. Avoid situations where one type or group consistently feels comfortable while others consistently feel stretched and challenged.

Planning a Constructive Learning Environment

Remember, learning is fundamentally relational. Your data analysis sessions are one of many situations where the effort to build relationships will be worth your investment of time and attention. The more that members of a group feel connected to one another, the more they can support one another through honest and challenging conversations. It's vitally important to remember this social side of learning. Be sure to review the strategies from Practice III on building relationships.

You want to create a space that simultaneously welcomes both risk-taking and the constructive critique of ideas and information. Maslow has noted that human beings have within them two powerful driving forces: one that strives for genuine growth and another (perhaps a stronger force in equity work) that clings understandably to personal safety. We know that a person cannot grow without admitting she does not know, an uncomfortable position for many. However, when given the option to choose growth or safety, many adults will instinctively choose safety.

Specific to equity learning, no one wants to feel stupid, ignorant, or insensitive, especially in front of peers. The threat of saying something offensive, or that may be met with disapproval, can cause people to choose not to participate or to participate in inauthentic, closed ways. For some, conversations about equity could send them running for the door. If you acknowledge this fact early and take time to reflect and prepare, you can be thoughtful about how you might proceed.

Safety is not the same as being comfortable while learning. As a leader, you should be prepared for times when, intentionally or not, someone makes a racist, culturally insensitive, or offensive statement. Be calm but attentive when this occurs. These challenging situations give you an opportunity to model how you can (1) be diligent about noticing insensitive and/or harmful language and actions; (2) maintain the dignity of a colleague who commits an honest error and support learning to use more sensitive, inclusive, accurate, and respectful language; and (3) acknowledge and work to correct any harm that's been done.

One way to preserve people's dignity while they are learning and making potentially hurtful mistakes is to develop a set of agreements for equity-focused conversations. This is not a fail-safe remedy, but it truly helps people find the courage to express themselves, and it reminds the group to be more understanding as learning develops. The facilitator (often the leader) should set the tone for encouraging colleagues to be compassionate and expect to be forgiven enough to keep the learning going forward. It is essential to make the learning space feel secure; but doing so must be balanced against taking healthy risks, because without both, individuals and the organization simply will not grow.

Sample Activities for Data Analysis and Professional Learning Sessions

Reflecting on Learning Preferences

Being mindful of the variety of ways people learn can truly help you manage challenges that might accompany individual and group resistance. Recall that just as change is necessary and natural, so are these challenges. Preparing participants with the information about paradigms and cognitive dissonance, demonstrating your curiosity and humility, and attending to various learning preferences will significantly increase your chances of engaging the most learners in the most learning.

- Provide the chart on learner types (Figure 13.1) to your participants by showing it on the screen or offering a handout. Have them reflect on the descriptions and consider how they take in new information under various circumstances, perhaps giving potential examples. Recognizing that each individual engages in each type at various times, have them draw a pie chart that depicts the percentages for their engagement in each of the four types.
- Have them then reflect on some pros and cons of their most preferred type and share that information with a neighbor or two.
- Provide time for the group to report out a few of the pro and con examples to the entire group. Ask, "What are some pros and cons that may challenge us and propel us as a data analysis and learning team?"

Agreements and Appreciative Interviews

The following activities involving agreements and appreciative interviews can help groups address equally important needs for growth and safety.

Part 1: Warm-up and Engagement

To begin, ask each participant to complete the sentence stem "For equity, I commit to _____." This is an open-ended question and participants can respond in any way that feels authentic to them. If you have a large number of participants, you can have them share their responses in pairs or small groups, and then solicit responses from the whole group.

Part 2: The Agreements

Using the steps that follow, introduce and discuss the proposed agreements (adapted from *Courageous Conversations About Race* [Singleton, 2015]):

- Speak my truth.
- Experience and embrace discomfort.

- Stay engaged.
- Stay curious and hopeful.
- Expect and work toward reasonable progress.
- Hold confidentiality.
- Do not "freeze" colleagues in time.

This activity works best when you provide adequate time (10 minutes per step) for participants to reflect on the agreements carefully and to share feelings.

- Step 1: Post the first two agreements on chart paper or a screen. Ask participants what they believe these agreements mean and why they are important.
- Step 2: Place the third and fourth items on the chart paper or screen. Ask participants why these agreements are necessary before discussing equity issues. Remind them to bring their full selves to this important work and seek first to learn, even if it makes them uncomfortable. Consider the following quote from Mahatma Gandhi to support productive dialogue: "As human beings, our greatness lies not so much in being able to remake the world . . . as in being able to remake ourselves." It reminds people that the issue they are addressing—in this case, equity—is a big issue, but they can start to do something about it if they focus on changing their behavior today.
- Step 3: The last three agreements are helpful in establishing a space that is safe but may not be comfortable for everyone. Include some dialogue and reminders about why it is important to hold confidence in this learning space, and publicly acknowledge that agreeing not to "freeze" people in their worst moments in the learning process can be helpful as well.

Part 3: Appreciative Interviews—Discovering and Building on the Root Causes of Success
This activity achieves two goals. First, it stimulates positive conversations on how to move forward when reviewing audit data, as this can be a sobering and discouraging experience. Second, it reminds participants to reflect on the pockets of hidden success with equity. Know and accept that some responses may not be very strong initially. The point is not to assess each one but to use them as opportunities to connect positively around the work and increase buy-in.

This activity is organized around the following prompt: "Tell a story about a time when you worked on an equity challenge with others and you were proud of what you accomplished. What made success possible? What did you have to overcome? What motivated you?"

Here are other guidelines for the activity:

- Arranging space and equipment:
    - Chair pairs facing one another (no need for tables)
    - Personal paper/device to take notes
    - Flip chart to record stories and assets/conditions

- Participation: All included with equal time
- Groups: Pairs at first, moving to foursomes
- Sequence of steps and time:
    - Review directions and provide the activity prompt. (4 minutes to reflect and write)
    - Instruct participants to pair with another participant, preferably someone they don't know well. In pairs, participants share their response. (6 to 9 minutes each; 12 to 18 minutes total)
    - In foursomes, each person retells their partner's story. Participants should carefully listen to the equity stories and note patterns in conditions/assets supporting success and connections to what they took away from the equity audit data. Remind colleagues to be focused. (15 minutes for groups of 4)
    - Collect and record insights and patterns for the whole group to see on a flip chart. Summarize the information. (15 minutes)

Conducting Collective Data Analysis

You have prepared the data for sharing and planned various ways to present the data. You have also prepared your participants to talk, work, and learn together to understand what that information is saying. As you plan for the data analysis portion of your sessions, remember how important it is to not rush to conclusions about the data and to use collective and dialectical processes to make meaning of the information. (Note: This part of the process can and should be conducted with your internal leadership team, as well as with multiple stakeholder groups.)

Part 1: Share Data

The first step is to share the data with participants and provide them with adequate time to review it. Again, remember that your process should help participants avoid moving too quickly to considering action when they haven't fully contemplated what the data could possibly mean. Your task is to construct a process to thoroughly and thoughtfully review the data, and to ask questions that broaden and deepen analysis and understanding.

Once participants have had time to individually consider the data, use the questions that follow as starting points for rich conversations about equity. Make time and space to go over the information—individually and together—and to wrestle with the questions and each participant's response to it. Remind participants that when they experience dissonance related to the data, they should resist the urge to deny their uncomfortable feelings. Instead, they should describe and discuss the information as it appears, without trying to explain it. Remember to use the practices in Practice III to help participants when their personal lenses conflict with the trends shown in the data.

Here are some questions to help you focus on the data and what the information is telling you:

- What is your emotional reaction to the data? Put plainly, how do you feel about the data? Are you surprised? Disappointed? Sad? Angry? Proud? Confused? Explain why you feel this way. What might your emotions reveal about your underlying paradigms?
- As factually as possible, describe what you see in the data. Refrain from inferences about why things are the way they are, or what should be done about it. What parts of the data do you think are most important? Why? What else might be important?
- What patterns do you see in the data?
- What meanings do you derive from the data? What other meanings are possible?
- What assumptions are contained within the meanings you have chosen?
- Are your assumptions accurate? Plausible? Credible? How do you know? How can you improve upon their accuracy and completeness?

Part 2: Identify the Equity Gaps

Now that you have worked to investigate initial observations and responses to the data, it is time to identify disproportionality. Again, do not act without thoughtfully identifying the problems or issues you are trying to solve. Here are some questions to guide the process:

- Where are you succeeding at proportional representation?
- Where are your equity gaps? (That is, where do you have disproportionality rather than proportional representation?)
- Create a list of the gaps or major issues the data present.

Part 3: Understand Service Delivery

Using the service delivery mapping process described in Chapter 12 and shown in Appendix B, you have created visual representations of your service delivery for special education, EL, interventions, and so on. Review which students are being removed for part or all of the day. Review how you are using your staff (teachers and paraprofessionals). Consider the ways in which your service delivery is fostering or inhibiting inclusion and belonging for students and staff. Examine to see if racial, socioeconomic, or other demographic patterns related to sociocultural identity exist about the kinds of services students are receiving. How do the data match state and federal rates?

Part 4: Understand the Environment

You have collected a variety of data to better understand your environment. Collate the data, and then disaggregate by roles and sociocultural identities/areas of difference. Review the data in a methodical way to discern key themes. Be sure to compare responses and themes by role and sociocultural identity to develop a fuller picture of the equity issues and strengths in the district. Across all these data sets, discern what these data are telling you.

Part 5: Assess Organizational Readiness

Through the organizational readiness data you've collected (see "Organizational Readiness Data Collection" in Appendix B), your goal is to discern the organizational assets, levers, challenges, and inhibitors related to equity change. Identify how you are currently using the school's/district's assets and creating conditions for success. What opportunities do you see to do more? What barriers and inhibitors are most prominent and powerful? You don't need to plan to address these issues yet. Your task at this point is just to identify these aspects of your organization's readiness.

A Framework for Designing Sustainable, Systemic, Equity-Focused Change

Once you have worked through a collaborative process to analyze your data, the next step is to consider how to turn your insights and conclusions into a set of strategic priorities, an initial action plan, and a framework for ongoing equity-focused improvement. The equity audit data and analysis will most likely point to several areas needing improvement. Don't be discouraged if the list initially looks and feels insurmountable. No district can fix everything at once, so avoid feeling overwhelmed. Reference this quote from James Baldwin to maintain your motivation and inspire others: "Not everything that is faced can be changed, but nothing can be changed until it is faced."

Remember, change is difficult, and change for equity is far more challenging than other changes because it requires high levels of personal investment in areas that cause tension and discomfort. However, we offer some steps here that can make the process easier. The work you did in Practice II laid the foundation for the team development you did in Practice III, which enabled you to create the strong, cohesive team needed to initiate and lead this change. Working together can make this load lighter and increase your chances for success.

The framework in Figure 14.1 can guide your planning for equity-focused, systemic change. As much as possible, continue to use inclusive, collaborative approaches to the work. Including a variety of stakeholders will ensure you reach strong decisions and develop reasonable, doable plans. Further, it will provide support for and skill in implementation. Once you have your planning team and process in place, work through the key components of this framework, using the instructions provided here.

FIGURE 14.1

Key Elements of an Equity-Focused Change Framework

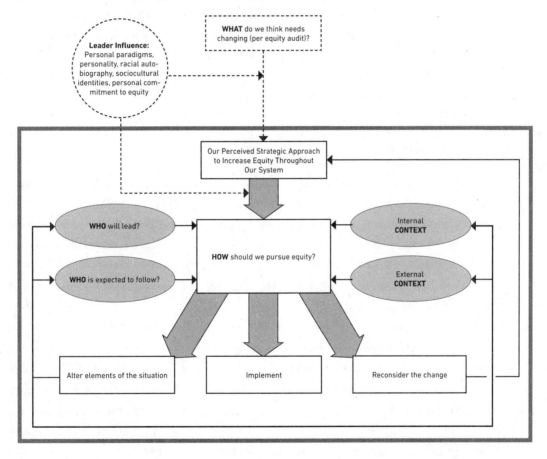

Source: From *Leading Change Management: Leadership Strategies That Really Work,* by D. M. Herold and D. B. Fedor, 2008, London: Kogan Page. Adapted with permission.

What We Think Needs Changing

The rectangle at the top of the change framework in Figure 14.1 highlights the first step: a focus on what needs to be changed. You can use various approaches, but early on be sure to consider what is important to internal context or mission. For example, what might be some high-priority items that would help your school or district work through inequities by race or culture? Which sociocultural identities are present on your staff and in your community? What related needs are pressing? Remember to avoid deficit-oriented thinking; in other words, rather than planning to "fix" students and families, consider how you can strengthen the school staff, make policies and practices more responsive and equitable, provide adequate resources based upon need, and so on.

To facilitate gathering ideas, you can use the inclusive dialogue activities outlined in Chapters 10 and 11. You can refer to the "Disruptive Practices" sections in Chapters 4 through 9 to identify action steps regarding specific sociocultural identities.

Leader Influence

Although "[t]he role of leaders is to provide direction, to motivate followers to exert the effort necessary to achieve organizational objectives, and to support or enable such efforts" (Herold & Fedor, 2008, p. 43), the way in which equity-focused leaders provide direction is to establish inclusive, collaborative processes that seek, respond to, and are informed by multiple and historically marginalized perspectives. Upholding this ideal is indispensable when encouraging your faculty and staff to work toward a new way of being as a learning organization.

As you create and rely on these inclusive processes, you must be conscious of how you design and influence what happens. You have a lens that you use to view the world; we hope at this stage you have developed, expanded, and clarified it via the exercises in previous chapters. Your lens (the circle in Figure 14.1) is influenced by your paradigms, personality, racial reflections, other sociocultural identities, and the degree and form of your personal commitment to equity. The first horizontal dashed line in the diagram indicates how you affect the process (indicated by the vertical dashed line) related to *what* your organization thinks must change, and the second horizontal dashed line shows you influencing strategy in *how* your organization will pursue equity. Remember that each person involved at each step is both influencing and being influenced by the process, via their own lenses. Being clear about your overall personal identity is an essential part of equity-focused leadership. Doing so involves important considerations, such as who you are as a leader, what your experiences are, how you make sense of them, and how you approach your work for equity. Review your work from Practice II to continue to refine and expand your awareness of yourself in the larger system.

Further, developing your followership has a great deal to do with how you connect with stakeholders. Avoid using your power and influence to be coercive; instead, work hard to be collaborative and continue to build relationships to support the development of your staff, teammates, and other stakeholders. As we've said before, equity work requires new and exciting learning, but it will likely challenge people's belief systems, which can feel personally upsetting—an important point to remember as you move forward. Draw on the processes for group dialogue you developed in Practice III to surface and work with individual and systems paradigms.

Remember, you influence the question of *how* you should pursue equity before, during, and after determining the proposed strategic approach. Throughout these stages, you should make your—and others'—thinking visible. In particular, we draw your attention back to paradigms—the lenses through which you view the world, including your values, beliefs, and assumptions. They are highly influenced by your various identities, experiences, and communities. As a leader, your self and systems paradigms will influence the process, thus you must be mindful about putting equity, inclusion, and critical analysis at the center.

Carrie and Layla—What Do We Think Needs Changing?

Carrie and Layla's audit revealed a range of places where inequity was lurking in their wealthy, high-performing district. The revelation did not fit with the historic identity of the district and the people who worked there, prompting much angst and discussion. Many people wanted to explain away the disparities around placement patterns for special education and reading intervention. Many reading intervention teachers, who had influence with their elementary colleagues, saw the current philosophy and implementation of reading intervention as the best and only way. Their individual paradigms and identities reinforced for them that reading intervention was working fine—even as the audit pointed out glaring issues. Some key stakeholders held a paradigm that the district was inclusive around special education, and they found it hard to accept the data pointing out large pockets of exclusion with racially discriminatory patterns.

It would have been easy to decide to tackle other issues. As superintendent, Carrie needed to provide a strong voice stating that these glaring inequities were unacceptable. Layla became the change facilitator, designing a process to hold the team accountable to engage in critical dialogue and to acknowledge what the data were telling them: reading intervention and special education required immediate attention. Both women, drawing on the racial aspects of their identities, pulled in personal narratives to give them strength and to show vulnerability as they passionately led these difficult dialogues. Rather than attack those who were resistant, Carrie and

Layla worked with the team to articulate a personal and professional vision of why the district needed to change. They designed a process to examine individual and collective paradigms that could prevent a more equitable view from gaining traction, and they worked with the team to discern the most effective means to improve their paradigms' accuracy and completeness, and to bring others into the conversation to help design and implement the change.

Discerning and Deciding on a Perceived Strategic Approach

When considering a perceived strategic approach, your team will be faced with the questions of *what* to change *when*. You may be tempted to offer a broad solution to an interlinked set of complex issues. Here we remind you of the systemic nature of inequity and offer a metaphor and a well-known joke: "How do you eat an elephant? One bite at a time." We recommend you pause, slow your thinking, and mull over the challenge of where and how to begin.

To lead systemic and social change, you must commit to creating and leading a *learning organization*. The routines and roles you learned and applied in Practice III are more critical now than ever. Working as a learning-oriented team will enable you to effectively identify change targets and develop effective action plans in response. With that view, you can begin to identify a problem—taking an approach that asks, "What problem should we fix right now?"

Although you might design for a few early wins to show progress and commitment and to build confidence, remember that the more important change processes focused on educational equity involve long-term and strategic planning. Many tools are available to help leaders design efforts that will produce meaningful, sustainable change over time. Here we provide two common processes: *theory of action* and *theory of change*.

Developing a Theory of Action and a Theory of Change

A *theory of action* (ToA) is a simple "if-then" statement that conveys the action to be taken and the expected result. One clear example is related to the standardized-testing movement, which suggests that *if* we regularly test students using standardized measures, *then* we will create more equitable outcomes. You can see that when we articulate a theory of action this clearly, it contains significant assumptions. Perhaps it assumes that with clear data to show disparities and student progress, teachers will change their practice; this in turn assumes that teachers know how to change their practice to create different outcomes and have been choosing not to use their most effective skills.

Another common ToA related to educational equity is that *if* we place schools in competition with one another, *then* they will perform better. Although many people believe it to be true, this ToA assumes several things: that all school professionals are best motivated through competition, that schools aren't already doing the best they know how, and that school professionals need some other form of motivation beyond what already exists. However, extrinsic motivation (such as competition and reward systems) has limited impact and works only in certain limited situations. In addition, we believe that, in general, school professionals are doing the best they can with their available resources and skills. We hope that as you look at the ToAs of common reform efforts, you can see design flaws that help you understand why these initiatives alone have not created equitable schools. The lesson here is to articulate and examine your ToA to improve the design of your change efforts.

However, you will be wise to extend your efforts to include a *theory of change* (ToC), a longer-term statement of the change you want to create. In crafting this theory, you look further ahead and identify the change you would like to create. Then you backward-map all the steps it will take to get there. You do so in the most comprehensive way possible, including implications related to policy; resources (time, budget, personnel, skill); process; infrastructure; organizational culture; and social and political support. You also do so with a critical mind, identifying all the underlying assumptions, something we see in the following vignette.

Carrie and Layla—What Will Be Our Perceived Strategic Approach?

Carrie and Layla's team identified a series of ToAs that drove their change:

- If we create more inclusive service delivery:
  - We will see better results.
- If we create an inclusive service delivery system for special education:
  - Students with disabilities will be more authentic members of their classroom communities.
  - Students with disabilities will learn more of the general education curriculum.
  - Students *without* disabilities will see their peers who have disabilities as full members of the community and become more comfortable and committed to accepting differences as normal and natural.
  - We will avoid the racially segregating patterns that happened with self-contained special education rooms.

— More teachers will become better equipped to meet the learning needs of *all* students.

• If we create more inclusive reading intervention models:

— We will avoid the stigma of pullout programs.

— Students who struggle with literacy will achieve more and will not get stuck in the perpetual loop of intervention.

Carrie and Layla's team identified a ToC for their work as well. One part of it was a conversion to a fully inclusive service delivery model for special education. They rethought the service delivery map that they had created as part of their initial audit (Figure 14.2) and used the same staffing and resources to create a more inclusive plan. Figure 14.3 shows the new, inclusive service delivery

FIGURE 14.2
Equity Audit Map of Existing Special Education Service Delivery

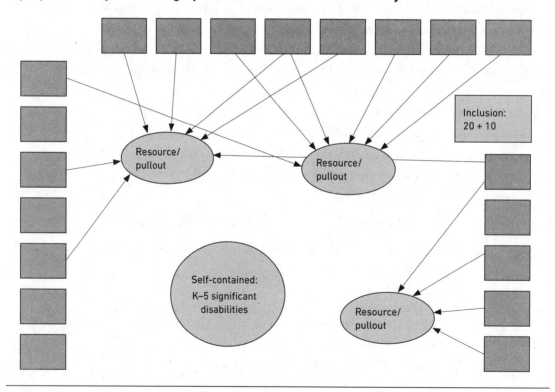

Outer rectangles = K–5 general education classrooms

"Resource/pullout" = Special education teachers pull out students from general education classrooms

"Inclusion: 20 + 10" = Classroom with general education and special education teacher team teaching 20 general education students and 10 special education students

"Self-contained" = Special education classroom where all students with significant disabilities receive instruction and spend most of their school day

Source: Used with permission from George Theoharis.

model for one of the district's schools. Teachers and administrators reconfigured the existing staff to form teams that collaboratively plan and deliver instruction to heterogeneous student groups. In this example, the teams comprise one special education teacher, two or three general education teachers, and a paraprofessional (not pictured).

Carrie and Layla's district team made a detailed and comprehensive plan to implement this new service delivery model, including a staff redeployment component, a two-year timeline, and specifications about changes needed in students' IEPs. Three particularly important components were to (1) engage affected staff to identify and provide for the necessary professional learning for them to feel confident and be capable in the new model; (2) engage affected students and their families to learn and address their concerns; and (3) develop and distribute strategic and effective communications as needed to increase school-community acceptance and support. By engaging stakeholders in the critical planning stages, the district improved the plan and developed deeper support and capacity for implementing it.

FIGURE 14.3
New Service Delivery Model for Special Education

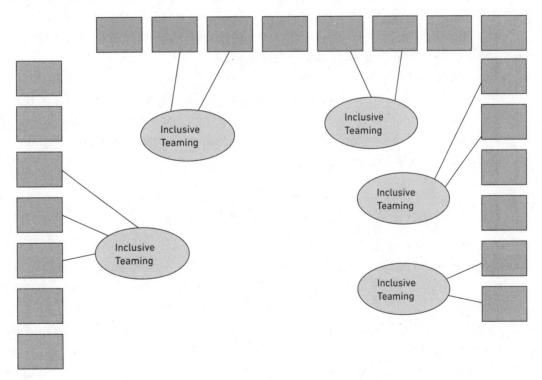

Rectangles = General education classrooms
"Inclusive Teaming" = 1 special education teacher, 2–3 general education teachers, 1 paraprofessional
Source: Used with permission from George Theoharis.

Carrie and Layla's ToC involved rethinking structures and the use of staff to create fully inclusive special education services. For all schools that move in a more inclusive direction, an essential component of this step is placing students into classrooms using the school's natural proportions of students with special education (or EL) as a guide. This means that if 13 percent of the students at the school have disabilities, then the student placement process should mirror that density of students with disabilities in general education classrooms throughout the school—adjusted for grade level natural proportion variations. Classrooms should not be created such that students with disabilities (and thus a need for additional support) are highly concentrated in some spaces and not others. Using natural proportions as a guide, it is important to strive for balanced, heterogeneous classes that mix abilities, achievement, behavior, and other learning needs. The concept of natural proportions needs to be carefully considered across all identity groups and dimensions of diversity. The idea is not to resegregate students based on one part of their identity—like disability or race. In a diverse school, you do not want to create one 3rd grade class that automatically clusters all the Black students or all the students with disabilities. But there are important considerations to keep in mind. For example, if a particular grade has only a couple of students who are learning English and speak Spanish as their home language, it can make sense to have those students in the same class so they have a language peer. Likewise, in a racially diverse school that had five sections of 7th grade, if there are only a handful of Black students, it is important to consider the social, identity, and academic support needs of those students to determine placement, keeping in mind their racial identity so students do not feel racially isolated.

How Should We Pursue Equity?

Returning to Figure 14.1, the question *"How* should we pursue equity?" is at the center. As you move to this step, your capacity is strengthened by the collaborative and inclusive approach you have developed along the way. You have built effective, cohesive, and critical equity leadership teams as described in Practice III. Further, you have carefully included many voices through engaging dialogues and activities. These actions are important because they represent ways to increase schoolwide discussion on the front end to improve your plans and increase support for, commitment to, and capacity in implementing them.

Consider, for example, Carrie and Layla's audit and how the interactions with colleagues and their data revealed reading intervention and special education issues as a top priority. They engaged in all four roles in the Equity-Focused Leadership Model: Educator, Leadership Practitioner, Equity Champion, and Decision Maker. Through these roles, their engagement with other stakeholders was vitally important.

Now, as you seek to discern *how* to pursue equity, we offer important considerations in two categories: (1) *who* to consider for aspects of the work and (2) the *context*. Each is divided into two parts, as shown in the four ovals in the diagram. These considerations highlight the political dynamics at play in your district. Your plans rely on an accurate and astute understanding of these dynamics.

The *Who*

The question of *who* involves both sides of the leadership coin: leaders and followers. In discerning who will lead, consider the willingness, capacity, and existing role of potential candidates. Remember, equity-focused leadership requires that individuals draw on racial and cultural identity skills built in Practices II and III. Those who have developed a strong skillset have the capacity for strong equity consciousness *and* the social capital, relationships, and networks to accomplish the change. They know how to work with colleagues in engaging and supportive ways to help them center equity in discussions and actions. They also know how to generate and harvest collective wisdom and support to design and implement change. They are willing and able to stay engaged and emphasize their equity-specific work.

The next question is who is expected to follow? In other words, who will need to change the way they learn, think, and do their work to effectively implement the change? We hope you have already engaged many of these people in getting to this point. Now you need to think through how you will engage the rest of those who are expected to change, including those who are willing and those who may feel and act resistant. As you proceed, consider four questions that followers most likely ponder before they engage in tasks related to change:

- Is this my work?
- Do I have the skillset to do this work?
- What will happen if I succeed (or fail)?
- How much do I value the consequences?

If people answer no to the first question, then they do not begin to engage in the change effort, or they do so with only minimal commitment and fidelity. Remember that people truly are the core of any successful (or unsuccessful) change effort. Thus it is important to set up multiple opportunities to engage with individuals and teams, taking time to get to know them and holding space for them to get to know each other and to contribute to understanding the problem and designing the change. At this point, it is crucial to design and facilitate effective processes, using the inclusive dialogue approaches described in Chapters 10 and 11. Remember, if activities are not carefully planned and the purpose not made clear early on, followers can easily become confused or frustrated. Further, return to Chapter 13 to review important information regarding adult learners. Last, consider how you will support faculty members when they feel they have failed to reach equity goals. How will they be motivated to stay with the process?

Again, meaningful change requires stakeholder involvement, commitment, and support. Be intentional about engaging as many stakeholders as possible throughout the effort. When you reach the stage of working with others to detail and implement the change, the people whose effort and commitment are needed are vital assets. Work with, support, and value them.

The Context

In this model, we use the term *context* to refer to the social, political, economic, relational, and other dynamics that surround you. *Internal context* refers to your organization, whether that is a school or a district. *External context* refers to everything that surrounds your internal context.

Regarding your internal context, what is happening in your school right now as you consider leading your faculty through equity-focused change? Although one could argue that it is better to conceptualize this type of change when it is integrated within current practices, be aware that the effort is still going to be perceived as change. As we have noted, the prospect of change evokes a range of emotions. Consider these questions relative to the internal context:

- Are people coming off another change effort? Did that effort succeed or fail?
- Can that event easily be considered an equity-focused change? What was learned?
- What other changes are going on?
- What is your overall SWOT (Strengths, Weaknesses, Opportunities, Threats) analysis for your internal context? What is your SWOT analysis specific to your planned equity-focused changes?

- How should you plan to accommodate all these dynamics and increase the commitment, fidelity, and success of your change?

After you have addressed these questions, consider answering the same ones for your external context.

Last, it bears repeating that you will be best informed about both the internal and external contexts if you *engage other stakeholders* in sharing their perspective and knowledge. Make sure you understand, from multiple perspectives, how others see the context and what their ideas are for working within it to support the change effort.

Putting It All Together

In this change model, your careful consideration of *what* should change (guided by an equity audit), your *strategy, who* should lead and *who* is expected to follow, and the internal and external *contexts* are all influenced by you as a *leader*. Additionally, these elements and your paradigms converge to consider the larger, central question as shown in Figure 14.1: *How* will we pursue equity?

The diagram provides three possible pathways (represented by the rectangles at the bottom). In the first pathway, you *implement* the steps that will lead to change and assess how things are going in this process. In the second option, if the agreed-upon path to change requires you to *alter* the elements of a plan or strategy, the model provides four points to consider how you might do so by looking at who leads and who follows, and the internal and external contexts. In some cases, the proposed path may be fraught with difficulty, and you may need to follow the third option: *reconsider* the change.

Reconsideration is a dicey option, not to be taken lightly. Consider carefully and deeply the reasons, rationale, and scope for reconsideration. Should you need to do so, loop back to questions about "perceived strategic approach," and as much as possible, re-engage with the stakeholders to problem solve your next steps, but be sure *not* to make a decision to dismiss equity-focused change altogether. In the end, the leader is ultimately responsible for determining whether the school is on an equity-focused and equity-advancing path; however, we remind you to engage in this process carefully. It is most important to stay connected with broad constituencies and make time and space for those who should most benefit from the plan—that is, those who have been historically marginalized—to inform and influence you.

Carrie and Layla—*How* Should We Pursue Equity?

We have seen how Carrie and Layla worked with stakeholders to articulate a theory of action and a theory of change that led to structural changes, among other things. Making those structural changes required a team approach. Carrie and Layla worked with their principals and special education administrator, who led the creation of the service delivery maps at each school. Carrie and Layla assumed leadership for the overall process and engaged various stakeholders as often as possible. For the individual school changes, they empowered principals and special education administrators to lead the redesign of service delivery with local school teams.

Carrie and Layla had hired an outside consultant to help with their equity audit, and this person became an important leader for parts of this process. As part of their ToC, they brought the consultant in to work with their school board on matters related to the audit results and subsequent changes.

Carrie and Layla had created a district team to help guide this process, and the team worked with a consultant to outline and deliver professional learning for teachers and paraprofessionals at each school across the change timeline. Their district team led various parts of their systemic change. One key aspect of their professional learning plan concerned collaboration and co-teaching. This aspect required district and school leadership to play key leadership roles in ensuring that staff received professional development, had the time and skill to collaborate, and implemented the changes in their daily work. Carrie and Layla realized that they could not mandate the way teams approached their collaborative teaching, but through school-based, shared implementation of designing and staffing the new service delivery, job-embedded professional learning, and supervision of key collaborative practices by the principals, they engaged a wide variety of staff in both leading and following.

1-2-4-ALL: A Process for Engaging Stakeholders in the Change Effort

A team activity called 1-2-4-ALL provides a dialogue format to engage everyone in simultaneously generating ideas about what to work on as part of the school's equity focus (Lipmanowicz & McCandless, 2013). Again, the leader will have to skillfully engage participants and stakeholders into areas they might want to avoid.

Group your participants at tables to begin a series of "rounds." With proper planning and skilled facilitation, one round can be completed in 12 minutes, an efficient way to generate and gather stakeholder input. Timing for each round is as follows:

- 1 minute to reflect alone
- 2 minutes to share in pairs
- 4 minutes to share in foursomes

- 5 minutes for groups to share in whole group

Have a recorder capture the top priorities reported out from each of the table groups. These priorities represent the focus of your work. The leader should offer commentary and insight on which ones to pursue, considering context and available team members to lead the work.

The 1-2-4-ALL activity can be a significant first step in building engagement for change. Note that it involves your stakeholders in collectively identifying a prioritized set of challenges to be addressed in search of possible solutions.

We suggest going through seven rounds of 1-2-4-ALL (although not all in one sitting or at one meeting or retreat). As outlined here, each round represents key areas for moving from the equity audit data toward systemic change:

- Round 1: The data analysis identified equity gaps and other inequities as revealed through service delivery, the overall environment, and organizational readiness. Given these findings, what will we address?
- Round 2: What, specifically, are we trying to achieve in addressing those gaps and inequities? For example, what is our goal for each issue we identify? And how will we know if we have made progress?
- Round 3: What are the causes of the gaps and inequities we've selected to address? (Think systemically. Do they connect to other gaps and inequities? If so, how? And what does that reveal about underlying causes?)
- Round 4: Create theories of action (ToA) by first asking, "What steps/structures/ practices do we think will address those gaps and their causes? How is that approach different from what we are already doing?" Then structure the theories of action using "if-then" statements, paying careful attention to the validity of the assumptions contained in each ToA. (See examples from Carrie and Layla.)
- Round 5: Create theories of change by asking, "What will it take to implement those theories of action systemically?" Consider changes in structure, policy, schedule, and staffing; sequence of actions; financial and other resources; professional learning; social, political, and organizational support; and so on.
- Round 6: How shall we pursue equity? For each theory of change, identify the following:
  —Which roles and individuals are expected to lead?
  —Who is expected to follow? What support will they need?

—What aspects of the internal and external context could support or complicate the changes related to the theories of action?

—What steps can be taken and resources used to best engage stakeholders and garner support for each proposed change action?

- Round 7: Alter elements of the situation/Implement/Reconsider:

   —What would be a careful and purposeful timeline and sequence for these changes? What professional development would be needed? What communication?

   —Are there aspects of the *context* and the *who* that need to be altered first?

   —Are there things that need to be reconsidered now to move toward greater equity?

To conclude this process, a designated work group should take each group's work and review them collectively to discern common themes and priorities. Then, draft an overall multiyear plan. Ideally, you would bring this plan back to your stakeholder group(s) for feedback and refinement. Throughout, remember to communicate your process and progress so that all those who will be involved and impacted can offer their input and ultimately, invest their energy and talent to implement the action plans and change.

* * * * *

Throughout this chapter, we have used Carrie and Layla's equity-change work as an example, but it is important to remember that this work is possible in many different contexts across the United States. Carrie and Layla's work represents just one way to consider how context matters, and it offers some important lessons for considering the major parts of change.

Many organizational change efforts fail for various reasons, including oversimplifying the solution or the change-design process. Equity change requires leaders to enlist the help of others while also preparing them for leadership in this process. It requires collective consciousness and awareness raising. It requires developing an effective strategy but being open to reflecting upon that strategy and modifying it along the way. The tools provided in this chapter offer a powerful approach to start this important process. Do not be afraid to refashion them for your context and purpose.

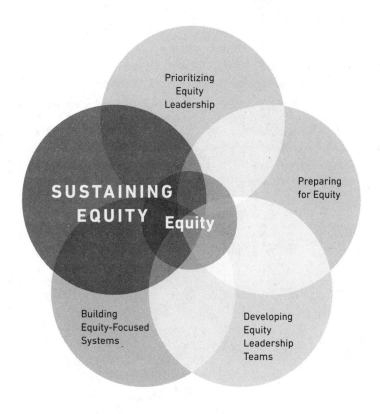

PRACTICE V
Sustaining Equity:
Preparing for the Long Haul

Looking Back and Planning Forward

In this final chapter, we offer a guide for establishing and embedding transformative systems that will continue to increase and sustain the equity in your system even as student demographics and school personnel evolve. We provide three things:

- An opportunity to review and reflect on the work you have done to this point
- A framework for planning your ongoing equity work to ensure that your district or school continues its forward momentum and institutionalizes an aspirational and practical commitment to equity
- A reminder of the courage and commitment it will take to sustain your equity leadership and the tremendous payoff you'll see in time

Pause and Reflect

Recall that in Practice I, we walked you through six sets of statistics related to inequity in schools. Our purpose was to provide information that would be common among your team and that you could use as a starting point for understanding the critical reasons school leaders must engage in equity leadership. We then laid out the five practices for building an equity-focused system. We emphasized two important reminders. First, equity work is long and deep, with no easy answers or simple solutions. Second, to do this work well, you need to open yourself and your team to the emotional and intellectual work that is required. We also asked you to reflect on yourself and the information presented.

Before we proceed, go back to wherever you have logged your reflections and consider these questions:

- How are you thinking about yourself as an equity leader now, after engaging with the book and this work? What is different now compared with when you began? What is the same?
- How has your reaction to data changed compared with when you began? In what specific ways has your meaning making improved? Where or how can it still improve?

As we took the initial steps of prioritizing equity leadership, we discussed the need to commit to a *transformative approach*, again emphasizing that equity leadership goes beyond technical and adaptive approaches. Through Ezra's story in Chapter 2, we reviewed the levels of systemic inequity to illustrate how inequity is historical, structural, institutional, and both individual and interpersonal. We used a set of stories to illustrate the ways people explain away the shortcomings of equity efforts when those efforts don't achieve their hoped-for outcomes; we asked if you saw yourself in any of these stories, and if so, to reflect on what that would mean for your equity leadership.

Review your reflections about how inequity is historical, structural, institutional, and individual/interpersonal by asking these questions:

- What do you know now that you did not know before?
- What do you see now in your school/district/community that you did not see before?
- What observations do you have about how your team has grown since your first group discussions?

Next, in Practice II, we tackled the emotional and intellectual nature of equity work. Specifically, we reminded you that the learning you will need to do to be an effective equity leader is long, deep, important, and never-ending, and we acknowledged that you are likely to react with strong emotions to some, if not all, of the information that you will encounter in learning to be an equity leader. We want you to be aware of this likelihood and develop strategies to deal with it, because you cannot let your emotions deter you.

Take a moment to reflect on all that you have learned about the emotional and intellectual nature of equity work by asking these questions:

- What stands out as the most challenging moment(s) of learning?
- What made it hard? How did you get through these difficult moments? What did you learn?
- How would you describe your overall emotional response to the learning and the work in this book? What have you learned about your emotions as an equity leader?

We also asked you to spend considerable time unpacking six aspects of sociocultural identity by learning about the category in general, the ways that discrimination and exclusion happen specific to that category, and finally, reflecting on your identity and experiences related to that category. We began with race, because we know that race is an identity category that has profound impacts for all Americans and at the same time, is often the hardest topic for White Americans to think and talk about. We hypothesized that by starting your work with race, you would establish a powerful foundation for the rest of this work. We emphasized that understanding your multifaceted identity, its relationship to your experiences of privilege and marginalization, and its impact on your leadership and relationships with others is something you must continually revisit as you seek to serve and grow as an equity-focused leader.

Review your reflections throughout this book related to your identity. How has your awareness grown, expanded, or deepened since you initially completed your reflections?

Then, in Practice III, we shifted from an individual focus to laying the foundations for team-based equity leadership. We affirmed that any time a team begins working together, individual and relationship-based histories are present and active. We asked that you think through how you might engage whole-heartedly and with commitment with your team, respecting and striving to heal the past while also continuing to press forward. We proposed a model based on routines (Chapter 10) and roles (Chapter 11) that you could use to develop a strong, effective, and cohesive equity-focused leadership team. Take a moment to reflect on your team's development by asking the following questions:

- What assets did your team develop using the Equity-Focused Leadership Team Model?
- What are the next, most important steps you need to take to strengthen your team to continually increase your consciousness and effectiveness related to equity?

Finally, in Practice IV, we provided you with a format for conducting an equity audit of your site or district. We suggested strategies to work with the data you collected to unearth the most significant equity-centered implications, and provided a format to design and articulate your equity-focused, organizational improvement plan.

You completed this work just before beginning this chapter. Take a moment to reflect on it, asking these questions:

- What excites you about your plan?
- What makes you nervous?
- What do you need to do or learn next to ensure the success of your plan?

Planning

Now that you have worked your way through most of the book and reflected on your growth up to now, your next task is to plan for ensuring that you persist. Doing so requires a framework to sustain your focus on, and movement toward, equity. The goal is to institutionalize—for yourself and your team—your aspirational and practical commitment to equity and the ongoing learning and effort it requires. In this section, we offer you a framework, accompanied by graphic organizers, for planning in 1-, 2-, 5-, and 10-year increments.

Return to the work you did on *theory of change* in Chapter 14. You articulated your equity plans by developing your theories of change and then back-mapping all the steps you need to get to the changed state. Your plan includes all the technical and practical considerations you need to address, such as policy, resources (time, budget, personnel, skill), process, and infrastructure implications. What do you think the long-term impact of this plan will be? Is this a 1-, 2-, 5-, or 10-year vision?

Although equity change certainly requires practical and technical considerations, you cannot afford to lose sight of the fact that you are working to change *social constructions*, which have long and deep foundations in your and others' underlying paradigms. Change that challenges paradigms is uncomfortable, unpopular, and controversial, and it can make relationships difficult at times, which is why we spent so much time and focus on building your equity leadership team earlier—so that you can consistently focus your equity lenses, support and challenge one another, and work together cohesively. In your leadership team, you can build your personal and collective capacity to surface, examine, and improve your personal and collective paradigms; you can also build your personal and collective resilience to engage in this process and the discomfort that comes along with it. This effort is a critical foundation of equity leadership.

In addition, the change you envision—if it will really move your system toward equity—will involve the redistribution of power and resources. You should not be afraid of this fact, but you must be prepared for the resulting resistance and political struggle. However, before we address that issue, remember that it is crucial that you not be too quick to label a person or group "the resistance." Few people appreciate being labeled, and this particular label makes quick enemies out of potential allies and concerned community members. As we noted earlier, people who are not allowed to be a part of the solution may eventually become part of the problem. In short, do what you can to involve as many people as possible in your change ideation and processes. Do whatever you can to build relationships with people who are committed to equity—and those whose commitment you question. Remember that levels of systemic inequality have contributed to collective paradigms that, in one form or another, believe the status quo is normal and natural. We all have learning to do! It is most often through relationships and experience that those who initially oppose equity initiatives eventually open their minds and hearts to the harsh and unjust realities that previously lay outside their awareness.

That said, some people will be deeply opposed to your equity work and will work hard to ensure its failure. This reality is characteristic of all organizational change, but even more potent and deep for equity-focused change. Remember to not take it as a personal attack. Your ability to move your organization forward in the face of this resistance will depend on three things: (1) the doability and the "should-doability" of the change; (2) your ability to *build strong coalitions* for the work you're doing within your organization and in the surrounding environment; and (3) your ability to *sustain* yourself, your team, and your vision over time and despite obstacles and pushback. If you have engaged in the group-learning and collective-planning processes presented earlier, then you now have a doable and "should-doable" plan for the near future—and a vision for the more distant future. What remains is the critical coalition-building and sustainability measures that will ensure your progress continues into the long term.

Building Coalitions

When we speak of coalitions, we are taking a lesson from the long history of progressive social change throughout the world. The most reliable way to shift social constructions, as is necessary in equity work, is through movement building. By *movement building*, we mean engaging as many others as possible who care about children, their learning, and equity, to support the work you are doing. This effort takes many forms, including building a network of leaders (a broad community of practice) and engaging local officials and organizations.

Broad Networks

A broad and supportive network of leaders is an essential aspect of building alliances for your work. Such a network comprises people with whom you might not be working through the activities in this book, but for whom educational outcomes are central to their goals. Some might be local leaders, and others might be from across the nation or globe, but all bring practical, emotional, and intellectual support to your equity leadership. These are the people you meet at conferences or meetings, who care about the same things that you do, and whom you come to trust over time. You share your experiences and perspectives, generate ideas together, learn from one another. These connections foster your personal and professional sustainability and help you improve your practice. For an example, you can read in the Conclusion of this book how the four of us as coauthors became this type of network for and with one another. We also belong to a larger community of scholar-practitioner-activists who gather biannually to learn from and with one another through structured sessions and informal conversation.

As you consider the importance of supportive networks, ask yourself these questions:

- Who is in your supportive network of leaders?
- Who would you like to develop a stronger relationship with?
- When and where are the settings that provide an opportunity to build your network and connect more deeply for learning and support?

Local Networks

Equity work benefits from support at the local level. Be sure to reach out to public officials and community organizations. Learn about their priorities, needs, values, mission, and upcoming action plans. Although you will begin with relationship building and getting to know one another and each other's work, your long-term relationship could take many different forms. You may find that you have intersecting or mutual interests that can be bolstered through a joint action plan, such as cohosting a community engagement series that creates strong connections between families, their schools, and other community services. You may find that you are duplicating efforts and by working more collaboratively, you can accomplish more with fewer resources. For example, how could you expand the learning opportunities available for parents and caregivers if you partnered with your community education department or perhaps even a local community college or university? Through these collaborations, you will learn from one another about community needs, resources, and priorities. Ultimately, through sincere and authentic connections, you will become allies who can support one another—whether publicly or privately—when the going gets tough.

Finally, become and stay closely connected to communities and groups that have been historically marginalized. We have seen committed equity leaders lose touch with these groups and take actions that were unpopular and undesirable for these groups. Don't assume you know more than the community about what the community needs; learn from the community. In the preceding example about offering learning opportunities for parents and caregivers, ask families what they most need and want. Perhaps they want classes in English learning or building a résumé or navigating the school system; or perhaps they want classes on monitoring their students' use of technology or preparing their students to engage in a job or college search. When you subsequently take action to address these needs, you'll have community support—if not from across the community, then in sufficient number to back you up.

Sustainability

When we talk about sustainability, we are referring to three components: you, your team, and your vision. Each requires many of the same commitments for sustainability: clear values, learning and reflection, continuous improvement, and care for the human nature of this work.

Clear Values

As we've said, equity work is moral work, grounded in our values. For example, our author team values antiracism and social justice in their most liberating forms. We also value people, relationships, factual information, multiple perspectives, and our imperfection, so we push each other to engage with challenging information and ideas that expand our understanding of antiracism and social justice, and we do so with care, honesty, and humility while avoiding self-righteousness.

It is important to clarify your personal values and the values you aim to advance in your work. You have values that drive you to engage in equity leadership and values about how you interact with others and care for your relationships. Finally, you have values about learning, education, and children. Write these down and put them in a place where you can revisit them regularly, to serve as inspiration.

In addition, you and your team members need to clarify your collective values. What values cause you to commit to equity-focused work? What are your values about how you should interact with one another and your constituencies to keep your focus on equity? What are your values about learning, education, and children? You can engage in a collective discussion exercise with your team—similar to those that we outlined in Practices III and IV—to identify and clarify your team values. Once you've articulated specific value statements, consider developing your understanding even further: for each value statement, discuss what the value looks like, sounds like, and feels like in action. Identify the times when each value is challenged or in conflict with other goals or values; discuss how you navigate and resolve those tensions. Keep your value statements and any other relevant information posted in your meeting space. Revisit them regularly as part of your meeting agenda.

Last, how does your work align with your values? Use your individual and team value statements to guide your decision making. Whether you are faced with action planning or resistance to your plans, use your values to help you make decisions. You can and should also use your value statements as a tool to evaluate your work. In other words, do the outcomes of your work align with your values, or do you need to do more?

Learning and Reflection

By now you know that equity leadership requires ongoing learning and reflection. Your learning plans should push you to consistently learn more about the systemic levels of inequity, injustice, and inequality, and the real ways these affect people with various sociocultural identities. They also should push you to learn how to deliver services better and build more authentic relationships across a variety of differences. To sustain your equity work, you will need to plan for this learning and reflection.

Specific to yourself, what is your learning edge regarding equity? Do you need to learn more about race, racism, white privilege, and white fragility? What about ableism? Gender diversity? Anti-Semitism and Islamophobia? Whatever your learning edge, how will you "go there" and gather more information, ensuring that you stay open to information that feels challenging or dissonant with your previous experiences or ways of seeing things? Make a plan in which you articulate your individual learning goals, along with a set of experiences or resources through which you can achieve those goals. Reflect on your learning and your reactions to it. Push yourself to stay open and curious.

Similarly, your team needs a learning and reflection plan. Earlier in the book we described some important qualities you need to develop in your group to become a high-quality, equity-focused learning community. Moving forward, work with your team annually to identify the most pressing learning goals for the year; then plan a schedule, experiences, and resources that will help you achieve those goals. Be sure to build in regular times for critical reflection.

You will also need to identify the learning that will enable you to do your equity work well. For example, if you plan to implement the type of inclusion models that are described in Chapter 14, how will you equip your staff with the skills and knowledge to do this effectively? If you plan to address disproportionality in discipline and suspensions, what skills, attitudes, and approaches must your staff develop to ensure that they are better equipped to build relationships with students and encourage positive, engaged behavior?

Continuous Improvement

The third aspect of sustainability is to create a culture of continuous improvement, built on an ongoing cycle of planning and evaluation, as discussed in previous chapters. To continuously improve, it's important to combine these processes. You identify a goal, develop a plan to achieve it, implement your plan, and then evaluate your work. Your evaluation should then inform your goal and plan moving forward. We've all done this before, but now you know that you need to ensure that your goals are focused on advancing equity, that your plans are doable and "should-doable," that your evaluations measure what really matters, and that you use them to inform your work moving forward.

At the individual level, identify the most important steps you need to take to better increase equity, make a doable plan for achieving those steps, evaluate when you are done and, finally, use your evaluation to inform your next plan. Remember that your individual action plans and evaluations should be connected to and supportive of your team plan and the overall organizational equity vision and plan.

Similarly, to ensure sustainability, plot your team plan on a timeline and incorporate regular evaluation activities to learn the degree to which you are making progress on your goals through your planned activities. Identify the specific work your team still has to do to support that plan, along with time for continuing evaluations.

The Human Nature of This Work

Last and perhaps most important, to sustain yourself, your team, and your vision, it is imperative that you remember and ground yourself in the human nature of this work. By "human nature," we mean four things:

Human beings are imperfect. You will be imperfect in how you engage in your equity work. You will say things that are unintentionally offensive or insensitive. You will do something that makes things worse rather than better. Others will be imperfect too. Hold yourself and others in high regard and maintain high expectations. In other words, forgive yourself and one another while simultaneously acknowledging harm that has been done and working to fix it.

Self-care is vital to sustainability. You will not completely fix the problems of inequity in the span of your career or your lifetime. To do something meaningful to improve conditions locally, you will be running a marathon. Get enough rest, eat well, exercise, have fun at work and outside of work, spend time with your loved ones, avoid self-destructive and addictive behaviors and substances. This advice is not meant as a sermon or a moral directive, but as a practical necessity. You cannot be good at your work if you don't feel good physically, mentally, and emotionally.

This is relationship-based work. Earlier we talked about the danger of the lone heroic leader. We remind you of that danger, but we also remind you to invest in the people around you and reach out to connect to new individuals, especially those you believe are different from you. Remember that relationships require care and time and are a vital space where your commitment to equity can be enacted. Because improving equity requires learning and learning is fundamentally relational, make the time and effort to care and tend to your relationships. Doing so is a key component of your work as an equity-focused educational leader.

This is people and heart work. At the end of the day, you must ask yourself whether you treated others—*all* others—as full human beings. If or when you fail to do so, acknowledge that your actions do not align with your equity values because you have indicated that *some* people are worthy of your respect and others aren't. You don't have to respect other people's values or actions when they grossly misalign with equity and social justice, but you can respect that those individuals are human beings. Do not act with the intolerance that you are fighting against.

As you consider how to sustain your work over time, you can use the table shown in Figure 15.1 to complete a plan that addresses each of the components of sustainability just described (clear values, learning and reflection, continuous improvement, and the human nature of this work). Consider completing a table for your 10-year vision and plan and then back-planning with tables for five years, two years, and one year.

FIGURE 15.1

Components of Sustainability

	Clear Values	Learning and Reflection	Continuous Improvement	The Human Nature of This Work
Self	Indicate when and how you will (1) revisit and update your personal values statement and (2) reflect on the degree to which your actions align with your values.	Articulate your learning goals and identify the experiences and resources you will engage with to achieve those goals. Set up a regular schedule for reflection.	Indicate the ways that you will work for greater equity in your daily work. Set up a schedule and plan for implementing that work, as well as for evaluating your efforts.	List the aspects of the human nature of the work that you personally most want to attend to. Identify how you will stay focused on those aspects and set a schedule to reflect on your progress.
Team	First, work as a team to articulate your values. Then, indicate when and how you will (1) revisit and update your values statement and (2) reflect on the degree to which the way you interact with one another aligns with your team values.	Articulate your learning goals and identify the experiences and resources you will engage with to achieve those goals. Set up a regular schedule for reflection.	Indicate the ways that your team will work for greater equity in your daily work. Set up a schedule and plan for implementing that work, as well as for evaluating your efforts.	List the aspects of the human nature of the work that, as a team, you most want to attend to. Identify how you will stay focused on those aspects and set a schedule to reflect on your progress.
Vision and Work	Articulate the values that guide your organization's educational equity commitment. Indicate when and how you will revisit your values statements to update them, and to reflect on the degree to which your action plans and outcomes align with your team and organizational values.	Identify learning needs in your organization related to the key activities and initiatives you will undertake in the coming year and create a plan to address these needs through learning opportunities. Set up a regular schedule for reflection.	List the primary organizational development/change goals and objectives, and articulate how they will increase equity throughout the system. Identify a clear plan and schedule for evaluation and subsequent planning.	Articulate your aspirations for uplifting and prioritizing the human aspects of your work and your organization. Generate ideas for what this will look like in practice. Identify the tensions and challenges in enacting this vision. Create a regular schedule and format for assessing your performance in this area.

Conclusion: A Final Word

To close, we want to first thank you. We are grateful that you chose to read and interact with this book and, more importantly, that in doing so you have demonstrated a strong commitment to educational equity. We are grateful to be in community with you and are encouraged by your engagement and leadership. As we expand our relationships with each other out to you through this book, we invite your feedback, questions, comments, and concerns. Please feel free to contact us!

Next, we want to remind you of the courage and commitment it will take for you to sustain your equity leadership over time. Equity leadership is not for the faint of heart. Over the course of your career as an equity leader, at one time or another you may end up questioning almost everything you ever learned or believed. You may become incredibly discouraged when you realize how powerful systems of inequality really are; how deeply entrenched old habits can be even when they are clearly inequitable; how much neglect and harm is inflicted upon children and their families who are marginalized; how much you did not or do not yet know about race and racism, ableism, homophobia, or religious intolerance; how few are the authentic relationships—or even encounters—you have with people who are different from you.

You must not give up. We believe we will not reach the end of this work in our life-times—but that doesn't excuse us from committing our best efforts, skills, and assets to move it forward. As Dr. Martin Luther King, Jr. said, "The arc of the universe is long, but it bends toward justice." Your job is to do as much as you can, as well as you can, for as long as you can. If we all do that, we will make significant progress.

And with that, we offer you a message of hope and support. We began our collaboration as a group of four individuals—across genders and races—who respected one another's work and knew one another only casually. Through our work together to write this book, we had countless hours of conversations in which we wrestled with and vetted these ideas with the goal that we would provide something meaningful and practical for you, the reader, and ultimately, the children and families you serve. We stepped cautiously, but persistently, into conversations about various areas of difference, but race and racism, sexism, and ableism were the most frequent. Through sharing ideas, pushing one another with differing perspectives, and keeping social justice at the fore, we not only learned a tremendous amount, we also grew to know one another in deep and meaningful ways that will now extend long beyond this publication.

We offer this testament as an example for you. As much as equity leadership is hard and never-ending, it is also rewarding beyond measure. Although it can be nerve-wracking to try to have an honest conversation about differences with someone who is different from you, it is liberating to be humble and learn from another perspective. As much as recognizing that the scope of inequality can be painful, it is empowering to know more truth.

Living and working in alignment with your deepest values for humanity and social justice can light a fire in the belly that will fuel you. Belonging to a diverse community that embraces its diversity and cares for the humanity and well-being of each and every one of its members—as well as those in other communities—is an experience like no other. It is only through connecting and expanding our networks that we can continue to move forward with strength, support, skill, and effectiveness.

Acknowledgments

This book is the result of a coming together of the four of us to think, teach, learn, work, build relationships, and imagine into the future with each other. We are each grateful for the others in this endeavor, and recognize how each of us is also the result of a coming together of our respective families, communities, and experiences who we wish to acknowledge here.

We extend enormous gratitude to Susan Hills, Joy Scott Ressler, Kathleen Florio, Liz Wegner, and the rest of the team at ASCD. We thank them for seeing the value in this project and for making this work better, more coherent and accessible, and a part of the present-day equity leadership discussion. Your ongoing support, encouragement, suggestions, and guidance have pushed us to make this work the best it could be and shown us how to do so. We thank you!

We are also indebted to our larger communities of scholars across the country and around the world, who have both supported this work and nourished us professionally. We are motivated and enriched to be in community with you. We are grateful for all of our colleagues who we have connected with through UCEA, AERA Div A, the LSJ-SIG, and AESA.

We thank our families—immediate, extended, and created—who keep us grounded in the importance of quality time, family traditions, school activities, and making the world a more just and equitable place. Thank you for your care and time to encourage, push, eat, drink, and play with us. This book is a direct result of your support and love.

* * * * *

Sharon expresses her deepest gratitude to her coauthors: Our partnership and this outcome are a pinnacle in my life; words cannot express what an unmatched privilege this has been. Thank you to the countless individuals who have been so generous in their mentorship and tangible support throughout my entire career, especially Bruce Kramer, who not only taught me so much about my work, but even through his passing, how to lead every day in every way. I am forever and joyfully indebted to the St. Paul-based Bush Foundation for believing in my leadership capacity long before I did; your generous funding of my leadership development has altered the trajectory of my life and certainly expanded my capacity to meaningfully improve our education system. I am grateful to my scholar-siblings and cherished friends, Rosa Rivera-McCutchen, Leslie Hazle Bussey, Latish Reed, Tanetha Grosland, Madeline Hafner, Hollie Mackey, Joanne Marshall, and countless others in the LSJ-SIG, AERA Div. A, and UCEA: You show me the way, and I thank you all for sharing your knowledge, wisdom, and beautiful spirits with me. I am ever thankful for my colleagues at the Midwest and Plains Equity Assistance Center at IUPUI, especially Seena Skelton, Tiffany Kyser, and Kathleen King Thorius: You have taught me—in practice and with love—how to speak up and invite others into the difficult but vital work we must do. My dear friend and teaching partner, Marisol Chiclana-Ayala: Your wisdom, knowledge, generosity, and friendship have not only made me a better professional, but more importantly, a better person. My graduate assistant, Amanda Steepleton: You have done so much to make me a better teacher and are always supportive, dependable, skillful, and wise; your work, loyalty, and courage to speak your truth in the last four years have been invaluable and are present in this book. I am grateful for my former colleagues in the K–12 system in Minnesota, as well as my colleagues and students at St. Catherine University; you all have helped me learn the best lessons about the joy and messiness of education, organizations, and social justice. And last, but not at all least, I thank my sister, Joann Gillis: You are the best part of our family; I thank you for being you and for loving me so completely and unconditionally.

Gretchen acknowledges her coauthors on this project, Sharon I. Radd, Mark Anthony Gooden, and George Theoharis. Special recognition to my colleagues in the Department of Educational Foundations and Leadership in the School of Education at

Duquesne University, especially Amy Olson, Connie Moss, and Rick McCown; to Miguel A. Guajardo, Francisco Guajardo, Chris Janson, Matt Militello, and Lynda Tredway and those educators planning and conducting Community Learning Exchanges (CLE); to my professors and colleagues from UNC Chapel Hill who shaped me and my work, George Noblit, Rhonda Jeffries, Paula Groves Price, and Sheryl Cozart; and to my former and current students who are also my teachers and friends, Jacqueline Roebuck Sakho, Tyra Good, Renee Knox, Tia Wanzo, Ramona Crawford, and Carol Schoenecker.

Mark truly appreciates the supportive and committed coauthors of this book for giving of themselves in significant ways. I am so much better for that. I also appreciate my dear colleagues at Teachers College, Columbia University and The University of Texas-Austin, Terrance Green, H. Richard Milner IV, Muhammad Khalifa, Linda Tillman, Michael Dantley, Philip T.K. Daniel, Khaula Murtadha, Kofi Lomotey, Michelle Young, April Peters-Hawkins, Noelle Witherspoon Arnold, Dana Thompson Dorsey, Ann O'Doherty, Sonya Douglass Horsford, Rosa Rivera-McCutchen, Terri Watson, Terah Venzant Chambers, Anjalé Welton, Daniel Spikes, Decoteau Irby, Bradley Davis, James Davis, Judy Alston, María Luisa González, Gary Crow, Jay Scribner, J. John Harris, Colleen Capper, Sheneka Williams, Cindy Reed, Andrea Rorrer, Mónica Byrne Jimenez, Victoria Showunmi, Larry Parker, Michelle Knight-Manuel, Yolanda Sealey-Ruiz, Erica Walker, and all of the UTAPP and SPA-NYC cohorts I was blessed to engage with as we learned about leadership.

George acknowledges his remarkable coauthors of this book for this wonderful professional collaboration. I am really proud to have done this together. I hold a deep sense of appreciation for the larger communities of scholars that have both supported this work and nourished me professionally. I am motivated and enriched to be in the company of Martin Scanlan, John Rogers, Tara Affolter, Steve Hoffman, Deb Hoffman, Jeff Brooks, Joanne Marshall, Frank Hernandez, Rebecca Lowenhaupt, Jessica Rigby, Sonya Douglass Horsford, Josh Bornstein, Rosa Rivera-McCutchen, Leslie Hazle Bussey, Michael Dantley, Madeline Hafner, Terah Venzant Chambers, Terri Watson, Dave DeMatthews, Nancy Parachini, Colleen Capper, and Terrance Green. I deeply appreciate my inspiring colleagues as well as equity team collaborators from Syracuse: Leela George, Courtney Mauldin, Marcelle Haddix, Beth Myers, Tom Bull, Christy Ashby, Nate Franz, Sara Gentile, Corey Williams, Meredith Devennie, and Ben Steuerwalt. I appreciate the wisdom and friendship I have gained from my many, many teaching and leadership students who have (knowingly or not) been a part of the development of this book for years.

Appendix A:
Sample Equity Audit

(*Note:* This audit is designed to collect school-level data but can be adapted to collect data across multiple schools or for an entire district, region, or state.)

Assessment of Proportional Representation, Access, and Outcomes
The prompts in Column 1 are designed to guide your collection of quantitative data related to specific areas of difference. For each question/prompt, review your data from the previous 12 months (unless otherwise indicated) and provide an accurate count in each column. You can add areas of difference for which you have data in the far right column. Add additional columns as needed. Some lines, such as "grade level" or "other labels," may require that you add additional lines. Modify the table as needed to best address your needs, population, and context.

Data collected by:

E-mail:

Site:

Phone:

Race: For the purpose of this audit, students who identify with more than one race should be counted as mixed race

	Sex/Gender			Race					Ethnicity		ESL Services		IEP or 504			SES		Other (name)
	Cisgender Female	Cisgender Male	Other	White/Euro-American	African/African American/Black	Asian American	Indigenous/American Indian	Mixed Race	Latino/a/x Y	Latino/a/x N	Yes	No	IEP-Y	504-Y	Neither IEP or 504	FRL Y	FRL N	
Number of teaching/certified staff in your school																		
Number of noncertified/paraprofessional staff in your school																		
Number of administrators in your school																		
Number of students in your district																		
Number of students in your school																		
Number of students in each grade level in your school																		

Students who would attend your school if they did not qualify for special services (Special Education, ESL, etc.) but who attend another school in order to receive those services. (Including students with significant disabilities who are sent to special education programs/schools.)								
Students who do not live in your attendance area who attend your school								
Students labeled "gifted" or receiving gifted services								
Students referred for special education services this year								
Students receiving Tier 2 and Tier 3 RTI								
Students labeled with a disability								
Students who receive special education services								
Students labeled ESL or bilingual								
Students with any other kind of label (include the label)								

continued

	Sex/Gender			Race — For the purpose of this audit, students who identify with more than one race should be counted as mixed race					Ethnicity		ESL Services		IEP or 504			SES		Other (name)
	Cisgender Female	Cisgender Male	Other	White/Euro-American	African/African American/Black	Asian American	Indigenous/American Indian	Mixed Race	Latino/a/x Y	Latino/a/x N	Yes	No	IEP-Y	504-Y	Neither IEP or 504	FRL Y	FRL N	
Students educated primarily in general education setting/classroom																		
Students educated primarily in pull-out programming																		
Students who attend an alternative school/setting																		
Students who received an in-school suspension last year																		
Students who received an out-of-school suspension last year																		
Students who received another form of discipline (office referrals, behavior intervention room, etc.). Identify what discipline data you are collecting.																		
Students expelled in the past year																		
Students placed in Alternative Interim Placement																		

Students with attendance issues and/or truancy										
Students meeting adequate achievement and/or testing proficient in reading										
Students meeting adequate achievement and/or testing proficient in math										
Students who graduated on time										
Students who dropped out of school										
Students who took ACT, SAT, or Advanced Placement exams										
Students who participate in advanced academic classes/curricular opportunities										
Students who participate in out-of-school-time academic clubs/activities										
Students who participate in school-time performing arts classes/opportunities										
Students who participate in out-of-school-time arts activities										

continued

	Sex/Gender			Race — For the purpose of this audit, students who identify with more than one race should be counted as mixed race					Ethnicity		ESL Services		IEP or 504			SES		Other (name)
	Cisgender Female	Cisgender Male	Other	White/Euro-American	African/African American/Black	Asian American	Indigenous/American Indian	Mixed Race	Latino/a/x Y	Latino/a/x N	Yes	No	IEP-Y	504-Y	Neither IEP or 504	FRL Y	FRL N	
Students who participate in out-of-school-time athletic activities																		
Students who participate in out-of-school-time leadership activities																		

Source: From *Leading for Social Justice: Transforming Schools for All Learners,* by E. M. Frattura and C. A. Capper, 2007, Thousand Oaks, CA: Corwin Press. Copyright 2007 by Corwin Press. Adapted with permission.

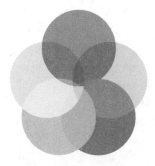

Appendix B: Tools for Environmental Scans

Building Observations, Part 1

The following questions aim to gather identity-related perceptions, experiences, and opinions from students, family members, and school and district staff. You can use various methods to gather the information, including interviews, surveys, discussions at staff meetings, and focus groups. Summarize your findings in Column 2 of the table.

How and where do students labeled "gifted" receive services (in class, out of class)? Are students labeled gifted proportionally represented in all classes/courses/learning experiences?	
How and where are students receiving Tier 2 and Tier 3 interventions served?	
How has the number of students referred for special services changed over time?	

continued

Of students referred for special services each year, what fraction/percentage were then identified to receive special services? Indicate which special services they received by previous identity category.	
How many students in your school openly and freely claim or express an LGBTQIA+ identity? What do you notice about how other students treat them? How staff treat them?	
How many staff in your school openly and freely claim or express an LGBTQIA+ identity? What do you notice about how students treat them? How other staff treat them?	
Does your school/district have policies that support differences in gender and sexual identity? If so, what are they?	
Does your curriculum provide instruction related to sexual orientation? If so, how and to what extent?	
If a group of students asked your school principal for permission to begin a gay/lesbian support group, what would be your principal's or district's likely response?	
Considering your school's library and media holdings, to what extent do students in your school have access to information about sexual orientation? What is the nature of this information?	

How does your school communicate symbolically about sexual and gender diversity? Are there gender-neutral bathrooms? What terms are used on forms intended for parents/guardians (e.g., do they ask for information for "mother" and "father")? Are gender nonconforming individuals and same-sex couples welcomed and comfortably included at school-sponsored activities (e.g., prom or homecoming)? How do you know? Are games organized as "boys versus girls"? Are there other examples of expecting heteronormativity or the gender binary? If so, what are they?	
Does your school/district have any active policies that address religious diversity and inclusion? If so, what are they?	
What, if any, religious-related practices does your school/district have in place?	
How and to what extent—for example, in what grades and in what subjects—does your district's curriculum provide instruction about religion?	
How does your school communicate symbolically about religious diversity? To what extent is the calendar tailored to Christian holidays versus holidays in other religions or no religious holidays at all? To what extent are a variety of religions represented in the symbols and rituals in your school, such as music at concerts, curriculum, and student activities?	

continued

If students approached your building principal and asked for space to host a Bible study group, how do you think your principal or district would respond? What if students requested space for an atheists club? An Islamic student club? A prayer room?	
How many students in your school openly and freely claim or express a religious identity other than Christianity? What do you notice about how other students treat them? How staff treat them?	
How many staff in your school openly and freely claim or express a religious identity other than Christianity? What do you notice about how students treat them? How other staff treat them?	

Source: From Leading for Social Justice: Transforming Schools for All Learners, by E. M. Frattura and C. A. Capper, 2007, Thousand Oaks, CA: Corwin Press. Copyright 2007 by Corwin Press. Adapted with permission.

Building Observations, Part 2

A 30-minute building walk-through can provide information you might not otherwise obtain. Record your observations in the spaces provided.

Site:	Data collected by:	
Date:	Start time:	End time:
Introductory notes (What is going on at the site today? What have you observed and heard before the formal walkthrough?)		
Instructions: As you walk through the school and observe the hallways, classrooms, meeting spaces, office spaces, and open/free spaces (e.g., lunchroom, entry), make a checkmark for each time you notice one of the described situations, and write a note about the evidence. When you have completed your walk-through, summarize what you saw.		

The walls of the school (hallways, classrooms, office, cafeteria, etc.) are decorated with items that reflect the diversity of the region, nation, and world, in terms of race, sex, gender diversity, dis/ability, and so on.	
The walls of the school (hallways, classrooms, office, cafeteria, etc.) are decorated with items that connect to the current students, including displays of student work, student achievements, and student possibilities.	
The students in any given space are representative of the entire student body (spaces are integrated in terms of race, sex, gender, dis/ability, etc.).	
In classrooms, students are positively engaged with one another and the teacher or in challenging work.	
The tone of conversations is respectful, kind, compassionate, and encouraging.	
Adults intervene effectively and respectfully when student behavior and interactions need redirection.	
Parents, guardians, other family members, and visitors are welcomed and greeted with respect (at entrance and office).	
Parents, guardians, other family members, and visitors are welcomed and treated with respect (in meetings, etc.)	
Notes/overall observations/conclusions: Review your notes as you record your reactions. What are your impressions? What conclusions do you have about why things are the way they are?	

Source: Used with permission from Sharon I. Radd.

Organizational Readiness Data Collection

Through surveys, focus groups, or interviews, gather information from employees at all levels of the building/district by asking the following questions:

- How is the organizational climate overall? The community climate? Where is there cohesiveness and where is there division?
- What are the attitudes toward change and learning?
- What is the attitude toward difference? Inclusivity?
- What are the varying definitions of equity? Whose job is it to work for equity? Who is equity for?
- Who is seen to be marginalized? Who is invisibly marginalized?
- What are the organizational structures and processes that support organizational learning and change? Which change strategies have been effective historically?
- What resources are available? What resource constraints exist? What factors influence resource distribution?
- What infrastructure is available? What infrastructure is lacking in the district? (*Note:* Here, *infrastructure* refers to a wide range of supports, processes, and structures.)
- Who are the formal and informal leaders? What are their priorities, commitments, and styles? Who makes things happen? Who can prevent things from happening?
- Who cares enough about equity to go the extra mile? Whose support and engagement are available and needed?
- What are the strengths of the building/district in terms of educational equity? What are the most pressing needs in the building related to equity? What are the barriers that inhibit the ability to achieve equity?
- How capable do you feel to advance equity through your work here? What would help you feel more capable?

Stakeholder Concerns

Instructions: Use the following components for surveys, focus groups, or interviews to gather information from parents, guardians, and other caregivers. Modify the questions and statements as needed to collect information from employees and other stakeholder groups. Be sure to take adequate precautions to ensure the rights of individuals to decline participation and to maintain the confidentiality (and even anonymity) of the data.

Background Information

With which racial or ethnic category(ies) do you primarily identify (e.g., White, Black/African American, African, Latino/a/x, Asian American, Indigenous, Hmong, Somali, Mexican)?

Does your child receive special education services?

Do you or your child identify as LGBTQIA+?

Does your child qualify for free or reduced lunch?

Do you practice or identify with a religion other than Christianity?

What is the primary language spoken in your family?

What grades are your children currently in?

What schools do your children attend?

Likert-Scale Statements

- My child feels included and a sense of belonging at school.
- *All* students are included and valued at my child's school.
- Through experiences at school, my child is learning to work and relate positively, constructively, and inclusively with others.
- I have seen or heard about racist, prejudicial, and excluding language and actions at my child's school.
- School staff notice and respond effectively when racist, prejudicial, and excluding words or actions happen at my child's school.
- I feel welcome and comfortable in my child's school and the district overall.
- My family's race, culture, and ethnicity are accurately and fully represented in the curriculum and the climate at my child's school.
- The school and district work well with a variety of learning styles and differences.
- The school and district value diversity and create a welcoming environment for everyone, inclusive of differences.
- The school and district are taking intentional and meaningful steps to provide a more equitable learning environment and education for all children.

Experiences and Perceptions (Open-Ended Responses)

- What is it like for you in this school (or district) as the parent/guardian/caregiver of a student? How would you describe the way that the school engages with you, and how do you respond?
- What are the messages, if any, that you get about diversity at this school?

- How would you describe the school regarding whether all students are included and belong? What examples can you give about who is included and how? What about who is excluded? Do you ever hear or see racist or prejudicial words or actions while at school? If so, can you describe? What is the staff knowledge of and response to these incidents?
- What is going wrong when a student doesn't succeed in the system? What do school personnel need to do to achieve better outcomes? What barriers exist to achieving better outcomes?
- The Strategic Framework for the school district states _____. Equity is defined as _____. The district goals related to equity include _____. What do these statements mean to you? In what ways are these ideas important? In what ways are they hard to accomplish?

Individual Reflection, Observations, and Conclusions on Data Collection

Review the data you have collected and record your reactions. What common themes do you see across the data? What contradictions or paradoxes do you notice when comparing data across data sets? In particular, take note of how various demographic groups see things similarly and differently. What do you think is going on? What do these data mean? What conclusions do you have about why things are this way? Be sure to include both strengths and areas for improvement in curriculum, instruction, and other learning opportunities as related to students with a variety of identities. Pay special attention to how students with marginalized identities are served. Consider and record potential concrete, specific actions for eliminating inequities related to each area of difference. Reference points like state or national average percentages for students in special education or students in restrictive settings can be useful in providing guideposts.

Service Delivery Mapping

Understanding service delivery is a key and often overlooked part of equity audits aimed at meeting the range of student needs. Service delivery mapping is a way for you and your team to create a clear and accurate picture of how your school provides services and uses human resources—how and where all relevant staff members work—for special education, EL, interventions, and so on. The resulting map is a visual representation of general education and other classrooms and spaces, and how they are used and staffed, including which staff pull students from which classrooms, which students learn in self-contained spaces, and where paraprofessionals provide support.

Figure AppB.1 is an example of a visual map of service delivery for special education. The model concentrated much of the provision of services in only some classrooms, while others lacked both students with disabilities and additional adult support. Some students were removed from the general education curriculum, instruction, and social interaction with general education peers for some or all of each school day. As this example illustrates, service delivery mapping can help your team see how spaces and human resources (including general education and special education teachers) are being used and which students have more access to the entire school, its services and opportunities, and which have less, giving them a bird's-eye view of services in their school.

FIGURE APPB.1

Equity Audit Map of Existing Special Education Service Delivery

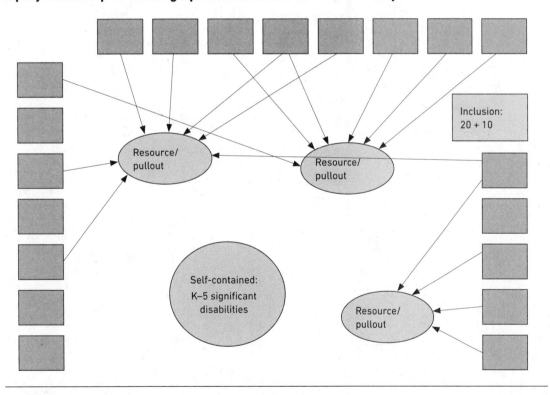

Outer rectangles = K–5 general education classrooms

"Resource/pullout" = Special education teachers pull out students from general education classrooms

"Inclusion: 20 + 10" = Classroom with general education and special education teacher team teaching 20 general education students and 10 special education students

"Self-contained" = Special education classroom where all students with significant disabilities receive instruction and spend most of their school day

Source: Used with permission from George Theoharis.

Remember that service delivery mapping can be used for other services in addition to special education. A map for EL services, for example, might show a configuration of spaces and distribution of staff similar to that shown in Figure AppB.1.

References

AAA Statement on Race. (1998). *American Anthropologist,100*(3), 712–713. Retrieved from www.jstor.org/stable/682049

ADL. (2018). *Anti-Semitic Incidents Surged Nearly 60% in 2017, According to New ADL Report.* Retrieved from https://www.adl.org/news/press-releases/anti-semitic-incidents-surged-nearly-60-in-2017-according-to-new-adl-report

American Association of University Women. (2018, Fall). The simple truth about the gender pay gap. Retrieved from https://aauw-tn.aauw.net/files/2018/12/simple-truth-one-pager.pdf

Argyris, C. (1985a). Interventions for improving leadership effectiveness. *The Journal of Management Development, 4*(5), 30.

Argyris, C. (1985b). *Strategy, change, and defensive routines.* Boston: Pitman.

August, D., & Hakuta, K. (1998). *Educating language-minority children.* Washington, DC: National Academies Press.

Badgett, M. V. L., Choi, S. K., & Wilson, B. D. M., (2019, October). *LGBT poverty in the United States: A study of differences between sexual orientation and gender identity groups.* Los Angeles: The Williams Institute. Retrieved from https://williamsinstitute.law.ucla.edu/wp-content/uploads/National-LGBT-Poverty-Oct-2019.pdf

Bailey, S. P. (2015, September 14). A startling number of Americans still believe President Obama is a Muslim. *Washington Post.* Retrieved from https://www.washingtonpost.com/news/acts-of-faith/wp/2015/09/14/a-startling-number-of-americans-still-believe-president-obama-is-a-muslim/

Banaji, M. R., & Greenwald, A. G. (2013). *Blindspot: Hidden biases of good people.* New York: Delacorte Press.

Beirich, H. (2019). The year in hate: Rage against change. Intelligence Report. Southern Poverty Law Center. Spring Issue 166. Retrieved from https://www.splcenter.org/fighting-hate/intelligence-report/2019/year-hate-rage-against-change

Benz, M. R., Lindstrom, L., & Yovanoff, P. (2000). Improving graduation and employment outcomes of students with disabilities: Predictive factors and student perspectives. *Exceptional Children, 66*(4), 509–529.

The BIPOC Project. (n.d.). A black, indigenous, and people of color movement. Retrieved from https://www.thebipocproject.org/

Blad, E. (2016). Feds urge schools to shield Muslim students from harassment. *The Education Digest*, *81*(9), 26.

Bornstein, J. (2014). Runners, biters, and chair throwers: Discourses of order and medicalization in inclusion. Dissertations - ALL. 72. https://surface.syr.edu/etd/72

Brenan, M. (2018, December 24). Religion considered important to 72% of Americans. Retrieved from https://news.gallup.com/poll/245651/religion-considered-important-americans.aspx

Brookfield, S. M. (2012). *Teaching for critical thinking: Tools and techniques to help students question their assumptions*. San Francisco: Jossey-Bass.

Bruner, J. (1990). *Acts of meaning*. Cambridge, MA: Harvard University Press.

Bryk, A. S., Sebring, P., Allensworth, E., Luppescu, S., & Easton, J. (2010). *Organizing schools for improvement: Lessons from Chicago*. Chicago: University of Chicago Press.

Cambron-McCabe, N. & Mccarthy, M. (2005). Educating school leaders for social justice. *Educational Policy, 19*(1), 201–222. doi: 10.1177/0895904804271609

Capper, C. A., & Young, M. D. (2015). The equity audit as the core of leading increasingly diverse schools and districts. In G. Theoharis & M. Scanlan (Eds.), *Leadership for increasingly diverse schools* (pp. 212–223). New York: Routledge.

Carroll, J. (2004, March 2). American public opinion about religion. Retrieved from https://news.gallup.com/poll/10813/religion.aspx

Carter, E. W., & Hughes, C. (2005). Increasing social interaction among adolescents with intellectual disabilities and their general education peers: Effective interventions. *Research and Practice for Persons with Severe Disabilities, 30*(4), 179–193.

Carter, P. L., & Welner, K. G. (Eds.). (2013). *Closing the opportunity gap: What America must do to give every child an even chance*. New York: Oxford University Press.

Cartiera, M. R. (2006). Addressing the literacy underachievement of adolescent English language learners: A call for teacher preparation and proficiency reform. *New England Reading Association Journal, 42*(1), 26.

Cartledge, G., Singh, A., & Gibson, L. (2008). Practical behavior-management techniques to close the accessibility gap for students who are culturally and linguistically diverse. *Preventing School Failure: Alternative Education for Children and Youth, 52*(3), 29–38.

Center for American Progress. (2020). Quick Facts About the Gender Wage Gap. Retrieved from https://cdn.americanprogress.org/content/uploads/2020/03/23133916/Gender-Wage-Gap-.pdf

Cigna. (2017). LGBT Health Disparities. Retrieved from https://www.cigna.com/individuals-families/health-wellness/lgbt-disparities

Cimera, R.E. (2010). Can community-based high school transition programs improve the cost-efficiency of supported employment? *Career Development for Exceptional Individuals, 33*(1), 4–12.

College Board. (2019). SAT results, class of 2019. Available at https://reports.collegeboard.org/sat-suite-program-results/class-2019-results

Collier, V. P. (1995). Acquiring a second language for school. *Directions in Language and Education, 1*(4), n4.

Collier, V. P., & Thomas, W. P. (2009). *Educating English learners for a transformed world*. Dual Language Education of New Mexico/Fuente Press.

Collier, V. P., & Thomas, W.P. (2012). What really works for English language learners: Research-based practices for principals. In G. Theoharis & J. Brooks (Eds.), *What every principal needs to know to create equitable and excellent schools* (pp. 155-173). New York: Teachers College Press.

Cosier, M. (2010). *Exploring the relationship between inclusive education and achievement: New perspectives*. Syracuse University.

Crawford, J. (2004). *Educating English learners: Language diversity in the classroom* (5th ed.). Los Angeles: Bilingual Educational Services

Cummins, J. (1979). Cognitive/academic language proficiency, linguistic interdependence, the optimum age question and some other matters. *Working Papers on Bilingualism*, No. 19.

Cummins, J. (2000). Putting language proficiency in its place: Responding to critiques of the conversational/academic language distinction. *English in Europe: The Acquisition of a Third Language*, *X*(x),54–83.

Darling-Hammond, L., LaPointe, M., Meyerson, D., Orr, M. T., & Cohen, C. (2007). Preparing school leaders for a changing world: Lessons from exemplary leadership development programs. School Leadership Study. Final Report. *Stanford Educational Leadership Institute*

De Brey, C., Musu, L., McFarland, J, Wilkinson-Flicker, Sl, Diliberti, M., Zhang, A., Branstetter, C., & Wang, X. (2019). Status and trends in education of racial and ethnic groups 2018 (NCES 2019-038). U.S. Department of Education. Washington, D.C.: National Center for Education Statistics. Retrieved from https://nces.ed.gov/ pubsearch/

De Jong, A. (2018, May 10). Protestants decline, more have no religion in a sharply shifting religious landscape (POLL). Retrieved from https://abcnews.go.com/Politics/protestants-decline-religion -sharply-shifting-religious-landscape-poll/story?id=54995663

Diamond, J. B. (2006). Still separate and unequal: Examining race, opportunity, and school achievement in "integrated" suburbs. *The Journal of Negro Education*, *75*(3), 495–505.

Dunbar-Ortiz, R. (2014). *An indigenous history of the United States*. Boston: Beacon Press.

Economic Policy Institute. (2020). Black-white wage gaps are worse today than in 2000. Retrieved from https://www.epi.org/blog/black-white-wage-gaps-are-worse-today-than-in-2000/

Engel v. Vitale, 370 U.S. 421, 82 S. Ct. 1261, 8 L. Ed. 2d 601 (1962).

Feldman, S., & Malagon, V.F. (2017). Unlocking learning: Science as a lever for January 2017 English learner equity. Retrieved from https://west.edtrust.org/wp-content/uploads/sites/3/2015/11/ Ed-Trust-West-Unlocking-Learning-Report.pdf

Ferguson, S. (2014, September 29). Privilege 101: A quick and dirty guide. Retrieved from https:// everydayfeminism.com/2014/09/what-is-privilege/

Ferri, B.A. (2015). Response to Intervention: Persisting concerns. In B. Amrhein & K. Ziemen (Eds.), *Diagnostics in the Context of Inclusive Education: Theories, Ambivalences, Operators, Concepts*, pp. 7-21. Klinkhardt, Germany: Verlag.

Fisher, M., & Meyer, L. H. (2002). Development and social competence after two years for students enrolled in inclusive and self-contained educational programs. *Research and Practice for Persons with Severe Disabilities*, *27*(3), 165–174.

Ford, D. Y., & Moore, J. L. (2013). Understanding and reversing underachievement, low achievement, and achievement gaps among high-ability African American males in urban school contexts. *The Urban Review*, *45*(4), 399–415.

Frattura, E. M., & Capper, C. A. (2007). *Leading for social justice: Transforming schools for all learners*. Thousand Oaks, CA: Corwin Press.

Freeman, D. (2004). Teaching in the context of English-language learners: What do we need to know. *Teaching immigrant and second-language students*, 7–20.

Gallup. (2016). Five key findings on religion in the U.S. Retrieved from https://news.gallup.com/ poll/200186/five-key-findings-religion.aspx

Gallup. (2018). Religion considered important to 72% of Americans. Retrieved from https://news .gallup.com/poll/245651/religion-considered-important-americans.aspx

Gallup. (2019). Religion. Retrieved from https://news.gallup.com/poll/10813/religion.aspx

GLAAD. (2016, October 26) GLAAD media reference guide–Lesbian/gay/bisexual glossary of terms. Retrieved from https://www.glaad.org/reference/lgbtq

GLAAD. (2019, December 7). GLAAD media reference guide–Transgender. Retrieved from https:// www.glaad.org/reference/transgender

Gooden, M.A. & O'Doherty, A. (2015). Do you see what I see? Fostering aspiring leaders' racial awareness. *Urban Education*. 50(2), 225–255.

Gorski, P. C. (2013). *Reaching and teaching students in poverty: Strategies for erasing the opportunity gap*. New York: Teachers College Press.

Green E. (2017). How much discrimination do Muslims face in America? *The Atlantic*. Retrieved from https://www.theatlantic.com/politics/archive/2017/07/american-muslims-trump/534879/

Habib, D., Habib, S., McNamara, B., Habib, I., Jones, K., Malfy, A., Orellana, N., Huff, E., Desgres, R., Jorgenson, C.M., Schuh, M.C., University of New Hampshire, Institute on Disability, & DH Photography. (2007). *Including Samuel* (video recording). DH Photography.

Hauslohner, A. (2017, July 26). Discrimination against Muslims is increasing in U.S., Pew study finds. Retrieved from https://www.washingtonpost.com/national/discrimination-against-muslims-is -increasing-in-us-pew-study-finds/2017/07/25/dfa52756-717a-11e7-9eac-d56bd5568db8_story .html?nonredirect=on

Haynes, J. (2007). *Getting started with English language learners: How educators can meet the challenge.* Alexandria, VA: ASCD.

Herold, D. M., & Fedor, D. B. (2008). *Leading change management: Leadership strategies that really work.* London: Kogan Page.

Hicks, M. A., Berger, J. G., & Generett, G. G. (2005). From hope to action: Creating spaces to sustain transformative habits of mind and heart. *Journal of Transformative Education, 3*(1), 57–75. https:// doi.org/10.1177/1541344604270924

Hughes, C., Harvey, M., Cosgriff, J., Reilly, C., Heilingoetter, J., Brigham, N., ... & Bernstein, R. (2013). A peer-delivered social interaction intervention for high school students with autism. *Research and Practice for Persons with Severe Disabilities, 38*(1), 1–16.

Human Rights Campaign (2020). Violence against the transgender and gender non-conforming community in 2020. Accessed 7/30/2020 at https://www.hrc.org/resources/violence-against-the -trans-and-gender-non-conforming-community-in-2020

Individuals with Disabilities Education Improvement Act of 2004, 20 U.S.C. 1401 et seq. (2004). (Reauthorization of the Individuals with Disabilities Education action of 1990).

Intersex Society of North America. (n.d.). What is intersex? Retrieved from https://isna.org/faq/ what_is_intersex/

Katznelson, I. (2006). When is affirmative action fair? On grievous harms and public remedies. *Social Research,* 541–568.

Kena, G., Aud, S., Johnson, F., Wang, X., Zhang, J., Rathbun, A., ... & Kristapovich, P. (2014). The Condition of Education 2014. NCES 2014–083. *National Center for Education Statistics.*

Kindler, A. (2002). *Survey of the states' limited English proficient students and available education programs and services, 2000–2001 Summary Report.* Washington, DC: National Clearinghouse for English Language Acquisition and Language Instruction Educational Programs.

Koball, H., & Jiang, Y. (2018). Basic facts about low-income children: Children under 9 years, 2016. A report published by the National Center for Children in Poverty, Columbia University Mailman School of Public Health. Retrieved from http://www.nccp. org/publications/pub_1194.html

Kosciw, J. G., Greytak, E. A., Palmer, N. A., & Boesen, M. J. (2014). The 2013 national school climate survey. New York: Gay, Lesbian, and Straight Education Network.

Kowalski, T. J. (2013). District diversity and superintendents of color. *Educational Leadership Faculty Publications.* 19. Retrieved from https://ecommons.udayton.edu/eda_fac_pub/19

Kurth, J. A., & Mastergeorge, A. M. (2010). Academic and cognitive profiles of students with autism: implications for classroom practice and placement. *International Journal of Special Education, 25*(2), 8–14.

Ladson-Billings, G. (2006). From the achievement gap to the education debt: Understanding achievement in the U.S. schools. *Educational Researcher, (35)*7, 3–12.

Layton, L. (2014, April 28). National high school graduation rates at historic high, but disparities still exist. Retrieved from https://www.washingtonpost.com/local/education/high-school-graduation- rates-at-historic-high/2014/04/28/84eb0122-cee0-11e3-937f-d3026234b51c_story.html

Lee v. Weisman, 505 U.S. 577, 112 S. Ct. 2649, 120 L. Ed. 2d 467 (1992).

Leithwood, K., & Riehl, C. (2005). What do we already know about educational leadership. *A New Agenda for Research in Educational Leadership, X*(x)12-27.

Lenski, S. D. (2006). Reflections on being biliterate: Lessons from paraprofessionals. *Action in Teacher Education, 28*(4), 104–113.

Lipmanowicz, H., & McCandless, K. (2013). Liberating structures: Including and unleashing everyone. Retrieved from http://www.liberatingstructures.com/1-1-2-4-all

McCabe, J., Fairchild, E., Grauerholz, L., Pescosolido,, B.A., & Tope, D. (2011). Gender in twentieth-century children's books: Patterns and disparity in titles and central characters. *Gender & Society, 25*(2):197–226. Retrieved from https://www.researchgate.net/publication/241647875_Gender_in_Twentieth-Century_Children%27s_Books

McLaughlin, B., & McLeod, B. (1996). Educating all our students: Improving education for children from culturally and linguistically diverse backgrounds (Vol. 1). Santa Cruz: University of California at Santa Cruz. *National Center for Research on Cultural Diversity and Second Language Learning.*

Meskill, C. (2005). Infusing English language learner issues throughout professional educator curricula: The training all teachers project. *Teachers College Record, 107*(4), 739–756.

Meyer, D., Madden, D., & McGrath, D. J. (2005). English language learner students in U.S. public schools: 1994 and 2000. *Education Statistics Quarterly*, 6(3), X.

Mezirow, J. (2000). Learning to think like an adult. *Learning as transformation: Critical perspectives on a theory in progress*, 3–33. San Francisco: Jossey-Bass.

Milner, H. R. (2010). What does teacher education have to do with teaching? Implications for diversity studies. *Journal of Teacher Education, 61*(1–2), 118–131. https://doi.org/10.1177/0022487109347670

Milner, H. R. (2012). But what is urban education? *Urban Education, 47*(3), 556–561. https://doi.org/10.1177/0042085912447516.

Montecel, M. R., Cortez, J. D., & Cortez, A. (2002). What is valuable and contributes to success in bilingual education programs. A paper presented at the 2002 Annual Meeting of the American Education Research Association, New Orleans, LA.

Mosca, C. (2006). How do you ensure that everyone in the school shares the responsibility for educating English language learners, not just those who are specialists in the field. *English language learners at school: A guide for administrators*, 109–110. Philadelphia: Caslon Publishing.

National Center for Education Statistics. (n.d.a). Back to school statistics. Retrieved from https://nces.ed.gov/fastfacts/display.asp?id=372

National Center for Education Statistics. (n.d.b). Trends in educational equity of girls & women: 2004. Conclusion. Retrieved from https://nces.ed.gov/pubs2005/equity/Section11.asp

National Center for Education Statistics. (2020a). Racial/ethnic enrollment in public schools. Retrieved from https://nces.ed.gov/programs/coe/indicator_cge.asp

National Center for Education Statistics. (2020b). Characteristics of public school teachers. Retrieved from https://nces.ed.gov/programs/coe/indicator_clr.asp#:~:text=In%20 2017%E2%80%9318%2C%20about%2079,1%20percent%20of%20public%20school

National Center for Education Statistics. (2020c). Characteristics of public school principals. Retrieved from https://nces.ed.gov/programs/coe/indicator_cls.aspl

National Center for Education Statistics. (2020d). English language learners in public schools. Retrieved from https://nces.ed.gov/programs/coe/indicator_cgf.asp

Obiakor, F. E. (2001). *It even happens in" good" schools: Responding to cultural diversity in today's classrooms*. Thousand Oaks, CA: Corwin Press.

Obiakor, F. E., & Utley, C. A. (2004). Educating culturally diverse learners with exceptionalities: A critical analysis of the Brown case. *Peabody Journal of Education, 79*(2), 141–156.

Parrish, T. (2002). Racial disparities in the identification, funding, and provision of special education. *Racial Inequity in Special Education, X*(x),1537.

Payne, R. (1996). *A framework for understanding poverty*. Highlands, TX: aha! Process, Inc.

Pew Research Center. (2008, July 16). Still think Obama is Muslim? Retrieved from https://www.pewresearch.org/fact-tank/2008/07/16/still-think-obama-is-muslim/

Pew Research Center. (2015, May 7). Christians decline as share of U.S. population; Other faiths and the unaffiliated are growing. Retrieved from https://www.pewforum.org/2015/05/12/americas-changing-religious-landscape/pr_15-05-12_rls-00/

Pew Research Center. (2017, July 26). U.S. Muslins concerned about their place in society, but continue to believe in the American dream. Retrieved from https://www.pewforum .org/2017/07/26/findings-from-pew-research-centers-2017-survey-of-us-muslims/

PFLAG NYC. (2020). Statistics you should know about gay & transgender students. (n.d.). Retrieved from http://www.pflagnyc.org/safeschools/statistics

Radd, S.I. & Kramer, B.H. (2016). Dis Eased: Critical consciousness in school leadership for social justice. *Journal of School Leadership, 26*(4), 580–606.

Reyes, A. (2006). Reculturing principals as leaders for cultural and linguistic diversity. In *Preparing quality educators for English language learners*. 155–176. New York: Routledge.

Riehl, C. J. (2000). The principal's role in creating inclusive schools for diverse students: A review of normative, empirical, and critical literature on the practice of educational administration. *Review of Educational Research, 70*(1), 55–81.

Riese, J. (2016). *Youth who are bullied based upon perceptions about their sexual orientation.* Center City, MN: Hazelden Publishing. Retrieved from http://www.violencepreventionworks.org/public/ bullying_sexual_orientation.page

Roberto, M.A. (2013). *Why great leaders don't take yes for an answer: Managing for conflict and consensus* (2nd ed.). Upper Saddle River, NJ: Pearson Prentice Hall.

Roberts, A. L., Austin, S. B., Corliss, H. L., Vandermorris, A. K., & Koenen, K. C. (2010). Pervasive trauma exposure among US sexual orientation minority adults and risk of posttraumatic stress disorder. *American Journal of Public Health, 100*(12), 2433–2441.

Rothenberg, P. (2002). *White privilege: Essential readings on the other side of racism.* New York: Worth Publishers.

Ruíz, R. (1984). Orientations in language planning. *NABE Journal, 8*(2), 15–34.

Sadker, D., Sadker, M., & Zittleman, K.R. (2009). *Still failing at fairness: How gender bias cheats girls and boys in school and what we can do about it.* New York: Scribner.

Samway, K. D., & McKeon, D. (2007). *Myths and realities: Best practices for English language learners.* Heinemann Educational Books.

Santa Fe Independent School Dist. v. Doe, 530 U.S. 290, 120 S. Ct. 2266, 147 L. Ed. 2d 295 (2000).

Scheurich, J. J., & Skrla, L. (2003). *Leadership for equity and excellence: Creating high-achievement classrooms, schools, and districts.* Thousand Oaks, CA: Corwin Press.

School District of Abington Township v. Schempp. (1963). 374 U.S. 203.

Schmidt, J., Shah, P., Vedantam, S., Boyle, T., Penman, M., & Nesterak, M. (2019, January 24). Creative differences: The benefits of reaching out to people unlike ourselves. Retrieved from https://www .npr.org/2019/01/24/687707404/creative-differences-the-benefits-of-reaching-out-to-people- unlike-ourselves

Senge, P. M. (1990). *The fifth discipline: The art and practice of the learning organization.* New York: Doubleday

Senge, P. M., Cambron-McCabe, N., Lucas, T., Smith, B., Dutton, J., & Kleiner, A. (2012). *Schools that learn (updated and revised): A fifth discipline fieldbook for educators, parents, and everyone who cares about education.* New York: Crown Business.

Shah, S. (2011, September 8). For Muslim students, life changed after Sept. 11. *Education Week.* Retrieved from https://www.edweek.org/ew/articles/2011/09/08/02muslim_ep.h31.html

Shaw, P. (2003). Leadership in the diverse school. *Multilingual education in practice, X*(x), 97–112.

Shields, C. M. (2004). Dialogic leadership for social justice: Overcoming pathologies of silence. *Educational administration quarterly, 40*(1), 109–132.

Silberman, M. L. (2006). *Training the active training way: 8 strategies to spark learning and change* (Vol. 12). Hoboken, NJ: John Wiley & Sons.

Singleton, G. E. (2015). *Courageous conversations about race: A field guide for achieving equity in schools.* Thousand Oaks, CA: Corwin Press.

Skrla, L., McKenzie, K. B., & Scheurich, J. J. (Eds.). (2009). *Using equity audits to create equitable and excellent schools.* Thousand Oaks, CA: Corwin Press.

Southern Poverty Law Center (2020, March). The year in hate and extremism 2019. Retrieved from https://www.splcenter.org/news/2020/03/18/year-hate-and-extremism-2019

Stevenson, B. (2014). *Just mercy: A story of justice and redemption.* New York: Spiegel & Grau.

Taliaferro, J. D., & DeCuir-Gunby, J. T. (2008). African American educators' perspectives on the advanced placement opportunity gap. *The Urban Review, 40*(2), 164–185.

Test, D. W., Mazzotti, V. L., Mustian, A. L., Fowler, C. H., Kortering, L., & Kohler, P. (2009). Evidence-based secondary transition predictors for improving postschool outcomes for students with disabilities. *Career Development for Exceptional Individuals, 32*(3), 160–181.

Theoharis, G., & O'Toole, J. (2011). Leading inclusive ELL: Social justice leadership for English language learners. *Educational Administration Quarterly, 47*(4), 646–688.

U.S. Census Bureau. (2017). 2016 American community survey 1-year estimates, poverty status in the past 12 months. Washington, D.C.: United States Census Bureau. Retrieved from https:// factfinder.census.gov

U.S. Census Bureau News. (2020, April 28). Quarterly residential vacancies and homeownership, fourth quarter 2019. (2020, January 30). Retrieved from https://www.census.gov/housing/hvs/files/currenthvspress.pdf

U.S. Department of Education. (2018). Academic performance and outcomes for English learners: Performance on national assessments and on-time graduation rates. Department of Education. Retrieved from https://www2.ed.gov/datastory/el-outcomes/index.html

U.S. Department of Education Office for Civil Rights. (2012). Civil rights data collection. Retrieved from https://ocrdata.ed.gov/

U.S. Department of Education Office for Civil Rights. (2014). Civil rights data collection data snapshot: School discipline. *Issue brief no. 1.*

U.S. Department of Education Office for Civil Rights. (2015). Civil rights data collection. Retrieved from https://ocrdata.ed.gov/

U.S. Department of Education Office for Civil Rights. (2016). 2013–2014 civil rights data collection: A first look. Retrieved from https://www2.ed.gov/about/offices/list/ocr/docs/2013-14-first-look.pdf

U.S. Department of Education Office for Civil Rights. (2018a). 2015-16 Civil rights data collection: School climate and safety. Retrieved from https://www2.ed.gov/about/offices/list/ocr/docs/school-climate-and-safety.pdf

U.S. Department of Education Office for Civil Rights. (2018b). 2015–2016 Civil rights data collection: STEM course taking. Available at https://www2.ed.gov/about/offices/list/ocr/docs/stem-course-taking.pdf

U.S. Department of Justice-Federal Bureau of Investigation. (2018). 2017 hate crime statistics; Incidents and offenses. Retrieved from https://ucr.fbi.gov/hate-crime/2017/topic-pages/incidents-and-offenses

Walquí, A. (2000). Contextual factors in second language acquisition. *ERIC Digest.*

Weiler, K. (2002). The dreamwork of autobiography: Felman, Freud, and Lacan. In *Feminist ngagements* (pp. 97–116). New York: Routledge.

Wenger, K. J., Lubbes, T., Lazo, M., Azcarraga, I., Sharp, S., & Ernst-Slavit, G. (2004). Hidden teachers, invisible students: Lessons learned from exemplary bilingual paraprofessionals in secondary schools. *Teacher Education Quarterly, 31*(2), 89–111.

West Virginia Bd. of Ed. v. Barnette, 319 U.S. 624, 63 S. Ct. 1178, 87 L. Ed. 1628 (1943).

Yell, M. L., Katsiyannis, A., Ryan, J. B., McDuffie, K. A., & Mattocks, L. (2008). Ensure compliance with the individuals with disabilities education improvement act of 2004. *Intervention in School and Clinic, 44*(1), 45–51.

Zehr, M. A. (2008). Hurdles remain high for English learners. *Education Week, 27*(39), 1–1.

Index

Note: The letter *f* following a page number denotes a figure.

About the Authors

Sharon I. Radd, EdD, is Program Director and Associate Professor of Organizational Leadership at St. Catherine University, Principal at ConsciousPraxis, and an Equity Fellow for the Midwest and Plains Equity Assistance Center at IUPUI. Prior to entering higher education, she spent 23 years in K–12 public education as a school social worker, building-level administrator, district administrator, and professional development provider. Awarded a 2006 Archibald Bush Leadership Fellowship and winner of the 2010 Social Justice Dissertation Award given by the American Educational Research Association LSJ-SIG, Radd specializes in the areas of leadership, educational equity, adult learning, organizational change, team functioning, and cross-cultural proficiency. Her research interrogating whiteness and re-envisioning equitable and more humane leadership has been published in the *Journal of School Leadership, Educational Policy*, the *Journal of Education Policy,* and *Urban Education;* she has several published book chapters, and is author of the blog, *A Woman with Heart.* As a professor, presenter, and consultant, Radd's work with leaders serves to hone their communication, conflict, and decision-making skills and enhance their leadership to restore more authentic relationships, more expansive and optimistic purposes for work, and more social justice in the public sphere. A captivated yoga student and open-heart surgery patient, she infuses yogic principles and insights about open-hearted living into all aspects of her work. Radd holds bachelor's degrees in sociology and social work; a master's in educational leadership; and from the University of St. Thomas, an EdD in Leadership. She lives in Minneapolis, Minnesota, and is grateful to share life and love with her husband, adult son, and two teenagers.

Gretchen Givens Generett, PhD, is Interim Dean and Professor of Foundations in the School of Education at Duquesne University. She holds the Noble J. Dick Endowed Chair in Community Outreach at Duquesne, where she has served as the Director for the University Council Educational Administration (UCEA) Center for Educational Leadership and Social Justice and the Associate Dean for Graduate Studies and Research. Generett is a qualitative researcher whose research is centered around educational stories that have been rendered invisible in America. At its core, her scholarship intermingles traditional sociology of education, African American studies, and feminist studies with more progressive concepts of justice that examine agency, empowerment, and action. Her research is published in numerous academic journals including *Urban Education, Urban Review, Educational Studies, Educational Foundations, The Western Journal of Black Studies, NER, the Middle School Journal, Journal of Transformative Education,* and the *International Journal of Qualitative Studies.* Her most recent scholarship is a book entitled *The American Dream for Students of Color: Barriers to Educational Success* coauthored with Dr. Amy M. Olson. Generett received her BA in English from Spelman College and her PhD in Foundations of Education from the University of North Carolina at Chapel Hill. She lives in Pittsburgh, Pennsylvania, with her husband William O. Generett, Jr., son William, and daughter Gabrielle.

Mark Anthony Gooden, PhD, is the Christian Johnson Endeavor Professor in Education Leadership and Director of the Endeavor Antiracist & Restorative Leadership Initiative (EARLI). He is the former Director of the Summer Principals Academy-NYC in the Department of Organization and Leadership at Teachers College, Columbia University. His research focuses broadly on culturally responsive school leadership with specific interests in antiracist leadership, the principalship, urban educational leadership, and legal issues in education. His research has appeared in a range of outlets including the *Journal of Negro Education,*

Urban Education, American Educational Research Journal, Educational Administration Quarterly, Teachers College Record, Review of Educational Research, Journal of School Leadership, and others. He is Past President of the University Council for Educational Administration (UCEA), a consortium of over 100 higher education institutions committed to advancing the preparation and practice of educational leaders and the 2017 recipient of the Jay D. Scribner Mentoring Award. Before entering higher education, he served as a secondary mathematics teacher and departmental chairperson in Columbus Public Schools. He transitioned into higher education as an assistant professor at the University of Cincinnati, where he also directed several leadership programs for seven years. Gooden went on to eventually rise to the role of Margie Gurley Seay Centennial Professor of Education at The University of Texas-Austin where he served as director of the principalship program for nearly nine years. He has spent two decades in higher education developing and teaching courses in culturally responsive leadership, race, law, and research methods and consulting with school districts, universities, and non-profit organizations by designing and delivering professional development courses/workshops in antiracist leadership, law, and community building. He earned his BA in mathematics from Albany State University (a historically Black college/university) and his MEd in Mathematics Education, a second master's, and a PhD in policy and leadership, all from The Ohio State University. He lives in New Rochelle, New York, with his lovely wife and his highly analytical and beautiful daughter.

George Theoharis, PhD, is a Professor in Educational Leadership and Inclusive Education at Syracuse University. He has extensive field experience in public education as a principal and teacher. At Syracuse, he previously served as Department Chair, Associate Dean for Urban Education Partnerships, and the Director of Field Relations. Theoharis teaches classes in educational leadership and elementary/early childhood teacher education. He coordinates the Inclusive Early Childhood and Special Education undergraduate program and the EdD program in Leading Equitable Schools. His interests, research, and work with K–12 schools focus on issues of equity, justice, diversity, inclusion, leadership, and school reform. His books titled *The School Leaders Our Children Deserve* (2009), *Leadership for Increasingly Diverse Schools* (2015), *What Every Principal Needs to Know to Create Excellent and Equitable Schools* (2012), and *The Principal's Handbook for Leading Inclusive Schools* (2014) focus on issues of leadership and creating more equitable schools. Theoharis'

work bridges the worlds of K–12 schools and higher education. As such, he writes for public audiences in outlets such as: *The School Administrator, Educational Leadership* (online), *The Principal, The Washington Post,* and *The Syracuse Post-Standard,* as well as writing for numerous academic journals. He coruns a summer institute in Syracuse for school leaders from around the country and globe focusing on issues of disability, equity, and inclusion and consults with leaders, schools, and districts around issues of leadership, equity, diversity, and inclusive reform around the United States and Canada. His PhD is in educational leadership and policy analysis from the University of Wisconsin–Madison. Theoharis, who grew up in Wisconsin in an activist family committed to making a more just world, lives in Fayetteville, New York, with his two awesome, teenage children and his father.

Related ASCD Resources

At the time of publication, the following resources were available (ASCD stock numbers appear in parentheses):

Cultural Competence Now: 56 Exercises to Help Educators Understand and Challenge Bias, Racism, and Privilege by Vernita Mayfield (#118043)

Culture, Class, and Race: Constructive Conversations That Unite and Energize Your School and Community by Brenda CampbellJones, Shannon Keeny, and Franklin CampbellJones (#118010)

What Every School Leader Needs to Know About RTI by Margaret Searle (#109097)

Leading an Inclusive School: Access and Success for ALL Students by Richard A. Villa and Jacqueline Thousand (#116022)

Aim High, Achieve More: How to Transform Urban School Through Fearless Leadership by Yvette Jackson and Veronica McDermott (#112015)

Turning High-Poverty Schools into High-Performing Schools, 2nd Edition by William H. Parrett and Kathleen M. Budge (#120031)

Building Equity: Policies and Practices to Empower All Learners by Dominique Smith, Nancy E. Frey, Ian Pumpian, and Douglas E. Fisher (#117031)

Educating Everybody's Children: Diverse Teaching Strategies for Diverse Learners, Revised and Expanded 2nd Edition, edited by Robert W. Cole (#107003)

Excellence Through Equity: Five Principles of Courageous Leadership to Guide Achievement for Every Student by Alan M. Blankstein and Pedro Noguera with Lorena Kelly (#116070)

The Innocent Classroom: Dismantling Racial Bias to Support Students of Color by Alexs Pate (#120025)

For up-to-date information about ASCD resources, go to www.ascd.org. You can search the complete archives of *Educational Leadership* at www.ascd.org/el.

ASCD myTeachSource®

Download resources from a professional learning platform with hundreds of research-based best practices and tools for your classroom at http://myteach-source.ascd.org/.

For more information, send an e-mail to member@ascd.org; call 1-800-933-2723 or 703-578-9600; send a fax to 703-575-5400; or write to Information Services, ASCD, 1703 N. Beauregard St., Alexandria, VA 22311-1714 USA.

WHOLE CHILD
TENETS

1 HEALTHY
Each student enters school healthy and learns about and practices a healthy lifestyle.

2 SAFE
Each student learns in an environment that is physically and emotionally safe for students and adults.

3 ENGAGED
Each student is actively engaged in learning and is connected to the school and broader community.

4 SUPPORTED
Each student has access to personalized learning and is supported by qualified, caring adults.

5 CHALLENGED
Each student is challenged academically and prepared for success in college or further study and for employment and participation in a global environment.

The ASCD Whole Child approach is an effort to transition from a focus on narrowly defined academic achievement to one that promotes the long-term development and success of all children. Through this approach, ASCD supports educators, families, community members, and policymakers as they move from a vision about educating the whole child to sustainable, collaborative actions.

Five Practices for Equity-Focused School Leadership relates to the **engaged**, **supported**, and **challenged** tenets.

For more about the ASCD Whole Child approach, visit **www.ascd.org/wholechild.**